METHODS IN MOLECULAR BIOLOGY

Series Editor
John M. Walker
School of Life and Medical Sciences
University of Hertfordshire
Hatfield, Hertfordshire, AL10 9AB, UK

For further volumes:
http://www.springer.com/series/7651

P-Type ATPases

Methods and Protocols

Edited by

Maike Bublitz

Department of Biochemistry, University of Oxford, Oxford, UK

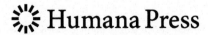

Editor
Maike Bublitz
Department of Biochemistry
University of Oxford
Oxford, UK

ISSN 1064-3745   ISSN 1940-6029 (electronic)
Methods in Molecular Biology
ISBN 978-1-4939-3178-1   ISBN 978-1-4939-3179-8 (eBook)
DOI 10.1007/978-1-4939-3179-8

Library of Congress Control Number: 2015952622

Springer New York Heidelberg Dordrecht London
© Springer Science+Business Media New York 2016
This work is subject to copyright. All rights are reserved by the Publisher, whether the whole or part of the material is concerned, specifically the rights of translation, reprinting, reuse of illustrations, recitation, broadcasting, reproduction on microfilms or in any other physical way, and transmission or information storage and retrieval, electronic adaptation, computer software, or by similar or dissimilar methodology now known or hereafter developed.
The use of general descriptive names, registered names, trademarks, service marks, etc. in this publication does not imply, even in the absence of a specific statement, that such names are exempt from the relevant protective laws and regulations and therefore free for general use.
The publisher, the authors and the editors are safe to assume that the advice and information in this book are believed to be true and accurate at the date of publication. Neither the publisher nor the authors or the editors give a warranty, express or implied, with respect to the material contained herein or for any errors or omissions that may have been made.

Printed on acid-free paper

Humana Press is a brand of Springer
Springer Science+Business Media LLC New York is part of Springer Science+Business Media (www.springer.com)

Preface

A lipid membrane separating the "inside" from the "outside" of a cell is one of the most crucial hallmarks of all living organisms. In order to control the cellular content and to exchange material and information with its surroundings, every cell possesses a certain set of transport proteins and gated channels embedded in both the plasma and — if present — the organellar membranes. These transport proteins mediate the maintenance and tight control of the compositional differences between the "inside" and "outside" worlds. The P-type ATPases are a large family of such membrane transport proteins: under the consumption of ATP, they "pump" ions across the lipid bilayer, thereby building up electrochemical gradients, which are then used for manifold other processes, such as nutrient transport, nerve excitation, or muscle contraction.

Since the discovery of the sodium pump in 1957 by Nobel laureate Jens Christian Skou, generations of researchers have been studying the members of this large and diverse family of enzymes, making use of very different scientific approaches, ranging from electrophysiology, enzyme kinetics, and cell culture to animal models, structural studies, and bioinformatics.

This volume of *Methods in Molecular Biology* provides a broad collection of protocols for many of the common experimental techniques used for the characterization of P-type ATPases. The book aims to provide comprehensive practical instructions for all researchers in the P-type ATPase field, from the protein biochemist to the mouse geneticist, covering the identification and classification, isolation, purification, *in vitro* characterization, knockout studies, as well as crystallization and structural analysis.

Since many of the methods are not exclusively applicable to P-type ATPases, the techniques described will probably also prove useful in a variety of biochemistry laboratories working with other ATPases or with membrane proteins in general.

I hope this book will become a firm piece of inventory in all laboratories studying P-type ATPases, and a helpful and reliable companion for every scientist working in this fascinating field.

Oxford, UK *Maike Bublitz*

Contents

Preface.. v
Contributors.. xi

1. An Introduction to P-type ATPase Research........................... 1
   Poul Nissen

PART I PROTEIN PRODUCTION, ISOLATION,
        PURIFICATION, AND STABILIZATION

2. Purification of Na,K-ATPase from Pig Kidney......................... 5
   Natalya U. Fedosova

3. Preparation of Ca^{2+}-ATPase 1a Enzyme
   from Rabbit Sarcoplasmic Reticulum................................. 11
   Jesper V. Møller and Claus Olesen

4. Isolation of H^+,K^+-ATPase-enriched Membrane Fraction
   from Pig Stomachs.. 19
   Kazuhiro Abe and Claus Olesen

5. Overproduction of P_{IB}-Type ATPases............................. 29
   Xiangyu Liu, Oleg Sitsel, Kaituo Wang, and Pontus Gourdon

6. Coordinated Overexpression in Yeast of a P4-ATPase
   and Its Associated Cdc50 Subunit: The Case of the Drs2p/Cdc50p
   Lipid Flippase Complex... 37
   *Hassina Azouaoui, Cédric Montigny, Aurore Jacquot,
   Raphaëlle Barry, Philippe Champeil, and Guillaume Lenoir*

7. The Plasma Membrane Ca^{2+}-ATPase: Purification
   by Calmodulin Affinity Chromatography, and Reconstitution
   of the Purified Protein.. 57
   Verena Niggli and Ernesto Carafoli

8. Expression of Na,K-ATPase and H,K-ATPase Isoforms
   with the Baculovirus Expression System............................. 71
   Jan B. Koenderink and Herman G.P. Swarts

9. Time-Dependent Protein Thermostability Assay....................... 79
   Ilse Vandecaetsbeek and Peter Vangheluwe

PART II ACTIVITY ASSAYS

10. Colorimetric Assays of Na,K-ATPase................................ 89
    Kathleen J. Sweadner

11 ATPase Activity Measurements by an Enzyme-Coupled
   Spectrophotometric Assay... 105
   Pankaj Sehgal, Claus Olesen, and Jesper V. Møller

12 Antimony-Phosphomolybdate ATPase Assay.......................... 111
   Gianluca Bartolommei and Francesco Tadini-Buoninsegni

13 ATPase Activity Measurements Using Radiolabeled ATP............ 121
   Herman G.P. Swarts and Jan B. Koenderink

14 Assaying P-Type ATPases Reconstituted in Liposomes 127
   Hans-Jürgen Apell and Bojana Damnjanovic

15 Coupling Ratio for Ca^{2+} Transport by Calcium Oxalate Precipitation........ 157
   Pankaj Sehgal, Claus Olesen, and Jesper V. Møller

16 Calcium Uptake in Crude Tissue Preparation 161
   Philip A. Bidwell and Evangelia G. Kranias

17 Measuring H^+ Pumping and Membrane Potential Formation
   in Sealed Membrane Vesicle Systems................................ 171
   *Alex Green Wielandt, Michael G. Palmgren, Anja Thoe Fuglsang,
   Thomas Günther-Pomorski, and Bo Højen Justesen*

18 Assay of Flippase Activity in Proteoliposomes
   Using Fluorescent Lipid Derivatives................................ 181
   Magdalena Marek and Thomas Günther-Pomorski

PART III IN VITRO FUNCTIONAL STUDIES

19 The Use of Metal Fluoride Compounds as Phosphate Analogs
   for Understanding the Structural Mechanism in P-type ATPases..... 195
   Stefania J. Danko and Hiroshi Suzuki

20 Phosphorylation/Dephosphorylation Assays......................... 211
   Hiroshi Suzuki

21 Tryptophan Fluorescence Changes Related to Ca^{2+}-ATPase Function 227
   Pankaj Sehgal, Claus Olesen, and Jesper V. Møller

PART IV LIGAND BINDING STUDIES

22 Determination of the ATP Affinity of the Sarcoplasmic
   Reticulum Ca^{2+}-ATPase by Competitive Inhibition
   of [γ-^{32}P]TNP-8N$_3$-ATP Photolabeling............................... 233
   *Johannes D. Clausen, David B. McIntosh, David G. Woolley,
   and Jens Peter Andersen*

23 Ca^{2+} Binding and Transport Studied with Ca^{2+}/EGTA Buffers
   and $^{45}Ca^{2+}$.. 261
   Pankaj Sehgal, Claus Olesen, and Jesper V. Møller

24 Assay of Copper Transfer and Binding to P_{1B}-ATPases................. 267
   Teresita Padilla-Benavides and José M. Argüello

Part V Electrophysiology

25 Voltage Clamp Fluorometry of P-Type ATPases 281
    Robert E. Dempski

26 Electrophysiological Measurements on Solid Supported Membranes 293
    Francesco Tadini-Buoninsegni and Gianluca Bartolommei

27 Electrophysiological Characterization of Na,K-ATPases Expressed
    in *Xenopus laevis* Oocytes Using Two-Electrode Voltage Clamping 305
    Florian Hilbers and Hanne Poulsen

Part VI Functional Studies by Cell Culture and Transgenic Animals

28 Functional Studies of Na^+,K^+-ATPase
    Using Transfected Cell Cultures 321
    Elena Arystarkhova and Kathleen J. Sweadner

29 HPLC Neurotransmitter Analysis 333
    Thomas Hellesøe Holm, Toke Jost Isaksen, and Karin Lykke-Hartmann

30 Behavior Test Relevant to $\alpha_2/\alpha_3 Na^+/K^+$-ATPase
    Gene Modified Mouse Models 341
    *Toke Jost Isaksen, Thomas Hellesøe Holm,
    and Karin Lykke-Hartmann*

31 Zebrafish Whole-Mount In Situ Hybridization Followed
    by Sectioning 353
    *Canan Doganli, Jens Randel Nyengaard,
    and Karin Lykke-Hartmann*

32 Whole-Mount Immunohistochemistry for Anti-F59
    in Zebrafish Embryos (1–5 Days Post Fertilization (dpf)) 365
    Canan Doganli, Lucas Bukata, and Karin Lykke-Hartmann

33 Cell-Based Lipid Flippase Assay Employing
    Fluorescent Lipid Derivatives 371
    *Maria S. Jensen, Sara Costa, Thomas Günther-Pomorski,
    and Rosa L. López-Marqués*

34 Transient Expression of P-type ATPases in Tobacco Epidermal Cells 383
    Lisbeth R. Poulsen, Michael G. Palmgren, and Rosa L. López-Marqués

Part VII Lipid Techniques

35 Lipid Exchange by Ultracentrifugation 397
    Nikolaj Düring Drachmann and Claus Olesen

36 Reconstitution of Na^+,K^+-ATPase in Nanodiscs 403
    *Jonas Lindholt Gregersen, Natalya U. Fedosova, Poul Nissen,
    and Thomas Boesen*

PART VIII   CRYSTALLIZATION

37 Crystallization of P-type ATPases by the High Lipid–Detergent (HiLiDe) Method 413
   Oleg Sitsel, Kaituo Wang, Xiangyu Liu, and Pontus Gourdon

38 Two-Dimensional Crystallization of the Ca^{2+}-ATPase for Electron Crystallography 421
   John Paul Glaves, Joseph O. Primeau, and Howard S. Young

39 Two-Dimensional Crystallization of Gastric H^+,K^+-ATPase for Structural Analysis by Electron Crystallography 443
   Kazuhiro Abe

PART IX   COMPUTATIONAL APPROACHES IN ANALYZING P-TYPE ATPASES

40 MD Simulations of P-Type ATPases in a Lipid Bilayer System 459
   Henriette Elisabeth Autzen and Maria Musgaard

41 Computational Classification of P-Type ATPases 493
   Dan Søndergaard, Michael Knudsen, and Christian Nørgaard Storm Pedersen

42 Molecular Modeling of Fluorescent SERCA Biosensors 503
   Bengt Svensson, Joseph M. Autry, and David D. Thomas

43 How to Compare, Analyze, and Morph Between Crystal Structures of Different Conformations: The P-Type ATPase Example 523
   Jesper L. Karlsen and Maike Bublitz

Index 541

Contributors

KAZUHIRO ABE • *Cellular and Structural Physiology Institute, Nagoya University, Nagoya, Japan; Department of Medicinal Sciences, Graduate School of Pharmaceutical Sciences, Nagoya University, Nagoya, Japan*

JENS PETER ANDERSEN • *Department of Biomedicine, Aarhus University, Aarhus C, Denmark*

HANS-JÜRGEN APELL • *Department of Biology, University of Konstanz, Konstanz, Germany*

JOSÉ M. ARGÜELLO • *Department of Chemistry and Biochemistry, Worcester Polytechnic Institute, Worcester, MA, USA*

ELENA ARYSTARKHOVA • *Laboratory of Membrane Biology, Massachusetts General Hospital, Boston, MA, USA*

JOSEPH M. AUTRY • *Department of Biochemistry, Molecular Biology and Biophysics, University of Minnesota, Minneapolis, MN, USA*

HENRIETTE ELISABETH AUTZEN • *Department of Molecular Biology and Genetics, Aarhus University, Aarhus C, Denmark; Centre for Membrane Pumps in Cells and Disease—PUMPkin, Danish National Research Foundation, Aarhus C, Denmark*

HASSINA AZOUAOUI • *Institute for Integrative Biology of the Cell (I2BC) – UMR 9198 CEA/CNRS/Université Paris-Sud, Gif-sur-Yvette Cedex, France; CEA, iBiTec-S/SB2SM, CEA Saclay, Gif-sur-Yvette Cedex, France*

RAPHAELLE BARRY • *Institute for Integrative Biology of the Cell (I2BC) – UMR 9198 CEA/CNRS/Université Paris-Sud, Gif-sur-Yvette Cedex, France; CEA, iBiTec-S/SB2SM, CEA Saclay, Gif-sur-Yvette, France; INSERM, U968, Paris, France; UMR_S 968, Institut de la Vision, Sorbonne Universités, UPMC Univ Paris, Paris, France; CNRS, UMR 7210, Paris, France*

GIANLUCA BARTOLOMMEI • *Department of Chemistry "Ugo Schiff", University of Florence, Sesto Fiorentino, Italy*

PHILIP A. BIDWELL • *Department of Pharmacology and Cell Biophysics, College of Medicine, University of Cincinnati, Cincinnati, OH, USA*

THOMAS BOESEN • *Department of Molecular Biology and Genetics, Aarhus University, Aarhus C, Denmark; Centre for Membrane Pumps in Cells and Disease—PUMPkin, Danish National Research Foundation, Aarhus C, Denmark; Danish Research Institute for Translational Neuroscience—DANDRITE, Nordic EMBL Partnership of Molecular Medicine, Aarhus C, Denmark*

MAIKE BUBLITZ • *Department of Biochemistry, University of Oxford, Oxford, UK*

LUCAS BUKATA • *Sanford-Burnham Medical Research Institute, La Jolla, CA, USA*

ERNESTO CARAFOLI • *VIMM, Venetian Institute of Molecular Medicine, University of Padova, Padova, Italy*

PHILIPPE CHAMPEIL • *Institute for Integrative Biology of the Cell (I2BC) – UMR 9198 CEA/ CNRS/Université Paris-Sud, Gif-sur-Yvette Cedex, France; CEA, iBiTec-S/SB2SM, CEA Saclay, Gif-sur-Yvette Cedex, France*

Contributors

JOHANNES D. CLAUSEN • *Department of Biomedicine, Aarhus University, Aarhus C, Denmark*

SARA COSTA • *Department of Plant and Environmental Sciences, Centre for Membrane Pumps in Cells and Disease (PUMPKIN), University of Copenhagen, Frederiksberg C, Denmark*

BOJANA DAMNJANOVIC • *Department of Biology, University of Konstanz, Konstanz, Germany*

STEFANIA J. DANKO • *Asahikawa Medical University, Asahikawa, Hokkaido, Japan*

ROBERT E. DEMPSKI • *Department of Chemistry and Biochemistry, Worcester Polytechnic Institute, Worcester, MA, USA*

CANAN DOGANLI • *Smith Cardiovascular Research Institute, University of California, San Francisco, CA, USA*

NIKOLAJ DÜRING DRACHMANN • *Centre for Membrane Pumps in Cells and Disease – PUMPkin, Danish National Research Foundation, Aarhus, Denmark; Syngenta, Jealott's Hill International Research Centre, Bracknell, United Kingdom*

NATALYA U. FEDOSOVA • *Centre for Membrane Pumps in Cells and Disease—PUMPkin, Danish National Research Foundation, Aarhus C, Denmark; Department of Biomedicine, Aarhus University, Aarhus C, Denmark*

ANJA THOE FUGLSANG • *Department of Plant and Environmental Sciences, Centre for Membrane Pumps in Cells and Disease (PUMPKIN), University of Copenhagen, Frederiksberg C, Denmark*

JOHN PAUL GLAVES • *Department of Biochemistry, University of Alberta, Edmonton, AB, Canada*

PONTUS GOURDON • *Department of Biomedical Sciences, University of Copenhagen, Copenhagen, Germany; Department of Experimental Medical Science, Lund University, Lund, Sweden*

JONAS LINDHOLT GREGERSEN • *Department of Molecular Biology and Genetics, Aarhus C, Denmark; Centre for Membrane Pumps in Cells and Disease—PUMPkin, Danish National Research Foundation, Aarhus C, Denmark*

THOMAS GÜNTHER-POMORSKI • *Department of Plant and Environmental Sciences, Centre for Membrane Pumps in Cells and Disease (PUMPKIN), University of Copenhagen, Frederiksberg C, Denmark*

FLORIAN HILBERS • *Danish Research Institute of Translational Neuroscience – DANDRITE, Nordic-EMBL Partnership for Molecular Medicine, Aarhus University, Aarhus C, Denmark; Centre for Membrane Pumps in Cells and Disease – PUMPKIN, Danish National Research Foundation, Department of Molecular Biology and Genetics, Aarhus University, Aarhus C, Denmark*

THOMAS HELLESØE HOLM • *Department of Biomedicine and Centre for Membrane Pumps in Cells and Disease-PUMPKIN, Danish National Research Foundation, Aarhus University, Aarhus University, Aarhus C, Denmark*

TOKE JOST ISAKSEN • *Department of Biomedicine and Centre for Membrane Pumps in Cells and Disease-PUMPKIN, Danish National Research Foundation, Aarhus University, Aarhus C, Denmark*

AURORE JACQUOT • *Institute for Integrative Biology of the Cell (I2BC) – UMR 9198 CEA/CNRS/Université Paris-Sud, Gif-sur-Yvette Cedex, France; CEA, iBiTec-S/SB2SM, CEA Saclay, Gif-sur-Yvette Cedex, France; UMR 5004, Biochimie et Physiologie Moléculaire des Plantes, Institut de Biologie Intégrative des Plantes-Claude Grignon, Montpellier, France*

MARIA S. JENSEN • *Department of Plant and Environmental Sciences, Centre for Membrane Pumps in Cells and Disease (PUMPKIN, University of Copenhagen, Frederiksberg C, Denmark*

BO HØJEN JUSTESEN • *Department of Plant and Environmental Sciences, Centre for Membrane Pumps in Cells and Disease (PUMPKIN), University of Copenhagen, Frederiksberg C, Denmark*

JESPER L. KARLSEN • *Department of Molecular Biology and Genetics, Aarhus University, Aarhus C, Denmark*

MICHAEL KNUDSEN • *Bioinformatics Research Centre (BiRC), Aarhus University, Aarhus C, Denmark*

JAN B. KOENDERINK • *Pharmacology/Toxicology 149, Radboud University Medical Center, Radboud Institute for Molecular Life Sciences, Nijmegen, The Netherlands*

EVANGELIA G. KRANIAS • *Department of Pharmacology and Cell Biophysics, University of Cincinnati, College of Medicine, Cincinnati, OH, USA*

GUILLAUME LENOIR • *Institute for Integrative Biology of the Cell (I2BC) – UMR 9198 CEA/CNRS/Université Paris-Sud, Gif-sur-Yvette Cedex, France; CEA, iBiTec-S/SB2SM, CEA Saclay, Gif-sur-Yvette Cedex, France*

XIANGYU LIU • *School of Medicine, Tsinghua University, Beijing, China*

ROSA L. LÓPEZ-MARQUÉZ • *Department of Plant and Environmental Sciences, Centre for Membrane Pumps in Cells and Disease (PUMPKIN), University of Copenhagen, Frederiksberg C, Denmark*

KARIN LYKKE-HARTMANN • *Department of Biomedicine and Centre for Membrane Pumps in Cells and Disease-PUMPKIN, Danish National Research Foundation, Aarhus University, Aarhus C, Denmark; Department of Molecular Biology and Genetics, Aarhus University, Aarhus C, Denmark; Aarhus Institute of Advanced Studies (AIAS), Aarhus University, Aarhus C, Denmark*

MAGDALENA MAREK • *Department of Plant and Environmental Sciences, Centre for Membrane Pumps in Cells and Disease (PUMPKIN), University of Copenhagen, Frederiksberg C, Denmark*

DAVID B. MCINTOSH • *Division of Chemical Pathology, Faculty of Health Sciences, Institute of Infectious Diseases and Molecular Medicine, University of Cape Town, Cape Town, South Africa*

JESPER V. MØLLER • *Department of Biomedicine, Aarhus University, Aarhus C, Denmark; Centre for Membrane Pumps in Cells and Disease, Danish National Research Foundation, Aarhus C, Denmark*

CÉDRIC MONTIGNY • *Institute for Integrative Biology of the Cell (I2BC) – UMR 9198 CEA/CNRS/Université Paris-Sud, Gif-sur-Yvette Cedex, France; CEA, iBiTec-S/SB2SM, CEA Saclay, Gif-sur-Yvette Cedex, France*

MARIA MUSGAARD • *Department of Biochemistry, University of Oxford, Oxford, United Kingdom*

VERENA NIGGLI • *Department of Pathology, University of Bern, Bern, Switzerland*

POUL NISSEN • *Department of Molecular Biology and Genetics, Aarhus University, Aarhus C, Denmark*

JENS RANDEL NYENGAARD • *Stereology and Electron Microscopy Laboratory, Centre for Stochastic Geometry and Advanced Bioimaging, Aarhus University Hospital, Aarhus University, Aarhus C, Denmark*

CLAUS OLESEN • *Department of Molecular Biology and Genetics, Aarhus University, Aarhus C, Denmark; Centre for Membrane Pumps in Cells and Disease, Danish National Research Foundation, Aarhus C, Denmark*

TERESITA PADILLA-BENAVIDES • *Department of Chemistry and Biochemistry, Worcester Polytechnic Institute, Worcester, MA, USA*

MICHAEL G. PALMGREN • *Department of Plant and Environmental Sciences, Centre for Membrane Pumps in Cells and Disease (PUMPKIN), University of Copenhagen, Frederiksberg C, Denmark*

CHRISTIAN NØRGAARD STORM PEDERSEN • *Bioinformatics Research Centre (BiRC), Aarhus University, Aarhus C, Denmark; Centre for Membrane Pumps in Cells and Disease—PUMPKIN, Danish National Research Foundation, Aarhus C, Denmark*

HANNE POULSEN • *Danish Research Institute of Translational Neuroscience – DANDRITE, Nordic-EMBL Partnership for Molecular Medicine, Aarhus University, Aarhus C, Denmark; Centre for Membrane Pumps in Cells and Disease – PUMPKIN, Danish National Research Foundation, Department of Molecular Biology and Genetics, Aarhus University, Aarhus C, Denmark*

LISBETH R. POULSEN • *Department of Plant and Environmental Sciences, Centre for Membrane Pumps in Cells and Disease (PUMPKIN), University of Copenhagen, Frederiksberg C, Denmark*

JOSEPH O. PRIMEAU • *Department of Biochemistry, University of Alberta, Edmonton, AB, Canada*

PANKAJ SEHGAL • *Department of Biomedicine, Aarhus University, Aarhus C, Denmark; Centre for Membrane Pumps in Cells and Disease, Danish National Research Foundation, Aarhus C, Denmark*

OLEG SITSEL • *Department of Molecular Biology and Genetics, Aarhus University, Aarhus C, Denmark*

DAN SØNDERGAARD • *Bioinformatics Research Centre (BiRC), Aarhus University, Aarhus C, Denmark; Centre for Membrane Pumps in Cells and Disease—PUMPKIN, Danish National Research Foundation, Aarhus C, Denmark*

HIROSHI SUZUKI • *Asahikawa Medical University, Asahikawa, Hokkaido, Japan*

BENGT SVENSSON • *Department of Biochemistry, Molecular Biology and Biophysics, University of Minnesota, Minneapolis, MN, USA*

HERMAN G.P. SWARTS • *Radboud University Medical Center, Nijmegen, The Netherlands*

KATHLEEN J. SWEADNER • *Laboratory of Membrane Biology, Massachusetts General Hospital, Boston, MA, USA*

FRANCESCO TADINI-BUONINSEGNI • *Department of Chemistry "Ugo Schiff", University of Florence, Sesto Fiorentino, Italy*

DAVID D. THOMAS • *Department of Biochemistry, Molecular Biology and Biophysics, University of Minnesota, Minneapolis, MN, USA*

ILSE VANDECAETSBEEK • *Department of Cellular and Molecular Medicine, Laboratory of Cellular Transport Systems, KU Leuven, Leuven, Belgium*

PETER VANGHELUWE • *Department of Cellular and Molecular Medicine, Laboratory of Cellular Transport Systems, KU Leuven, Leuven, Belgium*

KAITUO WANG • *Department of Biomedical Sciences, University of Copenhagen, Copenhagen, Denmark*

ALEX GREEN WIELANDT • *Department of Plant and Environmental Sciences, Centre for Membrane Pumps in Cells and Disease (PUMPKIN), University of Copenhagen, Frederiksberg C, Denmark*

DAVID G. WOOLLEY • *Division of Chemical Pathology, Institute of Infectious Diseases and Molecular Medicine, Faculty of Health Sciences, University of Cape Town, Cape Town, South Africa*

HOWARD S. YOUNG • *Department of Biochemistry, University of Alberta, Edmonton, AB, Canada*

Chapter 1

An Introduction to P-type ATPase Research

Poul Nissen

P-type ATPases — the word alone may not ring a bell, but adding ion pumps and lipid flippases it would immediately spur interest. Electrochemical gradients and asymmetric distributions of lipids across biomembranes are fundamental aspects of life, and are maintained by the members of this protein family. Considering also that P-type ATPases consume a very significant proportion of the energy in the cell — approximately 1/3 overall and some estimated 40–70 % in brain — it becomes immediately apparent that studies of the structure, function, and mechanism of P-type ATPases are of fundamental importance in the life sciences. For understanding life we need to investigate these proteins also.

The P-type ATPase field has a very long and proud tradition. Jens Chr. Skou identified the Na^+, K^+-ATPase enzyme as a membrane-associated ATPase activity responding to the amounts and ratios of Na^+, K^+, and Mg^{2+} in the medium. At a time where the concepts of biomembranes and protein structure were still in their infancy this was arguably a first note of an integral membrane protein. The P-type ATPases therefore became destined to pioneer many of the experimental approaches that are now being used for membrane protein purification and biochemical and biophysical characterization.

Another important point is that many P-type ATPases are naturally abundant in native tissues such as the sarcoplasmic reticulum Ca^{2+}-ATPase (SERCA) from muscle and the Na^+, K^+-ATPase from kidney (for native source isolation protocols, *see* Chaps. 2–4). Early on, SERCA was also sequenced, cloned, expressed, and functionally probed in mutant forms. The function of P-type ATPases can be approached in many ways, probing for example the overall ATPase activity, the transport of radioactive isotopes of, e.g., calcium and rubidium ions or as probed by fluorescent dyes (*see* **Part II**), or the partial reactions of phosphorylation and dephosphorylation (*see* **Part III**). Equally important, some of the members can be targeted by specific, high-affinity inhibitors, making it possible to assess very accurately the inhibitor-sensitive activity of P-type ATPases in a complex background — e.g., the ouabain-sensitive Na^+, K^+-ATPase activity or the thapsigargin-sensitive activity of SERCA (*see*, e.g., Chaps. 2, 10, 14, 19, 21, 23, 26, and 28).

P-type ATPases were also targeted from the very beginning for structural characterization, first by electron microscopy. Since the year 2000, complete structures have been determined by X-ray crystallography, and today more than 76 structures are deposited in the Protein Data Bank, representing for example more than 12 intermediate states of the SERCA transport cycle and key structures of Na^+, K^+-ATPase, H^+, K^+-ATPase, H^+-ATPase, and Cu^+- and Zn^{2+}-ATPases (*see* also Chap. 43).

To some extent the bounty of P-type ATPases and the pioneering approaches of earlier times also became a straitjacket. It is very difficult to introduce new methods when established approaches work so well — and how can one qualify suggestions of other cell biological functions when a key role as ion pumps can explain almost anything?

This is the challenge of the future of the P-type ATPase field. Not to say that the established methods have played out their game — not even close — but the field needs to embrace also new approaches, including single-molecule studies, computer simulations, and protein-protein interaction studies in native or near-native environments like lipid nanodiscs (*see*, e.g., Chaps. 36, 40, and 42). Also the use of advanced transgenic cell culture and animal models such as zebrafish and mouse (*see* **Part VI**) will continue to develop, and the studies of mutations associated with disease will translate research into clinical models or drug discovery. This book summarizes the state-of-the-art in many aspects of P-type ATPase research, which will hopefully also inspire the reader to identify the new opportunities that exist, and how current approaches tackle problems on the way.

Part I

Protein Production, Isolation, Purification, and Stabilization

Chapter 2

Purification of Na,K-ATPase from Pig Kidney

Natalya U. Fedosova

Abstract

The method of purification of Na,K-ATPase from pig kidney is based on a differential centrifugation and SDS-treatment of a microsomal preparation. The yield is 0.4 mg protein per 1 g tissue with the specific (ouabain-sensitive) activity of 25–28 μmol P_i/min per mg protein and nucleotide binding capacity of 3 nmol/mg. The protein/lipid ratio is 1/1 (mg/mg) with a protein purity of ~80 %.

Key words Na,K-ATPase, Kidney, Microsomal fraction differential centrifugation, SDS-treatment

1 Introduction

Na,K-ATPase, a member of the P-type ATPase superfamily, maintains transmembrane gradients of Na^+ and K^+, critical for cell homeostasis. The minimal functional unit of the Na,K-ATPase consists of a large catalytic α-subunit, defining its membership in the P-ATPase family (M_r 113 kDa), and a glycosylated β-subunit (M_r 55 kDa). A third regulatory subunit from the FXYD family is often present in a tissue-specific manner. The stoichiometry under normal conditions is 1α/1β/optionally 1FXYD.

The physiological significance of the Na,K-ATPase (Na-pump) has nourished the interest to investigation of its structure, function and regulation for over half a century. Protein chemistry research and in the recent years crystallization studies consume considerable amount of purified enzyme. Although present in most animal cells, the enzyme is particularly plentiful in secretory and excitable tissues. The best sources for purification are kidney, brain, salt glands, and electric organs. A collection of purification protocols from different sources is available in Methods in Enzymology [1]. Klodos et al. [2] described the large-scale preparation of partly purified Na,K-ATPase from pig kidney with precise estimation of person-hours of work and discussed the quality of the final preparation.

This chapter presents the above method [2] step by step, the protocol successfully used in our lab to supply material for functional as well as crystallization studies.

2 Materials

Prepare all solutions using ultrapure water and analytical grade reagents. All solutions must be stored at 4 °C. The volumes are sufficient for purification of the enzyme from 15 kg whole kidneys.

1. Buffer I: 30 mM Histidine, 250 mM sucrose; pH 7.3 (20 °C). Weigh 4.65 g L-histidine, 85.58 g sucrose, add 900 mL H$_2$O and adjust to pH 7.3 with 1 N HCl. Make up to 1 L with H$_2$O.

2. Buffer II: 25 mM Imidazole, 250 mM sucrose, 1 mM EDTA; pH 7.4 (20 °C). Weigh 10.21 g imidazole, 513.5 g sucrose, 1.752 g Na$_2$EDTA·2H$_2$O, add 5.5 L H$_2$O and adjust pH to 7.4 with 1 N HCl and make up to 6 L with H$_2$O.

3. Buffer III: 20 mM Histidine, 250 mM Sucrose, 0.9 mM EDTA; pH 7.0 (20 °C). Weigh 6.2 g Histidine, 171.15 g Sucrose, 0.526 g Na$_2$EDTA·2H$_2$O, add 1.5 L H$_2$O and adjust pH to 7.0 with 1 N HCl and make up to 2 L with H$_2$O.

4. 15 kg of fresh kidneys packed in plastic bags (maximum 10 kidneys per portion) at the slaughterhouse, immediately cooled, and transported on ice.

5. Stainless steel tissue press, with plates with 2 and 1.4 mm diameter holes.

6. Pestle homogenizer with Teflon pestle (Sartorius Potter S).

3 Methods

3.1 Kidney Dissection

1. Each kidney is cut longitudinally in half (Fig. 1).

2. Cut (dissect) the inner medulla away with a scalpel and discard it. This procedure gives access to the outer medulla (Fig. 1).

3. Dissect the outer medulla with the Rongeur tong (forceps) (Fig. 2); that is, take the pink and reddish tissue (Fig. 1) and place it in small beakers on ice with some amount of cold buffer I. The beakers with buffer are weighed beforehand in order to allow estimation of the amount of collected tissue. The volume of the buffer I and the amount of tissue per beaker should be approximately equal. Collect 330 g outer medulla tissue.

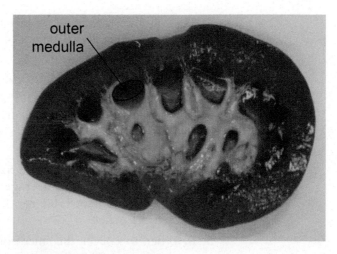

Fig. 1 Longitudinal cut over pig kidney

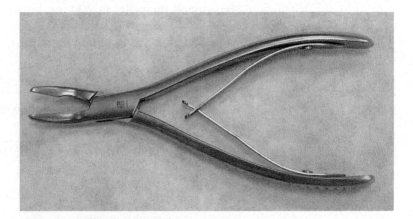

Fig. 2 Rongeur tong for outer medulla dissection

3.2 Disruption of Tissue

The following step is a disruption of the collected tissue by passage through a stainless steel tissue press.

1. Approximately 50 mL portions of tissue in buffer I are forced through first the plate perforated with 2.0 mm diameter holes and then through a plate with 1.4 mm diameter holes. The resulting suspension (approximately 500 mL) may be stored on ice in a refrigerator until the next day if necessary (*see* **Note 1**).

3.3 Preparation of Crude Microsomal Fraction by Differential Centrifugation

1. Add 500 mL of cold buffer II to the tissue suspension (500 mL) and homogenize in a pestle homogenizer with a Teflon pestle on ice at, e.g., 1500–2000 rpm, three times down and up (avoid foaming).

2. The homogenate (1 L) is diluted up to 3 L volume with buffer II; that is, the mixture contains 110 g tissue/L.

3. The homogenate is subjected to centrifugation at 3700×*g*-average for 20 min at 4 °C (e.g., Sorvall GSA-rotor 6000 rpm).

4. Supernatant (S1) is kept on ice while the pellets are resuspended in 1 L of buffer II, homogenized as before.

5. The second homogenate is diluted to 3 L with buffer II and centrifugation is repeated as in **step 3**. Keep supernatant (S2) on ice and discard the pellet.

4. Combine supernatant S1 and S2 (ca. 5 L) and subject it to centrifugation 7400×*g*-average for 20 min at 4 °C (e.g., Sorvall SS34 rotor 8500 rpm). Save the supernatant and discard the pellet.

5. In order to collect the microsomal fraction, centrifuge the supernatant at 38,000×*g*-average for 40 min at 4 °C (e.g., Sorvall SS34 rotor 20,000 rpm). Save the pellet.

6. Resuspend the pellet in 200 mL buffer II and homogenize as described above with five strokes down and up. The microsomal fraction can be stored for months at −80 °C (*see* **Notes 2 and 3**).

3.4 Titration with SDS

1. Prepare a stock solution of 1 % SDS w/v (highest purity available) in buffer II and make the following dilutions according to the scheme:

Nr.	SDS %	1 % SDS in buffer II, μL	Buffer II, μL
1	0	0	1000
2	0.1	100	900
3	0.2	200	800
4	0.3	300	700
5	0.35	350	650
6	0.4	400	600
7	0.45	450	550
8	0.5	500	500
9	0.55	550	450
10	0.6	600	400
11	0.65	650	350
12	0.7	700	300
13	0.75	750	250
14	0.8	800	200

2. After determination of the protein concentration in the microsomal preparation (*see* **Note 3**), thaw the second aliquot and dilute the microsomes to 4.6 mg protein/mL with buffer II.

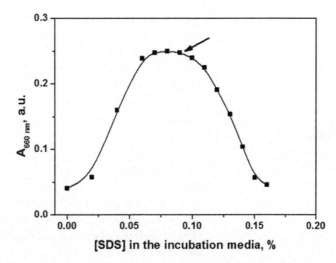

Fig. 3 Na,K-ATPase activity estimated colorimetrically from P_i release as function of SDS concentration in the incubation media. Optimal SDS concentration is marked by an *arrow*

3. Prepare 14 samples: To 80 μL microsomal fraction (4.6 mg/mL) add 20 μL of SDS solution in buffer II (according to the scheme above) and incubate the samples at 20 °C for 1 h on a water bath.

4. Measure Na,K-ATPase activity of each sample and plot the activity vs. final SDS concentration (Fig. 3). The Na,K-ATPase activity is defined as a difference in phosphate production in the absence or presence of 1 mM ouabain; *see* Chaps. 10–12 in this volume for protocols on colorimetric activity assays. Maximum of the bell-shaped curve, i.e., optimal SDS concentration for the activation of the microsomal fraction at the next step, is in the range of 0.08–0.1 % (*see* **Note 4**).

3.5 Treatment with SDS

1. Thaw the frozen microsomal preparation on a water bath at room temperature and dilute it with buffer II to a protein concentration of 4.6 mg/mL.

2. Add SDS dissolved in buffer II in the amount determined by the titration experiment and the relative volume as described above under continuous stirring (avoid foaming) (*see* **Note 5**).

Keep the suspension at 20 °C overnight.

3.6 Washing Procedure

1. Centrifuge the suspension at $127{,}000 \times g$-average for 50 min at 10 °C (e.g., Beckmann Ti 45 rotor at 40,000 rpm). Discard supernatant.

2. For washing—resuspend the pellet in 400 mL of buffer III using tight pestle homogenizer and repeat centrifugation (four cycles).

3. Suspend the final pellet in 25 mL of buffer II (to an approximate protein concentration of 5 mg/mL). Take a small aliquot for determination of the protein concentration and specific Na,K-ATPase activity. Store both the aliquot and the main batch at −20 °C.

4 Notes

1. Do not use a mixer or a blender. In our hands it decreased specific activity of the final enzyme preparation. However, other labs do use small blenders. In this case, the dissected tissue is drained on gaze, put it in the blender and blended with frozen buffer cubes. Ice cubes prevent foaming and therefore denaturation.
2. We freeze two aliquots (1 mL and 2 mL) of microsomal fraction for determination of protein concentration and optimal conditions for SDS-treatment conditions separately from the batch protein.
3. The protein concentration is measured according to Lowry et al. [3] against bovine serum albumin as a standard without correction for color factor.
4. In our hands, the shift of the bell-shaped curve towards higher SDS concentrations correlates with poor quality of the microsomal preparation and results in lower specific activity of the final preparation.
5. Add SDS very slowly under continuous stirring.

Acknowledgement

I thank Dr. Mikael Esmann for helpful suggestions. Ms. Birthe Bjerring Jensen, Anne Lillevang, and Angelina Damgaard are gratefully acknowledged for their expert advice and practical tips.

References

1. Methods in Enzymology 156, Section I. Preparation of Na$^+$,K$^+$-ATPase and Subunits. pp. 29–71
2. Klodos I, Esmann M, Post RL (2002) Large scale preparation of sodium-potassium ATPase from kidney outer medulla. Kidney Int 62: 2097–2100
3. Lowry OH, Rosebrough NJ, Farr AL, Randall RJ (1951) Protein measurement with the Folin phenol reagent. J Biol Chem 193:265–275

Chapter 3

Preparation of Ca^{2+}-ATPase1a Enzyme from Rabbit Sarcoplasmic Reticulum

Jesper V. Møller and Claus Olesen

Abstract

The SERCA calcium ATPase, probably the most well-investigated membrane protein both from a biophysical and structural view, can be purified from native source in substantial quantities, making it a favorable target when conducting biochemical experiments and structure determination, e.g., by X-ray crystallography.

Key words SR calcium ATPase, Native source purification, Membrane protein

1 Introduction

The Sarco/endoplasmic reticulum Ca^{2+}-ATPase (SERCA) is an active transporter of calcium ions from the cytoplasm to the intracellular sarco/endoplasmic reticulum (SR) store, thereby maintaining the Ca^{2+} gradient across the membrane of intracellular vesicles of all cells. This process is of particular importance in muscle cells in which Ca^{2+} triggers contraction by binding troponin C, allowing the myosin-actin cross-bridge interaction that facilitates the muscle contraction. The protein belongs to the P-type ATPase family with which the exploration started in 1957 with Jens Christian Skou's discovery of the Na/K-ATPase purified from crab legs, and subsequently a Ca^{2+}-ATPase was identified in rabbit fast twitch muscles by Hasselbach and Makinose (1959). SERCA functions by an ATP hydrolysis-driven transition between two major states: E1 with a high calcium affinity (Ca_2E1) where Ca^{2+}-ions from the cytosol are bound, and an E2 state with a low calcium affinity (H_nE2), where the Ca^{2+}-ions are released to the SR lumen. The two major states are further subdivided into a number of intermediate steps to account for the phosphorylation by ATP with occlusion of the calcium ions followed by their translocation.

Since its discovery the Ca^{2+}-ATPase has been subject to extensive investigations, which has been made possible for a number of reasons such as availability and stability both in membrane and detergent solubilized form. This has allowed detailed studies of the intermediate states by activity measurements and other assays. Furthermore, the ability to produce both tight and leaky SR vesicles and to reconstitute the ATPase in proteoliposomes have been important tools in the biophysical characterization of the Ca^{2+}-ATPase, in calcium transport/uptake and phosphorylation experiments [1].

In the following, we present a purification protocol for isolation of SR vesicles from skeletal muscles, where SERCA1a is the predominant isoform. The protocol is based on a series of differential centrifugation steps that originally was developed by Hasselbach and Makinose [2] and later refined by de Meis and Hasselbach [3]. These are detailed in the present protocol which also includes an up-to-date account with our own modifications.

The latter part of this chapter includes a description of a protocol for a mild deoxycholate (DOC) extraction based on a procedure by Meissner et al. [4] for purification of Ca^{2+}-ATPase from the SR preparation described above, in which loosely attached proteins like calsequestrin and M55 glycoprotein are removed from the microsomes. It is important to note that this procedure has the further advantage that it makes the microsomes leaky to Ca^{2+}, a circumstance that is often helpful to avoid complications arising from the formation of Ca^{2+} transmembrane gradients in the interpretation of activity experiments.

2 Materials

Prepare all solutions by using ultrapure water (18 MΩ cm sensitivity). Filtrate all solutions by 0.22 μm filters and keep all buffers and solutions ice-cold for every step of the purification procedure (very important).

2.1 Preparation of SR Vesicles

1. Two rabbits (approx. 4–5 kg each).
2. Balance.
3. Blender (a household type is adequate).
4. Hand-operated meat grinder.
5. Scalpels, scissors, tinfoil.
6. Measuring cylinder 250 mL.
7. Ice.
8. Sorvall centrifuge, Sorvall SLA-1500 rotor, and tubes.
9. Sorvall centrifuge, SS-34 rotor, and clear tubes.
10. Beckmann Coulter centrifuge, Ti-45 rotor, and tubes.
11. Beckmann Coulter centrifuge, Ti-70 rotor, and tubes.

12. Timer.
13. Automatic homogenizer apparatus with ice cooling of the homogenizing glass.
14. 5 mL Teflon homogenizer.
15. 30 mL Teflon homogenizer.
16. Buffer A: 3 L of 100 mM KCl, 2.5 mM K_2HPO_4, 2.5 mM KH_2PO_4, 2.0 mM EDTA.
17. Solution B: 150 mL of 1 M sucrose, 50 mM KCl.
18. Buffer C: 236 mL of 120 mL 2 M KCl, 8 mL 100 mM MgATP, pH 7.0, 108 mL water. The final solution is not pH adjusted. Leave the mixture on ice for 30 min to 1 h before use.
19. Solution D: 200 mL of 50 mM KCl.
20. Buffer E: 50 mL of 0.3 M sucrose, 5 mM Hepes, pH 7.4.

2.2 Preparation of Ca^{2+}-ATPase Membranes by DOC Extraction

1. Sodium deoxycholate (*see* **Note 1**).
2. Sorvall centrifuge, SS-34 rotor, and clear tubes.
3. Beckmann Coulter ultracentrifuge, Ti-45 rotor, and tubes.
4. Beckmann Coulter ultracentrifuge, Ti-70 rotor, and tubes.
5. Timer.
6. Plastic Pasteur pipet and glass rods.
7. Automatic homogenizer apparatus with ice cooling of the homogenizing glass.
8. 5 mL Teflon homogenizer.
9. 30 mL Teflon homogenizer.
10. Extraction buffer I: 0.3 M sucrose, 0.5 M KCl, 1 mM EDTA, 10 mM Tris, 0.01 mM $CaCl_2$, 1.25 mM $MgCl_2$, pH 7.9 (we prepare a large stock and keep it in the freezer in suitable aliquots for later use).
11. Extraction buffer II: 150 mL Extraction buffer I + DOC 0.5 mg/mL × 150 mL = 60 mg and DTT 0.5 mg/mL × 150 mL = 75 mg (freshly prepared).
12. Washing buffer: 60 mL of 5 mM TAPS (pH 7.5), 0.3 M Sucrose, 0.5 M KCl, 0.5 mM $MgCl_2$, 10 μM $CaCl_2$.
13. Buffer E: 50 mL of 0.3 M sucrose, 5 mM Hepes pH 7.4.

3 Methods

3.1 Preparation of SR Vesicles

1. Kill two fasting rabbits of approx. 4–5 kg by bleeding after a blow to the neck.
2. After removal of the skin, dissect the fast twitch muscles from the back and back limbs free with a scalpel and directly transfer them to the tinfoil placed on ice.

3. Carefully remove the fat tissue, fascia, etc. from the muscles with scissors, as they are being dissected from the rabbit. Be very careful not to contaminate the meat with blood and hair from the rabbit. The clean meat preparation is kept on ice until further use.

The following preparation is conducted in the cold room and all tubes and samples are continuously kept on ice!

4. Grind the meat twice using a hand driven meat grinder (to keep heating of the meat to an absolute minimum).
5. Mix portions of approx. 55 g of minced meat with 180 mL buffer A and then blend at full speed for 15 s followed by a 10-s break. Repeat this procedure three times (*see* **Note 2**).
6. Transfer and distribute the resulting rather viscous meat homogenate among plastic centrifuge flasks and centrifuge at 4 °C for 20 min and $6400 \times g$.
7. Filter the supernatant through 7–10 layers of gauze into a glass beaker placed on ice to further remove cell debris and fat.
8. Distribute the filtered sample in the plastic centrifuge flasks for medium speed centrifugation for 20 min at 4 °C at $9700 \times g$.
9. Once more, filter the supernatant through 7–10 layers of gauze into a glass beaker placed on ice to remove final remnants of cell debris.
10. Pour the supernatant into centrifuge tubes and centrifuge at 4 °C, and $47{,}800 \times g$ for 60 min to sediment the microsomal pellet (sarcoplasmic reticulum) (*see* **Note 3**).
11. Discard the supernatant and resuspend the pellet by homogenization using a Teflon homogenizer 25 times at low speed (400–500 rpm) in 150 mL of solution B.
12. Transfer the suspension into centrifuge tubes and centrifuge for 30 min at $4300 \times g$ and 4 °C.
13. Gently decant and save the supernatant and discard the pellet (*see* **Note 4**).
14. Mix the supernatant with KCl/ATP extraction buffer C and leave on ice or in the refrigerator for 30 min to 1 h (*see* **Note 5**).
15. The preparation is centrifuged in an ultracentrifuge for 90 min at $84{,}500 \times g$ and 4 °C.
16. Discard the supernatant and resuspend the pellet in 6×70 mL washing solution D followed by gentle homogenization 25 times with a Teflon homogenizer at approx. 500 rpm.
17. The homogenized supernatant is distributed into ultracentrifuge tubes and centrifuged for 60 min at $84{,}500 \times g$ and 4 °C.
18. Repeat **step 13**.
19. Discard the supernatant, resuspend the pellet and homogenize gently using a 5 mL Teflon homogenizer in 24 mL buffer E 25 times.

20. Transfer to cryo tubes in suitable aliquots (usually 1–5 mL), flash freeze in liquid nitrogen and store at −80 °C.

21. Determine the protein concentration, e.g., by the Lowry method [5] and subsequently measure the specific activity of the SR-Ca^{2+}-ATPase preparation (*see*, e.g., **Part II** in this book).

3.2 Preparation of Ca^{2+}-ATPase Membranes by DOC Extraction

This is a mild DOC extraction procedure for purification of Ca^{2+}-ATPase from the SR preparation protocol described above, in which the loosely attached proteins like calsequestrin and M55 glycoprotein are removed. It is important to note that if this procedure is done correctly, the DOC extraction will not solubilize the Ca^{2+}-ATPase but merely remove extrinsic proteins.

1. Mix the extraction buffer I with SR protein, DTT and DOC, aim at the following concentrations: 2.5 mg protein/mL, 0.5 mg DOC/mL, 0.5 mg DTT/mL (*see* **Note 6**).

2. Incubate the solution for 10 min on ice and then distribute into Ti-70 tubes and centrifuge for 75 min at 4 °C at $181,000 \times g$.

3. Resuspend the pellet in 147.9 mL of extraction buffer II, and spin for 75 min at 4 °C at $181,000 \times g$.

4. Wash the pellet in 60 mL washing buffer (20 mL/125 mg protein) and centrifuge for 75 min at 4 °C at $181,000 \times g$.

5. Resuspend the pellet in approx. 7.5 % of the starting volume in buffer E (approx. 7 mL).

6. Aliquot the microsome sample in cryo tubes in portions of 100 and 300 μL, flash freeze in liquid nitrogen, and store at −80 °C.

7. Determine the protein concentration by the Lowry method and then measure and calculate the specific activity of the preparation. Check the purification by SDS-PAGE (*see* Fig. 1) and compare with the activity of the SR preparation after permeabilization with A23187 (Ca^{2+}-specific ionophore) or addition of $C_{12}E_8$.

4 Notes

1. We obtain reproducible good results with sodium deoxycholate from Sigma Aldrich (D6750).

2. This relatively brutal maceration and homogenization in Subheading 3.1, **steps 4** and **5**, is important to ensure a sufficient disintegration of the muscle tissue such that the connective tissue is broken up and, more importantly, the myofibrils are finely cut, preventing them from binding various cell organelles, in particular SR and thereby cause these to co-sediment in the following low speed centrifugations steps (Subheading 3.1, **steps 6** and **8**) designed to remove the

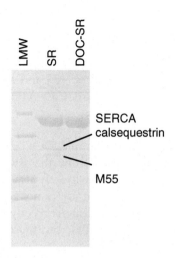

Fig. 1 SDS-PAGE of the SR and DOC-SR preparations

major part of the extracellular connective tissue, cell debris, nuclei and the majority of the myofibril and the mitochondria.

3. This will separate the major part of microsomal fraction from the soluble proteins.
4. This is an important and somewhat difficult step: The white/gray pellet is rather soft and loose and difficult to see when decanting, so be careful to avoid disturbing the sediment.
5. The KCl/ATP extraction procedure causes the actomyosine to dissociate and become soluble hence these components will not sediment and pellet in the following centrifugation.
6. As an aid to following the procedure, we have included an example based on a standard SR vesicle preparation result from the procedure described above. The starting point is an SR preparation which gave a yield of 28.44 mg protein/mL and had a total volume of 13 mL. 13 mL SR prep has a total protein content of (13 mL × 28.44 mg/mL) 369.72 mg. This is to be suspended in a total volume of (369.72 mg / 2.5 mg/mL) = 147.9 mL, i.e., by addition of (147.9 mL − 13 mL) = 134.9 mL extraction buffer I. If you prepare 150 mL extraction buffer I it should contain: (0.5 mg/mL × 147.9 mL × 150 mL / 134.9 mL = 82.2 mg DOC and (0.5 mg/mL × 147.89 mL × 150 mL) / 134.89 mL = 82.23 mg DTT in 150 mL.

To use a correct concentration of DOC is of the utmost importance as: adding too little DOC leads to insufficient purification and adding too much may lead to extensive inactivation. Nevertheless you should expect that even a successful ATPase purification leads to some (10–20 %) inactivation.

Acknowledgement

We would like to thank Nikolaj Drachmann for preparing the gel for Fig. 1 and Ingrid Dach for critical reading and commenting on this book chapter. A special thanks goes to our technicians Lisbeth Nielsen and Birte Nielsen.

References

1. Møller JV (2010) The sarcoplasmic Ca^{2+}-ATPase: design of a perfect chemi-osmotic pump. Q Rev Biophys 43(4):501–566
2. Hasselbach W, Makinose M (1963) Uber Den Mechanismus Des Calciumtransportes Durch Die Membranen Des Sarkoplasmatischen Reticulums. Biochem Z 339:94
3. de Meis L, Hasselbach W (1971) Acetyl phosphate as substrate for Ca^{2+} uptake in skeletal muscle microsomes. J Biol Chem 246:4759–4763
4. Meissner G et al (1973) Isolation of sarcoplasmic reticulum by zonal centrifugation and purification of Ca^{2+}-pump and Ca^{2+}-binding proteins. Biophysica Acta 298:246–269
5. Lowry OH et al (1951) Protein measurement with the Folin phenol reagent. J Biol Chem 193(1):265–275

Chapter 4

Isolation of H⁺,K⁺-ATPase-enriched Membrane Fraction from Pig Stomachs

Kazuhiro Abe and Claus Olesen

Abstract

Gastric H⁺,K⁺-ATPase is an ATP-driven proton pump responsible for the acid secretion. Here, we describe the procedure for the isolation of H⁺,K⁺-ATPase-enriched membrane vesicle fractions by Ficoll/sucrose density gradient centrifugation. Further purification by SDS treatment of membrane fractions is also introduced. These procedures allow us to obtain purified protein preparations in a quantity of several tens of milligrams, with the specific activity of ~480 μmol/mg/h. High purity and stability of H⁺,K⁺-ATPase in the membrane preparation enable us to evaluate its detailed biochemical properties, and also to obtain 2D crystals for structural analysis.

Key words Gastric proton pump, H⁺,K⁺-ATPase, Membrane protein, Purification

1 Introduction

The gastric H⁺,K⁺-ATPase is a membrane protein located in the parietal cells of the stomach. This enzyme stays in the cytoplasmic tubular membranes in the resting state and then moves to the microvilli of the expanded secretory canaliculus in the stimulated state of the parietal cell [1, 2]. The gastric H⁺,K⁺-ATPase catalyzes electroneutral exchange of H⁺ and K⁺ coupled with ATP hydrolysis, generating a million-fold proton gradient across the gastric parietal cell membrane, thereby acidifying the content of the stomach to a pH level as low as 1. This acidic environment of the gastric lumen is indispensable for digestion, and also constitutes an important defense mechanism against undesirable microorganism infection. However, too much acidification induces gastric ulcer or gastroesophageal reflux disease (GEAD), thus recent development in therapy of acid disease has relied heavily on the performance of drugs which inhibit the activity of gastric H⁺,K⁺-ATPase [3]. Here, we introduce the procedure for the isolation of the H⁺,K⁺-ATPase-enriched membrane fractions and an additional purification step using sodium dodecyl sulfate (SDS), based on the procedure developed by Chang et al. [4] and Yen et al. [5], respectively, with some modifications to increase the purity and activity of the preparation. Isolated membrane fraction (G1 fraction) is obtained as inside-out tight vesicles in which cytoplasmic domains of H^+,K^+-ATPase are facing outside. Because of this topology, isolated vesicles can be used for proton uptake measurements using a pH-sensitive dye [6] or direct measurement of solution pH by electrode [7]. On the other hand, due to inaccessibility of K^+ from the luminal side of the enzyme (inside of vesicle), apparent H^+,K^+-ATPase activity in this tight vesicle is kept low. Tight vesicles however can be leaked by the addition of K^+-ionophore such as gramicidin, by membrane permeabilization with alamethicin [8], mild detergent treatment [9], or lyophilization, resulting in increased specific H^+,K^+-ATPase activity of around 350 μmol/mg/h. These leaky vesicles have been used for detailed kinetic measurements of ATP hydrolysis and phosphorylation [10], binding measurements of inhibitors or antagonists which can be used as acid suppressants [11] and various biochemical experiments [12]. Further treatment of isolated tight vesicles with SDS can completely leak them, and at the same time impurities such as proteins susceptible to SDS that are weakly bound to the membrane can be removed. Giving increases in the maximum specific activity reaches up to around 480 μmol/mg/h. Purified preparations have been used for the determination of the ligand binding stoichiometry [13], biochemical characterization [9, 14], and structural analysis by 2D electron microscopy [15–17] (*see* also Chap. 39).

2 Materials

2.1 Isolation of Membrane Fractions from Pig Stomach

1. Pig stomachs: Whole stomachs are obtained from freshly slaughtered hogs. If possible, select stomachs where the fundus part is as smooth as possible hence this will easy the dissecting of parietal cells and save ample time (12 stomachs will usually give a yield around 30–50 mg of G1 fraction). Avoid freeze-thawing of the stomachs as the final yield will be significantly reduced.

2. Sucrose buffer A: Prepare 3 L of 0.25 M sucrose, 5 mM HEPES, 0.5 mM EGTA, pH 7.0 adjusted with Tris solution (*see* **Note 1**).

3. Saturated NaCl solution: 2 L of saturated NaCl solution, stir in the cold room until use.

4. Sucrose buffer B: 1 mM phenylmethylsulfonyl fluoride (PMSF), 1 mM benzamidine, 1 μg/mL pepstatin A, 2 μg/mL leupepsin, 2 μg/mL aprotinin in 720 mL of sucrose buffer A. Prepare just prior to use (alternatively *see* **Note 2**).

5. Ficoll solution: 7.5 % (w/v) Ficoll dissolved in sucrose buffer A. Dissolve 15 g of Ficoll PF-70 in 200 mL of sucrose buffer A. This solution is used for the separation of the H^+,K^+-ATPase membrane fraction by centrifugation. Alternatively, prepare the sucrose buffer in a beaker the day before and distribute the Ficoll powder onto top and let it dissolve overnight in the refrigerator.
6. 37 % sucrose solution: 37 % (w/v) sucrose, 5 mM HEPES, 0.5 mM EGTA, pH 7.0 adjusted with Tris in a volume of 200 mL.
7. Dilution buffer: 5 mM HEPES, 0.5 mM EGTA, pH 7.0 adjusted with Tris solution, in a volume of 500 mL.
8. A conventional electric food blender.
9. A Potter-Elvehjem homogenizer with a Teflon piston.

2.2 SDS Treatment

1. SDS stock solution: 1 % (w/v) SDS dissolved in water.
2. 25 % sucrose solution: 25 % (w/v) sucrose, 5 mM HEPES, 0.5 mM EGTA, pH 7.0 adjusted with Tris solution.
3. 35 % sucrose solution: 35 % (w/v) sucrose, 5 mM HEPES, 0.5 mM EGTA, pH 7.0 adjusted with Tris solution.
4. 60 % sucrose solution: 60 % (w/v) sucrose, 5 mM HEPES, 0.5 mM EGTA, pH 7.0 adjusted with Tris solution.
5. SDS treatment buffer: 10 mM PIPES, 3 mM ATP (Tris salt), 0.5 mM EGTA and 20 % glycerol, pH 7.25 adjusted with Tris solution.

3 Methods

Carry out all procedures on ice unless otherwise specified. Isolation procedures (Subheadings 3.1 and 3.2) are outlined and an SDS-PAGE of each isolated fraction is shown in Fig. 1.

3.1 Isolation of the Membrane Fractions

1. Open freshly slaughtered hog stomachs, by cutting their upper side, wash them with tap water to discharge the stomach content, and keep them on ice in 2 L sucrose buffer A until use.
2. Cut the reddish parts of stomachs (the fundus) (Fig. 2a) in pieces of suitable size for the following step (Fig. 2b).
3. Wash stomach pieces with saturated NaCl solution. Remove the sticky gastric mucosa during this step. Paper towel is helpful to wipe gastric surface.
4. Scrape off the surface cells using a butter knife (alternatively, use a filet knife and gently cut a very fine slice of parietal cells) and collect the red-to-white-colored sticky cell tissue in the 200 mL beaker on ice.

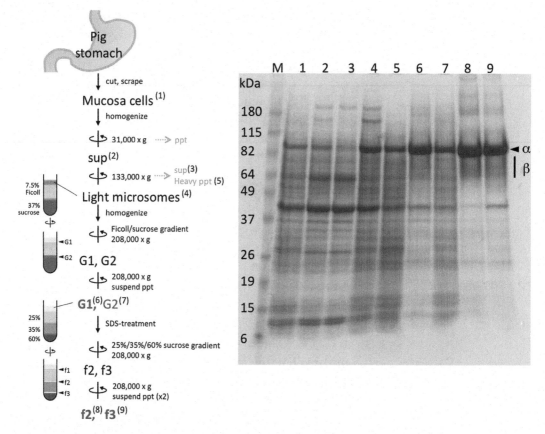

Fig. 1 Schematic overview of the isolation of the membrane fractions from pig stomachs (*left panel*). Cartoon drawings on the *left* indicate separation of membrane fractions by Ficoll/sucrose or sucrose stepwise density gradients. Numbers in parentheses correspond to the lane number in SDS-PAGE (*right panel*). The catalytic α-subunit (1034 amino acids) and heavily glycosylated β-subunit (290 amino acids with seven N-linked glycosylation sites) are indicated by a *black arrowhead* and a *black bar*, respectively. Each lane contains the same amount of protein (10 μg/lane) as determined by the Bradford method [18]. Molecular weights of size markers (M) are also indicated on the *left* (in kDa)

5. Suspend the cells collected in **step 4** in Subheading 3.1 ($200 \times g$) in 720 mL of sucrose buffer B containing protease inhibitors, and homogenize by the food blender in the cold room for 30 s × 6 times (each 30-s interval).

6. Centrifuge the cell suspension at $31,000 \times g$ for 40 min, to remove cell debris in the precipitate.

7. Filtrate supernatant through sterile gauze to remove fat and other floating cell debris.

8. Centrifuge the filtrated supernatant (about 720 mL + cell volume) at $133,000 \times g$ for 45 min.

9. Remove supernatant, and collect the upper (looser) red-white-colored precipitate (light microsomes) (Fig. 2c), by adding

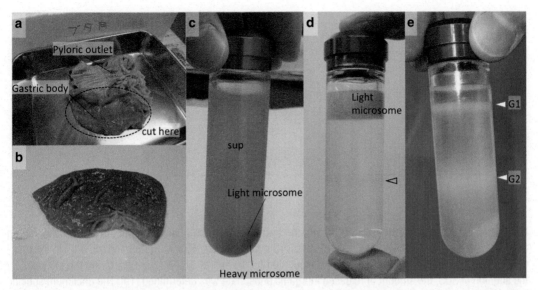

Fig. 2 Pig stomachs and isolated membrane fractions. Reddish parts of the stomach (**a**, indicated by *dotted circle*) were cut using a surgical knife, and washed by saturated NaCl solution. After removing the sticky mucosa (**b**), surface cells were collected using a butter knife. (**c**) After centrifugation in **step 8** in Subheading 3.1, microsome fractions were separated into two fractions (light and heavy microsomes). H$^+$,K$^+$-ATPase is enriched in the *red-white-colored* upper precipitate (light microsomes). (**d**) Ficoll/sucrose density gradient with homogenized light microsomes on the *top* of the layer (in **step 13** in Subheading 3.1). The *open arrowhead* indicates the interface between 7.5 % Ficoll solution (*upper layer*) and 37 % sucrose solution (*lower layer*). (**e**) After centrifugation, light microsomes were further separated into two fractions (G1, G2, indicated as *white arrowheads*) found at the interface between Ficoll and sucrose solutions

0.32 mL of sucrose buffer A into each centrifuge tube and gently vortex.

10. Again, add 0.32 mL of Sucrose buffer A to each centrifuge tube, and collect the suspended light microsomes.

11. Homogenize the resuspended light microsomes in a Potter-Elvehjem homogenizer with a teflon piston (use wide pipette tips created by manually cutting conventional pipette tips or a Pasteur single use pipette).

12. During centrifugation in **step 8** in Subheading 3.1, Ficoll/sucrose density gradient can be prepared. Dispense 10 mL of 7.5 % Ficoll/sucrose solution into the 26.3 mL polycarbonate centrifuge tube, then add 10 mL of 37 % sucrose solution to the bottom of the tube using a syringe with needle or a thin plastic tube (Fig. 2d).

13. The light microsomes collected in **steps 9–11** in Subheading 3.1 will have a total volume of approximately 40 mL. After homogenization, lay 2 mL of homogenate gently on the top of each Ficoll/sucrose stepwise gradient prepared in **step 12** in Subheading 3.1 (Fig. 2d).

14. Centrifuge for 70 min at 208,000 × g.

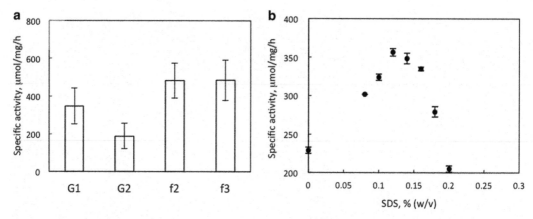

Fig. 3 Specific activities in the isolated membrane fractions (**a**). Each membrane fraction was suspended in 250 mM sucrose, 2.5 mM $MgCl_2$, 2.5 mM ATP/Tris, 10 mM CH_3COOK, 40 mM HEPES/Tris (pH 7.0), in the presence or absence of 100 μM SCH28080, and incubated for 10 min at 37 °C. After denaturing the enzymes by the addition of SDS, the amount of released inorganic phosphate was determined [19]. The maximum specific activities of G1 and G2 fractions treated by K^+-ionophore gramicidin were determined. Because specific activities in f2 and f3 were ionophore independent, neither ionophore nor detergent was added in the SDS-treated fractions. Data shown are means and standard deviations derived from 6 to 10 independent preparations. (**b**) A representative of the titration of SDS concentrations used for the large-scale treatment (**step 2** in Subheading 3.2). Data shown were means and standard errors from the triplicate measurement. In this case, an SDS concentration of 0.12 % is chosen for the following large-scale preparations. Specific activities were calculated as differences between the value in the presence or absence of SCH28080 for each sample

15. After centrifugation, membrane fractions are separated into mainly two fractions (Fig. 2e, *see* **Note 3**). One is in the upper layer of Ficoll solution (G1 fraction), and the other can be found at the interface between Ficoll and 37 % sucrose layer (G2 fraction).

16. Remove the upper solution carefully, and then collect the G1 fraction into a trap made of a glass flask and plastic tubes, using an aspirator (alternatively, use a plastic syringe with a wide aperture).

17. Then remove the G2 fraction as done for the G1 fraction.

18. Dilute both the G1 and G2 fraction in 200 mL using dilution buffer.

19. Precipitate each fraction by centrifugation at $208,000 \times g$ for 90 min.

20. Suspend the white-pink-colored precipitate in sucrose buffer A, and homogenize using a Teflon homogenizer.

21. Isolated G1 fractions show specific H^+,K^+-ATPase activity of around 350 μmol/mg/h (Fig. 3a). Both G1 and G2 fractions can be stored at −80 °C after flash freezing in liquid nitrogen.

3.2 Purification by SDS Treatment and Further Purification of the G1 Fraction

1. Before large-scale treatment with SDS, the suitable SDS concentration must be determined by a small-scale experiment (usually 20–100 μL).
2. Suspend 3 mg/mL G1 fraction in SDS treatment buffer, and add SDS to a final concentration of 0–0.2 % (w/v). After incubation for 10 min on ice, dilute the enzyme solutions 100 times with sucrose buffer A, and measure their ATPase activity [18] (*see* **Note 4** and Fig. 3b). Suitable SDS concentrations for the treatment are usually 0.12–0.16 % (w/v).
3. Treat the whole G1 fraction with SDS according to results obtained in **step 2** in Subheading 3.2 (usually 16–32 mL). Suspend the G1 fraction in the buffer without SDS, and add SDS solution drop by drop in the cold room while stirring.
4. After incubation for 5 min on ice, lay the SDS-treated G1 fraction (2 mL) on each step gradient (thus usually 8–16 tubes are required for one operation) consisting of three layers of 25 % (w/v), 35 % (w/v) and 60 % (w/v) sucrose solution in a 26.3 mL polycarbonate centrifugation tube.
5. Centrifuge for 90 min at 208,000 ×g.
6. The SDS-treated G1 is now separated into three fractions (Fig. 1). Collect f2 and f3 using an aspirator as in **step 16** in Subheading 3.1 (*see* **Note 5**).
7. Dilute each fraction with dilution buffer and centrifuge for 45 min at 208,000 ×g. For complete removal of SDS from these fractions, repeat this operation twice.
8. Resuspend the pellet in sucrose buffer A, and store at −80 °C.

4 Notes

1. Usually 2 L are needed for the collection and transportation of the stomachs from the slaughterhouse, and another 1 L for the preparation of the membrane fractions.
2. The purification can be performed without the use of any protease inhibitors. In this case it is extremely important to maintain the protein preparation and buffers ice cold throughout the purification process.
3. The G1 fraction contains the H^+,K^+-ATPase-enriched tubular membranes of the parietal cells, and the G2 contains the plasma membranes. The G1 fraction shows higher specific activity and purity than the G2 fraction (Fig. 3a).
4. Since the G1 fraction contains ion-impermeable, inside-out tubular vesicles in which most of H^+,K^+-ATPases are embedded with their cytoplasmic-side out, addition of SDS leaks tightly sealed vesicle membrane and increases specific

H^+,K^+-ATPase activity due to K^+ permeation to the vesicle inside. However, too much SDS denatures the enzyme. Therefore, SDS concentration that gives maximum activity has to be carefully determined in this step, and used for the following large-scale experiment.

5. Another membrane fraction can be found on the top of gradient (f1). However, this fraction contains denatured enzyme as it shows little H^+,K^+-ATPase activity.

Acknowledgement

K.A. thanks Drs. Kazuya Taniguchi and Shunji Kaya for their help in the development of isolation procedure of membrane fractions from pig stomach. This work was supported by Grants-in-Aid for Young Scientist (A) and Platform for Drug Design, Discovery, and Development from MEXT, Japan, to K.A. Also a special thanks to Jesper Vuust Møller and Ingrid Dach for discussion and comments on this protocol.

References

1. Ganser AL, Forte JG (1973) K^+-stimulated ATPase in purified microsomes of bullfrog oxynic cells. Biochim Biophys Acta 307:169–180
2. Shin JM, Munson K, Sachs G (2009) The gastric HK-ATPase: structure, function, and inhibition. Eur J Physiol 457:609–622
3. Sachs G, Shin JM, Vagin O, Lambrecht N, Yakubov I, Munson K (2007) The gastric H,K-ATPase as a drug target: past, present, and future. J Clin Gastroenterol 41:S226–S242
4. Chang H, Saccomani G, Rabon G, Schackmann R, Sachs G (1977) Proton transport by gastric membrane vesicles. Biochim Biophys Acta 464:313–327
5. Yen LA, Cosgrove P, Holt W (1990) SDS purification of porcine H,K-ATPase from gastric mucosa. Membr Biochem 9:129–140
6. Lee LC, Forte J (1978) A study of H^+ transport in gastric microsomal vesicles using fluorescent probes. Biochim Biophys Acta 508:339–356
7. Rabon EC, McFall TL, Sachs G (1982) The gastric [H,K]ATPase: H+/ATP stoichiometry. J Biol Chem 257:6296–6299
8. Montes MR, Spiaggi AJ, Monti JL, Cornelius F, Olesen C, Garrahan PJ, Rossi RC (2011) Rb^+ occlusion stabilized by vanadate in gastric H^+/K^+-ATPase at 25 °C. Biochim Biophys Acta 1808:316–322
9. Abe K, Tani K, Fujiyoshi Y (2010) Structural and functional characterization of H^+,K^+-ATPase with bound fluorinated phosphate analogs. J Struct Biol 170:60–68
10. Ljungström M, Vega FV, Mårdh S (1984) Effects of pH on the interaction of ligands with the $(H^+ + K^+)$-ATPase purified from pig gastric mucosa. Biochim Biophys Acta 769(1):220–230
11. Shin JM, Grundler G, Senn-Bilfinger J, Simon WA, Sachs G (2005) Functional consequences of the oligomeric form of the membrane-bound gastric H,K-ATPase. Biochemistry 44:16321–16332
12. Rabon EC, Bassilian S, Sachs G, Karlish SJD (1990) Conformational transitions of the H,K-ATPase studied with sodium ions as surrogates for protons. J Biol Chem 265:19594–19599
13. Abe K, Kaya S, Imagawa T, Taniguchi K (2002) Gastric H/K-ATPase liberates two moles of Pi from one mole of phosphoenzyme formed from a high-affinity ATP binding site and one mole of enzyme-bound ATP at the low-affinity site during cross-talk between catalytic subunits. Biochemistry 41:2438–2445
14. Dach I, Olesen C, Signor L, Nissen P, Le Maire M, Møller JV, Ebel C (2012) Active detergent-solubilized H^+,K^+-ATPase is a monomer. J Biol Chem 287:41963–41978

15. Abe K, Tani K, Nishizawa T, Fujiyoshi Y (2009) Inter-subunit interaction of gastric H^+,K^+-ATPase prevents reverse reaction of the transport cycle. EMBO J 28:1637–1643
16. Abe K, Tani K, Fujiyoshi Y (2011) Conformational rearrangement of gastric H^+,K^+-ATPase with an acid suppressant. Nat Commun 2:155
17. Abe K, Tani K, Friedrich T, Fujiyoshi Y (2012) Cryo-EM structure of gastric H^+,K^+-ATPase with a single occupied cation-binding site. Proc Natl Acad Sci U S A 109:18401–18406
18. Bradford MM (1976) A rapid and sensitive method for the quantitation of microgram quantities of protein utilizing the principle of protein-dye binding. Anal Biochem 7:248–254
19. Chifflet S, Torriglia A, Chiese R, Tolosa S (1988) A method for the determination of inorganic phosphate in the presence of labile organic phosphate and high concentrations of protein: application to lens ATPases. Anal Biochem 168:1–4

Chapter 5

Overproduction of P_{IB}-Type ATPases

Xiangyu Liu, Oleg Sitsel, Kaituo Wang, and Pontus Gourdon

Abstract

Understanding of the functions and mechanisms of fundamental processes in the cell requires structural information. Structural studies of membrane proteins typically necessitate large amounts of purified and preferably homogenous target protein. Here, we describe a rapid overproduction and purification strategy of a bacterial P_{IB}-type ATPase for isolation of milligrams of target protein per liter *Escherichia coli* cell culture, with a final quality of the sample which is sufficient for generating high-resolution crystals.

Key words Membrane proteins, P-type ATPase, Overproduction, Membrane preparation, Nickel affinity chromatography, Size-exclusion chromatography

1 Introduction

X-ray crystallography is a powerful tool to determine protein structures and thus understand protein function at an atomic level. While there has been a dramatic increase in the available structural information on soluble proteins, the progress in membrane protein (MP) structure biology still remains limited. As a consequence, an increase of the pace with which new MP structures are resolved is desirable, and this is not only important for basic science; about one third of genes across all kingdoms of life encode MPs, many disorders are directly related their malfunction, and roughly 30 % of all drugs act on MPs [1], and therefore new structures may also alleviate the drug discovery process for human disorders.

One of the major bottlenecks responsible for the significant underrepresentation of this class of proteins in structural databases stems from substantial difficulties in isolating the high yields of recombinantly produced MP required for structure determination projects (>10 mg). Overproduction is occasionally toxic to the host cells, or the biogenesis fails and results in aggregates gathered in inclusion bodies and/or insufficient amounts of protein retained in the membrane. Therefore, a common strategy is to assess homologous proteins and different hosts [2]. The most closely related system in evolution to the target source is often more likely to yield correctly folded and inserted MP, which reflects that proteins have been co-developed with the available cellular machinery for proper processing [3]. Nevertheless, the advantages of bacterial and yeast hosts include low costs, the rapidly obtained and high cellular yield, and easy manipulation. For these reasons *Escherichia coli*, *Lactococcus lactis*, *Saccharomyces cerevisiae*, and *Pichia pastoris* are frequently preferred initial choices. The yield of eukaryotic targets from such systems can sometimes be improved by e.g., using protein engineering for improved target stability [4], codon optimization [5], rare codon t-RNA-expressing plasmids [6], or co-synthesis of chaperones [7].

Chromatographic purification is also frequently necessary to obtain MP samples with sufficient homogeneity for crystallization, but this requires that the molecules are extracted from their natural membrane environment and kept in solution using detergents, which shield the lipophilic regions of the proteins when they have been solubilized. However, identifying a suitable detergent for solubilization and/or downstream applications, which can retain fold, subunits and/or lipids, as well as function and homogeneity, is a nontrivial empirical process which is hard to predict, and therefore detergents that have proven successful for other membrane proteins are tested initially. Investigating the protein's stability in the presence of different detergents greatly improves the probability of obtaining well-diffracting crystals suitable for structure determination. Such assessment may be done using for example either regular or fluorescence-detection size-exclusion chromatography [8], the latter being based on using for example a C-terminal GFP fusion construct as a reporter. The purification methods are essentially the same as for soluble proteins, and are employed based on separation according to different properties of the target such as charge, molecular weight/size and hydrophobicity or biorecognition. The main difference is that the detergent concentration is always maintained above the critical micelle concentration (CMC) and the frequent presence of stabilizing agents such as glycerol, as well as the typically lower efficiency of the purification steps due to strong interactions between the membrane proteins and the chromatographic media. Often, affinity purification is initially used to specifically capture the target and remove as much of the contamination as possible. For that reason, recombinant MPs are often produced with an affinity tag attached, with the most popular being the polyhistidine-tag, which facilitates its binding to immobilized nickel- or cobalt-containing media. It is common to employ size-exclusion chromatography as a final step to polish the sample and define the protein buffer for the crystallization experiments.

In this chapter, a rapid overproduction and purification strategy of a bacterial MP, the copper-transporting P_{IB}-type ATPase LpCopA from *Legionella pneumophila*, is provided, which allows for isolation of milligrams of target protein per liter of *Escherichia coli* cell culture, with a final quality of the sample which is sufficient for generating high-resolution crystals. It employs a T7-expression system vector and overproduction is induced at a low temperature. Solubilization and purification is performed in the presence of the detergent octaethylene glycol monododecyl ether, $C_{12}E_8$, and the protein is purified using nickel affinity and size-exclusion chromatography. Using this approach, we have obtained high-resolution crystals of two conformational states of LpCopA [9, 10].

2 Materials

All solutions are prepared with ultrapure ddH$_2$O water.

2.1 Overproduction

1. Expression strain, vector, and protein: *E. coli* BL21(DE3) derivative C43(DE3) or C41(DE3), pET22b vector and LpCopA from *Legionella pneumophila* (Uniprot Q5ZWR1, also known as lpg1024).

2. LB medium: Weigh 10 g of tryptone, 10 g of NaCl, and 5 g of yeast extract and dissolve in water to a final volume of 1 L, and autoclave (*see* **Note 1**).

3. LB agar: Weigh 10 g of tryptone, 10 g of NaCl, 5 g of yeast extract, and 15 g of agar, dissolve in water to a final volume of 1 L, and autoclave (*see* **Note 2**).

4. Ampicillin stock: Dissolve at 100 mg/mL in water and filter using 0.22 μm filters.

5. IPTG: Isopropyl β-D-1-thiogalactopyranoside. Dissolve at 1 M in water, store at −20 °C.

2.2 Purification

1. $C_{12}E_8$ detergent: Dissolve $C_{12}E_8$ (BL-8SY from Nikkol Chemicals) at 100 mg/mL.

2. Affinity chromatography medium: Ni Sepharose 6 Fast Flow (GE Healthcare).

3. Size-exclusion chromatography medium: Superose 6 10/300 (GE Healthcare).

4. Buffer A: 50 mM Tris–HCl, pH 7.6, 200 mM KCl, 20 % (v/v) glycerol. Filter using 0.22 μm filters.

5. Buffer B: 20 mM Tris–HCl, pH 7.6, 200 mM KCl, 20 % (v/v) glycerol, 5 mM βME (*see* **Note 3**), 1 mM MgCl$_2$. Filter using 0.22 μm filters.

6. Buffer C: 20 mM MOPS-KOH, pH 7.4 (*see* **Note 4**), 200 mM KCl, 20 % glycerol, 5 mM βME, 1 mM MgCl$_2$, 0.15 mg/mL $C_{12}E_8$. Filter using 0.22 μm filters.

7. Buffer D: As Buffer C but including 500 mM imidazole. Filter using 0.22 μm filters.
8. Buffer E: 20 mM MOPS-KOH, pH 6.8, 80 mM KCl, 20 % glycerol, 5 mM βME, 3 mM $MgCl_2$, 0.15 mg/mL $C_{12}E_8$. Filter using 0.22 μm filters.
9. PMSF: Phenylmethanesulfonylfluoride. Dissolve to 100 mM in ethanol (see **Note 5**).
10. High-pressure homogenizer.
11. Ultracentrifuge: See **Note 6**.
12. Pestle tissue grinder: Uses Teflon® pestle (Thomas Scientific).
13. Concentration tube: The most frequently used concentration tubes for LpCopA purification are with a 50 kDa molecular weight cutoff (Vivascience).

(Abbreviations: *βME* beta-mercaptoethanol, $C_{12}E_8$ octaethylene glycol monododecyl ether, *PI* Roche complete EDTA-free protease inhibitor).

3 Methods

3.1 Overproduction

1. Clone the target gene into a T7-expression vector such as pET22b, using appropriate restriction enzyme sites introduced to the DNA (through PCR or DNA synthesis). pET22b allows for inclusion of a C-terminal hexahistidine tag. However, LpCopA also binds to nickel-resins through its intrinsic heavy metal binding domain, and thus a construct without polyhistidine tag is recommended.
2. Transform (heat or electroporation) the plasmid into competent C43(DE3) cells, plate on LB agar plate with 100 μg/mL ampicillin, and incubate at 37 °C overnight.
3. Inoculate several colonies from the plate into 20 mL LB medium supplemented with 100 μg/mL ampicillin and incubate at 37 °C for 16 h with 200 rpm shaking.
4. Inoculate 20 mL of the pre-culture to 2 L LB medium supplemented with 100 mg/L ampicillin and shake with 200 rpm at 37 °C until the optical density at 600 nm (OD_{600nm}) reaches 0.6–0.8. This normally takes about 3 h.
5. Lower the temperature to 20 °C and add 1 mM IPTG (final concentration) to induce protein production (see **Note 7**).
6. Incubate with 200 rpm shaking at 20 °C for about 16 h.
7. Harvest the cells through centrifugation at $8000 \times g$ for 15 min.
8. Freeze the cells at −20 °C, or continue directly with the following step.

3.2 Cell Disruption and Membrane Isolation

The following is based on a preparation from 8 L of cell culture but may be scaled.

1. Cells are diluted to 20 mL Buffer A per liter of cell culture and thawed. Fresh βME (final concentration, $c_{final} = 5$ mM), fresh PMSF ($c_{final} = 1$ mM), and 1 Roche complete EDTA-free protease inhibitor (PI) tablet is added.

2. Add DNase I to $c_{final} = 2$ μg/mL before breakage. Break the cells with a high-pressure homogenizer by three runs at 15–20,000 psi (*see* **Note 8**).

3. Remove DNA and large aggregates from the lysate by centrifugation at $20,000 \times g$ for 45 min at 4 °C.

4. Isolate membranes by ultracentrifugation at $250,000 \times g$ (average) for 75 min to 5 h at 4 °C (the membrane yield increases with time) (*see* **Note 9**).

5. Membranes are resuspended using pestle tissue grinder in 15 mL/g membrane Buffer B and 1 tablet of PI (*see* **Note 10**).

6. If not used immediately, store membranes on ice at 4 °C (overnight) or freeze at −20 °C.

3.3 Solubilization and Nickel Affinity Chromatography

1. Solid $C_{12}E_8$ ($c_{final} = 1$ % w/v) is added to the homogenized membranes and solubilization performed at 4 °C on a spinning wheel for 60–90 min (*see* **Note 11**).

2. Use 1.5 mL Ni Sepharose beads per 1 g of membranes. Wash the beads with ultrapure water and equilibrate the beads with Buffer C in 50 mL falcon tubes.

3. Spin down unsolubilized membranes by ultracentrifugation of the membrane-detergent mixture at $200,000 \times g$ (average) for 75 min. Dilute with detergent-containing buffer (Buffer C) if required for balancing the centrifugation tubes.

4. Collect the supernatant and add 50 mM of imidazole, adjust the concentration of KCl from 200 mM with solid KCl ($c_{final} = 500$ mM), and filter (0.45 μm) (*see* **Note 12**).

5. Transfer the supernatant to the equilibrated Ni Sepharose beads and incubate on a spinning wheel at 4 °C for at least 1 h.

6. Pack the resin into an empty column. To save time the beads may be pelleted (do not exceed $500 \times g$), or the solution passed through the column with a syringe or peristaltic pump.

7. The column is transferred to a FPLC system (such as an ÄKTA from GE Healthcare) maintained at 4 °C and washed at 1 mL/min with Buffer C for 3–5 column volumes (when a stable baseline is reached).

8. Elute the target protein with a gradient increase of Buffer D and analyze selected fractions using SDS-PAGE.

3.4 Size-Exclusion Chromatography

1. Pool the fractions containing the protein and concentrate using Vivaspin20 tubes (Vivascience) with a 50 kDa cutoff by centrifugation at $5000 \times g$ until the concentration has reached 10–20 mg/mL (*see* **Note 13**).
2. Remove potential aggregates by spinning the sample at $10,000 \times g$ for 5 min.
3. Wash (with water) and equilibrate (with Buffer E) a Superose 6 10/300 GL column (*see* **Note 14**).
4. Run gel filtration column and inject no more than 10 mg of sample in no more than 500 µL per run.
5. Analyze selected fractions by SDS-PAGE.
6. Fractions containing target protein are pooled and the pooled volume noted. The pooled sample is concentrated using new Vivaspin20 tubes with a 50 kDa cutoff by centrifugation at $5000 \times g$ until the concentration has reached approximately 20 mg/mL. Note the final volume and calculate the concentration factor (useful for comparing the amount of concentrated detergent with other preparations).
7. The purified protein can be flash frozen by liquid nitrogen and stored at −80 °C freezer before use. The protein is able to produce well-diffracting crystals also after being frozen for more than 3 months.

4 Notes

1. Weigh the yeast extract and tryptone in a fume hood. Adding water before the LB powder components to the final container may help reduce the amount of airborne particles.
2. Agar is not soluble in cold water. The undissolved material can be autoclaved directly.
3. βME is unstable in solution, so add immediately before usage.
4. Once prepared, the MOPS buffer should be stored at 4 °C and protected from light to prevent the solution to turn yellow.
5. PMSF is toxic. Wash extensively with water if skin or eyes become exposed. PMSF is unstable in water, add immediately before usage.
6. Ultracentrifuges are spinning at very high speed and consequently the centrifuge tubes need to be carefully balanced and fully filled with solution to prevent damage to the tubes.
7. Different induction/expression temperatures should be tested for different proteins. In general, 20 °C works well for most P_{IB}-ATPases, including LpCopA.
8. Normally the cells need to pass through the high-pressure homogenizer 2–3 times. At the beginning of cell lysis, the cells

are viscous and light colored. At the end of cell breakage, the cell lysate should come out of the homogenizer in a nonviscous manner and darker-colored. Homogenizing generates heat; keep the solutions on ice during the procedure.

9. Normally every liter of *E. coli* cells results in ~0.5–0.8 g of membrane (wet weight). The membrane pellet looks like orange-brown wax.

10. Keep everything on ice to minimize heating during the membrane resuspension procedure.

11. Solid $C_{12}E_8$ may take some time to be fully dissolved; ensure that the $C_{12}E_8$ is a fine powder before the procedure is initiated.

12. High salt concentration may reduce nonspecific binding. Imidazole is toxic. Wash extensively with water if skin or eyes become exposed.

13. Prewash the concentrators with water and then detergent containing buffer. The protein solution should be fully re-mixed by pipetting every 10–15 min during concentrating. Otherwise the solution adjacent to the concentrator membrane may reach a high protein concentration locally, potentially resulting in protein aggregation and/or precipitation. Take care not to damage the concentrator membrane with the pipette tip.

14. The buffers and water need to be filtered using a 0.22 μm filter.

Acknowledgements

This work was supported by the Graduate School of Science and Technology at Aarhus University and by grants to P.G. from the Swedish Research Council and the The Lundbeck Foundation.

References

1. Hopkins AL, Groom CR (2002) The druggable genome. Nat Rev Drug Discov 1:727–730
2. Li M, Hays FA, Roe-Zurz Z, Vuong L, Kelly L, Ho CM, Robbins RM, Pieper U, O'Connell JD 3rd, Miercke LJ, Giacomini KM, Sali A, Stroud RM (2009) Selecting optimum eukaryotic integral membrane proteins for structure determination by rapid expression and solubilization screening. J Mol Biol 385:820–830
3. Grisshammer R, Tate CG (1995) Overexpression of integral membrane proteins for structural studies. Q Rev Biophys 28:315–422
4. Magnani F, Shibata Y, Serrano-Vega MJ, Tate CG (2008) Co-evolving stability and conformational homogeneity of the human adenosine A2a receptor. Proc Natl Acad Sci U S A 105:10744–10749
5. Elena C, Ravasi P, Castelli ME, Peiru S, Menzella HG (2014) Expression of codon optimized genes in microbial systems: current industrial applications and perspectives. Front Microbiol 5:21
6. Rosano GL, Ceccarelli EA (2009) Rare codon content affects the solubility of recombinant

proteins in a codon bias-adjusted Escherichia coli strain. Microb Cell Fact 8:41
7. Vogl T, Thallinger GG, Zellnig G, Drew D, Cregg JM, Glieder A, Freigassner M (2014) Towards improved membrane protein production in Pichia pastoris: general and specific transcriptional response to membrane protein overexpression. N Biotechnol 31(6): 538–552
8. Kawate T, Gouaux E (2006) Fluorescence-detection size-exclusion chromatography for precrystallization screening of integral membrane proteins. Structure 14:673–681
9. Gourdon P, Liu XY, Skjorringe T, Morth JP, Moller LB, Pedersen BP, Nissen P (2011) Crystal structure of a copper-transporting PIB-type ATPase. Nature 475:59–64
10. Andersson M, Mattle D, Sitsel O, Klymchuk T, Nielsen AM, Moller LB, White SH, Nissen P, Gourdon P (2014) Copper-transporting P-type ATPases use a unique ion-release pathway. Nat Struct Mol Biol 21:43–48

Chapter 6

Coordinated Overexpression in Yeast of a P4-ATPase and Its Associated Cdc50 Subunit: The Case of the Drs2p/Cdc50p Lipid Flippase Complex

Hassina Azouaoui, Cédric Montigny, Aurore Jacquot, Raphaëlle Barry, Philippe Champeil, and Guillaume Lenoir

Abstract

Structural and functional characterization of integral membrane proteins requires milligram amounts of purified sample. Unless the protein you are studying is abundant in native membranes, it will be critical to overexpress the protein of interest in a homologous or heterologous way, and in sufficient quantities for further purification. The situation may become even more complicated if you chose to investigate the structure and function of a complex of two or more membrane proteins. Here, we describe the overexpression of a yeast lipid flippase complex, namely the P4-ATPase Drs2p and its associated subunit Cdc50p, in a coordinated manner. Moreover, we can take advantage of the fact that P4-ATPases, like most other P-type ATPases, form an acid-stable phosphorylated intermediate, to verify that the expressed complex is functional.

Key words Membrane protein, P4-ATPase, Cdc50 protein, Co-expression, Yeast, *Saccharomyces cerevisiae*, Lipid transport, Phosphorylation

1 Introduction

P-type ATPases from the P4 subfamily (P4-ATPases) are prime candidates for lipid transport across eukaryotic cell membranes from the late secretory pathway, thereby establishing lipid asymmetry [1]. Additional proteins called Cdc50 proteins have been found to associate with P4-ATPases, and to be essential for both correct localization and catalytic function of P4-ATPases.

As for many other membrane proteins, Drs2p and Cdc50p are scarcely expressed in yeast membranes. In this chapter, we describe an approach, based on our previous experience with expression of the rabbit SERCA1a Ca^{2+}-ATPase in yeast [2, 3], to optimize the high-yield co-expression of Drs2p and Cdc50p in yeast membranes. To avoid imbalanced transcription which may occur when two genes are placed on different multi-copy plasmids, the *DRS2* and *CDC50* genes were cloned into the same expression plasmid, both under the control of a strong galactose-inducible promoter. To facilitate detection of Drs2p and Cdc50p as well as for purification of the complex, a biotin acceptor domain (BAD) and a decahistidine tag were added at the N-terminus of Drs2p and Cdc50p, respectively. After expression in rich medium, the recombinant Drs2p protein accounts for about 3 % of the total protein present in yeast "light" membranes obtained after differential centrifugation of the total cell lysate.

Expression of Drs2p and Cdc50p restores the ability of Δ*drs2* and Δ*cdc50* yeast cells, respectively, to grow at low temperature. A hallmark of P-type ATPases is the formation of a transient phosphorylated intermediate during their catalytic cycle. Thanks to the rather high expression level obtained for Drs2p, it is possible to detect phosphorylation of Drs2p on its catalytic aspartate residue upon incubation of the light membrane fraction with [γ-^{32}P]ATP. Formation of this [^{32}P]-labeled phosphoenzyme is inhibited by orthovanadate, a classical inhibitor of P-type ATPases, and co-expression of Cdc50p is essential for phosphorylation.

2 Materials

2.1 Yeast Culture

The *Saccharomyces cerevisiae* W303.1b/*GAL4* (*a, leu2-3, his3-11, trp1-1:: TRP1-GAL10-GAL4, ura3-1, ade2-1, canr, cir+*) yeast strain is used. The pYeDP60 plasmid was generously given by Denis Pompon (LISBP, Toulouse, France).

Ultrapure water is used, if not otherwise specified.

1. PLATE solution: 40 % PEG 4000, 100 mM Lithium-acetate, 10 mM Tris–HCl pH 7.5, 1 mM EDTA.

2. Denatured single-stranded carrier DNA (SSD): dissolve deoxyribonucleic acid sodium salt from salmon testis to 2 mg/mL in water and store at −20 °C. Just before use, denature SSD by heating at 100 °C for 5 min and chill on ice before yeast transformation.

3. S6AU minimal medium: 0.1 % (w/v) bactocasamino acids, 0.7 % (w/v) yeast nitrogen base, 2 % (w/v) glucose, 20 μg/mL adenine, 20 μg/mL uracil. Prepare a 5 % over-concentrated medium containing only bactocasamino acids and yeast nitrogen base. Sterilize for 20 min at 120 °C. After cooling, add sterile glucose, adenine, and uracil from a 40 % (w/v) stock solution (for glucose), and from a 10 mg/mL stock solution for adenine and uracil (*see* below).

4. S6AU agar minimal medium: prepare the medium as described for liquid S6AU medium but with the addition of 2 % (w/v) agar.

5. S6A minimal medium: 0.1 % (w/v) bactocasamino acids, 0.7 % (w/v) yeast nitrogen base, 2 % (w/v) glucose, 20 μg/mL adenine. Prepare a 5 % over-concentrated medium containing only bactocasamino acids and yeast nitrogen base. Sterilize for 20 min at 120 °C. After cooling, add sterile glucose and adenine from, respectively, 40 % (w/v) and 10 mg/mL stock solutions (*see* below).

6. S6A agar minimal medium: prepare the medium as described for liquid S6A medium but with the addition of 2 % (w/v) agar.

7. S5A minimal medium: 0.1 % (w/v) bactocasamino acids, 0.7 % (w/v) yeast nitrogen base, 2 % (w/v) galactose, 20 μg/mL adenine. Sterilize for 20 min at 120 °C before the addition of galactose and adenine. Preparation of this medium is identical to that of S6A except that glucose is replaced by galactose. Prepare a 40 % (w/v) galactose stock solution as described below.

8. S5AF agar minimal medium: 0.1 % (w/v) bactocasamino acids, 0.7 % (w/v) yeast nitrogen base, 2 % (w/v) galactose, 1 % (w/v) fructose, 20 μg/mL adenine, 2 % (w/v) agar. Prepare this medium as described for agar S6A medium but replace glucose with both galactose and fructose. After cooling, add galactose, fructose, and adenine from a 40 % (w/v) stock solution (for galactose and fructose), and from a 10 mg/mL stock solution for adenine.

9. Sterilize a 40 % (w/v) glucose stock solution by filtration through a 0.22 μm filter and add to the various media after sterilization and cooling to 40 °C.

10. Sterilize a 40 % (w/v) galactose stock solution by filtration through a 0.22 μm filter and add to the various media after sterilization and cooling to 40 °C.

11. Sterilize a 40 % (w/v) fructose stock solution by filtration through a 0.22 μm filter and add to the various media after sterilization and cooling to 40 °C.

12. 10 mg/mL adenine solution: prepare in 0.1 N HCl and filter through a 0.22 μm filter. Add then to the various media after sterilization and cooling to 40 °C.

13. 10 mg/mL uracil solution: neutralize with NaOH and filter through a 0.22 μm filter. Add then to the various media after sterilization and cooling at 40 °C.

14. An incubator shaker system equipped with a cooling system.

15. A spectrophotometer to follow yeast growth (*see* **Note 1**). Optical densities (OD) are measured at 600 nm.

16. YPGE2X rich medium: 2 % (w/v) Bacto™ Peptone, 2 % (w/v) yeast extract, 1 % (w/v) glucose, and 2.7 % (v/v) ethanol.

Sterilize the medium (without glucose and ethanol) directly in the Fernbach-type flask. After cooling to 28 °C, add glucose and ethanol, just before inoculation with yeast cells.

17. 96 % (v/v) ethanol stock solution.

2.2 Small-Scale Membrane Preparation for Expression Screens

1. TEPI buffer: 50 mM Tris–HCl pH 6.8, 5 mM EDTA, 20 mM NaN_3, 1 mM PMSF, 1× SIGMAFAST™ EDTA-free protease inhibitor cocktail (*see* below). The buffer can be prepared beforehand and stored at room temperature. Chill on ice, add NaN_3, PMSF, and protease inhibitor cocktail just before use.

2. Glass beads (0.5 mm diameter). They are reusable after decontamination with fungicidal solution, intensive rinsing with water, and sterilization with ethanol. After washing, dry beads by incubation in a Pasteur oven before storage at 4 °C.

3. Phenylmethanesulfonyl fluoride (PMSF): dissolve to 0.2 M in isopropanol. Prepare 10 mL beforehand that you may store protected from light at room temperature for several weeks.

4. 10 × concentrated SIGMAFAST™ EDTA-free Protease Inhibitor Cocktail (Sigma-Aldrich) stock solution: dissolve one tablet in 10 mL of TEPI buffer, and store at −20 °C (*see* **Note 2**).

5. 1 M NaN_3 stock solution in water.

6. Two tabletop centrifuges: the first one is a high-speed centrifuge used with a swinging bucket rotor for collecting yeast cells and the second one is an ultracentrifuge, for recovering yeast total membranes.

2.3 Large-Scale Membrane Preparation

1. Polycarbonate bottles with screw cap assemblies, each one inwardly covered with a liner.

2. Ultrapure water: store 2 L at 4 °C.

3. TEKS buffer: 50 mM Tris–HCl pH 7.5, 1 mM EDTA, 0.1 M KCl, 0.6 M sorbitol. Sterilize with a 0.22 μm filter and store at 4 °C.

4. TES buffer: 50 mM Tris–HCl pH 7.5, 1 mM EDTA, 0.6 M sorbitol. Sterilize with a 0.22 μm filter and store at 4 °C.

5. HEPES-sucrose buffer: 20 mM HEPES-Tris pH 7.4, 0.3 M sucrose, 0.1 mM $CaCl_2$. You may prepare this buffer beforehand provided it is stored at −20 °C.

6. Reagents for protein assay: bicinchoninic acid (BCA), 4 % $CuSO_4$, and bovine serum albumin (BSA) at 10 mg/mL in water.

7. 0.5 mm diameter glass beads and PMSF are as described in Subheading 2.2.

8. A Dounce homogenizer.

9. A "Pulverisette 6" planetary mill (Fritsch) for breaking yeast cells with glass beads.
10. 10 × concentrated SIGMAFAST™ EDTA-free Protease Inhibitor Cocktail stock solution: prepare freshly by dissolving one tablet in 10 mL TES buffer. For long-term storage, this solution may be frozen and kept at −20 °C for months.

2.4 SDS-Polyacrylamide Gel Electrophoresis (SDS-PAGE)

1. Separating gel solution: 375 mM Tris–HCl pH 8.8, 8 % acrylamide (w/v), 0.1 % SDS (w/v), 0.1 % ammonium persulfate (w/v), 0.06 % TEMED (v/v). Add ammonium persulfate and TEMED last, just before pouring.
2. Stacking gel solution: 125 mM Tris–HCl pH 6.8, 5 % acrylamide (w/v), 0.1 % SDS (w/v), 0.1 % ammonium persulfate (w/v), 0.1 % TEMED (v/v). Add ammonium persulfate and TEMED last, just before pouring.
3. 2 × concentrated SDS/urea sample buffer: 100 mM Tris–HCl pH 6.8, 1.4 M β-mercaptoethanol, 5 % SDS (w/v), 1 mM EDTA, 9 M urea, 0.01 % bromophenol blue (w/v).
4. Running buffer: 25 mM Tris, 250 mM glycine, 0.1 % SDS (w/v).
5. Coomassie blue staining solution: 0.05 % Coomassie blue R-250, 40 % (v/v) methanol, 10 % (v/v) acetic acid.

2.5 Western Blotting

1. TB buffer: 26 mM Trizma-base, 190 mM glycine, 10 % (v/v) methanol. Prepare 1 L per electrophoresis tank (for one or two blots). Stir for 2–3 h at 4 °C in water/ice bath for cooling, and store at 4 °C (*see* **Note 3**).
2. Polyvinylidene-difluoride (PVDF) membranes and chromatography paper.
3. Prestained molecular weight markers.
4. 10× concentrated PBS buffer: 27 mM KCl, 1.4 M NaCl, 0.13 M Na_2HPO_4, and 18 mM KH_2PO_4. Dissolve first in 800 mL water, adjust to pH 7.4 with NaOH, make up to 1 L with water, and filter-sterilize. Dilute tenfold before use.
5. PBS-T buffer: PBS buffer supplemented with 0.2 % Tween 20.
6. Avidin peroxidase conjugate, commercial solution for single use. Store 2 μL-aliquots at −70 °C. It will be referred to as biotin probe in the following protocol.
7. His-Probe-HRP peroxidase conjugate, commercial powder: reconstitute with 0.4 mL of water to a final concentration of 5 mg/mL. Store aliquots at −20 °C.
8. Enhanced chemiluminescent (ECL) reagents.
9. ECL detection system.
10. CCD-cooled camera.

2.6 Phosphorylation from [γ-³²P]ATP of Drs2p Embedded in Light P3 Membranes

1. Radiolabeled [γ-^{32}P]ATP (250 μCi, 3000 Ci/mmol): prepare a 10 μM stock solution by mixing [γ-^{32}P]ATP with non-radioactive ATP, to reach a specific radioactivity of about 0.25–1 mCi/μmol.
2. Trichloroacetic acid (TCA) and phosphoric acid (H_3PO_4): prepare a stock solution of 1 M TCA and 60 mM H_3PO_4 in water and store at 4 °C, protected from light.
3. A/E glass fiber filters (1 μm porosity), and cellulose GSWP filters (0.22 μm porosity).
4. Sodium orthovanadate (Na_3VO_4) stock solution: 100 mM in water.
5. Buffer A: 100 mM KCl, 5 mM $MgCl_2$, 50 mM Mops-Tris pH 7.
6. FILTER-COUNT liquid scintillation counting.
7. A scintillation counter.

3 Methods

We previously developed methods for expression of integral membrane proteins, especially for expression of the wild-type and mutated rabbit Ca^{2+}-ATPase SERCA1a [2–6]. For this purpose, we use the pYeDP60 plasmid, generously given by Dr. Denis Pompon [7]. The pYeDP60 plasmid contains a strong and inducible *GAL10-CYC1* hybrid promoter which is fully repressed by glucose and activated by galactose. In addition to the classical *URA3* selection marker, this plasmid contains an *ADE2* selection marker. This marker is of special interest since even rich culture media become rapidly limiting for adenine upon yeast growth. Thus, yeast cells auxotrophic for adenine and transformed with pYeDP60 can grow in rich medium with minimal plasmid loss.

A common limitation to the use of recombinant *GAL* promoters for overexpression of foreign proteins is the low abundance of the Gal4p protein, which triggers transcription of genes that are controlled by *GAL* promoters. To overcome this limitation, we integrated a hybrid gene in the yeast chromosome consisting of the *GAL10* promoter fused to the *GAL4* gene [2, 8, 9] (Fig. 1). Following the addition of galactose to the culture medium, the integrant strain synthesizes substantially more Gal4p protein as well as more of the target protein.

In the particular case of the flippase complex broached here, the challenge is to achieve the functional and coordinated overexpression of the two membrane-embedded partners, i.e., the P-type ATPase Drs2p and its associated subunit Cdc50p. Toward this goal, we constructed a plasmid which contained *DRS2* and *CDC50* open-reading frames, both under the control of the *GAL10-CYC1*

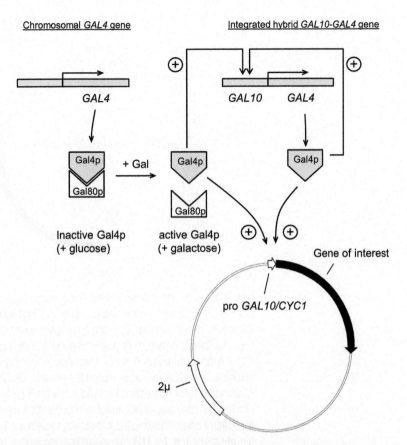

Fig. 1 Modification of the W303.1b yeast strain allowing regulated overproduction of the *GAL4* gene product. Gal4p is a positive regulatory factor required for the activation of transcription of genes under the control of *GAL* promoters. In the presence of glucose, the Gal80p protein antagonizes the action of the Gal4p protein. When glucose is entirely consumed and galactose is added to the medium, the interaction between Gal4p and Gal80p is disrupted and Gal4p may now activate *GAL* promoters. Chromosomal *GAL4* gene is constitutively expressed at very low levels in yeast cells (1–3 copies per cell). The W303.1b yeast strain has been genetically modified to integrate an additional copy of the *GAL4* gene fused to a *GAL10* promoter, allowing regulated overproduction of the *GAL4* gene product and an increased expression of foreign genes placed under the control of galactose-inducible promoters

inducible promoter [10, 11] (Fig. 2). The rationale for constructing this co-expression plasmid, rather than expressing the two proteins from different plasmids, was to avoid imbalanced transcription of *DRS2* and *CDC50* genes due to an unequal number of plasmids in each cell, a frequent behavior of 2 μ-based plasmids. To facilitate detection (and purification) of Drs2p and Cdc50p, a biotin acceptor domain (Bad), which is biotinylated in vivo in yeast, and a decahistidine tag were added at the N-terminus of Drs2p and Cdc50p, respectively.

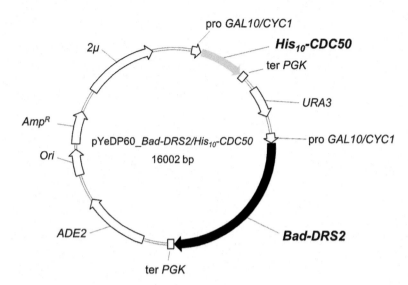

Fig. 2 Map of the plasmid used for co-expression of Bad-Drs2p and His$_{10}$-Cdc50p. The co-expression plasmid contains sequences encoding *CDC50* fused to an N-terminal decahistidine tag (*grey arrow*) and *DRS2* fused to an N-terminal Bad tag (*black arrow*), both under the control of a galactose-inducible *GAL10/CYC1* hybrid promoter. A short sequence encoding a tobacco etch virus (TEV) protease cleavage site was inserted between *DRS2* and *CDC50* open reading frames and their respective tags. Adapted from [10]. *ADE2* auxotrophy selection marker for adenine, *URA3* auxotrophy selection marker for uracil, *Ori* bacterial replication origin, *Amp^R* gene conferring resistance to ampicilline, *2 μ* yeast origin of replication, *ter PGK* phosphoglycerate kinase terminator sequence

Protein expression may be followed during the membrane preparation process, by Western blotting, using either a biotin probe coupled to horseradish peroxidase (HRP) (which binds to the biotinylated domain fused to Drs2p), or a histidine probe coupled to HRP (which binds to the Cdc50p decahistidine tag). The activity of the recombinant proteins may be followed by taking advantage of the fact that P-type ATPases undergo transient phosphorylation from ATP during their catalytic cycle.

3.1 Yeast Culture

3.1.1 Yeast Transformation

W303.1b/*GAL4* cells are transformed with the desired pYeDP60 plasmids using the lithium-acetate/single-stranded DNA/PEG method [12].

1. Grow a 24-h culture at 28 °C and 180 rpm in S6AU medium.
2. Pellet 2 mL of cells for 5 min and 3000×*g*, at room temperature. Carefully discard the supernatant.
3. Add 50 μL of 2 mg/mL denatured SSD and about 1 μg plasmid DNA to the pellet. Mix up by vortexing.
4. Add 500 μL PLATE solution as well as 20 μL sterile 1 M dithiothreitol. Mix the pellet up by vortexing.

5. Incubate overnight at 20 °C.

6. Centrifuge yeast cells at 450×*g* and resuspend the pellet in 100 µL selective S6A medium.

   Then, spread the whole volume onto a S6A plate and incubate for 3–5 days at 28 °C. As *S. cerevisiae* W303.1b/*GAL4* is auxotrophic for uracil, only those yeast cells transformed with pYeDP60 plasmids will have the ability to grow on this minimal medium.

3.1.2 Functional Complementation of Δ*drs2* and Δ*cdc50* Cold-Sensitivity

A convenient way to check the functionality of the various constructs is to take advantage of the fact that Δ*drs2* and Δ*cdc50* cells exhibit a cold-sensitive growth phenotype [13].

1. Spread yeast cells (transformed with the desired expression plasmid and stored at −80 °C as a glycerol stock) onto S6A plates and incubate at 28 °C for about 72 h.

2. Inoculate a few colonies into 5 mL S6A liquid medium and grow at 28 °C and 180 rpm to mid-exponential phase for 24 h in a shaking incubator.

3. Dilute cultures serially (to 0.02, 0.001, and 0.0002 OD_{600}) with a galactose- and fructose-containing S5AF medium (*see* **Note 4**). Spot 5 µL drops onto S5AF plates.

4. Incubate plates for 2–3 days at 28 °C or for 5–6 days at 20 °C. At the restrictive temperature of 20 °C, Δ*drs2* and Δ*cdc50* cells grow very slowly (Fig. 3).

3.1.3 Small-Scale Expression of Drs2p and Cdc50p in Minimal Medium

1. Spread yeast cells (transformed with the desired expression plasmid and stored at −80 °C as a glycerol stock) onto S6A plates and incubate at 28 °C for about 72 h.

2. Inoculate a few colonies into 5 mL S6A liquid medium and grow at 28 °C and 180 rpm to mid-exponential phase for 24 h in a shaking incubator.

3. Dilute cells grown in S6A into 20 mL S5A (containing 2 % galactose) to a final OD_{600} of 0.2, to induce protein expression. Grow cells for 17 h at 28 °C, 180 rpm. After 17 h growth, the OD_{600} should reach about 0.5.

4. Collect yeast cells (20 mL) by centrifugation at 1000×*g* for 10 min at 4 °C in 50-mL tubes (*see* **Note 5**). Wash pellets once with ice-cold TEPI buffer (containing fresh protease inhibitors, 1 mM PMSF and 20 mM NaN_3), and transfer to 1.5 mL tubes. Centrifuge the resuspended cells at 1400×*g* for 10 min at 4 °C.

5. Resuspend cell pellets in 100 µL ice-cold TEPI buffer and add glass beads so that they reach the meniscus. Break cells by vortexing vigorously for 20 min at 4 °C.

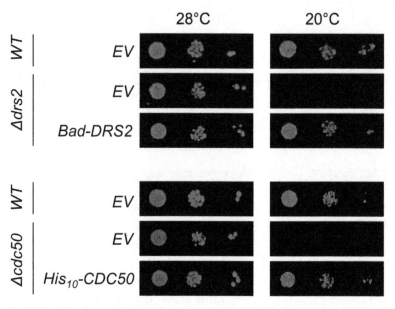

Fig. 3 Functional complementation of the temperature-sensitive phenotype of Δ*drs2* and Δ*cdc50* yeast cells. Yeast cells, either wild-type, Δ*drs2*, or Δ*cdc50* mutants, were transformed with plasmids bearing either *DRS2* tagged with a sequence coding for Bad, or *CDC50* tagged with ten histidines. Cells transformed with an empty vector (EV) were used as negative control. Serial dilutions of yeast cells were spotted on plates and incubated at 28 °C or at the restrictive temperature of 20 °C

6. Adjust volume to 1 mL with ice-cold TEPI buffer and spin down the lysate at $500 \times g$ for 5 min at 4 °C to remove nuclei and unbroken cells.

7. Transfer supernatants to 1.5 mL ultracentrifuge tubes and spin down at $72,000 \times g$ for 90 min at 4 °C (*see* **Notes 6** and **7**).

8. Remove the supernatant and add 200 μL 1× SDS/urea sample buffer to the pellet. Resuspend membrane pellets by vortexing for about 45 min at 4 °C. Incubate membrane samples for 10 min at 30 °C and analyze protein expression by Western blotting (*see* **Note 8**).

3.1.4 Large-Scale Expression of Drs2p and Cdc50p in Rich Medium

The procedure is described here for 4 L of yeast culture in glass Fernbach-type flasks, each flask containing 500 mL rich medium, but this protocol may be easily scaled up to 12 L.

1. Spread yeast cells (transformed with the desired expression plasmid and stored at −80 °C as a glycerol stock) onto S6A plates and incubate at 28 °C for about 72 h.

2. Inoculate a few colonies into 5 mL of S6A medium and grow at 28 °C for 24 h and 180 rpm, in a shaking incubator (the cell density reaches saturation, i.e., $OD_{600} \sim 3$, corresponding to $9-15 \times 10^7$ cells per mL).

3. Inoculate this pre-culture into 50 mL of S6A medium in order to reach a final OD_{600} of 0.1 and grow cells at 28 °C for 24 h and 180 rpm, in a shaking incubator (*see* **Note 9**). In the meantime, bring 4 L YPGE2X medium (without glucose and without ethanol) to 28 °C. That is, pour eight times 460 mL of sterile medium (which already contains the necessary amount of yeast extract and peptone for a final volume of 500 mL) in Fernbach flasks.

4. Prior to inoculation, add glucose and ethanol to the rich medium at a final concentration of 1 % (v/v) and 2.7 % (v/v), respectively. Add about 10 mL of the second pre-culture (at an OD_{600} of about 2–3) to each flask containing 460 mL YPGE2X, to reach an OD_{600} of about 0.05 (corresponding to $2-3 \times 10^6$ cells per mL). Grow cells at 28 °C and 130 rpm.

5. After 36 h growth, transfer the cultures into a water/ice bath for several minutes to allow cooling from 28 to 18 °C (*see* **Note 10**), and measure the OD_{600} (~10). When a temperature of 18 °C has been reached, add 2 % galactose powder (10 g/flask) to induce expression of the recombinant proteins. Transfer cultures back to the incubator at 18 °C and shake at 130 rpm.

6. 13 h after the first induction, start the second induction by adding 2 % galactose and grow the culture for a further 5 h at 18 °C and 130 rpm.

7. Harvest cells immediately after the second induction (*see* **Note 11**).

3.2 Membrane Preparation after Yeast Growth in Rich Medium

The procedure is described for 4 L of yeast culture (about 120–160 g yeast cells).

1. At the end of the culture, recover yeast cells in 4 centrifugation bottles (1 L each, *see* **Note 12**). Spin down for 10 min, at 4 °C and $4000 \times g$. Discard supernatants.
   All subsequent steps are performed at 4 °C in a cold room.

2. Wash yeast pellets with cooled water (with a volume equivalent to about 15 times the mass of yeast). Centrifuge resuspended cells for 10 min, at 4 °C and $4000 \times g$. After removing the supernatants, weigh the pellets. Usually, about 30–40 g of yeast are obtained for 1 L of culture.

3. Resuspend pellets in ice-cold TEKS buffer (2 mL TEKS per gram of yeast). Incubate the suspension for 15 min at 4 °C. At this stage, the cells may be transferred to a liner for easier resuspension. Centrifuge resuspended cells for 10 min, at 4 °C and $4000 \times g$.

4. Resuspend pellets with a small volume of TES buffer, equivalent to 1 mL per gram of yeast, so that the suspension eventually contains 1 mM PMSF and a 1× concentrated protease inhibitor cocktail. Using the liners, this step is easily carried out

by hand kneading. Pour the resulting suspension into agate pots (80–225 mL may be handled in 500 mL pots), and add 0.5 mm diameter glass beads (a volume equal to the total volume of the suspension). Break yeast cells with a planetary mill, by shaking for 3 min at 450 rpm first, followed by a 30 s pause, and an additional shaking for 3 min at 450 rpm in the opposite direction (*see* **Note 13**).

5. For recovery of the crude extract from the agate pots, connect a vacuum pump to a filtering flask (3 L Erlenmeyer, as a trap), that is itself connected to a 50 mL pipette. This allows efficient sucking of the crude extract from the beads into the trap.

6. Wash the glass beads in three successive steps with a total washing volume of 180–240 mL TES buffer supplemented with 1 mM PMSF and the 1× protease inhibitor cocktail (1.5 mL buffer per g of yeast). Control the pH of the total crude extract (the first extract plus the washes will result in a total volume of about 0.4 L): it must be around 7.0–7.5. If not, adjust the pH by adding NaOH (*see* **Note 14**).

7. Spin down the total crude extract in 500 mL bottles at $1000 \times g$ for 20 min, at 4 °C. Discard the pellet (P1), containing cell debris and nuclei (Fig. 4).

8. Centrifuge the corresponding supernatant (S1) in 500 mL bottles at $20,000 \times g$ for 20 min, at 4 °C. Remove the supernatant (S2) with great care as part of the pellet is not firmly attached to the bottom of the bottle. Resuspend the pellet (P2), corresponding to "heavy" membranes, with the rest of S2 supernatant, and save for further analyses (Fig. 4).

9. Centrifuge the S2 supernatant in 70 mL polycarbonate bottles at $125,000 \times g$ for 1 h, at 10 °C. Discard the supernatant (S3) (Fig. 4).

10. Resuspend the pellet (P3), corresponding to "light" membranes, at about 30–40 mg/mL of total protein in about 24–32 mL HEPES-sucrose buffer (0.2 mL per gram of yeast).

11. Homogenize P3 membranes with a Dounce homogenizer and store at −70 °C after snap freezing in liquid nitrogen (*see* Fig. 4: Western blot of the different fractions obtained in the course of the membrane preparation, using a biotin probe and a histidine probe).

3.3 Estimation of the Total Protein Concentration

Estimation of the total protein concentration is performed in 96-well plates, using the bicinchoninic acid (BCA) assay [14].

1. The total volume of the assay mixture is 25 μL, as follows: 10 μL of the sample to be quantified (diluted at least ten times in water, in triplicates), 10 μL of water, and 5 μL of 10 % SDS (w/v) so that final concentration is 2 %.

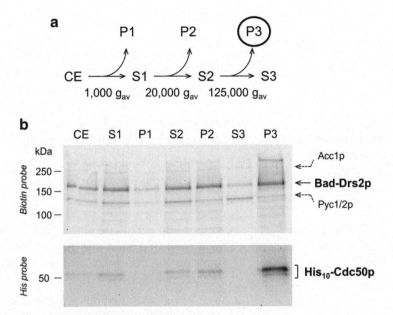

Fig. 4 Co-expression of Bad-Drs2p and His_{10}-Cdc50p in yeast membranes. (**a**) Diagram of the membrane fractionation procedure. CE: crude extract; S1 and P1: supernatant and pellet recovered after low-speed centrifugation; S2 and P2: supernatant and pellet recovered after medium-speed centrifugation; S3 and P3: supernatant and pellet recovered after high-speed centrifugation. (**b**) Western blot analysis of Drs2p and Cdc50p expression in various samples collected throughout membrane fractionation. After SDS-PAGE, proteins were electro-transferred to PVDF membranes and immunostained with either a biotin probe (*top gel*) or a histidine probe (*bottom gel*). In both cases, 1.5 μg total protein was loaded on gel. The biotin probe also weakly detects yeast proteins known to be biotinylated, Acc1p and Pyc1/2p (*dashed arrows*). Cdc50p appears as multiple bands corresponding to different glycosylation levels

2. For the standard curve, BSA is used. For reliability of the standard curve, the assay mixture must also contain 2 % SDS and the 10 μL of water are replaced with 10 μL of tenfold diluted TES or HEPES-sucrose buffer, depending on the sample tested. Establish the standard curve with BSA amounts comprised between 0 and 80 μg (10 μL of 0–8 mg/mL stock solution).

3. When SDS, the sample to be quantified, and water (or buffer) have been mixed together in a 1.5 mL tube for both the membrane samples and for the standard curve, add 225 μL of a 50/1 $BCA/CuSO_4$ mixture. After 30 min incubation at 37 °C (*see* **Note 15**), transfer 200 μL of each sample to 96-well plates. Prepare the $BCA/CuSO_4$ reagent just before use, to avoid premature oxidation.

4. Read the absorbance at 562 nm. The total protein concentration of the light membrane fraction (P3) is generally around

30–40 mg/mL and comparison, by Western blotting, of known amounts of the Bad-tagged SERCA1a Ca^{2+}-ATPase with membrane fractions containing Bad-Drs2p allows us to estimate that Drs2p contributes to about 3 % of the total membrane protein content in P3 membranes [11].

3.4 SDS-Polyacrylamide Gel Electrophoresis (SDS-PAGE)

Gels are prepared according to the Laemmli procedure [15], using glass plates with 0.75 mm-thick spacers.

1. Separating gel: prepare 8 % acrylamide gels by mixing 1 mL of 1.5 M Tris–HCl pH 8.8 with 1.07 mL of 30 % acrylamide/bis solution (29/1), 1.85 mL of water, 40 µL of 10 % SDS, 40 µL of 10 % ammonium persulfate and 2.4 µL of TEMED (add ammonium persulfate and TEMED last, just before pouring), for a total of 4 mL (one gel). Pour the gel (3.2 mL), leaving space for a stacking gel, and overlay with water (see **Note 16**). The gel polymerizes within about 30 min at 20 °C.

2. Prepare the stacking gel by mixing 250 µL of 1 M Tris–HCl pH 6.8 with 330 µL of 30 % acrylamide/bis solution, 1.38 mL of water, 20 µL of 10 % SDS, 20 µL of 10 % ammonium persulfate and 2 µL of TEMED (add ammonium persulfate and TEMED just before pouring), for a total of 2 mL (one gel). Remove the water on top of the separating gel, and pour an amount of stacking gel at least 1 cm-thick. Insert the comb immediately. The gel polymerizes within about 30 min at 20 °C.

3. Add the running buffer to the anode and cathode chambers of the gel unit.

4. Before running, mix samples with an equal volume of 2× concentrated SDS/urea sample buffer and heat at 30 °C for 10 min. Load the samples onto the gels and run for 90 min, at 110 V and 25 mA/gel.

3.5 Western blotting for Drs2p and Cdc50p Immunodetection

Samples that have been separated by SDS-PAGE may be transferred onto a PVDF membrane.

1. Soak the PVDF membrane briefly into pure methanol for maximum transfer efficiency.

2. Use prestained standards to check transfer efficiency.

3. Use TB buffer for transfer. In those conditions, the vast majority of the proteins (M_r ranging from about 20 to 200 kDa) are transferred from an 8 % Laemmli-type gel, in less than 1 h at 110 V. During transfer, set intensity to a maximum of 350 mA to prevent heating of the system. Along the same lines, place an ice cooling unit in the tank.

4. Following transfer, block unspecific binding sites on the PVDF membranes with PBS-T supplemented with 5 % dry milk, for 1 h at room temperature.

5. For detection of biotinylated proteins by the biotin probe, wash first membranes three times (10 min each) with PBS-T in order to remove any traces of milk (*see* **Note 17**). Add the probe (coupled to HRP) at a 1/25,000 dilution in PBS-T and incubate for 1 h at 20 °C.

6. For detection of histidine-tagged proteins, block the unspecific binding sites on PVDF membranes with PBS-T supplemented with 2 % BSA, for 1 h at room temperature. Add the histidine probe (coupled to HRP) at a 1/2000 dilution in PBS-T and incubate for 1 h at 20 °C.

7. In all cases, wash membranes three times with PBS-T buffer (10 min each), and two times with PBS (10 min each). The bound probes are revealed using ECL reagents and emitted light is detected using a CCD cooled camera (Fig. 4).

3.6 Phosphorylation from [γ-^{32}P]ATP of Drs2p Embedded in Light P3 Membranes

Drs2p belongs to the P-type ATPase family. Those ATPases are able to form a stable phosphorylated intermediate during their catalytic cycle. Formation of this intermediate may be measured after incubation with [γ-^{32}P]ATP followed by acid quenching, using a filtration protocol [16] (*see* Fig. 5: phosphorylation from [γ-^{32}P]ATP of Bad-Drs2p in yeast membranes, in the absence or presence of its associated subunit Cdc50p).

1. Pre-incubate 40 µL aliquots of P3 membranes, at 0.5 mg/mL total protein, on ice in buffer A, in the presence or absence of 1 mM orthovanadate.

2. Start phosphorylation by adding [γ-^{32}P]ATP (at 0.25–1 mCi/µmol) at a final concentration of 0.5 µM (add 2 µL of a 10 µM stock solution).

3. Stop the phosphorylation reaction after 30 s phosphorylation (once steady-state is reached) by adding 1 mL of 500 mM TCA + 30 mM H_3PO_4. Leave the samples on ice for at least 30 min after quenching, to allow aggregation of the precipitated proteins (*see* **Note 18**).

4. Filter the precipitated samples on either A/E glass fiber filters or GSWP nitrocellulose filters (pre-soaked with the quenching medium). Carefully rinse the filters with 50 mM TCA + 3 mM H_3PO_4.

5. Radioactivity is measured by scintillation counting. Place filters in a scintillation vial, add 4 mL FILTER-COUNT and shake until the filter is dissolved.

6. For the conversion of counts per minutes (cpm) to moles of phosphoenzyme formed, measure the radioactivity of various dilutions of the 10 µM [γ-^{32}P]ATP stock solution.

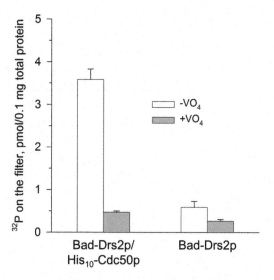

Fig. 5 Co-expression of Cdc50p is required for phosphorylation from [γ-^{32}P]ATP of Bad-Drs2p expressed in P3 membranes. P3 membranes containing either Drs2p co-expressed with Cdc50p or Drs2p expressed alone were used. Samples were incubated on ice in buffer A, in the absence or presence of 1 mM orthovanadate. After 30 s phosphorylation on ice with 0.5 μM [γ-^{32}P]ATP, samples were quenched and filtrated. Phosphorylation of Drs2p in P3 membranes is three- to fourfold higher than that in P2 membranes (for the same amount of Drs2p and Cdc50p), suggesting that the P3 fraction contains a higher proportion of active Drs2p

4 Notes

1. In most cases, true optical densities do not depend on the particular equipment used; they only depend on the optical cell path length. However, to monitor the growth of an organism, what is generally significant is the turbidity of the sample, which depends on its light scattering properties. Be aware that when measuring with an ordinary spectrophotometer, the loss of transmitted light will depend on the geometry of the equipment. For our yeast cells at a given stage, the measured OD could vary by a factor 2 depending on the spectrophotometer used. The optical density values reported throughout this manuscript are measured with a Novaspec II (Pharmacia Biotech).

2. Previously, we used the Complete EDTA-free protease inhibitor cocktail from Roche Diagnostics. However, in the case of Drs2p/Cdc50p expression, SIGMAFAST protease inhibitor cocktail proved more efficient at inhibiting unwanted protease activities. Another advantage of the SIGMAFAST protease inhibitor cocktail is that its composition is provided by the supplier.

3. If possible, prepare TB buffer the day before transferring, for efficient cooling. TB buffer is reusable 2–3 times and may be stored at 4 °C for about 1–2 weeks.

4. Fructose is added to the galactose-containing S5A medium to allow yeast growth at a fair rate, since galactose as the sole carbon source results in very poor yeast growth.

5. Using the swinging-bucket rotor is preferable in this case as it will allow yeast sedimentation to the bottom of the tube, and therefore limit loss of material when removing the medium after centrifugation.

6. We centrifuge at $72,000 \times g$ in a TLA-45 rotor because this is the maximum speed allowed for such a rotor. If available, you may alternatively use a TLA-55 rotor and in this case, you may centrifuge for 1 h at $100,000 \times g$.

7. The procedure for small-scale membrane preparation may be stopped after ultracentrifugation and removal of the supernatant. Membrane pellets may be stored at −80 °C after snap freezing in liquid nitrogen.

8. Drs2p is prone to aggregation when heated at temperatures that ensure denaturation (e.g., 2 min at 96 °C) before loading onto SDS-PAGE. This is the case even in the presence of urea, which has previously been shown to reduce significantly denaturation of the Ca^{2+}-ATPase SERCA1a when incubated at high temperatures. Therefore, we incubate samples at 30 °C for 10 min prior loading onto SDS-PAGE.

9. Both pre-cultures are carried out in Erlenmeyer-type flasks, with a volume of culture which is at most one-tenth of the volume of the flask. This ensures optimal ventilation of the yeast culture.

10. Lowering the temperature of the medium to 18 °C significantly increases the amount of properly folded and active protein. It is likely that at this temperature, the yeast metabolism is slowed down, thereby facilitating integration of newly synthesized protein in the ER membrane. It might be especially relevant in the case of overexpressed proteins.

11. If necessary, yeast cells can be frozen after removal of the culture medium and stored at −80 °C until their handling for membrane preparation.

12. Pay attention to the fact that depending on the geometry of your centrifugation tube (whether a sample has to migrate a few centimeters (as in preparative rotors) or a few millimeters (as, for example, in small tubes appropriate for TLA rotors) to reach the bottom of the tube), a given speed for a given period of time may lead to different results. That is why the volumes of centrifugation bottles are indicated.

13. Yeast breaking with the Pulverisette does occur at room temperature. Consequently, the culture will warm up. However, in our hands, thermal inertia of agate pots (the pots, the cells, and the beads are cold before starting shaking) allows the temperature to remain sufficiently low for preventing protein degradation.
14. Monitoring the pH is a critical step in the course of membrane preparation. Indeed, yeast breaking may result in the release of the vacuolar content and may contribute to acidifying the pH and therefore to increase the activity of vacuolar proteases.
15. To achieve good reliability for the protein assay, the period of time for color development should be long enough, so that highly diluted samples reach an OD_{562} significantly above background. Note that sorbitol contained in TES inhibits color development; thus, depending on the buffer used for membrane resuspension, you may need to adjust the time period for color development.
16. Alternatively, you may use isopropanol or ethanol for this purpose.
17. After blocking, traces of milk need to be carefully removed because of the presence of biotinylated proteins in milk. Hence, before incubation with the biotin probe, the PVDF membrane is washed three times with PBS-T. In addition, the biotin probe itself is diluted into PBS-T in the absence of milk.
18. We would like to mention here that for those phosphorylation measurements that are performed in the presence of detergent, the question arises as to whether the acid-denatured but previously detergent-solubilized Drs2p would be retained efficiently by the filters with relatively large pores (1 μm). In many cases, the answer is yes, provided that the acid-denatured proteins are kept on ice during a period of time sufficient for aggregation and therefore retention by the first filter. However, for a number of detergents, using two filters on top of each other for filtration revealed that the bottom filter contained 10–20 % of the radioactivity found on the first filter, suggesting that in those cases part of the radioactivity-carrying acid-quenched material had not been fully retained by the first filter.

Acknowledgments

We are grateful to Dr. Miriam-Rose Ash for critical reading of the manuscript and for insightful discussions. We also thank Dr. Marc le Maire and Dr. Christine Jaxel for discussions. This work was supported by grants from the Agence Nationale de la Recherche (ANR

Blanc program, MIT-2M), the Ile de France region (Domaine d'Intérêt Majeur Maladies Infectieuses, DIM Malinf), and the French Infrastructure for Integrated Structural Biology (FRISBI, ANR-10-INSB-05-01).

References

1. Lopez-Marques RL, Theorin L, Palmgren MG et al (2013) P4-ATPases: lipid flippases in cell membranes. Pflugers Arch 466:1227–1240
2. Lenoir G, Menguy T, Corre F et al (2002) Overproduction in yeast and rapid and efficient purification of the rabbit SERCA1a Ca(2+)-ATPase. Biochim Biophys Acta 1560:67–83
3. Cardi D, Montigny C, Arnou B et al (2010) Heterologous expression and affinity purification of eukaryotic membrane proteins in view of functional and structural studies: the example of the sarcoplasmic reticulum Ca(2+)-ATPase. Methods Mol Biol 601:247–267
4. Jidenko M, Lenoir G, Fuentes JM et al (2006) Expression in yeast and purification of a membrane protein, SERCA1a, using a biotinylated acceptor domain. Protein Expr Purif 48:32–42
5. Marchand A, Winther AM, Holm PJ et al (2008) Crystal structure of D351A and P312A mutant forms of the mammalian sarcoplasmic reticulum Ca(2+)-ATPase reveals key events in phosphorylation and Ca(2+) release. J Biol Chem 283:14867–14882
6. Clausen JD, Bublitz M, Arnou B et al (2013) SERCA mutant E309Q binds two Ca(2+) ions but adopts a catalytically incompetent conformation. EMBO J 32:3231–3243
7. Pompon D, Louerat B, Bronine A et al (1996) Yeast expression of animal and plant P450s in optimized redox environments. Methods Enzymol 272:51–64
8. Schultz LD, Hofmann KJ, Mylin LM et al (1987) Regulated overproduction of the GAL4 gene product greatly increases expression from galactose-inducible promoters on multi-copy expression vectors in yeast. Gene 61:123–133
9. Pedersen PA, Rasmussen JH, Jorgensen PL (1996) Expression in high yield of pig alpha 1 beta 1 Na, K-ATPase and inactive mutants D369N and D807N in Saccharomyces cerevisiae. J Biol Chem 271:2514–2522
10. Jacquot A, Montigny C, Hennrich H et al (2012) Stimulation by phosphatidylserine of Drs2p/Cdc50p lipid translocase dephosphorylation is controlled by phoshatidylinositol-4-phosphate. J Biol Chem 287:13249–13261
11. Montigny C, Azouaoui H, Jacquot A et al (2014) Overexpression of membrane proteins in *Saccharomyces cerevisiae* for structural and functional studies: a focus on the rabbit Ca^{2+}-ATPase Serca1a and on the yeast lipid "flippase" complex Drs2p/Cdc50p. Springer, New York, NY
12. Gietz RD, Schiestl RH, Willems AR et al (1995) Studies on the transformation of intact yeast cells by the LiAc/SS-DNA/PEG procedure. Yeast 11:355–360
13. Chen CY, Ingram MF, Rosal PH et al (1999) Role for Drs2p, a P-type ATPase and potential aminophospholipid translocase, in yeast late Golgi function. J Cell Biol 147:1223–1236
14. Smith PK, Krohn RI, Hermanson GT et al (1985) Measurement of protein using bicinchoninic acid. Anal Biochem 150:76–85
15. Laemmli UK (1970) Cleavage of structural proteins during the assembly of the head of bacteriophage T4. Nature 227:680–685
16. Hatori Y, Hirata A, Toyoshima C et al (2008) Intermediate phosphorylation reactions in the mechanism of ATP utilization by the copper ATPase (CopA) of Thermotoga maritima. J Biol Chem 283:22541–22549

Chapter 7

The Plasma Membrane Ca^{2+} ATPase: Purification by Calmodulin Affinity Chromatography, and Reconstitution of the Purified Protein

Verena Niggli and Ernesto Carafoli

Abstract

Plasma membrane Ca^{2+} ATPases (PMCA pumps) are key regulators of cytosolic Ca^{2+} in eukaryotes. They extrude Ca^{2+} from the cytosol, using the energy of ATP hydrolysis and operate as Ca^{2+}-H^+ exchangers. They are activated by the Ca^{2+}-binding protein calmodulin, by acidic phospholipids and by other mechanisms, among them kinase-mediated phosphorylation. Isolation of the PMCA in pure and active form is essential for the analysis of its structure and function. In this chapter, the purification of the pump, as first achieved from erythrocyte plasma membranes by calmodulin-affinity chromatography, is described in detail. The reversible, high-affinity, Ca^{2+}-dependent interaction of the pump with calmodulin is the basis of the procedure. Either phospholipids or glycerol have to be present in the isolation buffers to keep the pump active during the isolation procedure. After the isolation of the PMCA pump from human erythrocytes the pump was purified from other cell types, e.g., heart sarcolemma, plant microsomal fractions, and cells that express it ectopically. The reconstitution of the purified pump into phospholipid vesicles using the cholate dialysis method will also be described. It allows studies of transport mechanism and of regulation of pump activity. The purified pump can be stored in the reconstituted form for several days at 4 °C with little loss of activity, but it rapidly loses activity when stored in the detergent-solubilized form.

Key words Plasma membrane Ca^{2+} ATPase, Calmodulin, Affinity chromatography, Reconstitution, Phosphatidylserine, Phosphatidylcholine, Liposome, Ca^{2+} transport, Ca^{2+}-H^+ exchange

1 Introduction

Plasma membrane Ca^{2+} ATPases (PMCA pumps) extrude Ca^{2+} from the cytosol in exchange for protons, using the energy of ATP hydrolysis. Four major gene products of the pump have been identified and a number of splice isoforms have also been described. Isoforms 1 and 4 are ubiquitously expressed; isoforms 2 and 3 show a restricted tissue expression. The PMCA is organized in the plasma membrane with ten transmembrane helices and three large cytosolic domains. The Ca^{2+}-sensor calmodulin activates the pump by displacing a C-terminal autoinhibitory domain from binding sites situated in the vicinity of the active center. Acidic phospholipids also activate the pump in the absence of calmodulin. They bind to a stretch of predominantly basic residues in the loop between transmembrane domains 2 and 3, but also to the calmodulin-binding domain located in the tail of more than 100 residues emerging from transmembrane domain 10 [1]. Interestingly, the tail containing the calmodulin-binding domain is instead N-terminal in plant PMCAs [2].

1.1 Purification of the PMCA with Affinity Chromatography

The purification of the PMCA has enabled structure–function analysis of the pump, and has led to insights into regulation of its function and on the mechanisms of Ca^{2+} transport. It has also been instrumental in its cloning. The Ca^{2+}-dependent, reversible, high-affinity interaction of the PMCA with calmodulin has been exploited to devise the rapid isolation procedure that has resulted in the isolation of the functional pump with a high degree of purity. As detailed in Fig. 1, this procedure involves an affinity support with covalently linked calmodulin. Detergent-solubilized, purified plasma membranes (erythrocyte ghosts in the original protocol) depleted of endogenous calmodulin are applied to this column in

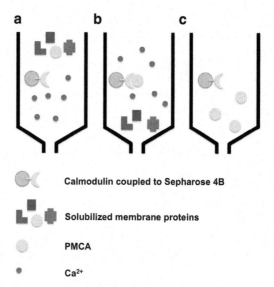

Fig. 1 The principle of calmodulin-affinity chromatography. (**a**) Detergent-solubilized plasma membrane proteins are applied to a column packed with calmodulin coupled to Sepharose 4B, in the presence of 0.1 mM $CaCl_2$. Calmodulin interacts reversibly with the PMCA molecules in the presence of Ca^{2+}. (**b**) Proteins binding calmodulin with high affinity, such as the PMCA, will be retained on the column, as they interact with the immobilized calmodulin. Non-binding proteins are eluted by a washing of the column with a Ca^{2+}-containing buffer. (**c**) Proteins binding specifically and reversibly to calmodulin, such as the PMCA, can now be eluted from the column by exchanging the Ca^{2+}-containing buffer for a Ca^{2+}-free buffer containing EDTA

the presence of Ca^{2+}. Non-bound proteins are removed by washing the column with a Ca^{2+}-containing buffer. Release of the PMCA bound to the immobilized calmodulin in the column is then easily achieved by washing the column with a buffer not containing Ca^{2+}. Erythrocytes are especially suited as starting material for the procedure: pure plasma membranes can be easily obtained from them as they lack organelles and nucleus. Moreover, the PMCA appears to be the main calmodulin-binding transmembrane protein in erythrocyte membranes. Studies of calmodulin binding to calmodulin-depleted erythrocyte membranes indeed indicate the presence of a single class of high-affinity Ca^{2+}-dependent binding sites ($K_d = 5$ nM). Independent estimates of the number of PMCA molecules in erythrocyte membranes correspond to the number of calmodulin-binding sites [3], i.e., the pump represents less than 0.1 % of the total membrane protein [4]. The affinity chromatography procedure using a calmodulin column described here purifies the PMCA pump in quantities sufficient for functional and biochemical studies. The purified erythrocyte pump is enriched at least 500-fold, based on its specific ATPase activity as compared to that of erythrocyte membranes [4]. Care must be taken to keep the protein active during purification, as the detergent-solubilized protein loses its activity. This loss can be (partially) prevented by the presence of phospholipids during the purification procedure or by the addition of 20 % glycerol, as detailed in Subheading 3.

1.2 Reconstitution of the PMCA into Phospholipid Vesicles

Properties of the purified PMCA such as the regulation of its function and the transmembrane transport of Ca^{2+} can only be analyzed in the pump reconstituted into phospholipid bilayers of a composition which mimics the natural environment of the pump. Obviously, transport activities cannot be studied on the solubilized enzyme, and detergents may modify the activity of the pump. Moreover, as already mentioned, the pump is much more stable when reconstituted in phospholiposomes than in the solubilized state. Reconstitution of the pump involves the addition of phospholipids to the detergent-solubilized protein and the removal of the detergent by dialysis. This results in the incorporation of the pump into unilamellar liposomes (Fig. 2), which can be used for a number of functional studies. For example, the impact of transmembrane electric potentials on Ca^{2+} transport can be studied by first generating liposomes with a predetermined K^+ concentration gradient across the bilayer, and by then adding the K^+ ionophore valinomycin to the liposomes to create a potential across the membrane. Niggli et al. [5] have used this protocol, and observed that valinomycin did not stimulate Ca^{2+} transport when added to PMCA-containing liposomes containing 130 mM KCl, suspended in a medium containing 130 mM NaCl, suggesting an electroneutral process. They also observed that the PMCA catalyzed a Ca^{2+}-H^+ exchange: when reconstituted in liposomes containing high con-

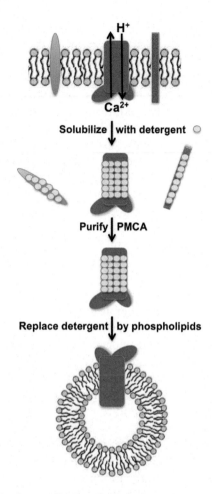

Fig. 2 The principle of reconstitution of membrane proteins (the PMCA pump) into liposomes. In a first step, purified plasma membranes are solubilized by detergent, followed by affinity purification on calmodulin columns (*see* Fig. 1). In a second step, purified phospholipids are added to the solubilized purified protein, resulting in mixed micelles containing detergent, protein, and lipids. This mixture is then placed into a dialysis bag with pores too small for the passage of PMCA and phospholipid micelles, but large enough to allow the passage of detergent. Upon removal of the detergent by dialysis against a large volume of detergent-free medium, the phospholipid micelles will rearrange to form unilamellar liposomes consisting of a phospholipid bilayer with the polar lipid headgroups facing the outside and the lumen of the vesicles and the hydrophobic non-polar fatty acid tails buried in the bilayer. The PMCA is now embedded in this bilayer as in the cell plasma membrane, allowing functional studies

centrations of buffer (200 mM HEPES and 100 mM MES, pH 7.2), the pump ejected H^+ to a medium containing low amounts of buffer (4 mM HEPES). Measurements of pH changes accompanying Ca^{2+} transport show an initial extrusion of H^+ that exceeded that produced directly by ATPhydrolysis, with a ratio of about 2 "extra" H^+ per ATP hydrolyzed. The concept that the

PMCA exchanges protons for Ca^{2+} is now well established, but the Ca^{2+}/H^+ stoichiometry, and thus the matter of electroneutrality or electrogenicity of the operation of the PMCA pump is still debated. Waldeck et al. [6] have also used proteoliposomes obtaining data comparable to those of Niggli et al. [5]. Thomas [7] has shown that in voltage-clamped snail neurons, that do not have other plasma membrane Ca^{2+} ejection systems, the recovery of internal Ca^{2+} and surface pH after a brief depolarization or Ca^{2+} injection is not slowed by hyperpolarization, consistent with a 1 Ca^{2+}:2 H^+ stoichiometry of the PMCA operation. By contrast, Hao et al. [8] using a proteoliposome protocol, have observed that the uptake of 1 Ca^{2+} was accompanied by the ejection of only 1 "extra" H^+ from the lumen of the vesicles, implying that the operation of the PMCA was partially electrogenic. The Ca^{2+}/H^+ stoichiometry has also been measured by others with inconclusive results: the plausible suggestion has been put forward that the slippage of protons could explain the partially electrogenic behavior of the PMCA observed with some protocols in erythrocytes and in proteoliposomes [9]. Finally, the reconstitution of the purified pump in liposomes has permitted to study the role of phospholipids in the regulation of the activity of the PMCA [10].

2 Materials

2.1 Isolation of Human Erythrocyte Plasma Membranes

1. Whole human blood (fresh or recently outdated) from healthy donors obtained from blood transfusion centers, for example 500 mL from which 0.1–0.2 μg purified PMCA can be obtained.
2. KCl-Tris buffer: 130 mM KCl, 10 mM Tris–HCl, pH 7.4.
3. Hemolysis buffer: 1 mM EDTA, 10 mM Tris–HCl, pH 7.4.
4. Hepes buffer: 10 mM Hepes, pH 7.4.
5. Hepes-NaCl-Magnesium-Calcium buffer: 130 mM NaCl, 0.5 mM $MgCl_2$, 0.05 mM $CaCl_2$, 10 mM HEPES, pH 7.4.

2.2 Purification of the PMCA from Erythrocyte Plasma Membranes by Calmodulin-Affinity Chromatography

1. Calmodulin Sepharose 4B (GE Healthcare Life Sciences).
2. Column (e.g., Tricorn 10/50 column, GE Healthcare Life Sciences).
3. Peristaltic pump (e.g., Peristaltic pump P-1, Amersham).
4. Automated fraction collector (e.g., FRAC-920, Amersham).
5. UV detector (e.g., Model 280 UV detector from Spectrum Laboratories Inc.).
6. Phosphatidylcholine and phosphatidylserine: e.g., from bovine spinal cord, grade I (e.g., from Lipid Products, Surrey, U.K.).

7. Phosphatidylcholine and phosphatidylserine stock solutions (20 mg/mL): evaporate the organic solvents under nitrogen, then resuspend the dried lipids in binding buffer by vortexing and sonication in a bath sonicator.

8. Binding buffer: 130 mM NaCl, 10 mM Hepes, pH 7.4, 0.4 % Triton X-100, 0.05 % phospholipids, 0.1 mM $CaCl_2$, 1 mM $MgCl_2$, 2 mM dithiothreitol.

9. Washing buffer: 130 mM NaCl, 10 mM Hepes, pH 7.4, 0.05 % Triton X-100, 0.05 % phospholipids, 0.1 mM $CaCl_2$, 1 mM $MgCl_2$, 2 mM dithiothreitol (sonicate for at least 20 min with a Branson-type tip sonifier in the pulsed mode (50 %) to properly resuspend the phospholipids in the presence of low concentration of detergent).

10. Elution buffer: 130 mM NaCl, 10 mM Hepes, pH 7.4, 0.05 % Triton X-100, 0.05 % phospholipids, 2 mM EDTA, 1 mM $MgCl_2$, 2 mM dithiothreitol (sonicate for at least 20 min with a Branson-type tip sonifier in the pulsed mode (50 %) prior to use).

11. $CaCl_2$ stock solution: 1 M in H_2O.

12. $MgCl_2$ stock solution: 1 M in H_2O.

2.3 Reconstitution of the Purified PMCA into Phospholipid Vesicles

1. Phosphatidylcholine and phosphatidylserine (*see* Subheading 2.2, **item 7**).

2. Asolectin (soybean phospholipids) (Sigma).

3. Cholate buffer: 80 mM potassium cholate, 100 mM KCl, 20 mM Hepes, pH 7.2 at 4 °C, 1 mM $MgCl_2$, 0.05 mM $CaCl_2$.

4. Triton-phospholipid-glycerol buffer: 0.05 % Triton X-100, 0.05 % phospholipid, 130 mM NaCl, 10 mM Hepes, pH 7.4, 3 mM $MgCl_2$, 0.05 mM $CaCl_2$, 2 mM EDTA, 2 mM dithiothreitol, 5 % glycerol.

5. Dialysis buffer: 130 mM NaCl or KCl, 0.05 mM $MgCl_2$, 20 mM Hepes (pH 7.2 at 4 °C), 1 mM dithiothreitol.

6. Dialysis tubing with 10,000 Da cut-off (e.g., G2 slide-A-lyzer dialysis cassette from Thermoscientific 0.25–0 75 mL).

7. Branson type sonifier.

3 Methods

3.1 Isolation of Calmodulin-Deficient Human Erythrocyte Plasma Membranes

All operations described in this section are carried out at 4 °C.

1. Centrifuge human blood from healthy donors, usually supplied in citrate-glucose buffer, at $5800 \times g$ for 10 min.

2. Wash the pelleted erythrocytes twice with 5 volumes of a KCl-Tris buffer and resuspend the washed packed erythrocytes in 10 volumes of hemolysis buffer, which contains EDTA. This

step results in the osmotic lysis of the erythrocytes and in the removal of calmodulin, since the buffer does not contain Ca^{2+}.

3. Centrifuge at $24,000 \times g$ for 35 min.
4. Wash the resulting erythrocyte plasma membranes ("ghosts") once in hemolysis buffer to remove all traces of calmodulin and then wash them 2–3 times with Hepes buffer, and finally 1–2 times with a Hepes-NaCl-Magnesium-Calcium buffer.
5. Resuspend the ghost membranes (ca. 5 mg protein/mL) in the Hepes-NaCl-Magnesium-Calcium buffer. They can be stored at −80 °C for at least 1 month without significant loss of PMCA activity. See **Note 1** for the preparation of plasma membranes from cells different from erythrocytes.

3.2 Purification of the PMCA from Erythrocyte Ghosts by Calmodulin-Affinity Chromatography

All operations described in this section are carried out at 4 °C.

1. Solubilize the purified ghost membranes, resuspended at 5 mg protein/mL in the Hepes-NaCl-Magnesium-Calcium buffer (see Subheading 3.1) by adding the detergent Triton X-100 (1 mg/mg ghost protein), then incubate for 10 min at 4 °C.
2. Centrifuge the solubilized membranes at $100,000 \times g$ for 35 min (4 °C) to remove non-solubilized material.
3. Add 0.5 mg/mL phosphatidylcholine or phosphatidylserine (final concentration) from the 20 mg/mL stock solutions to the supernatant. Concerning the rationale for the addition of lipids, and the choice of lipids, see **Note 2**.
4. Add $CaCl_2$ from a 1 M stock solution to the solubilized membranes to a final concentration of 0.1 mM.
5. Apply approximately 150 mg solubilized membranes to a calmodulin Sepharose 4B column (about 2 mL bed volume) equilibrated in binding buffer. For the preparation of the column and the flow rates, see **Note 3**.
6. Connect the column outlet to a peristaltic pump. The outlet of the pump is connected to a UV detector, the outlet of which is connected to a fraction collector. Collect fractions of ca. 0.5 mL.
7. After application of the ghost membranes, wash the column with 2–3 bed volumes of binding buffer, and then with 2–3 bed volumes of washing buffer which contains less Triton X-100 (0.05 %). The washing of the column with the Ca^{2+}-containing washing buffer should be continued until no further protein (as monitored by absorbance at 280 nm) is eluted.
8. Elute the PMCA pump with elution buffer (same composition as wash buffer except that Ca^{2+} is omitted and EDTA has been added). The chelation of Ca^{2+} disrupts the interaction of the PMCA enzyme with calmodulin in the column. This results in a transient increase of UV absorption of the eluate (see Fig. 1 in [4]).

9. Pool the fractions containing the PMCA (as monitored by absorption at 280 nm).

10. Add 2 mM $MgCl_2$ and 0.05 mM $CaCl_2$ (final concentrations) to the pooled fractions in order to stabilize the PMCA. If not immediately used, the purified enzyme can be stored at −80 °C in the presence of 5 % (v/v) glycerol.

The ATPase activity of the purified PMCA can be determined spectrophotometrically with a coupled enzyme assay as described [11]. The purified erythrocyte PMCA hydrolyzes ATP at a rate of approximately 4 μmol/mg protein/min, as measured in the presence of detergent [4]. The purity of the isolated pump should be evaluated by SDS-PAGE as described [12]. The PMCA protein runs as a single band of about 138 kDa in a 7.5 % PAGE [12], corresponding to the molecular masses of PMCA isoforms 1 and 4, which are expressed in human erythrocytes (138,755 and 137,920 Da respectively, calculated from the amino acid sequence, see [13]).

Following the successful purification from erythrocytes, the PMCA pump has been isolated from plasma membranes from various sources (see **Note 4**). For purity, activity, and yield of the enzyme see **Notes 5–7**.

3.3 Reconstitution of the Purified PMCA into Phospholipid Vesicles

3.3.1 Cholate Dialysis

In the original protocol, reconstitution of the PMCA pump has been performed starting either from Triton-X-100-phospholipid mixtures or cholate/phospholipid mixtures. Triton X-100 was then removed using a Bio-Beads SM-2 column; cholate was removed by slow dialysis. The method involving Triton X-100 is fast, however it generates multilamellar vesicles. Part of the enzyme may thus be occluded within vesicle layers. Moreover, although Ca^{2+} uptake can be measured, the vesicles are somewhat leaky, and thus not suitable for the study of countertransport and possible electrogenicity of the pump. The cholate dialysis method is slower, but yields ion-impermeable monolamellar vesicles. In addition, it allows the manipulation of buffer composition inside and outside the vesicles. Only the cholate dialysis method is therefore described in detail.

1. Place phospholipids (50 mg) dissolved in diethyl ether in a glass tube and dry them under a stream of nitrogen to generate a thin layer. Phospholipids originally dissolved in chloroform/methanol are dried under nitrogen, redissolved in diethyl ether, and dried again. Concerning the choice of phospholipids and the optimal lipid/protein ratio, see **Note 8**.

2. Disperse the dried lipids in 1 mL of cholate buffer by first vortexing in order to remove them from the wall of the tube, and then sonicate them to clarity with a Branson-type sonifier in

the pulsed mode (50 %) under a stream of nitrogen, and cooling them in an ice–water mixture.

3. Combine 0.35 mL of the PMCA purified as described above (Subheading 3.2) and resuspended in Triton-phospholipid-glycerol buffer at the concentration of 0.1–0.2 mg protein/mL, with 0.1 mL of the cholate/lipid mixture, and then mix gently on a vortex.

4. Place the mixture into a slide-A-lyzer dialysis cassette and dialyze against 500 mL (about 1000 volumes) of dialysis buffer at 4 °C for 16 h, with several buffer changes.

5. Store the liposomes at 4 °C (do not freeze them). The liposomes are monolamellar vesicles consisting of a single lipid bilayer enclosing a volume of approximately 0.4 µL/µmol of lipid. One µmol of lipid corresponds to approximately 2×10^{14} liposomes. One liposome contains on the average not more than one molecule of PMCA. The average diameter of the liposomes is in the range of 300–450 Å, less than 10 % of them being larger than 1000 Å [11].

Concerning the stability of the reconstituted enzyme and the sidedness of incorporation of the PMCA protein, *see* **Notes 9** and **10**.

3.3.2 Manipulation of Buffers inside and Outside the Lipid Vesicles

Different buffers may be required inside and outside the lipid vesicles. For example when studying the mechanism of Ca^{2+} transport, the presence of a high concentration of HEPES and of oxalate (e.g., 200 mM HEPES and 40 mM oxalate) inside the vesicles will minimize the increase in pH and free Ca^{2+} within them resulting from the transport activity which inhibits the enzyme. Low concentrations of HEPES (e.g., 4 mM) and omission of oxalate outside of the vesicles are required for Ca^{2+} transport studies [5].

1. To manipulate the inside buffer, dialyze the mixture of purified PMCA, lipid, and cholate (*see* Subheading 3.3.1) against 100 volumes of a buffer containing 18 mM K^+ or Na^+ cholate and the salt and pH buffer composition chosen for the inside of the liposomes for 30–45 min at 4 °C.

2. Change the dialysis buffer to 500 volumes of a buffer containing the same components as the first, but without cholate. The dialysis must be continued overnight.

3. Dialyze the mixture containing the closed bilayer structures for 2–4 h against 500 volumes of a buffer containing the components derived for the medium outside the liposomes. This last step can be accelerated by filtration on a Sephadex G-50 column, where the vesicles elute in the void volume. However, the gel filtration procedure will result in a two- to threefold dilution of the vesicle suspension.

4 Notes

1. The procedure described in Subheading 3.1 has also been used to prepare "ghosts" from erythrocytes obtained from patients with sickle cell anemia [14]. Large amounts of plasma membranes can be obtained from erythrocytes without contamination by membranes of cell organelles, but contamination by other membranes may be a problem when starting from other cell types. More complex procedures must then be used, e.g., involving density gradient centrifugation, as described for instance for the purification of the PMCA from heart sarcolemma [15].

2. The presence of phospholipids during the isolation procedure is important to keep the PMCA in a native conformation. In the presence of the zwitterionic phosphatidylcholine, the enzyme has a lower basal activity, which can be fully stimulated by the addition of calmodulin. At variance with zwitterionic phospholipids, acidic phospholipids activate the PMCA in a manner alternative to calmodulin [10]. Once the sequence of the pump became available, it was shown that they interact with the pump at a basic sequence stretch in the loop between transmembrane domains 2 and 3, and also with the basic calmodulin-binding domain [1]. Calmodulin-affinity chromatography still works with PMCA solubilized in the presence of the acidic phospholipid phosphatidylserine, suggesting that the affinity of calmodulin for the calmodulin-binding domain is greater than that of acidic phospholipids [10]. If required, isolation of the pump can also be performed with a buffer lacking lipids and containing instead 20 % (w/v) glycerol to stabilize the PMCA [16]. In this protocol, Triton X-100 is replaced by 0.5 mg/mL polyoxyethylene 10 laurylether ($C_{12}E_{10}$) and HEPES by 20 mM K-MOPS (3-(N-morpholino)-propanesulfonic acid). The activation of the pump by acid phospholipids is now attracting increasing interest as a native modulation of the pump, possibly as important as calmodulin [17].

3. The preparation of the calmodulin-affinity column starting from CNBr-activated Sepharose 4B (preactivated medium used for coupling proteins to Sepharose without an intermediate spacer arm) and bovine brain calmodulin has been described in detail elsewhere [4, 11]. Calmodulin Sepharose 4B is now commercially available from GE Healthcare Life Sciences. For the preparation of the column and for obtaining optimal flow rates (which depend on the geometry of the column), the instructions of the manufacturer must be followed. For a column with a diameter of 1 cm, the flow rate should be approximately 40 mL/h.

4. Application of the calmodulin-affinity procedure to the isolation of the PMCA from different cell sources: After the purification of the pump from erythrocytes, the calmodulin-affinity chroma-

tography protocol has been used to purify the PMCA pump from calf heart sarcolemma and from microsomal fractions obtained from plant cells, such as radish seedlings and *Arabidopsis* [2, 15, 18]. Interestingly, at variance with all animal PMCAs in which the domain is located canonically in the C-terminal tail, the PMCA from *Arabidopsis* (and from other plant sources) contains the calmodulin-binding domain at the N-terminus [2]. The calmodulin-affinity chromatography procedure has also been applied to purify the pump ectopically expressed in a number of cells, including yeast [19, 20]. The case of liver is particular: a plasma membrane Ca^{2+} pump was originally partially purified using concanavalin A-ultrogel chromatography, followed by DE-52 chromatography [21]. The pump was claimed to have properties different from those of the erythrocyte enzyme, including the sensitivity to some hormones.

5. Concerning the yield, 150 µg of enzyme could be purified from 156 mg of human erythrocyte membranes [4]; 300 µg of enzyme from 100 mg calf heart sarcolemma [15], 20 µg from 5.7 mg radish seedling microsomes [18], and 50 µg from 4 mg *Arabidopsis* microsomes [2]. For the isolation of larger quantities of PMCA (in the range of 2–4 mg), see the modifications described by Penniston et al. [22].

6. Concerning the purity of the enzyme: SDS-PAGE analysis of the purified erythrocyte pump shows a major band at around 138 kDa, with varying amounts of minor impurities such as bands of 90 kDa (the latter may correspond to band 3 or to glycophorin). These impurities can be removed, if required, by prolonging the washing time of the column to 18 h with a buffer containing 0.05 % Triton X-100, 300 mM KCl, 10 mM Tes-TEA, pH 7.4, 1–50 µM $CaCl_2$, 2 mM dithiothreitol, and 0.1 % phosphatidylcholine [12]. A band of mass higher than 200 kDa is very frequently also visible: it has been shown to be a dimer of the enzyme not dissociated by SDS. In non-erythrocyte plasma membranes, calmodulin-binding membrane proteins different from the PMCA may also contaminate the gels. For instance, calmodulin is a Ca^{2+}-sensing subunit of a variety of ion channels in a wide variety of species and tissues [23]. Calmodulin-affinity chromatography can for example be used for the partial purification of an epithelial K^+ channel [24]. Impurities may also correspond to proteolytic fragments of the PMCA; if required, protease inhibitors can be added to the isolation buffers [22].

7. The purified enzyme is rather unstable in detergents, such as Triton X-100. Up to two-thirds of the total activity applied to the calmodulin-affinity column is lost during purification of Triton-solubilized plasma membranes, whereas up to 98 % of the total PMCA protein is recovered. As mentioned, the enzyme becomes much more stable after reconstitution into

liposomes (*see* **Note 9**). Upon storage, the purified PMCA has a tendency to form aggregates, which persist in SDS, but aggregation does not result in loss of activity [11]. Actually, the dimerization of the enzyme, which occurs through its calmodulin-binding domain [25], even increases its activity [26].

8. When choosing lipids for the reconstitution of the PMCA, a mixture composed of 90 % phosphatidylcholine and 10 % phosphatidylserine results in optimal calmodulin stimulation of the reconstituted enzyme (five- to sixfold). However, these proteoliposomes are more permeable than those prepared with asolectin. Asolectin consists of about equal proportions of phosphatidylcholine, phosphatidylinositol and cephalin, with about 24 % saturated fatty acids, 14 % monounsaturated fatty acids, and 62 % poly-unsaturated fatty acids. Asolectin vesicles are impermeable to metal ions, protons, glucose, ATP, etc. The use of asolectin is thus required for transport studies which demand tightly coupled vesicles. However, the ATPase reconstituted in asolectin liposomes is already stimulated to maximal activity by the acidic phospholipids and unsaturated fatty acids present in asolectin, and thus does not respond to calmodulin. The lipid-to-protein ratio in the procedure described here corresponds to approximately 7000 molecules of phospholipid per molecule of PMCA. Assuming that a monolamellar liposome contains about 3500 phospholipid molecules, statistically one liposome would thus contain one molecule of ATPase or less, which would result in minimal disturbance of the permeability of the lipid bilayer.

9. ATPase activity of the reconstituted PMCA: The protein-containing lipid vesicles can be recovered without loss from the dialysis tubing, and the reconstituted enzyme is rather stable at 4 °C, losing not more than 10 % of its initial activity after 4 days. The enzyme should thus be stored in reconstituted form at 4 °C (the liposomes should not be frozen as this alters their structure). The reconstituted PMCA pump will accumulate Ca^{2+} into the lumen of the liposomes in exchange for protons against a concentration gradient in the presence and hydrolysis of ATP (*see* above). In the absence of Ca^{2+} ionophores, the Ca^{2+}-dependent ATPase activity will decrease with time during the measurement since the increase in Ca^{2+} in the lumen of the liposomes increases the Ca^{2+}gradient across the bilayer which the enzyme must overcome. Dissipation of this Ca^{2+} gradient with the Ca^{2+} ionophore A23187 results in the stimulation of the ATPase activity [11]. The increase indicates that the liposomes are impermeable to Ca^{2+} and that the ATPase activity is tightly coupled to Ca^{2+} transport.

10. Sidedness of the reconstituted enzyme: when adding Ca^{2+} and ATP to impermeable proteoliposomes, only the PMCA molecules that have the ATPase domain facing the outside of the ves-

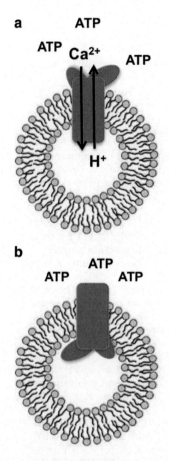

Fig. 3 The use of the reconstituted PMCA for functional studies. The lipid–protein ratio used in this Methods article for reconstitution results in the presence of maximally one PMCA molecule per liposome. This ensures maximal tightness of the liposome, when using asolectin as the lipid mixture. When using other lipids such as phosphatidylcholine and phosphatidylserine, the vesicles are more permeable. Concerning the sidedness of incorporation of the PMCA, either the ATPase domain will be exposed on the outside (**a**) or be buried in the inner space (**b**) of the liposome. Only the PMCA molecules with the ATPase domain on the outside (**a**) will be accessible to ATP added to the liposomes, and therefore only these molecules will show active Ca^{2+} transport. This requires impermeable vesicles that do not allow diffusion of ATP to their lumen. According to the measurement of ATPase activity in the absence or presence of detergent, about 80 % of the ATPase domain of the reconstituted PMCA faces the outside as depicted in (**a**)

icles will be active (Fig. 3). In order to assess the fraction of PMCA molecules with the ATPase domain buried in the interior of the liposomes, the effect of detergents on ATPase activity can be tested. The activity increases by about 20 % by the treatment of PMCA reconstituted by cholate dialysis with detergents. This suggests that about 80 % of the total PMCA pump has the domain that contains the active site (which in cells protrudes into the cytoplasm) exposed on the outside of the liposomes [11].

References

1. Brini M, Cali T, Ottolini D, Carafoli E (2013) The plasma membrane Ca^{2+} pump in health and disease. FEBS J 280:5385–5397
2. Bonza MC, Morandini P, Luoni L (2000) At-ACA8 encodes a plasma membrane-localized Ca^{2+}-ATPase of Arabidopsis with a calmodulin-binding domain at the N Terminus. Plant Physiol 123:1495–1505
3. Niggli V, Ronner P, Carafoli C, Penniston JT (1979) Effects of calmodulin on the (Ca^{2+}-Mg^{2+})-ATPase partially purified from erythrocyte membranes. Arch Biochem Biophys 198:124–130
4. Niggli V, Penniston JT, Carafoli E (1979) Purification of the (Ca^{2+}-Mg^{2+})-ATPase from human erythrocyte membranes using a calmodulin affinity column. J Biol Chem 254:9955–9958
5. Niggli V, Sigel E, Carafoli E (1982) The purified Ca^{2+} pump of human erythrocyte membranes catalyzes an electroneutral Ca^{2+}-H^+ exchange in reconstituted liposomal systems. J Biol Chem 257:2350–2356
6. Waldeck AR, Xu AS, Roufogalis BD, Kuchel PV (1998) Measurements of Ca^{2+} and H^+ transport mediated by A23187 and reconstituted plasma membrane Ca^{2+}-ATPase. Eur Biophys J 27:255–262
7. Thomas RC (2009) The plasma membrane calcium ATPase (PMCA) of neurons is electroneutral and exchanges 2 H^+ for each Ca^{2+} or Ba^{2+} ion extruded. J Physiol 587:315–327
8. Hao L, Rigaud JL, Inesi G (1994) Ca^{2+}/H^+ countertransport and electrogenicity in proteoliposomes containing erythrocyte plasma membrane Ca-ATPase and exogenous lipids. J Biol Chem 269:14265–14275
9. Niggli V, Sigel E (2008) Anticipating antiport in P-type ATPases. Trends Biochem Sci 33:156–160
10. Niggli V, Adunyah ES, Penniston JT, Carafoli E (1981) Purified (Ca^{2+}-Mg^{2+})-ATPase of the erythrocyte membrane: reconstitution and effect of calmodulin and phospholipids. J Biol Chem 256:395–401
11. Niggli V, Zurini M, Carafoli E (1987) Purification, reconstitution and molecular characterization of the Ca^{2+} pump of plasma membranes. Methods Enzymol 139:791–808
12. Graf E, Verma AK, Gorski JP, Lopaschuk G, Niggli V, Zurini M et al (1982) Molecular properties of Ca^{2+} pumping ATPase from human erythrocytes. Biochemistry 21:4511–4516
13. Stauffer TP, Guerini D, Carafoli E (1995) Tissue distribution of the four gene products of the plasma membrane Ca^{2+} pump. J Biol Chem 270:12184–12190
14. Niggli V, Adunyah ES, Cameron BF, Bababunmi EA, Carafoli E (1982) The Ca^{2+}-pump of sickle cell plasma membranes. Purification and reconstitution of the ATPase enzyme. Cell Calcium 3:131–151
15. Caroni P, Zurini M, Clark A, Carafoli E (1983) Further characterization and reconstitution of the purified Ca^{2+}-pumping ATPase of heart sarcolemma. J Biol Chem 258:7305–7310
16. Filomatori CV, Rega AF (2003) On the mechanism of activation of the plasma membrane Ca^{2+}-ATPase by ATP and acidic phospholipids. J Biol Chem 278:22265–22271
17. Lopreiato R, Giacomello M, Carafoli E (2014) The plasma membrane calcium pump: new ways to look at an old enzyme. J Biol Chem 289:10261–10268
18. Bonza C, Carnelli A, De Michelis MI, Rasi-Caldogno F (1998) Purification of the plasma membrane Ca^{2+}-ATPase from radish seedlings by calmodulin-agarose affinity chromatography. Plant Physiol 116:845–851
19. Heim R, Iwata T, Zvaritch E, Adamo HP, Rutishauser B, Strehler EE, Guerini D, Carafoli E (1992) Expression, purification and properties of the plasma membrane Ca^{2+} pump and of its N-terminally truncated 105-kDa fragment. J Biol Chem 267:24476–24484
20. Cura CI, Corradi GR, Rinaldi DE, Adamo HP (2008) High sensibility to reactivation by acidic lipids of the recombinant human plasma membrane Ca^{2+}-ATPase isoform 4xb purified from *Saccharomyces cerevisiae*. Biochim Biophys Acta 1778:2757–2764
21. Lotersztain S, Hanoune J, Pecker F (1981) A high affinity calcium-stimulated magnesium-dependent ATPase in rat liver plasma membranes. J Biol Chem 256:11209–11215
22. Penniston JT, Filoteo AG, McDonough CS, Carafoli E (1988) Purification, reconstitution and regulation of plasma membrane Ca^{2+}-pumps. Methods Enzymol 157:340–351
23. Saimi S, Kung C (2002) Calmodulin as an ion channel subunit. Annu Rev Physiol 64:289–311
24. Klaerke DA (1995) Purification and characterization of epithelial Ca^{2+}-activated K^+ channel proteins by calmodulin. Kidney Int 48:1047–1056
25. Vorherr T, Kessler T, Hofmann F, Carafoli E (1991) The calmodulin-binding domain mediates the self-association of the plasma membrane Ca^{2+} pump. J Biol Chem 266:22–27
26. Kosk-Kosicka D, Bzdega T (1988) Activation of the erythrocyte Ca^{2+} ATPase by either self-association or interaction with calmodulin. J Biol Chem 263:18184–18189

Chapter 8

Expression of Na,K-ATPase and H,K-ATPase Isoforms with the Baculovirus Expression System

Jan B. Koenderink and Herman G.P. Swarts

Abstract

P-type ATPases can be expressed in several cell systems. The baculovirus expressions system uses an insect virus to enter and express proteins in Sf9 insect cells. This expression system is a lytic system in which the cells will die a few days after viral infection. Subsequently, the expressed proteins can be isolated. Insect cells are a perfect system to study P-type ATPases as they have little or no endogenous Na,K-ATPase activity and other ATPase activities can be inhibited easily. Here we describe in detail the expression and isolation of Na,K-ATPase and H,K-ATPase isoforms with the baculovirus expression system.

Key words Na,K-ATPase, H,K-ATPase, Isoforms, Baculovirus, Insect cells, Sf9 cells, Membrane isolation

1 Introduction

The recombinant baculovirus expression system can be used to express Na,K-ATPase, H,K-ATPase, and other P-type ATPpases in insect cells. Basis for this system is the *Autographa californica* multiple nuclear polyhedrosis virus, which infects insect larvae of the fall armyworm *Spodoptera frugiperda* (Sf9 cells). The genes (α and β subunit) that will be expressed are first cloned into a donor plasmid downstream of the baculovirus promoters. This donor plasmid is then introduced into *E. coli* cells harboring the baculovirus genome as a shuttle vector (bacmid) and a transposition helper vector. Upon site-specific transposition between the donor vector and the bacmid, recombinant bacmids are selected and isolated. Subsequently, insect cells are transfected with these bacmids and the recombinant baculoviruses are harvested [1]. These recombinant viruses can be used for production of recombinant proteins. The membrane fractions of insect cells expressing the recombinant proteins can be isolated and Western blot analysis will reveal the expression patterns.

All Na,K-ATPase ($\alpha 1$, $\alpha 2$, $\alpha 3$, and $\alpha 4$) and the gastric and nongastric H,K-ATPase α-subunits have an apparent molecular mass of about 100 kDa. The β-subunits possess a carbohydrate-free and a core-glycosylated form. Recombinant P-type ATPases can be expressed easily in large quantities with low background ATPase activity in this system [2]. The isolated P-type ATPase can be studied with biochemical methods like Western blotting, ATPase activity, phosphorylation, and ligand binding.

2 Materials

1. The H,K-ATPase and Na,K-ATPase β subunits are placed downstream of the p10 promoter and the H,K-ATPase and Na,K-ATPase α subunits downstream of the polyhedrin promoter of the pFastbacdual vector (Fig. 1) (*see* **Note 1**) (Life Technologies, Breda, The Netherlands) [1].
2. As mock, a baculovirus expressing only β subunit or a non-ATPase protein.
3. Enzyme buffer: 0.25 M sucrose, 2 mM EDTA, and 50 mM Tris-acetate pH 7.0.
4. DH10Bac competent cells (Life Technologies, Breda, The Netherlands).

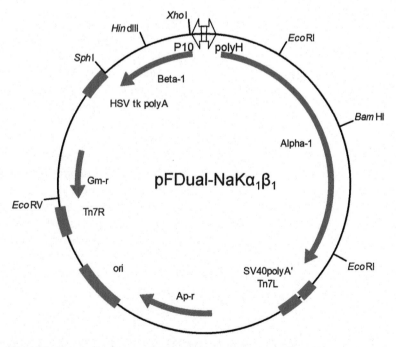

Fig. 1 Pfastbacdual vector with Na,K-ATPase α and β subunits

5. LB medium (Luria Broth).
6. LB medium with kanamycin 50 μg/mL kanamycin, 7 μg/mL gentamicin, 10 μg/mL tetracycline.
7. LB agar plates with 50 μg/mL kanamycin (Kan), 7 μg/mL gentamicin (Gen), 10 μg/mL tetracycline (Tet), 100 μg/mL X-Gal, and 40 μg/mL isopropyl β-D-1-thiogalactopyranoside (IPTG).
8. Resuspension buffer: 300 μL 0.1 μg/μL RNase solution (Life Technologies) in TE pH 8.0 (15 mM Tris–HCl and 10 mM EDTA).
9. Lysis buffer: 1 % SDS, 0.2 M NaOH.
10. KAc solution: 3 M potassium acetate, pH 5.5
11. Isopropanol. *p.a.*
12. 70 % ethanol.
13. TE buffer: 15 mM Tris–HCl and 10 mM EDTA, pH 8.0.
14. Cellfectin reagent (Life Technologies).
15. SF900 II insect cells medium (Life Technologies).
16. Shaker flasks (Corning) with vented cap for culturing insect cells.
17. Xpress medium (Lonza).
18. Xpress-Plus medium: Xpress medium supplemented with 10 % FBS (Greiner, heat inactivated).

3 Methods

Carry out all procedures at room temperature unless otherwise specified.

3.1 Transformation to DH10Bac-Competent Cells

Competent DH10bac *E. coli* cells harboring the baculovirus genome (bacmid) and a transposition helper vector, are transformed with the pFastbacdual transfer vector containing different cDNAs encoding P-type ATPase subunits.

1. DH10Bac-competent cells (per transformation one tube with 100 μL) are thawed on ice.
2. Add 5 μL (±1 μg) recombinant pFastbacdual vector and leave on ice for 30 min.
3. Heat shock the mixture in 42 °C water bath for 45 s and place immediately on ice for 2 min.
4. Add 900 μL LB medium and incubate for 4 h at 37 °C in shaker (~150 rpm).

5. Plate 5 µL and 40 µL on LB agar plates (Kan, Gen, Tet, Xgal, IPTG).
6. The rest of the transformed DH10Bac cells are centrifugated for 1 min at $15,000 \times g$, medium is reduced to 50 µL, and resuspended cells are plated on LB-agar plates (Kan, Gen, Tet, X-gal, IPTG).
7. The plates are incubated for 48–72 h at 37 °C.

3.2 Selection of the Transformed DH10Bac Cells

1. The colonies are screened for white and blue staining: the white colonies have the plasmid transposed into the bacmid DNA.
2. White colonies (big, round, and white) are streaked on LB-agar plates (Kan, Gen, Tet, X-gal, and IPTG) to ensure that they are truly white.
3. After 48–72 h at 37 °C a clear white colony is picked and cultured in 4 mL LB medium (Kan, Gen, Tet) and incubated overnight at 37 °C.

3.3 Isolation of Recombinant Bacmid DNA (>100 kb)

1. 1.5 mL DH10Bac culture is centrifuged for 1 min at $15,000 \times g$.
2. The supernatant is decanted and the cells are resuspended in resuspension buffer.
3. Next, 300 µL lysis buffer is added slowly while mixing carefully.
4. After incubation for 5 min at RT the solution becomes transparent.
5. 300 µL KAc is added slowly while mixing carefully.
6. After incubation for 5–10 min on ice the reaction tube is centrifugated, at 4 °C at $15,000 \times g$ for 10 min.
7. The supernatant is transferred to reaction tubes containing 800 µL isopropanol (2-propanol).
8. The solutions are mixed and incubated for 5–10 min on ice.
9. Next, the sample is centrifugated at $15,000 \times g$, RT for 15 min.
10. The supernatant is decanted and 500 µL 70 % ethanol is added.
11. The sample is centrifugated again at $15,000 \times g$, RT for 5 min.
12. The supernatant is removed and the pellet is air-dried for 5–10 min.
13. Finally, the DNA (pellet) is dissolved in 40 µL TE buffer and stored at −20 °C.

3.4 Transfection of Sf9 Cells

Insect Sf9 cells cultured in Xpress-Plus medium are transfected with recombinant bacmids using Cellfectin reagent [3].

1. 12 µL Cellfectin reagent is added to 200 µL Sf-900 II.
2. 10 µL bacmid DNA is added to 200 µL Sf-900 II.

3. These mixtures (1 and 2) are combined (mixed again) and incubated at RT for 30–60 min.

4. 2 mL Sf9 cells in Sf-900 II SFM (1.0×10^6) cells/mL are transferred to a 25 cm² cell culture flask.

5. The transfection mixture is added to the cells and the cells are incubated at 27 °C for 5 h.

6. The medium containing the transfection mix is removed and 4 mL Xpress-Plus culture medium is added.

7. The cells are cultured at 27 °C for 48–72 h.

8. The flask is knocked to detach the cell.

9. The cells and medium are collected and centrifugated at $4000 \times g$, RT, for 5 min.

10. The supernatant (medium containing the recombinant baculoviruses; P1) is collected and stored at 4 °C.

3.5 Production of Recombinant Viruses

1. 0.05 mL of the harvested recombinant baculoviruses (P1) is used to infect a new batch of Sf9 cells (multiplicity of infection ~ 0.1; see **Note 2**). Five days after infection, the amplified viruses are collected (P2).

2. The virus stocks are stored at 4 °C (see **Note 3**).

3.6 Expression of Recombinant Proteins in Sf9 Insect Cells

1. Adapt Sf9 insect cells from the Xpress-Plus culture (T175) to Xpress-medium (protein free medium) for 4–6 days at 27 °C in shaker flasks (speed 100 rpm) (see **Note 4**).

2. Keep the cell density between around 0.8–1.5×10^6 cells/mL. Avoid low cell densities and subculture the cells daily before the start of an expression experiment.

3. Add to each 500 mL shaker flask containing 100 mL Sf9 cells, 5 mL virus suspension from the P2 stock (see **Note 5**), to increase expression 1 % ethanol should be added [4].

4. Incubate the cells on a shaker for 3 days at 27 °C (speed 100 rpm).

3.7 Membrane Isolation

1. Harvest the Sf9 cells by centrifugation in 50 mL tubes at $2000 \times g$ for 5 min at RT.

2. Discard the supernatant (remove as much as possible) and keep the pellet on ice. At this stage the pellet can be stored at −20 °C.

3. Resuspend the pellet in 15 mL ice-cold enzyme buffer.

4. Sonicate the cells twice for 30 s at 65–70 W on ice (use a 5 mm Ø tip).

5. Centrifuge the disrupted cells at $10,000 \times g$ for 30 min at 4 °C.

6. Centrifuge the supernatant at $100,000 \times g$ for 60 min at 4 °C.

7. Resuspend the pelleted membranes in 2 mL of the enzyme buffer.

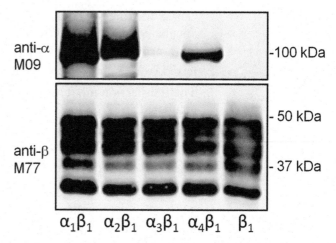

Fig. 2 Expression of recombinant Na,K-ATPase isoforms (*see* **Note 6**). Na,K-ATPase α1, α2, α3, or α4 in combination with the Na,K-ATPase β1 subunit are produced with the baculovirus expression system. The membrane fractions of the Sf9 cells are isolated and Western blot analysis reveals that in all samples the Na,K-ATPase β1-subunits is detected with the antibodies M77 [2], also in the mock-infected preparation (β1). This antibody recognizes both a carbohydrate-free (±30 kDa) and different core-glycosylated forms of the β-subunit, depending on the degree of glycosylation. The polyclonal antibody (M09) [5] raised against the Na,K-ATPase α1-subunit also recognizes the α2 subunit and to a lesser extent the α4 subunit. The Na,K-ATPase α3 is hardly recognized

8. Pass the suspension 20 times through a Potter-Elvehjem homogenizer on ice.
9. Store the final membrane fraction at 4–12 mg protein/mL at −20 °C.
10. Determine the protein concentration.

3.8 Analysis of Expression

1. The samples can be analyzed by Western blotting (Fig. 2).
2. P-type ATPase activities can be analyzed in ATPase activity assays [6] (*see* chapters in **Part II**). In addition, the enzyme expression and function may be studied by binding of the specific Na,K-ATPase inhibitor ouabain (*see* Ref. 6) (Fig. 3).

4 Notes

1. It is important to realize that the different subunits (α and β) are produced by one virus. The use of a Gateway (Life Technologies) destination vector with constant β-subunit gives the possibility to combine it with different catalytic α-subunits. By performing mutagenesis on the pEntry-α-subunit vector the mutated α-subunit can easily be cloned into the destination vector.

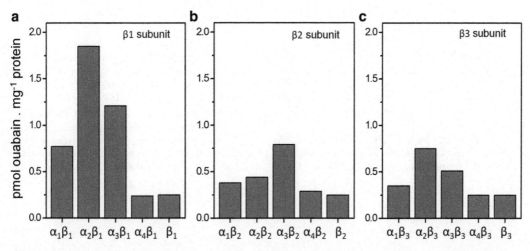

Fig. 3 Ouabain binding of recombinant Na,K-ATPase isoforms. Na,K-ATPase α1, α2, α3, or α4 in combination with the Na,K-ATPase β1, β2, and β3 subunits are produced with the baculovirus expression system. The membrane fractions of the Sf9 cells are isolated and ouabain binding capacity of the membranes fraction is shown. Figures 2 and 3A can be compared, showing expression analyses with different methods

2. We do not determine virus titers (generally 10^8 pfu/mL) as this is quite laborious and experience from the past shows that directly using the volumes in this protocol gives good results.

3. Viral stocks can be stored at 4 °C for many years. In our experience the titer of these stocks slowly reduces and we generally do not use stocks that are older than 2 years for protein production.

4. Sf9 insect cells are cultured without CO_2.

5. For production of the ATPase subunits $1.0-1.5 \times 10^6$ cells/mL are infected at a multiplicity of infection of 1–3 in Xpress medium.

6. Many different recombinant baculoviruses expressing Na,K-ATPase [1], gastric [7] or non-gastric H^+,K^+-ATPase [2], or Na,K-ATPase of *Drosophila melanogaster* [8] have been used in the past.

References

1. Koenderink JB, Hermsen HP, Swarts HG, Willems PH, De Pont JJ (2000) High-affinity ouabain binding by a chimeric gastric H, K-ATPase containing transmembrane hairpins M3–M4 and M5–M6 of the alpha 1-subunit of rat Na, K-ATPase. Proc Natl Acad Sci U S A 97(21):11209–11214

2. Swarts HG, Koenderink JB, Willems PH, De Pont JJ (2005) The non-gastric H, K-ATPase is oligomycin-sensitive and can function as an H, NH4-ATPase. J Biol Chem 280(39): 33115–33122

3. Luckow VA, Lee SC, Barry GF, Olins PO (1993) Efficient generation of infectious recombinant baculoviruses by site-specific transposon-mediated insertion of foreign genes into a baculovirus genome propagated in Escherichia coli. J Virol 67(8):4566–4579

4. Klaassen CH, Swarts HG, De Pont JJ (1995) Ethanol stimulates expression of functional H,K-ATPase in SF9 cells. Biochem Biophys Res Commun 210(3):907–913
5. Koenderink JB, Geibel S, Grabsch E, De Pont JJ, Bamberg E, Friedrich T (2003) Electrophysiological analysis of the mutated Na, K-ATPase cation binding pocket. J Biol Chem 278(51):51213–51222
6. De Pont JJ, Swarts HG, Karawajczyk A, Schaftenaar G, Willems PH, Koenderink JB (2009) The non-gastric H,K-ATPase as a tool to study the ouabain-binding site in Na,K-ATPase. Pflugers Arch 457(3):623–634
7. Swarts HG, Klaassen CH, de Boer M, Fransen JA, De Pont JJ (1996) Role of negatively charged residues in the fifth and sixth transmembrane domains of the catalytic subunit of gastric H,K-ATPase. J Biol Chem 271(47):29764–29772
8. Dalla S, Swarts HG, Koenderink JB, Dobler S (2013) Amino acid substitutions of Na,K-ATPase conferring decreased sensitivity to cardenolides in insects compared to mammals. Insect Biochem Mol Biol 43(12):1109–1115

# Chapter 9

Time-Dependent Protein Thermostability Assay

Ilse Vandecaetsbeek and Peter Vangheluwe

Abstract

Membrane protein purification often yields rather unstable proteins impeding functional and structural protein characterization. Low protein stability also leads to low purification yields as a result of protein degradation, aggregation, precipitation, and folding instability. It is often required to optimize buffer conditions through numerous iterations of trial and error to improve the homogeneity, stability, and solubility of the protein sample demanding high amounts of purified protein. Therefore we have set up a fast, simple, and high-throughput time-dependent thermostability-based assay at low protein cost to identify protein stabilizing factors to facilitate the handling and characterization of membrane proteins by subsequent structural and functional studies.

Key words Membrane protein, Protein stability, Thermostability assay, Melting temperature, Fluorescent dye

1 Introduction

With the exception of the Ca^{2+} pumps of the fast twitch skeletal muscle (SERCA1a [1]) and the heart muscle (SERCA2a [2]) and the Na^+/K^+ ATPase from kidney outer medulla tissue [3], most other P-type ATPases display a low native tissue expression. Detailed functional and structural studies of these pumps therefore require purification from a recombinant expression system. However, many membrane proteins, once purified from an environment devoid of their potential natural stabilizing factors, show loss of function and reduced stability in standard sample buffer conditions [4]. Identifying components essential for recovering protein activity and integrity is therefore a crucial, but time consuming step for subsequent functional characterization and structure determination. Structural studies are especially demanding, because the protein sample has to remain stable under non-native experimental conditions.

Several sophisticated methods can be employed for analyzing protein stability behavior, such as size exclusion chromatography or light scattering experiments. In general, these methods demand large quantities of purified protein and are not adapted for high-throughput analysis. Therefore, fluorescence-based protein thermostability screens were developed that allow high-throughput screening of many conditions in a 96-well format [5–8]. These assays use fluorescent dyes to probe changes of the protein's folding state. These dyes exhibit different fluorescence properties as a function of their local environment. The most commonly used dye is Sypro Orange which has the ability to bind to exposed hydrophobic regions of a protein, which allows quantifying protein denaturation/unfolding. However, Sypro Orange is less compatible with membrane proteins because the detergents used to solubilize membrane proteins interfere with the analysis, creating a hydrophobic environment due to micelle formation. This dye-detergent interaction does not allow correct measurement of the unfolding temperature of the sample, limiting the analysis to water-soluble proteins [9]. On the other hand, the thiol-specific fluorophore N-[4-(7-diethylamino-4-methyl-3-coumarinyl)phenyl]maleimide (CPM) has been proven to be a better fluorescent dye for stability profiling of membrane proteins [6, 8, 10]. CPM is almost non-fluorescent in its unbound state, but becomes fluorescent upon binding to free sulfhydryl groups [11]. As transmembrane segments often contain cysteines, cysteine accessibility is a good measure of protein unfolding of membrane proteins [8]. CPM is therefore an ideal sensor for the overall integrity of the folded state of membrane proteins [6, 8, 10]. An alternative dye to investigate membrane protein stability, Proteostat©, was developed by Enzo Life Sciences. As compared to Sypro Orange, Proteostat© is compatible with detergents and monitors protein aggregation rather than protein unfolding [4].

A fluorescence-based protein stability screening can basically be performed in two ways: temperature or time dependent. In the temperature-dependent method, the protein sample is gradually heated (i.e., stepwise by increasing temperature with 1 °C). This method allows determining the "melting" temperature (T_m), which is the temperature at which a half maximal fluorescence is observed (maximal fluorescence at the highest temperature). The higher the melting temperature, the more stable is a protein [5, 7]. In the time-dependent method, the protein sample is incubated at a fixed elevated temperature (e.g., 40 °C) and the fluorescence is measured over a period of time. The "half-life" time ($t_{1/2}$) is then determined, which corresponds to the time to reach 50 % of the maximum fluorescence. A stabilized protein results in a longer $t_{1/2}$ value [7, 8]. By varying the buffer components one can screen for favorable conditions, i.e., stabilizing factors, such as ligands, lipids, additives, or detergents. Here, we present a time-dependent thermostability-based approach to screen in a high-throughput assay for potential stabilizing factors such as changes in pH, buffer type, ionic strength, ions, detergents, lipids, ligands, and additives.

2 Materials

All buffers and solutions must be prepared with ultrapure water and stored at 4 °C.

1. 96-Well black plate, clear bottom (*see* **Note 1**) (96W Microplate, Black/µClear, or equivalent).
2. 96-Well normal transparent plate (96W Microplate, µClear, or equivalent).
3. Aluminum foil.
4. 50 mL reagent reservoirs (Corning Costar™ or equivalent).
5. 96-Well plate fluorescence reader with temperature control and bottom reading (Flexstation, Molecular Devices™ or equivalent).
6. Multichannel pipets.
7. Transparent cover sticker for 96-well plates.
8. N-[4-(7-dimethylamino-4-methyl-3-coumarinyl)phenyl] maleimide (CPM) (Molecular Probes™, Life Technologies™): Stock solution of 4 mg/mL in dimethylsulfoxide (DMSO) (stored at −20 °C or −80 °C) (*see* **Note 2**).
9. CPM buffer: 100 mM MOPS pH 7, 100 mM NaCl, 0.03 % DDM (*see* **Note 3**).

3 Methods

3.1 Determination of the CPM:Protein Ratio

Determine the optimal concentrations of CPM and protein to measure fluorescence of your protein of interest within the detectable fluorescence range.

1. Prepare a desired CPM dilution series (e.g., six different concentrations) in a transparent 96-well plate (plate 1): example: 100, 250, 500, 750, 1000, 1500 ng/well final concentration. Prepare for each dilution a volume sufficient for ten reactions at 10 µL (for eight different protein dilutions + 2, *see* **step 3**). Thus, transfer a total volume of 10 × 10 µL CPM dilution/well (i.e., 100 µL in each well of row A) (*see* **Note 4**).
2. Cover plate 1 with aluminum foil.
3. Prepare a desired protein dilution series (e.g., eight different concentrations) in plate 2 (example: 0, 0.5, 1, 2.5, 5, 10, 15, 20 µg/well). Dilute the protein in CPM buffer in a total volume corresponding to eight reactions, (i.e., the number of CPM dilutions to test + 2, *see* **step 1**) of 10 µL protein/well (i.e., 80 µL in each well of column 1).

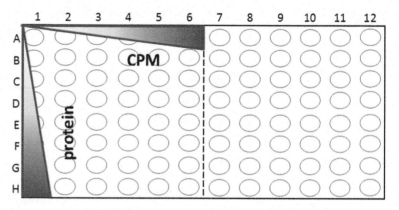

Fig. 1 Schematic representation of final dilutions of CPM and protein in plate 3

4. Keep plate 2 on ice.
5. Add 130 μL CPM buffer to the wells of plate 3, which should be a black 96-well plate.
6. Transfer 10 μL from column 1 of plate 2 to the first six columns of plate 3 with a multichannel pipet (Fig. 1).
7. Start the reaction by transferring 10 μL of the CPM dilutions of row A in plate 1 to the eight rows of plate 3 (Fig. 1) with a multichannel pipet (protect the plates from the light as much as possible).
8. Keep plate 3 on ice, covered by aluminum foil.
9. Seal plate 3 with a transparent cover sticker to avoid evaporation.
10. Put plate 3 in the fluorescence plate reader at a constant temperature of 42 °C (or 37 °C in case of very unstable proteins).
11. Take recordings every minute with a 5 s shaking interval. Detect fluorescence at 463 nm (excitation wave length is 387 nm) until a stable, maximal fluorescence signal is obtained (can range from several minutes to a few hours).
12. The fraction of non-denatured protein at each time point is calculated as follows. F_{max} is the maximal fluorescence measured in 1 well of the entire plate [8] and $Ft = x$ is the fluorescence at a given time point x:

$$\% \text{ of folded protein} = 100 - \left(\frac{(Ft = x)}{(F_{max})} \times 100 \right)$$

13. The relative unfolding rate ($t_{1/2}$) based on cysteine accessibility over time is then calculated by a single exponential decay curve and plotted. An example of recombinant protein of the Golgi Ca^{2+} pump hSPCA1a is depicted in Fig. 2.

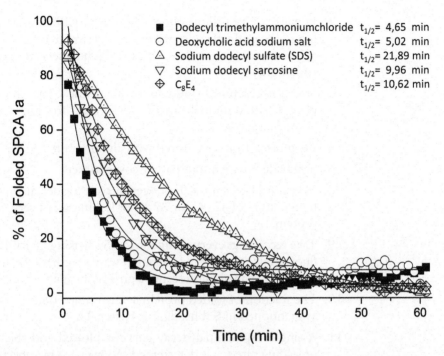

Fig. 2 Example of the time-dependent denaturation in different detergents for the purified SPCA1a, plotted by a single exponential decay curve

14. The higher the $t_{1/2}$ value of a certain condition is, the more this condition stabilizes your protein. Keep in mind that false positive results can occur, e.g., due to interference of the tested compound with the CPM dye. Positive results can always be verified by the fluorescence-detection size-exclusion chromatography based thermostability assay [12]. However, this is a protein and time-consuming assay and is therefore not desired as a first screening method but merely as a second verification assay.

3.2 Screening for Ligands/Detergents/Additives/Lipids to Improve Thermostability

The 96-well plates are always kept on ice unless stated otherwise.

1. Add 15 μL of a screening solution in which additives, ligands, lipids, or detergents are varied (*see* **Note 5**), to plate 4, a black 96-well (*see* **Note 6**).
2. Add 115 μL of the CPM buffer to plate 4 (*see* **Note 7**).
3. Dilute your protein to the required protein concentration in CPM buffer (optimal dilution was determined in Subheading 3.1) in the wells of one column of a transparent 96-well plate. Transfer 10 μL of protein dilution to all wells of plate 4 by means of a multichannel pipet.

4. Prepare a CPM-fluorophore dilution in CPM buffer according to the optimal concentration as determined in Subheading 3.1 and transfer it into a 50 mL reagent reservoir (protected from light).
5. Start the reaction by adding 10 μL of the CPM dilutions to plate 4 with a multichannel pipet (protect from light sources as much as possible).
6. Keep plate 4 on ice covered with aluminum foil.
7. Seal plate 4 with a transparent cover sticker.
8. Measure plate 4 in a fluorescence plate reader at a constant temperature of 42 °C (or 37 °C or less in case of very unstable protein).
9. Take recordings every minute with a 5 s shaking interval. Detect fluorescence at 463 nm (excitation wave length is 387 nm) until a stable, maximal fluorescence signal is obtained.
10. The fraction of folded protein at each time point is calculated as mentioned in Subheading 3.1, **step 12**.
11. A single exponential decay curve is plotted and the relative unfolding rate $t_{1/2}$ is determined. A lower $t_{1/2}$ in a specific condition is indicative for a stabilizing effect of the supplied ligand, detergent, lipid, or any other additive.

4 Notes

1. It is crucial to use black, non-transparent 96-well plates with clear bottom to prevent any fluorescence signal to leak to neighboring wells.
2. CPM is light sensitive, so aliquots should be stored in brown glass vials and working solutions and plates should be protected from light sources as much as possible and covered with aluminum foil.
3. It is possible to adapt the CPM buffer to buffer components of interest, but success cannot be guaranteed. When adapting the buffer, use the following guidelines: 20–100 mM buffer pH 7–7.5, 100 mM monovalent salt, maximal 3.5 critical micelle concentration (CMC) of detergent.
4. A CPM stock solution is made in DMSO and hence not soluble at 4 °C. This stock solution has to be put at room temperature prior to further handling.
5. Commercial screening kits available for optimizing crystallization conditions can be used for fast and simple thermostability screening. Also homemade screening solutions can be used.
6. In case lipids are added to the reaction, dry the lipids in a glass vial with N_2-gas and dissolve them in 130 μL CPM buffer.

7. Keep in mind to recalculate your CPM buffer so that you have the same final mM concentrations as mentioned in **item 9** of Subheading 2. The final volume of CPM buffer will be 135 μL (10 μL CPM in CPM buffer + 10 μL protein dilutions in CPM buffer + 115 μL CPM buffer) in a total reaction volume of 150 μL (135 CPM buffer (with CPM dye and protein) + 15 μL test substrate).

Acknowledgements

This work was supported by the Interuniversity Poles of Attraction of the Belgian Science Policy Office (P7/13) and the Flanders Research Foundation (FWO G044212N, G0B1115N and 1514514N) and the KU Leuven (OT/13/091).

References

1. Olesen C, Picard M, Winther AM, Gyrup C, Morth JP, Oxvig C, Moller JV, Nissen P (2007) The structural basis of calcium transport by the calcium pump. Nature 450:1036–1042

2. Yao Q, Chen LT, Bigelow DJ (1998) Affinity purification of the Ca-ATPase from cardiac sarcoplasmic reticulum membranes. Protein Expr Purif 13:191–197

3. Morth JP, Pedersen BP, Toustrup-Jensen MS, Sorensen TL, Petersen J, Andersen JP, Vilsen B, Nissen P (2007) Crystal structure of the sodium-potassium pump. Nature 450:1043–1049

4. Boivin S, Kozak S, Meijers R (2013) Optimization of protein purification and characterization using Thermofluor screens. Protein Expr Purif 91:192–206

5. Ericsson UB, Hallberg BM, Detitta GT, Dekker N, Nordlund P (2006) Thermofluor-based high-throughput stability optimization of proteins for structural studies. Anal Biochem 357:289–298

6. Alexandrov AI, Mileni M, Chien EY, Hanson MA, Stevens RC (2008) Microscale fluorescent thermal stability assay for membrane proteins. Structure 16:351–359

7. Abts A, Schwarz CKW, Tschapek B, Smits SHJ, Schmitt L (2011) Rational and irrational approaches to convince a protein to crystallize. In: Kolesnikov N, Borisenko E (eds) Modern aspects of bulk crystal and thin film preparation. ISBN p. Chapter 22 497–528

8. Sonoda Y, Newstead S, Hu NJ, Alguel Y, Nji E, Beis K, Yashiro S, Lee C, Leung J, Cameron AD, Byrne B, Iwata S, Drew D (2011) Benchmarking membrane protein detergent stability for improving throughput of high-resolution X-ray structures. Structure 19:17–25

9. Forneris F, Orru R, Bonivento D, Chiarelli LR, Mattevi A (2009) ThermoFAD, a Thermofluor-adapted flavin ad hoc detection system for protein folding and ligand binding. FEBS J 276:2833–2840

10. Fan J, Heng J, Dai S, Shaw N, Zhou B, Huang B, He Z, Wang Y, Jiang T, Li X, Liu Z, Wang X, Zhang XC (2011) An efficient strategy for high throughput screening of recombinant integral membrane protein expression and stability. Protein Expr Purif 78:6–13

11. Ayers FC, Warner GL, Smith KL, Lawrence DA (1986) Fluorometric quantitation of cellular and nonprotein thiols. Anal Biochem 154:186–193

12. Hattori M, Hibbs RE, Gouaux E (2012) A fluorescence-detection size-exclusion chromatography-based thermostability assay for membrane protein precrystallization screening. Structure 20:1293–1299

Part II

Activity Assays

Chapter 10

Colorimetric Assays of Na,K-ATPase

Kathleen J. Sweadner

Abstract

The Na,K-ATPase is a plasma membrane enzyme that catalyzes active ion transport by the hydrolysis of ATP. Its activity in vivo is determined by many factors, particularly the concentration of intracellular sodium ions. It is the target of the cardiac glycoside class of drugs and of endogenous regulators. Its assay is often an endpoint in the investigation of physiological processes, and it is a promising drug target. As described in this unit, its enzymatic activity can be determined in extracts from tissues by test tube assay using a spectrophotometer or ^{32}P-ATP. The protocols in this chapter measure inorganic phosphate as the end product of hydrolysis of ATP.

Key words Na,K-ATPase, Activity assay, Cardiac glycosides, Drug assay, ATP hydrolysis

1 Introduction

There are several ways to measure Na,K-ATPase activity. Measuring the hydrolysis of ATP with colorimetric reagents, as described here and Chap. 12, is appropriate for homogenates, membrane fractions, purified enzyme, and solubilized enzyme preparations where both sides of the membrane are accessible. To be active, the enzyme must have access to ATP, Mg^{2+}, and Na^+ at the cytoplasmic surface and K^+ at the extracellular surface simultaneously. An alternative that is very good for measuring activity as it would occur in physiological conditions but that is difficult to scale up, is to measure pump current by patch clamp electrophysiology [1], or atomic absorption spectrometry [2], methods that are not covered here. In membrane preparations ATP hydrolysis can also be measured continuously in a thermostated cuvette by coupling the hydrolysis of ATP to the oxidation of NADH [3] (*see* Chap. 11), but that assay is usually used only when time-dependent changes are studied.

This chapter has three related protocols. Two are the basic ATPase assay with alternatives for the color development step, the second of which can be adapted for ^{32}P-ATP (*see* **Note 1**).

The third is a modification of the sample preparation step that is appropriate for highly active and difficult-to-fractionate preparations like brain. The assays are performed in test tubes and require common equipment. The basic steps are

Hydrolysis of ATP → ADP + P_i

Stop the reaction by addition of acid and molybdate

P_i + molybdate → visible color (blue or yellow depending on oxidation state)

The basic method requires the least equipment and fewest steps, and is suitable for enzyme sources with a reasonable level of activity. Typically these are tissues that perform a significant amount of ion transport, such as transporting epithelia or the nervous system, but the assay will also work for many cultured cells and expression systems.

Sensitivity to inhibition by cardiac glycosides, such as ouabain, strophanthidin, and digoxin, defines the Na,K-ATPase. Except in highly purified preparations of enzyme, there will be some other ATPase activities present. These produce a background level of ATPhydrolysis that must be subtracted from the total ATP hydrolysis. The procedure is to measure activity in parallel in the presence and absence of the inhibitor. Like K^+ ions, the inhibitor must have access to the extracellular surface.

There are different isoforms of Na,K-ATPase α, β, and FXYD subunits in mammals, each from a different gene. The protocols here are appropriate for all of them. The Na,K-ATPases can differ by several orders of magnitude in their affinities for cardiac glycosides, however. The inhibitor concentration should be appropriate for the species and tissue to be studied, and if it is not known, the affinity should be measured.

The assay will yield values for maximal Na,K-ATPase activity. It is important to understand how to interpret the activity level if the objective is to understand physiology. The reaction mixture described here is saturating for the enzyme's binding sites, but it is not physiological because in the cell the enzyme would see different concentrations of Na^+ and K^+ at the inside and outside surfaces. Intracellular Na^+ is normally rate-limiting in vivo. Na^+ and K^+ also have competitive interactions at their binding sites on the enzyme that are not accurately modeled by the test tube assay. Activity in vivo can be affected greatly not only by changes in the concentration of the ligands, but also by trafficking of the enzyme to and from the plasma membrane. Finally, the pump is voltage-dependent, and this differs between isoforms. This assay estimates the maximal activity the preparation could have if enzyme were fully active. The activity measured will not report the real activity in the tissue right before it was homogenized, because the assay overrides the ion concentrations that occurred in vivo. It also may differ from the activity in the tissue because the Na,K-ATPase is highly regulated.

Regulatory modifications of Na,K-ATPase activity will be detected only to the extent that they are stable in the conditions of sample preparation and assay.

Test tube ATPase assays, on the other hand, lend themselves well to measuring inhibition or activation by pharmacological compounds. However, a solid knowledge of Na,K-ATPase kinetic properties is important. For example, a drug might be preincubated with the enzyme for a period of time prior to the assay to allow occupation of binding sites. Because the Na,K-ATPase undergoes large conformation changes during the course of a single turnover, and because drugs might be able to bind in only one conformation, the binding rate or apparent affinity could vary with the conditions. The cardiac glycosides are a classic example, binding with highest affinity in the K^+-bound state and much less well in the Na^+-bound state. Consequently a thorough characterization of drug interactions would include many parameters (*see* **Note 2** for more considerations about drug discovery). In particular, a drug that acted at the extracellular surface or in the ouabain site would be best studied either with highly purified enzyme, or with the SDS protocol below, to eliminate the need to run the reaction with and without ouabain.

2 Materials

2.1 Materials and Solutions for All Protocols

Materials for all protocols:

1. Safety glasses are needed at all steps (*see* **Note 3**, safety).
2. A detergent to open sealed vesicles (*see* Subheading 3.1).
3. Pipettors.
4. Water bath.
5. Vortexer.
6. Disposable glass test tubes, 13 × 100 mm.
7. Stopwatch.
8. Spectrophotometer.
9. Cuvette for 1 mL volume.
10. ATPase sample containing 2–6 mg protein/mL, if possible.

Stock solutions for all protocols:

Use Milli-Q or equivalent water in all solutions.

1. Reaction mixture: 140 mM NaCl, 20 mM KCl, 4 mM $MgCl_2$, 3 mM ATP (Tris salt), 30 mM histidine HCl (*see* **Note 4**). Adjust the final pH to 7.2 with 1 M Tris base. The reaction mixture should be stored frozen and kept on ice to reduce background hydrolysis of ATP. Discard if there is an increase in

background inorganic phosphate in assay blanks. MgCl$_2$ is hygroscopic and should be stored in a dry atmosphere.

2. Ouabain stock solution (*see* **Note 3, safety**) in water or dimethylformamide (*see* **Note 5**).

3. 10 N sulfuric acid stock solution (*see* **Note 3, safety**), stable at room temperature (*see* **Note 6**).

4. 5 % ammonium molybdate, 5 g per 100 mL water (*see* **Note 7**). Stable for at least 6 months at room temperature, but discard if a precipitate forms.

5. Quenching solution: 1 N H$_2$SO$_4$, 0.5 % ammonium molybdate. Mix eight parts of water, one part of 10 N H$_2$SO$_4$, and one part of 5 % ammonium molybdate. Each tube will need 0.6 mL of quenching solution. The solution needs to be prepared freshly.

6. Reducing solution: 25 mg of Fiske–Subbarow reducing powder/500 µL water (*see* **Note 8**). The solution needs to be prepared freshly.

7. Inorganic phosphate standard solution: 10 mM KH$_2$PO$_4$. Stable at room temperature as long as securely capped to prevent evaporation (*see* **Note 9**).

2.2 Materials for the Yellow Method

1. Benchtop centrifuge with holders for 13 × 100 mm glass test tubes.

2. Water-saturated isobutanol. Pour some isobutanol into a glass-stoppered flask. Add about ¼ volume of water. Shake well, and let the bottle stand on the shelf. It will separate. The upper phase is water-saturated isobutanol. Stable at room temperature, but keep securely closed (*see* **Note 10**).

3. For optional radioactive assay: Scintillation counter, γ-^{32}P-ATP, and an isobutanol- and water-compatible scintillation fluid for ^{32}P-ATP assay. Perkin Elmer publishes a guide to the solvent compatibilities of its many scintillation fluids.

2.3 Materials for the SDS Method

1. Bovine serum albumin (BSA), >96 % pure.
2. Sodium dodecyl sulfate (SDS), crystalline (*see* **Note 11**).
3. Teflon-glass homogenizer.
4. Fiske–Subbarow reducing powder (Sigma-Aldrich).
5. SET Buffer used for homogenizing and storing the tissue samples: 250 mM sucrose, 20 mM Tris base, 1 mM EDTA pH 7.2.
6. 25 mM histidine HCl, pH adjusted to 7.2 with Tris base. Stock solutions with histidine buffer should be stored frozen to prevent growth of microorganisms.
7. 4.0 mg/mL BSA in the 25 mM histidine buffer.
8. 0.33 mg/mL BSA in the 25 mM histidine buffer.
9. 2.3 mg/mL SDS in the 25 mM histidine buffer.

3 Methods

3.1 Strategic Factors in All Na,K-ATPase Assays

Buffer choice: The buffer used to prepare the enzyme should not interfere with the assay. There must be no phosphate buffer in the enzyme sample. Sucrose at 250 mM or 315 mM buffered with 20 mM Tris base or other non-phosphate buffer with 1 mM EDTA works well. The EDTA chelates calcium and heavy metals and retards growth of microorganisms (*see* **Note 12**).

Protein concentration: If possible, make the sample preparations 2–6 mg protein/mL. There is a limit to how much protein can be added to an assay tube without interfering with the assay. Too much protein (20 μg) will aggregate and absorb the colored reaction product. Too much volume (more than 15 % of the reaction mixture) will affect the assay by diluting the reactants.

Detergent treatment, an important step when assaying membrane proteins: If the preparation has not already been treated with detergent, it may be necessary to do this before the assay to open sealed vesicles that form during tissue homogenization. This should be checked experimentally because vesicle formation may cause sample-to-sample variability. Mild detergents can be used to open sealed vesicles. Sodium deoxycholate is a bile salt that circulates in the body. It has been used in many studies to fully expose renal medulla and cultured cell Na,K-ATPase for assay without solubilizing it [4] (*see* **Note 13**). Other detergents that preserve Na,K-ATPase activity well are $C_{12}E_8$ or $C_{12}E_9$ (octa- or nonaethylene glycol mono n-dodecyl ether), non-ionic detergents that solubilize it in a form stable enough for crystallization. In contrast, Triton X100 can reduce the activity. The use of SDS with BSA to buffer the SDS is described below. For any new source of Na,K-ATPase, it is important to experimentally determine an acceptable detergent and detergent-to-protein ratio for maximal activation. Figure 1 shows a typical preliminary experiment in which a new tissue source (a plasma membrane preparation from mouse liver) was pretreated with four different concentrations of a detergent (SDS in this case) to determine the optimum pretreatment conditions.

3.2 Basic Fiske–Subbarow Assay Method

1. Wear safety glasses and protective clothing.
2. Turn on the water bath and thaw solutions. Frozen reaction mix: mix thoroughly when thawed because ATP separates. Enzyme samples: thaw at room temperature or 37 °C. Keep reaction mix and enzyme on ice as soon as thawed.
3. Set up and label tubes. The procedure is almost always to assay activity in parallel in the presence and absence of ouabain. A typical single assay would have two tubes without ouabain, two tubes with ouabain, and a set of tubes for an inorganic phosphate standard curve. When using highly purified enzyme

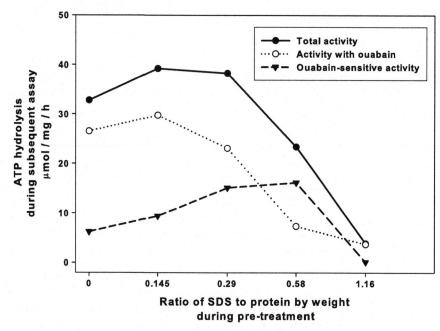

Fig. 1 Optimizing Na,K-ATPase activity measurement by controlled detergent pretreatment. Preincubating with detergent (30 min at room temperature) was done at 1.4 mg total protein/mL and the indicated SDS/protein ratios in a buffer containing 20 mM histidine, 1 mM EDTA, and 3 mM ATP. Mg^{2+} was omitted, preventing ATP hydrolysis until samples were diluted into assay reaction mixture for the timed assay. It can be seen that total activity increased slightly with the two lowest detergent concentrations, and declined with higher concentrations. With zero detergent, ouabain inhibited only about 20 % of the total ATP hydrolysis activity. The increase with detergent is due to opening sealed membrane vesicles, allowing access to both sides. Ouabain inhibits bigger fractions of the total with increasing detergent largely because unrelated ATP-hydrolyzing enzymes are inactivated. The ouabain-sensitive Na,K-ATPase activity (the total minus the activity with ouabain) was optimal at 0.58 mg SDS/mg protein, and at this point ouabain inhibited about 70 % of the total activity. At the highest SDS concentration tested, almost all activity was lost

with no contaminating ATP hydrolases, ouabain tubes are not necessary (*see* **Note 14**).

4. Pipette 400 μL of reaction mixture into each tube. Add ouabain to appropriate tubes and vortex for 1 s.

5. Prepare a standard curve from a 10 mM stock solution of KH_2PO_4.

μmol P_i	μL of 10 mM P_i stock
0 (blank)	0
0.05	5
0.1	10
0.15	15
0.2	20

The assay is reliable for 0.05–0.2 μmol of P_i per tube; each tube has 1.2 μmol of ATP (see **Note 15**).

6. Turn on the visible lamp of the spectrophotometer (see **Note 16**).
7. Pre-warm the assay tubes at 37 °C for 5 min.
8. Prepare a dilution of enzyme, if necessary. Start the reaction by adding enzyme to each tube at 30-s intervals, and vortex each tube for 1 s before returning it to the water bath (see **Note 17**). A typical assay might use 2 μL of a 5 mg/mL sample.
9. After a fixed time at 37 °C (e.g., 20 min), stop each reaction by adding 0.6 mL of quenching solution at 30-s intervals. Vortex for 3 s.
10. As soon as all tubes are quenched, add 10 μL of reducing solution to each tube, vortexing for 3 s immediately after addition.
11. Let the samples sit at room temperature for 30 min. The blue color will develop slowly, however pH-dependent hydrolysis of ATP will make the color continue to increase, so consistent timing is important.
12. Read OD 700 nm (see **Note 18**).
13. Discard liquid waste into a labeled bottle in a hazardous waste depot. Discard test tubes in appropriate waste containers.
14. Average the values for the replicates without ouabain and the replicates with ouabain. Determine the μmol/tube using the standard curve, which is normally linear. Calculate how many μmol of ATP were hydrolyzed per hour per volume of sample assayed. From the known concentration of protein added per tube, calculate the specific activity: μmol/h/mg protein or nmol/min/mg protein. Subtract +ouabain values, which represent unrelated ATP-hydrolyzing activities, from the −ouabain values. The difference is the ouabain-sensitive Na,K-ATPase activity (see **Note 19**).

3.3 Yellow Method: Na,K-ATPase Assay for Samples with High Lipid Content or for Use of ^{32}P-ATP

The blue color in the basic protocol appears only because the phosphomolybdate complex is chemically reduced. Without this step it is pale yellow. The extinction coefficient of the yellow form is greatly increased by extracting the phosphomolybdate complex into an organic phase. This step also accommodates large amounts of lipid that would otherwise precipitate and interfere with color measurement. There are some tricks to this method (see **Note 20**).

First, follow **steps 1–12** of the basic Fiske–Subbarow assay method, omitting **step 9** (no reducing solution needed).

13. As soon as all tubes are quenched, add 1.4 mL of water-saturated isobutanol to each tube. Vortex for 15 s each (see

Note 21). Test tubes smaller than 13 × 100 mm are too small for the vortexing step.

14. Cap the tubes with aluminum foil and centrifuge for 10 min at medium speed in a swinging-bucket tabletop centrifuge. This speeds phase separation and minimizes contamination of the upper phase with droplets of lower phase.

15a. From each tube remove the top 1.0 mL of isobutanol and read the absorbance of this organic phase at 380 nm in a spectrophotometer.

15b. Alternatively, if using ^{32}P-ATP, remove the top 1.0 mL of isobutanol to a scintillation vial and add a water and isobutanol-compatible scintillation fluid for counting (*see* **Note 22**).

16. Discard waste and glass tubes appropriately.

17. Follow the calculation instructions from **step 16** in the basic Fiske–Subbarow assay method. Use a standard curve for colorimetric determination, but base the calculations on measured specific activity of the ^{32}P-ATP if radioactivity is used. Validate the assay solutions and conditions with a colorimetric standard curve even if using ^{32}P-ATP.

3.4 SDS Method: Na,K-ATPase Assay of Crude Homogenates of Brain and Other Tissues After SDS Treatment

This protocol differs from the basic Fiske–Subbarow assay only in the way the samples are prepared. A relatively strong detergent, SDS, is used to modify crude homogenates in two ways. First, it opens sealed vesicles formed during homogenization, a common problem in brain because of the fine structure of the tissue. Second, it reduces the background of other ATP-hydrolyzing enzyme activities, by capitalizing on the resistance of Na,K-ATPase to this detergent. To prevent damage caused by SDS, the protocol calls for a timed exposure to the detergent, followed by dilution. Excess BSA is added in both steps to compete for SDS binding as first suggested by Forbush [5]. (*See* **Note 23** for more detail on the rationale.)

The extra steps are performed after thawing the enzyme samples, and the additional stock solutions can be made in advance and stored frozen. The extra steps entail adding fixed volumes of the BSA and SDS stocks together, then an equal volume of sample (diluted appropriately with its buffer), incubating 10 min at room temperature, then diluting the detergent by a factor of 4 with more BSA-containing solution.

In the protocols above, the buffer used for preparing tissue samples was left up to the investigator other than the caveat that phosphate buffers cannot be used. For the SDS method, samples should be prepared in advance as follows:

1. Homogenize tissue in SET buffer, using a motor-driven Teflon-glass homogenizer. For each gram of wet tissue weight, homogenize in 10 mL of buffer. Assay protein concentration so that the SDS/protein ratio can be carefully controlled (*see* **Note 24**). Store at −70 °C.

2. Follow **steps 1–7** of the basic Fiske–Subbarow assay.

3. Substitute the following longer procedure for **step 8**. However, do all the calculations in advance. This protocol will produce working sample aliquots large enough to assay in triplicate, i.e., three tubes without ouabain and three tubes with ouabain, plus extra left over (*see* **Note 25**).

4. Keeping the volume constant and the amount of protein and detergent constant is important for optimal results. Because protein concentrations of samples will vary, the procedure compensates for this to ensure uniformity of the final mixtures. In advance, calculate what volume of each sample (X) will contain 108 μg of protein, which is 13.5 μg per assay tube. Then calculate the volume (Y) of SET buffer so that $Y = 50\ \mu L - X$.

5. At room temperature, for each sample to be assayed add 25 μL of 4.0 mg/mL BSA-histidine to a 1.8 mL Eppendorf tube.

6. Then add 25 μL of 2.3 mg/mL SDS-histidine; vortex for 1 s. SDS will interact with BSA.

7. Add the calculated YμL of SET to each tube; vortex for 1 s.

8. After setting up all samples, start the timed detergent treatment. Add XμL of homogenized sample to each tube at 30 s intervals. Vortex for 1 s. Adding the sample last ensures that it is not exposed to a higher-than-intended concentration of SDS. Extended vortexing with SDS present will result in some enzyme inactivation.

9. Incubate at room temperature for 10 min. At 5 min, start pre-warming the assay tubes at 37 °C (pre-warming the reaction mixture tubes is **step 9** of the basic Fiske–Subbarow assay).

10. Dilute the 100 μL of detergent-treated sample by adding 300 μL of 0.33 mg/mL BSA-histidine and vortex for 1 s.

11. Start the ATPase reaction by adding 50 μL of sample to each tube of 400 μL of reaction mixture at 30 s intervals, and vortex each tube for 1 s before returning it to the water bath at 37 °C.

12. Carry on with the assay from **steps 9** to **14** of the basic Fiske–Subbarow assay.

4 Notes

1. The basic Fiske–Subbarow protocol is derived from ATPase assays first developed in the 1950s and the experience of many investigators [6]. The yellow method is not used often because the color development step is more complex, but in the case of liposome work, it may be the only assay that works. By utilizing ^{32}P-ATP, the yellow method can in principle greatly extend

the sensitivity of the assay to preparations with low activity or limited material. In the end, sensitivity is limited by background: how much ATP is already hydrolyzed in the purchased powder or ^{32}P-ATP.

2. Colorimetric assays are suitable to characterize Na,K-ATPase inhibitors or activators. It is straightforward to add pharmacological compounds to this assay to assess their ability to inhibit or activate the maximal hydrolysis of ATP. An example is the inhibitory effect of a cGMP-dependent protein kinase inhibitor, Rp-8-pCPT-cGMP, which is a nucleotide analog. After preincubation with the enzyme, it inhibited Na,K-ATPase activity even after dilution into assay reaction mixture with 3 mM ATP [7]. For drug development, however, it is important to bear in mind some strategic considerations. If a compound interfered with ouabain binding, for example, which it could do either directly, or by favoring the Na$^+$-binding (E1) over the K$^+$-binding (E2) conformation of the enzyme, the ability of ouabain to be used to subtract out unrelated enzyme activities could be compromised. In this case use of purified enzyme or the SDS protocol would be preferred if the candidate compound is not affected by detergent (such as by partitioning into detergent micelles). If a compound acted by favoring one conformational state over another, its most interesting actions might be revealed only at sub-saturating concentrations of Na$^+$ and K$^+$, which would also be more physiologically relevant. Voltage-dependent drug effects would not be seen in open membrane preparations. Drug studies might include comparing preincubations in Na$^+$ alone or K$^+$ alone and in turnover conditions (Na$^+$, K$^+$, Mg^{2+}, and ATP), with the understanding that a new steady state might be obtained quickly once added to the complete assay reaction mixture, depending on the affinity and dissociation rate of the compound. It is common to find quantitative differences in affinities measured for Na,K-ATPase between labs for this reason, so strict attention to detail and comprehensive reporting of methods is recommended. The drug affinity and dissociation rate will also dictate whether the concentration of the drug needs to be as high in the assay tube as in a preincubation. It is possible to perform the preincubation in an incomplete reaction mixture and start the reaction by addition of ATP and the missing ligand. However, some drugs, like the cardiac glycosides themselves, might bind best when the enzyme's active site is in the phosphorylated state, i.e., in Na$^+$, Mg^{2+}, and ATP. Familiarity with the reaction mechanism and kinetics of the Na,K-ATPase, which is too complex to cover here, is strongly recommended before undertaking serious pharmacological investigation.

3. Protective clothing and safety glasses should be used at all steps. Ouabain is toxic. Strong acids are used. Frequent vortexing increases the risk of splashing.

4. When preparing the assay solutions, note the formula weights and water content given on the bottle, if any, and use that for the calculations. It is essential to use ATP that is vanadium-free, since vanadate is an inhibitor of the Na,K-ATPase. ATP (and samples) prepared from muscle may contain vanadate. Store ATP desiccated at −20 °C. It is all right to use the Na^+ salt unless it is necessary to vary the sodium concentration in the assay. The Tris salt tends to hydrolyze spontaneously at a higher rate. $MgCl_2$ is hygroscopic and should be stored in a dry atmosphere. Buffers: Phosphate buffers cannot be used! Histidine is thought to chelate heavy metals that might inhibit activity. Tris is not optimal (inhibits somewhat). HEPES has been used successfully. In practice, imidazole can be contaminated with breakdown products; if it does not have pure white crystals, do not use it. ATP and Mg^{2+} form a chelate complex, so with equal concentrations there is very little free Mg^{2+}. Mg^{2+} alone has a binding site on the enzyme. Excess Mg^{2+} produces kinetic effects [8], but makes it possible to neglect the small amounts of EDTA introduced to the assay from the sample's buffer.

5. Ouabain has limited solubility in water. Cardiac glycoside affinities differ from species to species and isoform to isoform, so knowing the appropriate concentration to use is essential information. If the enzyme has very high affinity, 3 µM (final concentration) ouabain may be sufficient and a 0.24 mM stock solution can be prepared in water. Rat or mouse preparations with α2 or α3 isoforms require 10 µM or 30 µM ouabain (final) from a 2.4 mM aqueous stock. If the enzyme is of low affinity, like rodent α1, a 240 mM stock solution in dimethyl formamide is convenient because 5 µL per tube yields 3 mM, a concentration sufficient to inhibit any Na,K-ATPase. Higher concentrations may have non-specific solvent effects. Store ouabain stocks frozen. Cardiac glycosides are light-sensitive, so keep them out of direct sunlight, and discard them if they are accidentally left out overnight. Other cardiac glycosides can be used but most are less soluble in water than ouabain. Check whether the same final concentration of dimethyl formamide alone inhibits activity in the preparation.

6. Preparation of the 10 N sulfuric acid stock should be performed in a fume hood. Never add water to concentrated acid as it will boil and splash explosively. Carefully and very slowly add 27.8 mL concentrated H_2SO_4 to 50 mL water in a Pyrex beaker equipped with a stir bar on a magnetic plate. It will get hot. Use a disposable glass pipette and keep its painted mark-

ings out of the acid. Use a beaker pre-marked for 100 mL, then bring the total volume to 100 mL with water. Store in a bottle with a ground glass stopper. Rinse the pipette before discarding.

7. If ammonium molybdate does not initially dissolve despite extensive stirring, add ammonium hydroxide, one drop at a time, until it all dissolves.

8. For the record, the Fiske–Subbarow powder recipe is 0.2 g 1-amino-2-naphthol-4-sulfonic acid, 1.2 g Na bisulfite, 1.2 g Na sulfite. Mix thoroughly (mortar and pestle) and store dry.

9. KH_2PO_4 is preferred to K_2HPO_4 as a phosphate standard because the solid is less hygroscopic. If the powder appears to have already absorbed water (i.e., it is caked), heat some in an oven for an hour in a glass dish before weighing.

10. If this isobutanol preparation step is omitted, during the assay the upper phase will absorb a fraction of the lower phase from the assay tube, including some ATP.

11. When working with SDS, if it is a fine powder, wear a mask and work in a fume hood to prevent inhalation and alveolar collapse. If it does not appear crystalline (glossy plates), recrystallize it from hot 70 % ethanol, rinse the crystals with ethanol on a Buchner funnel with filter paper, and dry thoroughly.

12. Interference from buffers can be tested by including aliquots of them in an inorganic phosphate standard curve, and comparing it to a standard curve without additives run in parallel. Be aware of other buffer components that might have unanticipated inhibitory effects, such as kinase inhibitors that sometimes also inhibit other ATP-binding enzymes. The vanadate in some phosphatase inhibitor cocktails is an example. If a drug being tested absorbs in the visible range, it could increase absorbance in the colorimetric assay, and this should be tested.

13. Vilsen's method is as follows: Treat crude membranes at a protein concentration of 0.25–0.45 mg/mL with 0.65 mg/mL sodium deoxycholate in 2 mM EDTA, 20 mM imidazole for 30 min at 20 °C. This mixture may then be used directly in the assay.

14. The number of samples that can be assayed at once is limited by operator skill. At two tubes/min, only 40 tubes can be run in a 20-min assay, i.e., only ten different samples if reactions with and without ouabain are run. The assays should be designed so that all comparisons are run together.

15. It is important to allow the reactions to consume no more than 20 % of the ATP in each tube. Published papers showing a saturation of ATPase activity with increasing concentration

of protein are almost certainly due to simply running out of ATP, and of course a reaction consuming the ATP would not be linear with time.

16. It is recommended to zero the spectrophotometer against water and record the absorbance of the blank tubes, rather than zero using a blank tube. This will enable you to detect if your reaction mixture stock solution has elevated hydrolyzed ATP (in which case you should prepare a fresh solution) or if an error has been made that gives unusual baseline absorbance.

17. Maintain order by shifting each tube over one place in the rack when enzyme is added. Blot the wet tube bottoms on a paper towel before vortexing to avoid spattering water. If the preparation of enzyme and the reagents are free of heavy metals, inhibitory compounds, and degradative enzymes, activity should be linear with time for 30 min or more. Sometimes, however, activity is not linear, so reaction times should be kept as short as possible. Only activity in the near-linear portion of the reaction time curve can be used for reproducible quantification.

18. Pasteur pipettes can be used to fill and empty the cuvette, but if it is a quartz cuvette (softer than glass) it should be protected from scratches by using a plastic long-tipped pipette tip like a gel tip.

19. Specific activity for Na,K-ATPase is usually expressed as μmol ATP hydrolyzed/mg protein/h or nmol/mg protein/min, so it will be necessary to determine the protein concentrations of the samples. Specific activity might be 3.0 for a cell culture homogenate; 50 for a brain homogenate; 3000 for highly purified enzyme from renal medulla.

20. Yellow method tricks. First, if using a spectrophotometer, the cuvette must be dry because moisture will form droplets in the isobutanol that will cause light scattering and unreliable readings. This assay can be difficult to perform in humid environments. Second, if using ^{32}P-ATP, note that the radioactive ATP will remain in the lower phase, but any contamination of upper phase with lower phase will cause high readings. Contamination with humidity is not a problem though, as long as the scintillation fluid tolerates some water. Finally, there are old protocols that call for use of an isobutanol/benzene mixture. Use of benzene is not necessary and not recommended because of the toxicity of its vapor.

21. The long vortex step is to maximize extraction into the organic phase. The standard curve tubes should show an obvious gradient of yellow color in the upper, isobutanol phase, while the lower phase is colorless.

22. The radioactive assay can be performed with small amounts of radioactivity added to the normal reaction mixture, such that hydrolysis of 20 % of the ATP produces ~1000 cpm by scintillation counting. Follow institutional and government policy for personnel protection and safety procedures for working with radioactivity. For radioactive assay, all test tubes should be capped before any vortexing step.

23. Quantitative analysis of Na,K-ATPase in brain is complicated by difficulties with membrane fractionation. While it is possible to isolate membrane, synaptosome, and axolemma fractions from brain, in practice a large fraction of the Na,K-ATPase sediments with the nuclear fraction during differential centrifugation. This results from the unique fine structures of neuronal and glia processes, and the complex matrix and junction elements that organize them into stable structures. Since attempts to fractionate tissue samples complicate any quantitative analysis of activity, such as when studying activity in mutant mice, an assay was developed that can be performed with crude brain homogenates.

    The method utilizes a detergent, SDS, normally thought to be denaturing. Actually it is well-established that SDS promotes α-helical structure, and it is used as a membrane protein solvent in NMR structure determination. The Na,K-ATPase assay method builds on important prior observations. First, P. Jørgensen showed that the Na,K-ATPase is remarkably resistant to SDS denaturation under carefully controlled conditions, and he developed it into a purification method that is still the standard for renal medulla [9]. SDS leaves the Na,K-ATPase embedded in the lipid bilayer while inactivating and removing contaminating proteins, including other ATP-hydrolyzing enzymes. Second, Forbush [5] showed that SDS could be used to open sealed vesicles without Na,K-ATPase inactivation when an excess of BSA was present to "buffer" the SDS. The detergent is expected to interact with BSA both by binding at its fatty acid binding sites (as monomeric detergent) and to denatured protein (as micellar detergent). The Forbush method was adapted here, reducing the concentration of BSA to a level that is tolerated by the ATPase assay of the basic protocol, and adjusting volumes and incubations to maximize the detection of activity. The result is ouabain-sensitive ATPase activities that are 20–25 % higher than obtained in the same samples stimulated with sodium deoxycholate. Ouabain-resistant activity (due to other proteins) can be reduced to almost zero. Similar to the Jørgensen purification method, however, it is possible to use too much SDS and inactivate some of the enzyme. Trial and error demonstrated that optimal activation occurred when there was still some residual ouabain-insensitive activity, at SDS to protein ratio of 0.575. Activity was still detected at ratios as high as 0.8. For comparison, the ratio of bound SDS to fully denatured protein (such as during gel electrophoresis) is 1.4. In crude brain homogenates, of course, there are many proteins, in addition to the added BSA, competing for the SDS, and it is probably the unusual resistance of the Na,K-ATPase to SDS that makes the method successful. The detergent-to-protein ratio described in this unit was optimized for mouse brain homogenates. If the method is adapted for other tissues, lipid differences may require optimization of the detergent/protein ratio. Like the other protocols, the SDS protocol is intended to measure the maximum activity that the enzyme can achieve, and is suitable for questions of an enzymological nature. While pump activity in vivo will be controlled largely by the availability of substrates (Na^+, K^+ and ATP), the assay saturates the substrate sites so that intrinsic activity, V_{max} regulation, and enzyme kinetics can be assessed.

24. Note that samples (brain homogenates) with less than 2.7 mg/mL protein will not be sufficiently concentrated for this protocol without adjustments.

25. If assaying triplicates, three sets of six tubes can be run in one assay.

Acknowledgement

This work was supported by NIH grants HL036271, NS045283, EY014390, NS050696, and NS081558. The author is grateful to all of the laboratory members who mastered and refined this methodology.

References

1. Poulsen H, Khandelia H, Morth JP, Bublitz M, Mouritsen OG, Egebjerg J, Nissen P (2010) Neurological disease mutations compromise a C-terminal ion pathway in the Na,K-ATPase. Nature 467:99–102

2. Dürr KL, Tavraz NN, Spiller S, Friedrich T (2013) Measuring cation transport by Na,K- and H,K-ATPase in *Xenopus oocytes* by atomic absorption spectrophotometry: an alternative to radioisotope assays. J Vis Exp (72):e50201

3. Scharschmidt BF, Keeffe EB, Blankenship NM, Ockner RK (1979) Validation of a recording spectrophotometric method for measurement of membrane-associated Mg- and NaK-ATPase activity. J Lab Clin Med 93:790–799

4. Vilsen B (1992) Functional consequences of alterations to Pro328 and Leu332 located in the 4th transmembrane segment of the α-subunit of the rat kidney Na^+,K^+-ATPase. FEBS Lett 314:301–307

5. Forbush B III (1983) Assay of Na,K-ATPase in plasma membrane preparations: increasing the permeability of membrane vesicles using sodium dodecyl sulfate buffered with bovine serum albumin. Anal Biochem 128:159–163

6. Esmann M (1988) ATPase and phosphatase activity of Na^+,K^+-ATPase: molar and specific activity, protein determination. Methods Enzymol 156:105–115

7. Ellis DZ, Nathanson JA, Sweadner KJ (2000) Carbachol inhibits Na^+-K^+-ATPase activity in choroid plexus via stimulation of the NO/cGMP pathway. Am J Physiol 279:C1685–C1693
8. Arystarkhova E, Donnet C, Asinovski NK, Sweadner KJ (2002) Differential regulation of renal Na,K-ATPase by splice variants of the γ subunit. J Biol Chem 277:10162–10172
9. Jørgensen PL (1974) Purification and characterization of $(Na^+ + K^+)$-ATPase. III. Purification from the outer medulla of mammalian kidney after selective removal of membrane components by sodium dodecylsulphate. Biochim Biophys Acta 356:36–52

Chapter 11

ATPase Activity Measurements by an Enzyme-Coupled Spectrophotometric Assay

Pankaj Sehgal, Claus Olesen, and Jesper V. Møller

Abstract

Enzymatic coupled assays are usually based on the spectrophotometric registration of changes in NADH/NAD$^+$ or NADPH/NADP$^+$ absorption at 340 nm accompanying the oxidation/reduction of reactants that by dehydrogenases and other helper enzymes are linked to the activity of the enzymatic reaction under study. The present NADH-ATP-coupled assay for ATPase activity is a seemingly somewhat complicated procedure, but in practice adaptation to performance is easily acquired. It is a more safe and elegant method than colorimetric methods, but not suitable for handling large number of samples, and also presupposes that the activity of the helper enzymes is not severely affected by the chemical environment of the sample in which it is tested.

Key words Coupled assay, Ca^{2+}-ATPase, ATP hydrolysis, Enzymatic spectrophotometry, NADH, SERCA

1 Introduction

A common way to determine ATPase activity is to measure the liberation of inorganic phosphate arising from hydrolysis of ATP by colorimetric methods, based on ANSA (anilino-naphthalene sulfonate) reaction after ammonium phosphate molybdate precipitation [1], staining with Baginski reagents [2], or malachite green [3]. Such methods are convenient for collecting large sets of data, e.g., for high-throughput screening of substances as inhibitors of enzyme activity with a plate reader. Main drawbacks of the colorimetric tests of ATPase activity are the occurrence of large background signals arising from the non-specificity of the assays, nonenzymatic ATP hydrolysis, and the variations that may occur in the enzymatic hydrolysis rates during screening with ATP as a function of time. When it comes to following the details of the reaction colorimetric methods are less versatile than NADH- or NADPH-ATPase coupled assays where activities can be calculated

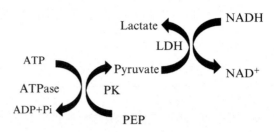

Fig. 1 The assay is based on a reaction in which the regeneration of hydrolyzed ATP is coupled to the oxidation of NADH. Following each cycle of ATP hydrolysis, the regeneration system consisting of phosphoenolpyruvate (PEP) and pyruvate kinase (PK) converts one molecule of PEP to pyruvate by which ADP is reconverted to ATP. The pyruvate in the subsequent reaction is reduced to lactate by L-lactate dehydrogenase (LDH) together with oxidation of one NADH molecule to NAD+. The assay measures the decrease of NADH concentration as a function of time, which equals the rate of ATP hydrolysis, from the decrease in optical density at 340 nm. The constant regeneration of ATP allows monitoring the ATP hydrolysis rate over the entire course of the assay as long as NADH is present

without standard curves and where from the spectrophotometric traces it is possible at a glance to detect changes in activity that may arise from e.g., destabilization caused by detergents or other effects that adversely affect enzyme activity during the reaction period. The method described here for determination of Ca^{2+}-ATPase activity by enzymatic spectrophotometry has been used over the past 40 years [4–6]. A particular feature is that the accumulation of ADP during the assay is avoided by regenerating the formed ADP to ATP by coupling it with a phosphoenolpyruvate catalyzed kinase reaction (PK in Fig. 1). The pyruvate formed by this reaction is then converted to lactate by NADH and lactate dehydrogenase in the following reaction step shown in Fig. 1, and the resulting formation of NAD+ is registered spectrophotometrically by monitoring the decrease in NADH absorbance at 340 nm.

These reactions allow a specific determination of ATP hydrolysis rates under conditions where the ADP concentration remains essentially zero. But in the implementation of this method it is necessary to check that the method is performed in such a way that the ATPase reaction is the rate determining step. This means that the linked reactions should proceed with the same rate as the ATPase reaction under conditions where the concentrations of ADP and pyruvate remain constant and very close to zero. In practice, it is the pyruvate kinase reaction that is the critical factor that can result in an insufficient rate of NADH oxidation, especially at low nucleotide concentrations. This situation can be revealed by a pyruvate kinase test by which it is shown that adding more PEP-kinase does not result in an increased NADH oxidation rate. In the procedure presented below LDH readily converts PEP to lactate at such a high rate that it will not represent a rate-limiting step.

2 Materials

2.1 Equipment

1. Spectrophotometer with UV attachment, e.g., Shimadzu UV 1800, with recorder and software, Kyoto, Japan.
2. Water bath with accurate temperature controller connected to the spectrophotometer for thermo-equilibration of the cuvette compartment and all solutions at 23 ± 0.1 °C, except for ATPase and other enzymes that are stored on ice.

2.2 Buffers and Reagents

Use all reagents in analytical grade.

1. TES/KCl/Mg^{2+} buffer: 10 mM TES, 1 mM Mg^{2+}, and 100 mM KCl, adjusted to pH 7.5 with NaOH.
2. Basic Ca^{2+} containing assay medium: Add 0.1 mM Ca^{2+} to the TES/KCl/Mg^{2+} buffer.
3. MgATP stock solution: 100 mM adenosine 5′-triphosphate (Na_2HATP) and 100 mM Mg^{2+}, solubilized by the TES/KCl/Mg^{2+} buffer and adjusted to pH 7.5 with NaOH.
4. MgPEP stock solution: 50 mM Mg^{2+} and 50 mM Phospho(enol) pyruvic acid tri(cyclohexyl-ammonium) salt solubilized in the TES/KCl/Mg^{2+} buffer and adjusted to pH 7.5 with KOH.
5. PK solution: Pyruvate kinase (PK) from rabbit muscle (Sigma), around 500 U (37 °C)/mg protein, solubilized in TES/KCl/Mg^{2+} buffer at 5 mg protein/mL.
6. NADH stock solution: 15 mM nicotinamide adenine dinucleotide, solubilized in de-ionized water.
7. L-Lactate dehydrogenase (LDH), from rabbit muscle (Roche), approx. 1100 U (37 °C)/mg.
8. Ca^{2+}-ATPase to be assayed: Suspend the DOC purified Ca^{2+}-ATPase membranes (20–25 mg protein/mL), prepared according to Chap. 3 in the basic Ca^{2+} containing assay medium at 1 mg protein/mL.

3 Methods

3.1 Spectrophotometric Activity Assay

1. Turn on the UV spectrophotometer and stabilize the lamp for about 1 h; set the UV spectrophotometer at 340 nm for the absorbance measurements.
2. Suspend Ca^{2+}-ATPase at a protein concentration of 1 mg/mL in the TES/KCl/Mg^{2+} buffer, supplemented with either

0.1 mM Ca^{2+} or 1 mM EGTA (for studies with E2 inhibitors) (*see* **Notes 1** and **2**).

3. Transfer 3 mL of the basic assay medium (10 mM TES, pH 7.5, 100 mM KCl, 1 mM MgCl$_2$, and 0.1 mM CaCl$_2$) to a thermostated (23 ± 0.1 °C) quartz cuvette, add 10 μg of the Ca^{2+}-ATPase sample to be assayed, followed by 1 mM Mg-PEP, 0.175 mM NADH, 65 U PK, and 82.5 U LDH.

4. Immediately before the activity measurement add 5 mM MgATP (pH 7.5) solution to the cuvette. This addition will give rise to an instantaneous decrease in the light absorption baseline at 340 nm, from about 1.1 to 0.9. This arises due to the presence of ADP always present as a contaminant in the commercial ATP, which becomes phosphorylated to ATP by the PEP and PK present, while the pyruvate concomitantly formed leads to NADH oxidation (Fig. 1) and the decrease in light absorption at 340 nm.

5. When the light absorbance has become stabilized (which takes less than 1 min), add 10 μg of the Ca^{2+}-ATPase suspension with gentle mixing. Close the lid of the cuvette compartment and record the evolution of ATP cleavage by measurement of the decrease in NADH absorbance at 340 nm over a period of at least 5 min. The decrease in absorbance may be slightly slower in the start, but should become perfectly linear with time after 1 min and continue until all the NADH has been used up.

3.2 Calculations

1. Calculate the activity of Ca^{2+}-ATPase in terms of μmol/min/mg from the decrease in of absorbance per min of NADH at 340 nm (ΔOD), by using the following formula:

$$\text{Activity} = \frac{\Delta OD \times \text{Vol.}_{\text{Act buf}}(3000\,\mu L)}{6.22 \times \text{Vol.}_{\text{ATPase}}(10\,\mu L) \times [\text{ATPase}]\left(\frac{1\,\text{mg}}{\text{mL}}\right)}$$

where ΔOD is the difference in absorbance at a respective time interval (min^{-1}), Vol.$_{\text{Act buf}}$ is the total volume of activity buffer present in the cuvette, Vol.$_{\text{ATPase}}$ and [ATPase] is the volume and concentration of the Ca^{2+}-ATPase aliquot added to the activity medium in the cuvette, [ATPase] is the stock concentration of Ca^{2+}-ATPase used in the activity reaction mixture, and 6.22 is the millimolar absorption coefficient of NADH in mol^{-1} cm^{-1}.

4 Notes

1. The presence of Ca^{2+} is essential to safeguard against inactivation during hour-long measurements. For unpurified preparations be sure that the SR membrane is permeabilized, e.g., by addition of low concentration of ionophore.

2. To study the effect of inhibitors on Ca^{2+}-ATPase activity, these can be added in known concentrations to the 3 mL activity medium, or in some cases added to the 1 mg/mL ATPase sample before the activity measurement for pre-equilibration of inhibitor with the ATPase before the measurement. Thus, it is often advisable with E2 reacting inhibitors to pre-equilibrate for 3–6 min in the presence of 1 mM EGTA to allow sufficient time for many of inhibitors to interact fully with the Ca^{2+}-ATPase before the activity measurement.

References

1. Fiske CH, Subbarow Y (1925) Colorimetric determination of phosphorus. J Biol Chem 66:375–400
2. Baginski ES, Foa PP, Zak B (1967) Determination of phosphate: study of labile organic phosphate interference. Clin Chim Acta 15:155–158
3. Lanzetta PA, Alvarez LJ, Reinach PS, Candia OA (1979) An improved assay for nanomole amounts of inorganic phosphate. Anal Biochem 100:95–97
4. Warren GB, Toon PA, Birdsall NJ, Lee AG, Metcalfe JC (1974) Reconstitution of a calcium pump using defined membrane components. Proc Natl Acad Sci U S A 71:622–626
5. Møller JV, Lind KE, Andersen JP (1980) Enzyme kinetics and substrate stabilization of detergent-solubilized and membraneous $(Ca^{2+}+Mg^{2+})$-activated ATPase from sarcoplasmic reticulum. Effect of protein-protein interactions. J Biol Chem 255:1912–1920
6. Rayan G, Adrien V, Reffay M, Picard M, Ducruix A, Schmeetz M, Urbach W, Taulier N (2014) Surfactant bilayers maintain transmembrane protein activity. Biophys J 107:1129–1135

Chapter 12

Antimony-Phosphomolybdate ATPase Assay

Gianluca Bartolommei and Francesco Tadini-Buoninsegni

Abstract

Hydrolytic activity is an important functional parameter of enzymes like adenosinetriphosphatases (ATPases). It is measured to test enzyme functionality, but it also provides useful information on possible inhibitory effects of molecules that interfere with the hydrolytic process. Here, we describe a molybdenum-based protocol that makes use of potassium antimony (III) oxide tartrate and may be valuable in biochemical and biomedical investigations of ATPase enzymes as well as in high-throughput drug screening. This method has been successfully applied to native and recombinant ATPases.

Key words ATPases, Enzymes, Hydrolytic activity, Molybdenum blue, Antimony-phosphomolybdate, Colorimetric assay, Calibration curve, UV/Vis spectrophotometry

1 Introduction

The enzymes known as adenosinetriphosphatases (ATPases) produce inorganic phosphate (P_i) by cleavage of the γ-phosphate of ATP. For these enzymes, P_i detection is useful to evaluate the rate of P_i production and the related enzymatic activity, a very important functional parameter.

Different phosphate detection methods have been optimized during years to be used both in enzymology [1–5] and in environmental analysis [6–8]. These methods are usually based on the chemistry of molybdenum. In fact, it is well known that phosphate and molybdic acid form a complex that can be reduced to produce a deep-blue-colored complex called molybdenum blue [2]. Classical experimental protocols for P_i detection involve the use of ammonium heptamolybdate in acid environment, together with a reducing agent having a critical role in determining the stability of the reduced complex and affecting the spectroscopic properties of the produced molybdenum blue species [5].

A new protocol involving potassium antimony (III) oxide tartrate has recently been optimized in order to be used in enzymology

[9]. Potassium antimony (III) oxide tartrate reacts with ammonium heptamolybdate in an acid medium with diluted solutions of phosphate to form an antimony-phosphomolybdate complex. This complex can be reduced to an intensely blue-colored compound by the employed reducing agent.

The procedure here described consists in the preparation of a coloring solution, the determination of a calibration curve with a series of standard solutions of P_i and, finally, the spectrophotometric quantification of P_i released by the ATPase protein at subsequent time intervals.

We have applied the method to native and recombinant ATPases and demonstrated its validity, sensitivity, and versatility [9].

2 Materials

Prepare all solutions using water obtained by a purification system (resistivity should increase up to 18 MΩ cm) possibly provided with a sterile filter that eliminates bacterial content and a system that reduces total organic carbon. Store each solution as indicated (*see* **Note 1**).

The coloring solution is prepared mixing four stock solutions and diluting with water.

The composition of the buffer solution depends on the ATPase to be tested.

2.1 Coloring Solution

1. Acid solution: 2.5 M H_2SO_4. Dilute 14 mL of concentrated H_2SO_4 to 100 mL with water. Use basic safety rules to handle the acid. Store at room temperature (RT).

2. Ammonium molybdate solution: 24 mM ammonium heptamolybdate tetrahydrate. Weigh 0.75 g for 25 mL of solution. Store at RT and protect from light (*see* **Note 2**).

3. Reducing agent solution: 0.3 M ascorbic acid. Weigh 1.35 g of L(+)-ascorbic acid for 25 mL of solution. Store at 4 °C and protect from light.

4. Tartrate solution: 4 mM potassium antimony (III) oxide tartrate trihydrate. Weigh 68 mg for 25 mL solution. Store at RT.

5. Coloring solution: mix stock solutions and add water according to Table 1 to obtain the following final concentrations: 125 mM H_2SO_4, 0.5 mM ammonium molybdate, 10 mM ascorbic acid, and 40 µM tartrate (*see* **Note 3**).

2.2 Buffer Solutions for Incubation

The composition of the buffer solution is strictly related to the ATPase under investigation (*see* **Notes 4** and **5**). In order to determine a net activity, three different incubation buffers should be prepared:

Table 1
Preparation of coloring solution. The table reports volume percentages of each component (stock solution or water) with respect to the total volume of solution. As an example, the volumes needed to prepare 50 mL of coloring solution (sufficient for 55 test tubes) are shown (*see* **Note 3**)

Component	% with respect to the total volume	Amount of stock solutions to prepare 50 mL
2.5 M H_2SO_4	5 %	2.5 mL
24 mM ammonium molybdate	2.1 %	1.05 mL
0.3 M ascorbic acid	3.3 %	1.65 mL
4 mM tartrate	1 %	0.5 mL
H_2O	88.6 %	44.3 mL

1. A control buffer (**CB**), that does not contain the transported cation (*see* **Note 6**).
2. An inhibition buffer (**IB**), that contains a specific inhibitor for the ATPase under study (*see* **Note 7**).
3. A measurement buffer (**MB**), that contains the transported cation (*see* **Note 8**).

2.3 Other Solutions

1. Protein suspension: ATPase proteins are usually stored at −80 °C in their resting buffer. Depending on protein concentration, it may be necessary to prepare a diluted protein suspension to perform the measurements (*see* **Note 9**).
2. Phosphate solution: 10 mM, 1 mM and 0.1 mM KH_2PO_4. Weigh 34.0 mg of KH_2PO_4 to prepare 25 mL of 10 mM stock solution. Prepare 1 mM and 0.1 mM by 1:10 serial dilution (*see* **Note 10**). Store all solutions at 4 °C.
3. ATP solution: 100 mM Na_2-ATP hydrate. Weigh 55.1 mg of ATP salt for 1 mL of solution. Divide in small samples of 20–30 μL and store at −20 °C (*see* **Note 11**).
4. Citrate solution: 10 % (w/w) tri-sodium citrate dihydrate in water. Weigh 10 g of citrate and solubilize with 100 mL of water. Store at +4 °C (*see* **Note 12**).

3 Methods

Carry out all procedures at room temperature, but keep solutions on ice (especially if containing the protein) during the time of the experiment.

3.1 Determination of the Calibration Curve

1. Prepare at least five standard solutions of P_i from stock solutions in disposable glass test tubes. Standard solution concentrations depend on protein activity, but values below 0.4 mM can usually be employed. After addition of the required amount of P_i, make up to 100 μL with water in each test tube (*see* **Note 13**).

2. Add 900 μL of coloring solution prepared as explained above (Subheading 2.1) to each test tube. Color development starts immediately and stops after a few minutes, remaining stable for several hours [9].

3. Measure absorbance at 850 nm (*see* **Note 14**) for each sample and plot the absorbance versus P_i concentration. Linear fitting of experimental data should provide a straight line passing through the origin with a slope of about 0.020 $nmol^{-1}$ [9] (*see* **Note 15**).
Figure 1 shows a typical calibration curve.

3.2 Measurements with the ATPase Protein

1. Prepare an appropriate number of disposable glass test tubes on a rack, depending on the number of time samples to be taken. Figure 2 shows an array of test tubes for five sampling times. Each buffer should be prepared as a triplicate, in order to perform a statistical analysis of results.

2. Add 900 μL of coloring solution to each test tube.

Perform measurements in the order **CB**, **IB** (if present), and **MB** (*see* **Note 16**):

3. Prepare the first triplicate buffer series in eppendorf tubes and incubate at 37 °C for 3 min in order to obtain a homogeneous temperature.

4. Add 1 mM ATP to each eppendorf tube to start the reaction (*see* **Note 17**).

5. Take 100 μL aliquots from each eppendorf tube at subsequent times (*see* **Note 18**) and add the aliquot to the corresponding glass test tube containing the coloring solution.

6. Repeat **steps 3–5** with the other triplicate series.

7. Measure absorbance at 850 nm for each sample (*see* **Note 19**).

3.3 Data Elaboration

1. For each series, calculate the mean of the three absorbance values obtained at each sampling time (*see* Fig. 2) and the corresponding standard deviation.

2. Determine the net absorbance values by subtracting the average absorbances obtained with **CB** from those relative to **MB** and to **IB**, if used.

3. Use the fitting parameters (*see* **Note 20**) of the calibration curve to convert absorbance values into nanomoles of produced P_i and divide by protein amount (mg) (*see* **Note 21**).

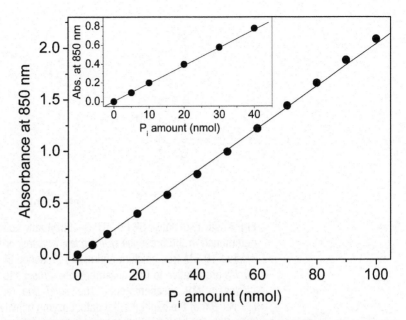

Fig. 1 Calibration curve. An example of a calibration curve is shown. Here several points are reported, but for a routine experiment 5–6 standard solutions are enough to obtain a good regression (*see* inset). The *solid line* represents the linear fitting ($Y = A + BX$) of all experimental data: $A = -0.008 \pm 0.001$; $B = 0.02048 \pm 0.00005$ nmol^{-1} ($R = 0.99953$). Error bars, representing the SEM of at least three independent measurements, are masked by symbols. Modified from [9]

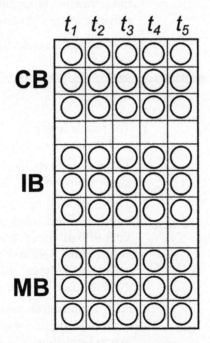

Fig. 2 Array of test tubes. The scheme shows an example of array of test tubes on a rack. The three series relative to **CB**, **IB,** and **MB** are shown, each as a triplicate, and five sampling times (t_1 to t_5) are considered (*see* Subheading 3.2, **step 5**)

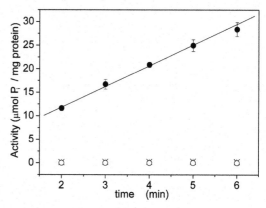

Fig. 3 Activity of native SR Ca-ATPase. Enzymatic activity of native SR Ca-ATPase determined in the presence (*full circles*, *crosses*) or in the absence (*empty circles*) of 10 μM free calcium and in the presence of 1 μM TG (*crosses*). *Empty circles* are relative to **CB** measurements, *crosses* to **IB** measurements and *full circles* to **MB** measurements. The *solid line* represents the linear fitting ($Y = A + BX$) of the experimental data. Enzyme activity is given by the slope, i.e., 4.4 ± 0.3 μmol P_i/(mg min) (corresponding to about 8.8 μmol P_i/(mg min) with respect to SR Ca-ATPase only [9]). Error bars represent the standard deviations of three measurements. Modified from [9]

4. Plot values (calculated as indicated above) versus time (minutes) and perform a linear fitting: the slope gives directly the ATPase activity expressed in nmol P_i/(min mg protein) (*see* **Note 22**).

Figure 3 shows the results for native SR Ca-ATPase (*see* **Note 23**).

4 Notes

1. To store solution both at room temperature and at 4 °C, plastic Falcon-type tubes can be used.

2. To protect from light, use dark containers or simply a container wrapped with aluminum foil.

3. The total volume of coloring solution depends on the total number of test tubes to be employed in the experiment (consider 900 μL for each test tube, *see* Subheading 3.1, **step 2** and Subheading 3.2, **step 2**). This solution should be clear and colorless, without any precipitate, and should be prepared fresh for each experiment.

4. Such a solution should be buffered at a pH for optimal activity of the ATPase under investigation. Moreover it should contain magnesium ions, a general cofactor for ATPases, and should have an ionic strength within physiological values. Depending

on the buffer type (*see* Subheading 2.2), the transported cation(s) maybe present or not.

5. The amount of buffer to be prepared depends on the number of samples that are collected to obtain the time dependence of P_i production. Usually, 4–5 samples are sufficient to obtain a good regression. As explained in Subheading 3.2, **step 5**, aliquots of 100 μL are taken for each sample. Therefore, at least 400 μL are necessary for four samples, but it is safer to use a slightly higher amount of solution, e.g., 450 μL. With five samples, a total volume of 550 μL can be prepared.

6. In the case of Na,K-ATPase **CB** is Na^+-free.

7. For example, thapsigargin in the case of Ca-ATPase or ouabain in the case of Na,K-ATPase. Measurements with **IB** are useful to confirm that the observed activity is related to the ATPase protein.

8. As an example, these are the solutions used with the SR Ca-ATPase:

   - **CB**: 80 mM KCl, 3 mM $MgCl_2$, 25 mM MOPS (pH 7.0 by Tris), 5 mM sodium azide, 2 mM EGTA, 2 μM A23187, 10 μg/mL of total protein.

   - **IB**: 80 mM KCl, 3 mM $MgCl_2$, 25 mM MOPS (pH 7.0 by Tris), 5 mM sodium azide, 0.2 mM EGTA, 0.2 mM $CaCl_2$ (10 μM free calcium), 2 μM A23187, 1 μM TG, 10 μg/mL of total protein.

   - **MB**: 80 mM KCl, 3 mM $MgCl_2$, 25 mM MOPS (pH 7.0 by Tris), 5 mM sodium azide, 0.2 mM EGTA, 0.2 mM $CaCl_2$ (10 μM free calcium), 2 μM A23187, 10 μg/mL of total protein.

   It is worth noticing that for SR vesicles prepared according to [10], SR Ca-ATPase is approximately 50 % of the total protein content [11].

   Add protein as the last component and keep it on ice during the time of the experiment. A good strategy to improve measurement reproducibility is to prepare a "global" buffer (**GB**) that contains all common components of **CB**, **IB**, and **MB**. Components present at different concentrations are added successively where needed. Therefore, in the case of SR Ca-ATPase **GB** should contain KCl, $MgCl_2$, MOPS, sodium azide, A23187, and SR vesicles suspension. $CaCl_2$, EGTA, and TG are then added at the desired amount where required.

9. The amount of protein has to be adjusted in order to use the minimal quantity of protein yielding a detectable activity. For example, considering an activity of about 9 μmol P_i/(mg min) for the Sarcoplasmic Reticulum (SR) Ca-ATPase (*see* legend to Fig. 3), in 4 min 18 nmol of P_i are produced by 0.5 μg of protein.

This amount of P_i is sufficient to generate an intense color easily detectable by the spectrophotometer (*see* Fig. 1).

10. To prepare standard solutions for the calibration curve, 1 mM and 0.1 mM stock solutions are usually employed. The more diluted stock solution is useful to prepare standard solutions at low P_i concentration.

11. Use a fresh ATP sample each time.

12. The sodium citrate solution should be used if the ATPase under study has a slow turnover or it is present at low concentrations or both (*see* Subheading 3.2).

13. A typical calibration curve can be prepared using the following concentrations: 0, 0.1, 0.2, 0.3, and 0.4 mM P_i. For example, the 0.1 mM standard solution is prepared by adding 10 μL of the 1 mM stock solution to the test tube followed by 90 μL of water. Note that the amount of stock solution in microliters corresponds numerically to the nanomoles of phosphate.

14. The wavelength range of the spectrophotometer should extend at least to 900 nm. To increase reproducibility, use a thermostatic cell holder.

15. Absorbance measurements can be performed during the period of color stability. Since measurements are performed in the visible range of the spectrum, disposable plexiglass semimicro cuvettes can be used as an alternative to quartz semimicro cuvettes. Use just one cuvette and proceed from more diluted to more concentrated samples: add the first solution to the cuvette, perform the absorbance measurement and replace the first solution with the subsequent solution using a Pasteur pipette. Be careful to remove completely the first solution from the cuvette before adding the subsequent solution.

16. As a general rule, the order should follow increasing concentrations of produced P_i and, therefore, a gradually more intense blue color.

17. For example, add 4.5 μL of 1 mM ATP to 450 μL of incubation buffer. To perform three measurements in parallel within each series, ATP additions to the eppendorf tubes should be staggered at 20 s intervals. The same time offset is maintained when taking the aliquots. The addition of ATP marks the zero-time point.

18. For example, for native SR Ca-ATPase aliquots can be taken after 2, 3, 4, 5, and 6 min. For recombinant SR Ca-ATPase longer time intervals are required, e.g., 5, 10, 20, 30, and 40 min [9].

19. If the ATPase under investigation has a slow turnover or is present at low concentration (or both), the duration of experiment increases significantly. As a consequence, if the time elapsed between addition of the aliquot to the coloring solution and absorbance measurement ("elapsed time") is relatively long (1 h or more), there may be a chemical interference. Using high ATP concentrations (e.g., 5 mM) further promote the interference. In fact, under these conditions the P_i produced by acid-ATP hydrolysis becomes significant, and adds to protein-released P_i. For long elapsed times, addition of 20 μL 10 % (w/w) sodium citrate to each test tube might be helpful. This compound acts as a stabilizing agent by chelating molybdenum, thus preventing the detection of nascent (acid-released) inorganic phosphate [4, 9]. The addition of citrate should be carried out after a complete color development that occurs after at least 10 min. It is worth noticing that the presence of citrate slightly interferes with color development for P_i concentration higher than 0.4 mM [9].

20. In particular, use the slope parameter with its error. On the other hand, the intercept parameter is very close to zero and the associated relative error could be quite high, therefore it should not be used.

21. The protein amount is that contained in the 100 μL aliquot. For example, in the case of native SR Ca-ATPase the protein concentration in the incubation buffer is usually 10 μg total protein/mL (*see* **Note 8**), which corresponds to 1 μg total protein.

22. Plot data points together with the error bars. The linear fitting provides the final value of the activity (the slope) with an uncertainty value that includes all sources of errors for the method.

23. This procedure can also be used to determine an inhibition curve (activity vs inhibitor concentration). It is just necessary to increase the number of triplicate buffer series using a series for each point of the curve (that is for each concentration of inhibitor). It is convenient to perform measurements by following an increasing amount of P_i (*see* **Note 16**).

Acknowledgement

This work was supported by Ente Cassa di Risparmio di Firenze (2009.0749) and the Italian Ministry of Education, University and Research (PRIN Project 20083YM37E).

References

1. Fiske CH, Subbarow Y (1925) The colorimetric determination of phosphorus. J Biol Chem 66:375–400
2. Ernster L, Lindberg O (1956) Determination of organic phosphorus compounds by phosphate analysis. Methods Biochem Anal 3:1–22
3. Baginski ES, Foa PP, Zak B (1967) Determination of phosphate: study of labile organic phosphate interference. Clin Chim Acta 15:155–158
4. Lanzetta PA, Alvarez LJ, Reinach PS, Candia OA (1979) An improved assay for nanomole amounts of inorganic phosphate. Anal Biochem 100:95–97
5. Katewa SD, Katyare SS (2003) A simplified method for inorganic phosphate determination and its application for phosphate analysis in enzyme assays. Anal Biochem 323:180–187
6. D'Angelo E, Crutchfield J, Vandiviere M (2001) Rapid, sensitive, microscale determination of phosphate in water and soil. J Environ Qual 30:2206–2209
7. U.S. Environmental Protection Agency (2005) Standard operating procedure for particulate-phase total phosphorous by persulfate oxidation digestion (LG209, Revision 06)
8. Matula J (2011) Determination of dissolved reactive and dissolved total phosphorus in water extract of soils. Plant Soil Environ 57:1–6
9. Bartolommei G, Moncelli MR, Tadini-Buoninsegni F (2013) A method to measure hydrolytic activity of adenosinetriphosphatases (ATPases). PLoS One 8, e58615
10. Eletr S, Inesi G (1972) Phase changes in the lipid moieties of sarcoplasmic reticulum membranes induced by temperature and protein conformational changes. Biochim Biophys Acta 290:178–185
11. Inesi G, Toyoshima C (2004) Catalytic and transport mechanism of the Sarco-(Endo) Plasmic reticulum Ca^{2+}-ATPase (SERCA). In: Futai M, Wada Y, Kaplan JH (eds) Handbook of ATPases: biochemistry, cell biology, pathophysiology. Wiley-VCH, Weinheim, pp 63–87

Chapter 13

ATPase Activity Measurements Using Radiolabeled ATP

Herman G.P. Swarts and Jan B. Koenderink

Abstract

ATP provides the energy that is essential for all P-type ATPases to actively transport their substrates against an existing gradient. This ATP hydrolysis can be measured using different methods. Here, we describe a method that uses radiolabeled [γ-^{32}P]ATP, which is hydrolyzed by P-type ATPases to ADP and ^{32}P$_i$. Activated charcoal is used to bind the excess of [γ-^{32}P]ATP, which can be separated from the unbound ^{32}P$_i$ by centrifugation. With this method, a wide range (0.1 μM–10 mM) of ATP can be used. In addition, we also describe in detail how ATP hydrolysis is translated into ATPase activity.

Key words ATPase activity, Adenosine triphosphate, ATP, ADP, P$_i$, Charcoal, H,K-ATPase, Na,K-ATPase

1 Introduction

The ATPase activity of an enzyme is defined as the amount of ATP hydrolyzed in time per mg protein. The γ-phosphate of ATP is cleaved off and the products formed are ADP and P$_i$ (inorganic phosphate). The activity is often measured as the amount of inorganic phosphate produced and can be determined with different techniques. The most commonly applied assay uses ammonium heptamolybdate in a sulfuric acid environment and FeSO$_4$ to form a blue complex with the liberated P$_i$ [1] (*see* Chap. 3.1). In addition, a malachite green method [2], or an ATP/NADH [3] coupled assay with phosphoenolpyruvate (PEP) and pyruvate kinase are used (*see* Chap. 11). Although all these techniques have their own advantages, they are all limited in the ATP range that can be used in these assays.

Here, we present a method in which radiolabeled ATP ([γ-^{32}P]ATP) is used as substrate and where the reaction products are ADP and ^{32}Pi (radiolabeled inorganic phosphate). When the ATPase reaction is stopped the unconverted [γ-^{32}P]ATP can be absorbed easily by activated charcoal and after centrifugation of the charcoal slurry, the unbound reaction product ^{32}P$_i$ can be determined in the clear supernatant. Charcoal is a form of carbon processed to have small pores with a high surface area available for adsorption. It is widely used in food technology, water decontamination, and also poisoning treatment [4].

2 Materials

2.1 Enzyme Preparations

1. Enzyme Buffer: 0.25 M sucrose, 2 mM EDTA, and 50 mM Tris-Acetate pH 7.0.
2. Sf9 membranes from expression studies with gastric [5], non-gastric H,K-ATPase [6], mammalian Na,K-ATPase α1β1 [7], α2β1 [8], α3β1 [9] or α4β1, or Na,K-ATPase of *Drosophila melanogaster* [10] can be used. It is, however, very well possible to use isolated P-type ATPases from animal tissues or other expression systems. The samples are stored at −20 °C in Enzyme Buffer at 4–12 mg protein/mL (*see* Chap. 8).
3. Dilutions of the membrane samples to 0.4 mg/mL in the above-mentioned Enzyme Buffer (prepared just before use).
4. Enzyme solution prepared according to Table 1.

2.2 ATPase Activity Assay

1. 10 % Charcoal in 6 % TCA: Take 50 g of charcoal and put this in a large beaker (>1 L), then add water slowly and mix with a magnetic stirrer. Let the mixture settle for half an hour and decant the light particles and the foam. Repeat this 2–5 times. Finally mix the slurry with 60 mL 50 % TCA and adjust the volume to 500 mL with water.
2. [γ-^{32}P]ATP, 3000 Ci/mmol and 10 mCi/mL (=3.3 μM) from Perkin Elmer.
3. A dilution of the 3.3 μM [γ^{32}P]ATP to ±2 μCi/mL ($\approx 4 \times 10^6$ cpm/mL).
4. 20 mM MgATP: 212 mg per 20 mL and bring to pH 7.0 with Tris. Store at −20 °C.
5. A dilution of part of the 20 mM MgATP to 1 mM MgATP.
6. ATP solution prepared according to Table 2.

Table 1
Enzyme dilutions are adjusted to the ATP concentrations used (Table 2)

	1	2	3	4	5	6
[enzyme][a]	0.08	0.22	0.55	1.42	3.13	8.00
Enzyme Buffer	396	389	372	329	244	0
0.4 mg/mL membranes	4	11	28	71	156	400

[a][enzyme] in experiment in μg per 100 μL

Table 2
Pipetting scheme for the ATP solutions

mM ATP[a]	1	2	3	4	5	6
	0.006	0.025	0.100	0.400	1.600	6.400
1 mM ATP-Mg	16	63	250			
20 mM ATP-Mg				50	200	800
Diluted [γ^{32}P]ATP	100	100	100	100	100	100
Water	884	838	650	850	700	100
Total (μL)	1000	1000	1000	1000	1000	1000

[a][ATP] in experiment

Table 3
Example of pipetting scheme to prepare the media (all volumes are in μL) (*see* **Note 4**)

Stock		Experiment
500 mM Tris-Acetate pH 7.0	750	50 mM Tris-Ac pH 7.0
100 mM MgCl$_2$	60	0.8 mM MgCl$_2$
100 mM Tris-N$_3$	75	1.0 mM Tris-N$_3$
10 mM EGTA	75	0.1 mM EGTA
100 mM EDTA	15	0.2 mM EDTA
1000 mM KCl	75	10 mM KCl
Water	1950	
Total	3000 μL	

7. 500 mM Tris-Acetate pH 7.0: Dissolve 12.1 g Tris-base, set pH 7.0 with acetic acid and adjust the volume to 200 mL with water.

8. 100 mM MgCl$_2$: 203 mg MgCl$_2$ · 6 H$_2$O in 10 mL of water.

9. 100 mM Tris-N$_3$: dissolve 1.3 g of NaN$_3$ in 100 mL of water (200 mM). Convert Dowex 50W-4 (Fluka, Switzerland) to Tris form: Mix the NaN$_3$ solution with 100 mL of Dowex-Tris, exchange and centrifuge and filter (0.2 μm). The supernatant is 100 mM Tris-N$_3$ (*see* **Note 1**).

10. 10 mM EGTA: 190 mg EGTA per 50 mL of water. Adjust to pH 7.0 with Tris.

11. 100 mM EDTA: 1.46 g EDTA-H$^+$ per 50 mL of water. Adjust to pH 7.0 with Tris.

12. 1000 mM KCl: 7.45 g KCl per 100 mL of water (*see* **Note 2**).

13. Assay medium in the desired volume. *See* Table 3 and **Note 3**.

14. Liquid Scintillation Analyzer.

3 Methods

Carry out all procedures at room temperature unless otherwise specified. Preferably use reactions in duplicate or triplicate and repeat these for three different protein isolations (*see* **Note 4**).

3.1 ATPase Assay

1. Pipette for each reaction 40 μL of assay medium in a 1.5 mL reaction tube and place on ice.
2. Add 20 μL of diluted enzyme (membranes) solution (0.08–8.0 μg).
3. Add 40 μL of ATP solution every 10 s and mix.
4. Place the reaction tube immediately at 37 °C to *start* the reaction.
5. Incubate for 30 min.
6. *Stop* the reaction by adding 500 μL of the charcoal-slurry (*see* **Note 5**) and place on ice (0 °C).
7. Repeat this every 10 s; so each reaction time is 30 min.
8. Leave the mixture at least 10 min on ice to allow proper binding.
9. Mix again and centrifuge for 10 s at RT in a table centrifuge ($3000 \times g$).
10. Pipette 150 μL of the clear supernatant in a liquid scintillation counting vial.
11. Add 3 mL of scintillation fluid.
12. Analyze the radioactivity (counts per minute) for 4 min in the Liquid Scintillation Counter.
13. Also pipette 40 μL of the original [γ^{32}P]ATP solutions directly in a counting vial, add 3 mL scintillation fluid, analyze and determine the specific radioactivity of the [γ^{32}P]ATP solution used in cpm per pmol ATP.
14. Also include a blank, without protein, to determine the non-enzymatic [γ^{32}P]ATP hydrolysis and the (*see* **Note 6**) [γ^{32}P]ATP fraction that does not bind to charcoal.

3.2 Calculations

The ATPase activity is expressed as μmol Pi/mg protein/h or as μmol ATP hydrolyzed/mg protein/h.

1. When the ATPase activity is calculated, the following factors should be accounted for:
   - dilution 600/150, only 150 μL is analyzed from the $(100+500)$ μL,
   - incubation time is generally 30 min,

- amount of (μg) protein used per sample,
- from pmol to μmol and μg to mg protein,
- specific radioactivity of the [γ^{32}P]ATP in cpm per pmol ATP.

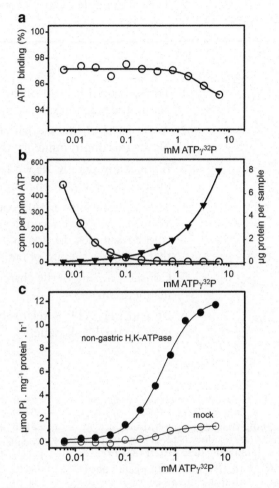

Fig. 1 (**a**) Binding of [γ^{32}P]ATP to charcoal in the absence of protein. In the absence of protein, 97 % of the [γ^{32}P]ATP binds to charcoal, only at very high ATP concentrations the binding capacity of the charcoal will decrease. The 3–5 % radioactivity at very low ATP concentrations that is not bound represents the ^{32}P$_i$ present in the [γ^{32}P]ATP stock solution and the spontaneous hydrolysis of [γ^{32}P] ATP at acidic pH during the experiment. (**b**) The specific radioactivity of the [γ^{32}P] ATP and the amount of protein in the experiment. When the specific radioactivity is kept constant, large amounts of radioactivity have to be used. To avoid that too much or too little [γ^{32}P]ATP is consumed, the protein concentration in the experiment has to be varied. Generally, a maximum of 10–25 % [γ^{32}P]ATP is converted to ensure that the hydrolysis is linear in time. (**c**) ATP dependence of the K$^+$-ATPase activity of non-gastric H,K-ATPase and mock membranes. If the activity of the mock is subtracted from that of the non-gastric H,K-ATPase, the specific ATP dependence of the non-gastric H,K-ATPase can be calculated (not shown). By varying the content of the media the K$^+$-dependence, as well inhibitory effect of inhibitors, like ouabain (Na,K-ATPase), SCH28080 (gastric H,K-ATPase), or vanadate (P-type ATPase) can be studied

$$\frac{\text{cpm sample} - \text{cpm blank}}{\text{cpm/pmol}} \times \frac{600}{150} \times \frac{60}{\text{time}} \times \frac{1}{\mu\text{g protein}} \times 10^3$$

2. Typical example of an ATPase activity assay (Fig. 1):

4 Notes

1. TrisN$_3$ is used instead of the sodium salt to reduce undesired Na$^+$ in the reaction mix.
2. Avoid using chemicals that contain (minor) amounts of K$^+$ or Na$^+$ as this might influence the ATPase activity.
3. If you study, e.g., K$^+$ dependence or specific inhibitors, prepare media containing a concentration range of these compounds.
4. This assay can use a wide concentration range of ATP (up to 10 mM). Moreover, samples contaminated with phosphate or colored samples do not influence the reaction. A laboratory that facilitates the use of radiolabeled compounds is required.
5. Constant mixing of the charcoal slurry is very important (use a magnetic stirrer).
6. At acid pH, ATP is relatively rapidly hydrolyzed, which will increase the blank value.

References

1. Swarts HGP, Moes M, Schuurmans Stekhoven FMAH, De Pont JJHHM (1992) Vanadate-sensitive phosphatidate phosphohydrolase activity in a purified rabbit kidney Na,K-ATPase preparation. Biochim Biophys Acta 1107:143–149
2. Carter SG, Karl DW (1982) Inorganic phosphate assay with malachite green: an improvement and evaluation. J Biochem Biophys Methods 7(1):7–13
3. Penefsky HS, Bruist MF (1984) Adenosinetriphosphatases. In: Bergmeyer HU (ed) Methods of enzymatic analysis, vol 4. Verlag Chemie, Weinheim, pp 324–328
4. Jones J, McMullen MJ, Dougherty J, Cannon L (1987) Repetitive doses of activated charcoal in the treatment of poisoning. Am J Emerg Med 5(4):305–311
5. Swarts HG, Klaassen CH, de Boer M, Fransen JA, De Pont JJ (1996) Role of negatively charged residues in the fifth and sixth transmembrane domains of the catalytic subunit of gastric H$^+$,K$^+$-ATPase. J Biol Chem 271:29764–29772
6. Swarts HG, Koenderink JB, Willems PH, De Pont JJ (2005) The non-gastric H,K-ATPase is oligomycin-sensitive and can function as an H$^+$, NH4$^+$-ATPase. J Biol Chem 280(39):33115–33122
7. Koenderink JB, Swarts HG, Willems PH, Krieger E, De Pont JJ (2004) A conformation-specific interhelical salt bridge in the K$^+$ binding site of gastric H,K-ATPase. J Biol Chem 279:16417–16424
8. Weigand KM, Swarts HG, Russel FG, Koenderink JB (2014) Biochemical characterization of sporadic/familial hemiplegic migraine mutations. Biochim Biophys Acta 1838(7):1693–1700
9. Weigand KM, Messchaert M, Swarts HG, Russel FG, Koenderink JB (2014) Alternating hemiplegia of childhood mutations have a differential effect on Na$^+$,K$^+$-ATPase activity and ouabain binding. Biochim Biophys Acta 1842(7):1010–1016
10. Dalla S, Swarts HG, Koenderink JB, Dobler S (2013) Amino acid substitutions of Na,K-ATPase conferring decreased sensitivity to cardenolides in insects compared to mammals. Insect Biochem Mol Biol 43(12):1109–1115

Chapter 14

Assaying P-Type ATPases Reconstituted in Liposomes

Hans-Jürgen Apell and Bojana Damnjanovic

Abstract

Reconstitution of P-type ATPases in unilamellar liposomes is a useful technique to study functional properties of these active ion transporters. Experiments with such liposomes provide an easy access to substrate-binding affinities of the ion pumps as well as to the lipid and temperature dependence of the pump current. Here, we describe two reconstitution methods by dialysis and the use of potential-sensitive fluorescence dyes to study transport properties of two P-type ATPases, the Na,K-ATPase from rabbit kidney and the K^+-transporting KdpFABC complex from *E. coli*. Several techniques are introduced how the measured fluorescence signals may be analyzed to gain information on properties of the ion pumps.

Key words Reconstitution, Lipids, Detergents, Dialysis, BioBeads, Vesicles, Voltage-sensitive fluorescence dyes, Electrogenic ion transport, Pump current, Substrate dependence, Membrane potential

1 Introduction

Several P-type ATPases have been reconstituted functionally in liposomes. This has become a well-established technique since decades and is used to study the transport functions and kinetic properties of several members of this family of ion pumps [1-9]. Various methods have been introduced to reconstitute P-type ATPases in liposomes. It turned out that dialysis methods are favorable because they are so mild that enzyme and transport activity of the reconstituted ion pumps are almost completely retained. Dialysis may be performed using dialysis tubings or BioBeads. Both methods are introduced in the following. Although each method is applied to a specific ATPase in this chapter, each approach may be used respectively also for other ATPases.

The advantage of the use of liposomes is on one hand that purified ATPases are integrated in lipid bilayer membranes without the background transport activity of other proteins. On the other hand, the lipid composition of the membrane may be varied deliberately and is well defined during the experiments. The pump activity can be triggered by addition of ATP, and the substrates in the aqueous compartments on both sides of the membrane can be freely chosen.

The terms "liposomes" and "vesicles" have been used likewise in the literature to refer to unilamellar structures in which transport proteins are integrated into the lipid bilayer, typically in both orientations, cytoplasmic side out and extracellular side out.

When ion pumping is induced by the addition of ATP to the liposome suspension, those ATPases are activated whose cytoplasmic side faces the liposome outside. Inversely reconstituted proteins remain inactive because ATP is unable to permeate through the liposome membrane. Therefore, the condition has to be kept in mind that in experiments performed with liposomes the outside corresponds to the cytoplasmic side of the membrane and the inner volume of the liposomes corresponds to the extracellular phase of the cell.

A consequence of this condition is that in measurements performed with reconstituted liposomes, the composition of the aqueous phase in which the liposomes are prepared represents the extracellular medium. It cannot be easily changed during the experiments. In contrast, the electrolyte solution to which the liposomes are added represents the cytoplasm and it can be ad hoc composed as desired.

To study and analyze ATPase-induced currents across the liposomal membranes, various techniques have been applied. In the beginning, radioactive isotopes, $^{22}Na^+$ and $^{42}K^+$ (or its congeners, $^{86}Rb^+$ and $^{137}Cs^+$) were used in the case of Na,K-ATPase to analyze pump-mediated fluxes [6, 7, 10–13]. Depending on the transport direction, uptake or release of the isotopes has been detected as function of time and substrate concentrations. This allowed the calculation of pump fluxes and the turnover rate, if the number of active pumps reconstituted in a liposome was known.

Another important experimental technique introduced to study electrogenic activities of the ATPases is the application of potential-sensitive fluorescence dyes. The term electrogenic means that the pump transfers net charge across the membrane while running through the pump cycle and thus modifies the electric potential difference across the membrane. For this purpose so-called slow-response dyes are frequently used which have characteristic time constants of 270 ms or longer [14] but produce fluorescence changes in the order of 1 %/mV change of the membrane potential. Well established dyes in studies with P-type ATPases are: cationic carbocyanines, such as $DiIC_1(5)$ [7] or $DiSC_3(5)$ [9, 13, 15], and anionic oxonol dyes, especially Oxonol VI [16–18].

In the following paragraphs (1) the preparation of liposomes with two different P-type ATPases, the Na,K-ATPase and the K^+-transporting KdpFABC complex from *E. coli* is presented, (2) the execution of fluorescence experiments with potential sensitive dyes is described, and (3) approaches of data analysis of the results of fluorescence experiments are introduced.

2 Materials

Prepare all solution with quartz double-distilled (DD) water. Sterile filtration (pore size 0.2 μm) is performed with all solutions used in fluorescence measurements to remove all contaminations by dust particles and bacteria, and the filtrated solutions are degassed. All solutions are kept at 4 °C unless indicated otherwise. All reagents are of (at least) analytical grade. To reduce the leak conductance of the liposome membranes, all salts are chosen to be sulfates instead of chlorides because the permeability of the sulfate anion is orders of magnitude lower than that of chloride. (Control experiments were performed to show that the enzyme activity of the ATPases was not affected when chloride was exchanged by sulfate.)

1. For dialysis, a tubing of regenerated cellulose is used with a molecular weight cut-off of ~14,000 Da. It turned out that Visking dialysis tubing 8/32, MWCO 12,000–14,000 (Serva Electrophoresis GmbH) is the most efficient choice.

2. Bio-Beads SM-2 (Bio-Rad) are used as adsorbent.

2.1 Solutions for Experiments with Na,K-ATPase Liposomes

1. Standard buffer: 30 mM imidazole, 1 mM ethylenediaminetetraacetic acid (EDTA), 5 mM $MgSO_4$, pH 7.2. To prepare 1 L of buffer weigh 2.04 g imidazole, 0.29 g EDTA, 0.60 g $MgSO_4$, and transfer all components into a 1-L beaker glass. Add 950 mL DD water. Stir with a magnetic stirrer until all components are dissolved. Adjust pH to 7.2 with 1 M H_2SO_4. (To check pH, immerse the pH electrode each time only for periods as short as possible to avoid significant contaminations by K^+ leaking through the ceramic junction of the reference electrode.) Fill the pH-adjusted buffer into a 1-L graduated flask and make up to 1 L with DD water. Store at 4 °C.

2. Buffer H: Standard buffer supplemented with 1 mM L-cysteine. Prepare as described for the Standard buffer, and add 0.12 g L-cysteine before dissolving all components in DD water.

3. Cholate buffer: Buffer H with 1 % (w/v) sodium cholate, 70 mM K_2SO_4, and 5 mM Na_2SO_4, pH 7.2. Prepare as Buffer H, to obtain 1 L add 10 g sodium cholate, 12.20 g K_2SO_4 and 0.71 g Na_2SO_4 before dissolving all components in DD water.

4. Dialysis buffer 1: Standard buffer with 70 mM K_2SO_4, and 5 mM Na_2SO_4, pH 7.2. Prepare as described for the Standard buffer, add 12.20 g K_2SO_4 and 0.71 g Na_2SO_4 before dissolving all components in 1 L DD water.

5. Dialysis buffer R: Standard buffer with 0.5 mM K_2SO_4, and 74.5 mM Na_2SO_4, pH 7.2. Prepare as described for the

Standard buffer, add 0.09 g K_2SO_4 and 10.58 g Na_2SO_4 before dissolving all components in 1 L DD water.

6. Cholate-containing lipid stock solution for liposome preparation: 20 mg lipid (e.g., dioleoyl phosphatidylcholine (PC 18:1) as powder, not as chloroform solution, obtained from Avanti Polar Lipids, Inc., Alabaster, AL) are dissolved in 1 mM ethanol (spectroscopic quality). 10 mg sodium cholate are dissolved in 500 μL ethanol (spectroscopic quality). Both components are mixed in a 50 mL round-bottom flask. (The flask has to be extremely well cleaned and dried before use.) The ethanol has to be removed completely by vacuum evaporation under rotation in a water bath at 40 °C for at least 1 h. Finally, a faint, clear film is formed on the flask wall. This film is dissolved in 1 mL DD water by smooth shaking. (Do not vortex to avoid foam formation!) The obtained clear solution is divided in 100 μL aliquots in 0.5 mL Eppendorf vessels and stored frozen at –24 °C. The samples may be stored for months before use. The final concentrations are 20 mg/mL PC 18:1 and 10 mg/mL sodium cholate. (The lipid contents may be checked with the Phospholipids C test from Wako Diagnostics, Wako Life Sciences, Inc. according to their provided protocol.)

7. 25 mg/mL *E. coli* polar lipid extract dissolved in chloroform (Avanti Polar Lipids, Alabaster, USA). Store contents of the opened glass ampule in a tightly sealed glass vial at –24 °C up to 3 months upon opening.

8. Oxonol VI stock solution (1 mM): A small sample (1–2 mg) of the dye Oxonol VI (Molecular Probes, Inc., Eugene, OR) is weighed into a small glass flask and an appropriate volume of ethanol (spectroscopic quality) is added to obtain a 1 mM stock solution. The molar mass of Oxonol VI is 316.4 g. If e.g., 1.5 mg of Oxonol VI are in the glass flask, 4.74 mL of ethanol (spectroscopic quality) are added. The flask with the stock solution is wrapped with aluminum foil to keep it dark. It is tightly sealed and stored at –24 °C (to avoid evaporation of ethanol). Before withdrawal of aliquots from the stock solution, it has to be warmed up to room temperature before opening to avoid condensation of water in the flask.

9. Measurement buffer: Standard buffer with 70 mM Na_2SO_4, and 5 mM K_2SO_4, pH 7.2 (unless the Na^+ or K^+ concentration dependence is studied). Prepare as described for the Standard buffer, add 0.87 g K_2SO_4 and 9.94 g Na_2SO_4 before dissolving all components in 1 L DD water.

2.2 Solutions for Experiments with KdpFABC Liposomes

1. Working buffer: 25 mM imidazole, 1 mM ethylenediaminetetraacetic acid (EDTA), 5 mM $MgSO_4$, pH 7.2. To prepare 1 L of buffer weigh 1.70 g imidazole, 0.29 g EDTA, and 0.60 g

MgSO$_4$, and transfer all components into a 1 L beaker glass, add 950 mL DD water and stir with a magnetic stirrer until all components are completely dissolved. Adjust pH to 7.2 with 1 M H$_2$SO$_4$. (While adjusting pH, immerse the electrode only for short periods of time to avoid K$^+$ leaking from the reference electrode.) Transfer the pH-adjusted buffer into a 1 L graduated flask and fill up to 1 L with DD water. Store at 4 °C.

2. Dialysis buffer 2: Working buffer with 70 mM K$_2$SO$_4$. Prepare as described for the Working buffer, add 12.20 g K$_2$SO$_4$ before dissolving all components in DD water.

3. Detergent stock solution mixture: 10 % β-DDM and 10 % C$_{12}$E$_8$ in Dialysis buffer 2. To prepare 1 mL of detergent mixture solution measure 0.10 g β-DDM and 0.10 g of C$_{12}$E$_8$ and add to 0.5 mL of the Dialysis buffer 2. Stir with a magnetic stirrer until a clear solution is obtained, and fill up to 1 mL with Dialysis buffer 2. Store at –24 °C.

4. DiSC$_3$(5) stock solution: To prepare a solution of 100 mM DiSC$_3$(5) (3,3′-dipropylthiadicarbocyanine iodide) weigh 0.055 g of DiSC$_3$(5) and dissolve in 1 mL of DMSO. Store in a black Eppendorf vessel tightly sealed with parafilm at 4 °C. For fluorescence experiments prepare a 0.3 mM DiSC$_3$(5) solution by 300-fold dilution of the stock solution in DMSO. To prepare the diluted solution, the stock solution has to be warmed up to room temperature before opening to avoid condensation of water in the flask. The effective concentration of the DiSC$_3$(5) solution can be checked spectroscopically. The molar extinction coefficient is 258,000 cm^{-1} M^{-1}.

5. Dialysis buffer R: Standard buffer with 0.5 mM K$_2$SO$_4$, and 74.5 mM Na$_2$SO$_4$, pH 7.2. Prepare as described for the Standard buffer, add 0.087 g K$_2$SO$_4$ and 10.58 g Na$_2$SO$_4$ before dissolving all components in 1 L DD water.

6. Measurement buffer L: Working buffer with 0.14 mM K$_2$SO$_4$. Prepare as described for the Working buffer, add 24 mg K$_2$SO$_4$ and 19.86 g Na$_2$SO$_4$ before dissolving all components in 1 L DD water. Use pH-indicator strips instead of a pH electrode to avoid K$^+$ contaminations from the electrode.

2.3 Other Solutions

1. Na$_2$-ATP stock solution: To obtain a 1 M ATP solution, 0.551 g of Na$_2$-ATP are weighed into a 1 mL graduated flask and DD water added up to 1 mL. ATP is rapidly dissolved. With a μL droplet on a paper strip pH indicator one can see that the pH is extremely acidic. Additions of ATP in the order of 2.5 μL do not change significantly pH in a buffered solution. If, however, pH adjustment is desired, add only 900 μL DD water to the dry substance of ATP, add 5 μL aliquots of 1 M NaOH and check pH after each addition by applying μL samples on pH-indicator

paper until you obtain a value in the range between 6 and 7. Finally, add DD water to obtain 1 mL total volume. Split the volume in aliquots of 10 μL in separate Eppendorf vessels and keep them frozen at –24 °C until used. When thawed for use, keep on ice and discard it at the end of the day. At pH > 6 auto-hydrolysis of ATP is considerable after a day.

2. Mg-ATP stock solution: To prepare a 0.5 M Mg-ATP stock solution weigh 0.25 g of Mg-ATP and dissolve in 1 mL of DD water. Stir with a magnetic stirrer until the compound is dissolved. If the measurements require a constant pH in the cuvette and traces of Na^+ ions are not perturbing, the pH of the ATP solution should be adjusted to ~6–7. In this case, dissolve the salt in 900 μL DD water, then add 5 μL aliquots of 1 M NaOH and check pH after each added drop using the indicator paper. Finally, add DD water up to 1 mL. Distribute the volume in aliquots of 10 μL in separate Eppendorf vessels and keep them frozen at –24 °C until used. When thawed for use, keep on ice and discard at the end of the day. At pH > 6 auto-hydrolysis of ATP is considerable.

3. Sodium orthovanadate solution for P-type ATPase inhibition: *The substance and its solution are toxic! Apply every appropriate precaution to avoid skin contact and ingestion!* To prepare 2 mL of a 200 mM Na_3VO_4 solution, weigh 109 mg of the dry substance into a small glass beaker and dissolve it in 1.9 mL DD water. A clear solution is obtained. Add 40 μL of 10 M NaOH to the solution to obtain pH ~12 (check with 3 μL solution pipetted on a paper strip pH indicator). Warm up the solution to 95 °C for 5 min and let it cool down to room temperature afterwards. Then add about 80 μL 25 % HCl to obtain pH ~7.5. The solution becomes clear and of bright orange color. Warm up the solution again to 95 °C for 10 min. During this time the solution becomes colorless and its pH rises to ~8. When cooled to room temperature, add approx. 8 μL 25 % HCl to obtain pH 7.5. The solution becomes orange colored. Warm up again to 95 °C for 10 min. The solution becomes clear and colorless, and after cooling pH remains at 7.5. The solution shall be divided in small aliquots of 50–100 μL. These aliquots of stock solution can be kept frozen at –24 °C for long times. Concentration of 0.1–2 mM *o*-vanadate in the cuvette will inhibit P-type ATPases almost completely.

4. Valinomycin stock solution: To prepare a 1 mM stock solution, weigh 11.1 mg valinomycin dry substance ($M = 1111.3$ g/mol) into a 10 mL graduated flask and add ethanol (spectroscopic quality) up to 10 mL. The stock solution is stored well sealed at –24 °C to avoid evaporation of the ethanol. For use in the fluorescence experiments diluted solutions may be used. For this purpose, small aliquots are taken from the stock at room

temperature and diluted by a factor of 20 to obtain a working solution of 50 μM. Use e.g., a micropipette to transfer 5 μL of the stock solution into a small Eppendorf vessel and add 95 μL ethanol (spectroscopic quality). Keep the vessel closed and on ice during the experimental session to minimize evaporation.

5. K_2SO_4 stock solution for membrane-potential calibration: To prepare 50 mL of a 3 M K_2SO_4 solution weigh 26.14 g of K_2SO_4 and dissolve in 40 mL of DD water. Fill the solution into a 50-mL graduated flask and make up to 50 mL with DD water. pH adjustment is not necessary. Split the volume in aliquots of 1 mL in separate Eppendorf vessels and keep them frozen at −24 °C until used.

3 Methods

3.1 Na,K-ATPase Liposomes

1. Enzyme preparation: Na,K-ATPase is prepared from outer medulla of rabbit kidneys according to procedure C of Jørgensen [19] (*see* also Chap. 2). This method yields purified enzyme in the form of open membrane fragments that consist of about 0.6 mg phospholipids and 0.2 mg cholesterol per mg protein [19, 20]. More than 98 % of the ATPase activity can be inhibited by ouabain, and specific ATPase activity is between 1800 and 2300 μmol P_i per hour and mg protein at 37 °C. The ATPase activity is determined by the pyruvate/lactate dehydrogenase assay [21] or the Malachite green test [22] (*see* also Chaps. 10–12), and the protein concentration by the Lowry method [23]. Typical protein concentrations at the end of the purification procedure are 3–5 mg/mL.

2. Determine the amount of enzyme to be solubilized: The amount of Na,K-ATPase applied for reconstitution controls the number of pump molecules reconstituted in the liposomes, and thus, the magnitude of the ion flux. Using an enzyme with a specific activity of 2000 μmol P_i per hour and mg protein at 37 °C, a reasonable flux for Oxonol VI experiments is obtained when one starts with an aliquot of purified membrane fragments that contains 300 μg Na,K-ATPase.

3. The enzyme suspension (about 100 μL) is centrifuged in an appropriate centrifuge for small volumes at ~160,000 × g for 15 min.

4. Aspirate the supernatant with a micropipette and discard it (note the volume and calculate the volume of the pellet!).

5. Add an aliquot of the cholate buffer to the pellet so that the protein concentration is 2 mg/mL (*see* **Note 1**).

6. To homogenize the protein suspension, pipet the contents up and down 10 times with a micropipette in the centrifuge tube.

7. Let the cholate react with the membrane fragment for 10 min at room temperature in the centrifuge tube.
8. Centrifuge the suspension again under the same conditions as given above.
9. After centrifugation, three phases are visible in the centrifugation tube: a pellet of undissolved membrane fragments, a slightly turbid layer floating above the pellet, and a major clear supernatant. Only both liquid layers are collected together (~140–145 μL). They contain the solubilized Na,K-ATPase. The pellet has to be avoided and can be discarded.
10. Liposome formation: Mix all of the suspension of solubilized Na,K-ATPase with 100 μL cholate-containing lipid stock solution, vortex the mixture for several seconds.
11. Transfer the mixture into a ca. 150 mm long pretreated dialysis tubing (*see* **Note 2**). The tubing is closed at both ends by a clamp or a knot.
12. The tubing is immersed into at least 200 mL Dialysis buffer 1 and dialyzed for 72 h or longer under continuous gentle stirring (magnetic stirrer) at 4 °C. Longer periods of dialysis or an exchange of the Dialysis buffer 1 may slightly decrease the leak conductance of the liposomes. The enzyme activity is not affected by the duration once the liposomes are formed.
13. After dialysis, the suspension is removed from the tubing by a micropipette with a long pipette tip.
14. The volume is measured to check possible concentration or dilution of the liposome suspension. If the volume remained constant, the lipid concentration in the liposome preparation is ~8 mg/mL. The liposome suspension is kept in a 500 μL Eppendorf vessel.
15. To determine the specific enzyme activity of the reconstituted ion pumps, enzyme activity and protein concentration may be determined (*see* **Note 3**). The specific activity of the reconstituted Na,K-ATPase is typically in the order of 500–600 μmol P_i per hour and mg protein at 37 °C when PC 18:1 is used as lipid (*see* **Note 4**).
16. Keep the liposomes on ice for about 3 h before use. This treatment decreases the leak conductance of the liposome membrane.
17. Liposome suspensions are stored at 4 °C (or on ice) and may be used for several days without significant loss of enzyme activity. Never freeze liposome suspensions because the formation of ice crystals affects the liposome suspension severely!

Detailed studies have shown that the liposome suspension consists of liposomes of different diameter and different number of ATPases reconstituted per liposome. This liposome heterogeneity affects the time course of the pump-induced fluxes across the membrane [7, 24, 25], which has to be taken into account for detailed discussion of the experiments. In the mentioned references, explicit calculations for fluorescence experiments are presented that may be used for further application.

3.2 Na,K-ATPase Liposomes for Membrane-Potential Calibration

1. Perform **steps 1–11** exactly as described in Subheading 3.1.
2. The tubing is immersed into at least 200 mL Dialysis buffer R and dialyzed for 72 h or longer under continuous gentle stirring (magnetic stirrer) at 4 °C. Longer periods of dialysis or an exchange of the Dialysis buffer 1 may slightly decrease the leak conductance of the liposomes.
3. Follow **steps 13–17** as described in Subheading 3.1 (*see* **Note 5**).

3.3 KdpFABC Liposomes

1. Enzyme preparation: The C-terminally His_{14}-tagged KdpFABC complex is expressed in *E. coli* as described previously [15]. The enzyme complexes are solubilized with β-DDM, and further purified using affinity chromatography. Subunit composition is analyzed on 12.5 % SDS-PAGE gel. The protein concentration is determined by the Lowry method [23], and typical protein concentrations at the end of the purification procedure are 2–5 mg/mL. ATPase activity is measured using the Malachite-green assay [22], and the specific activity is typically ~0.53 µmol P_i mg^{-1} min^{-1} in the presence of saturating K^+ and ATP concentrations at 37 °C. The amount of protein needed for a single measurement is ~1 µg.
2. To produce a liposome suspension for 1 day of measurements, take 120 µL of 25 mg/mL chloroform solution of *E. coli* lipids and vacuum-dry it under rotation in a glass flask.
3. 240 µL Dialysis buffer 2 are mixed with 60 µL of the Detergent stock solution mixture in an Eppendorf vessel to obtain 300 µL Dialysis buffer 2 supplemented with 2 % β-DDM and 2 % $C_{12}E_8$.
4. The thin lipid film obtained at the wall of the glass surface is dissolved in the 300 µL supplemented Dialysis buffer 2 to a final lipid concentration of 10 mg/mL.
5. The lipid/detergent mixture is sonicated under a nitrogen atmosphere (to prevent lipid peroxidation) and kept at room temperature until a clear solution is obtained, typically 2–3 h are sufficient.
6. Dilute the detergent-solubilized KdpFABC to a final concentration of 2 mg/mL with Dialysis buffer 2.
7. Mix 300 µL of 2 mg/mL KdpFABC and 300 µL of 10 mg/mL lipid/detergent solution to obtain a protein/lipid ratio of 1:5 (w/w) (*see* **Note 6**).

8. Add to 200 mg of the pretreated Bio-Beads to remove the detergent (see **Note 7**).

9. Leave the mixture in a sealed reaction vessel overnight under constant rotation at 40 rpm at 4 °C.

10. Finally, the liposome suspension is separated from the sedimented beads using capillary tips.

11. To minimize the leak conductance of the liposome membrane, the liposome suspension is kept on ice for about 3 h before starting the experiments. It is stored on ice during the experiments and is used only during the same day, since already after 24 h a leak conductance of the *E. coli*-lipid membranes is significantly increased.

The liposome diameter can be checked by dynamic light scattering (see **Note 8**). The ATPase activity of the reconstituted KdpFABC can be determined in the absence of detergents using, e.g., the malachite green assay. Under this condition only the activity of the inside-out oriented fraction of the KdpFABC complexes is measured due to the membrane impermeability to ATP. A typical ATPase activity is ~0.28 μmol P_i mg^{-1} min^{-1} in the presence of 70 mM K_2SO_4 and 5 mM $MgSO_4$ at pH 7.2 and 37 °C. It has to be taken into account that approximately 50 % of the reconstituted pumps is oriented inside-out and contributes to the enzyme activity. Therefore, the apparently reduced enzyme activity is in good agreement with the twofold larger activity obtained from the detergent-solubilized enzyme.

3.4 Choice of the Fluorescent Dye

When ion fluxes across the liposome membrane have to be detected, a choice of the fluorescent dye has to be made with respect to the direction in which net charge is transported. If the liposome interior is charged up positively by the pump, as in the case of the Na,K-ATPase, an anionic dye, e.g., Oxonol VI, has to be chosen. If the interior is charged up negatively, as in the case of the KdpFABC complex, a cationic dye is needed, e.g., $DiSC_3(5)$. Due to the detection mechanism, the sensitivity of the dyes is rather low when used for experiments in which the respective opposite membrane potential is generated.

To minimize noise and disturbances of the fluorescence signals, the cuvettes and the micropipette tips have to be flushed prior use carefully with compressed clean air to make sure that all fluff and dust particles are removed that frequently adhere the surfaces due to electrostatic attraction.

3.5 Oxonol VI Experiments

1. Experiments are performed in a fluorescence spectrophotometer with a thermostatically controlled cuvette holder equipped with a magnetic stirrer. Fluorescence cuvettes with a light-path length of 1 cm are used. A typical temperature for the experiments

is 20 °C. The excitation wavelength is set to 580 nm (slit width 20 nm) and the emission wavelength to 660 nm (slit width 20 nm) with an integration time of 1 s.

2. Take a bucket, fill it with crunched ice and have in it ready for use during the session:

    (a) Na,K-ATPase containing liposomes prepared in Dialysis buffer 1 (*see* Subheading 3.1).

    (b) About 100 µL of the 30 µM Oxonol VI solution in ethanol, prepared daily from stock solution (*see* **Note 9**).

    (c) 40 µL Na_2-ATP solution (250 mM). It is obtained by thawing an Eppendorf vessel containing 10 µL Na_2-ATP stock solution (1 M), add 30 µL DD water and vortex the mixture.

3. Take about 50 mL of the Measurement buffer, perform sterile filtration into a small glass flask that has been made dust free by blowing it with compressed air before filling.

4. 1 mL of measurement buffer is filled into a fluorescence cuvette together with a stirring magnet, placed into the cuvette holder and thermally equilibrated. (Depending on the thermal coupling between cuvette and cuvette holder this may need 3–10 min.)

5. Start fluorescence recording before addition of Oxonol VI.

6. Add 1 µL Oxonol VI solution and wait until a constant fluorescence signal is obtained (cf. Fig. 1).

7. Add an appropriate amount of the Na,K-ATPase containing liposome suspension to obtain ~20 ng lipid in the cuvette (at 8 mg/mL lipid concentration it would be 2.5 µL). Wait until a constant fluorescence signal is obtained.

8. Add 1 µL of the ATP solution to the ion pumping (*see* **Note 10**).

9. This process is observable by a rapid increase of the fluorescence (Fig. 1). To obtain detailed information of the transport properties of the Na,K-ATPase, such experiments can be performed at varying substrate conditions such as the Na^+ and K^+ concentrations in the measurement buffer (or inside the liposome preparations), ATP concentration, buffer pH, and temperature. Examples may be found in various publications, e.g., [17, 18, 26].

10. Fluorescence recording may be stopped after the elevated fluorescence level is constant for a few minutes.

3.6 DiSC$_3$(5) Experiments

1. Experiments are performed in a fluorescence spectrophotometer with a thermostatically controlled cuvette holder equipped with a magnetic stirrer. Fluorescence cuvettes with a light-path length of 1 cm are used. A typical temperature for the

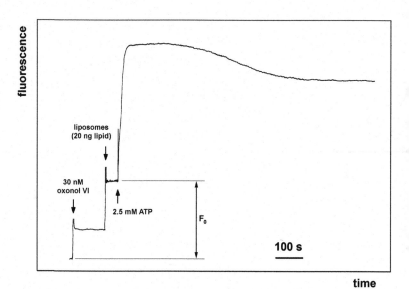

Fig. 1 Oxonol VI experiment with Na,K-ATPase-containing liposomes. To the thermally equilibrated 1 mL of measurement buffer, consecutively 1 μL Oxonol VI solution and 3 μL liposome suspension were added. The resulting fluorescence level is termed "F_0" and used as reference to normalize the fluorescence which is measured in arbitrary units. Pump activity was started by addition of 2.5 μL Na_2-ATP solution. After ~50 s a plateau level was reached that was stable for ~180 s before inset of a transition to a lower constant level that remained for at least 30 min. This transition was the consequence of an intra-liposomal depletion of K^+ ions which caused a change of the Na,K-ATPase function from the normal Na-K mode to the so-called Na-only mode with a significantly lower pump rate. In this experiment the time resolution was 1 s, i.e., the fluorescence signal was averaged for 1 s per each point recorded. The temperature was 18 °C

experiments is 18 °C. The fluorescence is excited at 650 nm (5 nm slit), and emission is detected at 675 nm (5 nm slit) with an integration time of 1 s.

2. Take a bucket, fill it with crunched ice and have in it ready for use during the session:
   (a) KdpFABC containing liposome suspension prepared in Dialysis buffer 2 (*see* Subheading 3.3).
   (b) About 50 μL of the 300 μM $DiSC_3(5)$ solution in DMSO.
   (c) 20 μL Na_2-ATP solution (500 mM). It is obtained by thawing an Eppendorf vessel containing 10 μL Na_2-ATP stock solution (1 M), add 10 μL DD water and vortex the mixture.
   (d) 20 μL valinomycin solution (1 mM in ethanol).
   (e) 50 μL of the 200 mM Na_3VO_4 solution.

3. Take about 50 mL of the Dialysis buffer 2, perform sterile filtration into a small glass flask that has been made dust free by blowing it with compressed air before filling.

4. 1 mL of Dialysis buffer 2 is filled into a fluorescence cuvette together with a stirring magnet, placed into the cuvette holder and thermally equilibrated at 18.0 ± 0.1 °C. (Depending on the thermal coupling between cuvette and cuvette holder this may need 3–10 min.)

5. Start fluorescence recording before addition of the liposome suspension.

6. Add an aliquot of the liposome suspension corresponding to the final lipid concentration of 80 µg/mL.

7. Shortly thereafter (~30 s) an aliquot of 1 mM $DiSC_3(5)$ is added, to obtain a dye concentration of 300 nM in the cuvette (*see* **Note 11**). After an initial, fast fluorescence increase a stable signal is achieved, typically within less than a minute.

8. Add 1 µL of 0.5 M Mg-ATP solution to trigger the pump activity.

    Extrusion of positive charge by the enzyme complexes generates an inside-negative potential, reflected in a fluorescence decrease (Fig. 2a), which approaches exponentially a steady-state level at which the pump current is compensated by the leak current due to the passive membrane conductance and the electric-potential gradient. Follow the signal until a constant fluorescence level is obtained.

9. Add 1 µL of the valinomycin solution (final concentration is 1 µM) (*see* **Note 12**). This potassium ionophore causes a stepwise fluorescence increase to a level that represents the Nernst potential determined by the actual K^+ concentration inside and outside the liposomes, according to Eq. 1.

10. Alternatively, add 5 µL of the *o*-vanadate solution to a final concentration of 0.1 mM of the inhibitor. The result of the pump inhibition is an exponential increase of the fluorescence signal (Fig. 2b).

    The time course can be used to evaluate the leak conductance of the liposome membrane. Extracting the time constant, τ, of the exponential increase allows the determination of the specific leak conductance of the membrane (*see* **Equation 3** in Subheading 3.9). The average value of the specific leak conductance of *E. coli*-lipid membrane was determined to be 21 ± 1 nS/cm^2 at pH 7.2. This parameter varied only insignificantly between different liposome preparations, and different pH of Dialysis buffer 2.

11. Fluorescence recording may be stopped after the fluorescence level is constant for a few minutes.

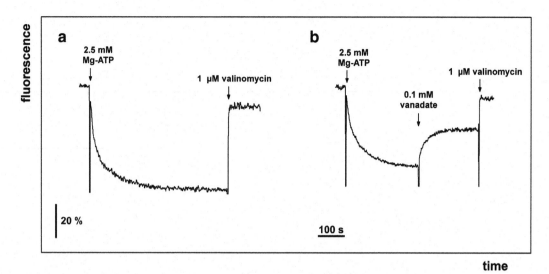

Fig. 2 Electrogenic pump activity of KdpFABC containing liposomes detected by the membrane-potential indicator DiSC$_3$(5). Recording of the fluorescence began after the equilibration of an aliquot of liposomes (80 μg/mL) and 1 mM DiSC$_3$(5) in 1 mL of Dialysis buffer 2. (**a**) The initial K$^+$ concentration was 140 mM inside and outside. Addition of ATP triggers K$^+$ pumping out of the liposomes. The increasing inside-negative potential is reflected by a fluorescence decrease. Addition of valinomycin induces a K$^+$-selective conductance of the liposome membrane that short-circuits the pump activity, and the resulting steady-state fluorescence level attains to the Nernst potential (Eq. 1), which is controlled by the ratio of K$^+$ concentrations inside and outside the liposomes. (**b**) When the ATP-induced steady state was reached, a non-saturating concentration of the inhibitor *o*-vanadate was added, and the membrane potential was reduced to a new steady-state level according to the reduced pump activity and the leak conductance of the membrane. Addition of valinomycin again leads to a fluorescence level representing the Nernst potential

3.7 Calibration of the Membrane Potential

1. Introductory remarks: The mechanism of the slow-response dyes is a redistribution of the dye molecules between the inside and outside (or to be more precise: particularly between inner and outer layer) of the lipid bilayer. Because of this fact the fluorescence response of the dyes depends on various parameters such as the total dye concentration, the ratio of lipid and dye concentration in the cuvette, and the liposome size. Therefore, it is not possible to use a fixed calibration curve that relates fluorescence response and membrane potential for all liposome preparations. A reliable method to calibrate the membrane potential was introduced in 1974 by Hoffman and Laris [27].

   The principle exploited by their method is that liposomes with a defined intravesicular K concentration, $[K^+]_{in}$, are added to a buffer of the K$^+$ concentration $[K^+]_{out}$. In the presence of 1 μM valinomycin within seconds an electric potential, V, arises which corresponds to the K$^+$ equilibrium (or Nernst) potential, E_K, calculated according to

$$V = E_K = \varphi_{in} - \varphi_{out} = \frac{RT}{F} \cdot \ln \frac{[K^+]_{out}}{[K^+]_{in}} \quad (1)$$

where $R = 8.31$ J/(mol K) is the gas constant, T is the absolute temperature, $F = 96{,}485$ C/mol the Faraday constant.

2. Choice of the sign of voltage: When inside-negative potentials shall be calibrated, liposomes should be prepared in 140 mM K^+, whereas for inside-positive potential liposomes should be prepared in 1 mM K^+ plus 149 mM of an inert cation such as $Tris^+$ to maintain the same ionic strength (*see* **Note 13**). The K^+ concentration on the outside will be varied between 1 mM and 140 mM.

   To perform calibration experiments of inside-positive membrane potentials with Oxonol VI, *see* below **steps 3–12**. To perform calibration experiments of inside-negative membrane potentials with $DiSC_3(5)$, *see* below **steps 13–22**. The analysis of the calibration experiments is described in Subheading 3.9, **step 2**.

3. Oxonol experiments are performed according to Subheading 3.5. In detail:

4. Take a bucket fill it with crunched ice and have in it ready for use during the session:

   (a) Na,K-ATPase containing liposomes are prepared for membrane-potential calibration (*see* Subheading 3.2).

   (b) About 100 µL of the 30 µM Oxonol VI solution in ethanol, prepared daily from stock solution (*see* **Note 9**).

   (c) 200 µL of a 50 µM valinomycin solution (take 10 µL of the 1 mM stock solution transfer it into an Eppendorf vessel and add 190 µL ethanol of spectroscopic quality).

   (d) 1 mL of 3 M K_2SO_4 solution.

   (e) 1 mL of 300 mM K_2SO_4 solution which is obtained by transferring 100 µL of the 3 M K_2SO_4 solution into an Eppendorf vessel, adding 900 µL DD water and vortexing the solution.

5. Take about 50 mL of the Dialysis buffer R, perform sterile filtration into a small glass flask that has been made dust free by blowing it with compressed air before filling.

6. 1 mL of Dialysis buffer R is filled into a fluorescence cuvette together with a stirring magnet, placed into the cuvette holder and thermally equilibrated (3–10 min).

7. Start fluorescence recording before addition of Oxonol VI.

8. Add 1 µL Oxonol VI solution and wait until a constant fluorescence signal is obtained.

9. Add an appropriate amount of the Na,K-ATPase containing liposome suspension to obtain ~20 ng lipid in the cuvette (at 8 mg/mL lipid concentration it would be 2.5 μL). Wait until a constant fluorescence level is obtained.

10. Add 1 μL of the valinomycin solution. (In some cases the addition of valinomycin causes a small change in the fluorescence intensity despite that fact that inside and outside electrolytes of the liposomes have the same composition and the membrane potential is 0. This is an artifact, and the F_0 level for normalization has to be taken after addition of valinomycin.)

11. Aliquots of the K_2SO_4 solutions are added which cause a rapid fluorescence increase because, according to Eq. 1, the elevated K^+ concentration induces an inside-positive potential (*see* **Note 14**). Between additions, an appropriate waiting time has to be chosen to be able to determine resulting fluorescence level.

12. Fluorescence recording is stopped after the final fluorescence level is obtained and remains stable.

13. $DiSC_3(5)$ experiments are performed accordingly to Subheading 3.6. In detail:

14. Take a bucket fill it with crunched ice and have in it ready for use during the session:

    (a) KdpFABC containing liposomes (*see* Subheading 3.3).

    (b) About 50 μL of the 300 μM $DiSC_3(5)$ solution in DMSO.

    (c) 50 μL valinomycin of the 1 mM stock solution.

    (d) 1 mL of 3 M K_2SO_4 solution.

    (e) 1 mL of 300 mM K_2SO_4 solution which is obtained by transferring 100 μL of the 3 M K_2SO_4 solution into an Eppendorf vessel, adding 900 μL DD water and vortexing the solution.

15. Take about 50 mL of the Measurement buffer L, perform sterile filtration into a small glass flask that has been made dust free by blowing it with compressed air before filling.

16. 1 mL of Measurement buffer L is filled into a fluorescence cuvette together with a stirring magnet, placed into the cuvette holder and thermally equilibrated.

17. Start fluorescence recording before addition of the liposome suspension.

18. Add an aliquot of the liposome suspension corresponding to the final lipid concentration of 80 μg/mL.

19. Shortly thereafter (~30 s) an aliquot of 1 mM $DiSC_3(5)$ is added, to obtain a dye concentration of 300 nM in the cuvette. After an initial, fast fluorescence increase a stable signal is achieved, typically within less than a minute.

20. Add 1 µL of the 1 mM valinomycin solution. Wait until a steady-state fluorescence is reached (*see* **Note 15**).

21. Aliquots of the K_2SO_4 solutions are added which cause a rapid fluorescence increase because, according to Eq. 1, the elevated K^+ concentration reduces the inside-negative potential. Between additions, an appropriate waiting time has to be chosen to be able to determine resulting fluorescence level (*see* **Note 16**).

22. Fluorescence recording is stopped after the last fluorescence level is obtained.

3.8 Inhibition of P-Type ATPases by o-Vanadate

1. Introductory remarks: P-type ATPases can be effectively inhibited from the cytoplasmic side by *o*-vanadate, which acts as a transition-state analog of phosphate. *o*-Vanadate binds in the presence of Mg^{2+} to the dephosphorylated E_2 state with a high affinity in the micromolar concentration range. Since in the case of ATPases reconstituted in liposomes, only those pumps that face the outside with their cytoplasmic side are activated by ATP, they may be easily inhibited by *o*-vanadate. While the binding affinity for *o*-vanadate in isolated enzyme preparations is in the order of 1–10 µM, it was found that the (apparent) binding affinity of the enzymes reconstituted in liposomes is significantly lower. Therefore, it is important to determine the precise half-inhibition constant, K_I, which can be derived from the enzyme activity in the presence of various *o*-vanadate concentrations between 0 and 5 mM.

2. Determine the ATPase activity by the pyruvate/lactate dehydrogenase assay for the Na,K-ATPase or the Malachite green test for the KdpFABC in the presence of different *o*-vanadate concentrations.

3. Plot the enzyme activity against the applied *o*-vanadate concentration.

4. Fit the dependence of the enzyme activity, E_A, on the *o*-vanadate concentration, c_{van}, by a Michaelis–Menten curve which provides K_I as half-inhibiting constant.

$$E_A(c_{van}) = E_A^{max} \cdot \frac{K_I}{K_I + c_{van}} \quad (2)$$

5. For the inhibition experiments with liposomes, a concentration of $\sim 30 \cdot K_I$ should be chosen. Under this condition $\sim 97\%$ of the enzyme is inhibited and the residual pump activity is no longer significant.

6. Perform an Oxonol VI experiment as described in Subheading 3.5, **steps 1–8**. After a constant fluorescence level is obtained, add a maximal-inhibiting amount of *o*-vanadate (e.g., 2 mM).

The fluorescence signal will decay exponentially to $V = 0$. When the fluorescence is constant, the recording is stopped. An example is shown in Fig. 3.

3.9 Analysis of Fluorescence Experiments

1. Normalization: The amplitude of the fluorescence signals is an extensive state variable, i.e., it depends on the amount of dye and liposomes added to the cuvette. Since the added volumes are in the order of µL, small but nevertheless significant variations of the additions may occur between different experiments and thus affect the fluorescence amplitude. Therefore, the fluorescence traces have to be normalized to allow a direct comparison between various experiments. This is achieved by relating the fluorescence changes to the specific initial experimental condition, when only the dye and the liposomes are in the cuvette. The corresponding fluorescence level is termed F_0 (see Fig. 1). The normalized fluorescence signal, $F(t)_{norm}$, is then calculated by:

$$F(t)_{norm} = \frac{F(t) - F_0}{F_0} \qquad (3)$$

By this normalization method, the fluorescence level is 0 before the addition of ATP (or any other substrate).

2. Calibration of the membrane voltage: The fluorescence signals in experiments obtained according to Subheading 3.7 are normalized (see Eq. 3). The normalized fluorescence levels of each K^+ concentration is plotted against the calculated respective Nernst potential (Eq. 1). Examples of such a voltage calibration are shown for Oxonol VI in Fig. 4 and for $DiSC_3(5)$ in Fig. 5. These plots allow the conversion of fluorescence into voltage.

   In the presence of a lipid concentration of 20 µg/mL and 30 nM Oxonol VI, the calibration curve is almost linear in the voltage range between 0 and 100 mV [17]. With a linear calibration curve the membrane potential has not to be read from a best-fit curve through the experimental data but can be directly calculated. For the experimental data of Fig. 4, it would be (according to the equation of a line):

$$V = \frac{\Delta F / F_0 + 0.02}{0.0115} \, mV \qquad (4)$$

   Negative membrane potentials: In the case of the $DiSC_3(5)$ experiments, linearity of the calibration curve is obtained in the voltage range between −40 and −120 mV (Fig. 5). At low voltages ($V > -30$ mV), the sensitivity of this dye is poor.

3. Determination of the leak conductance: The leak conductance of liposome membranes can be easily determined when the

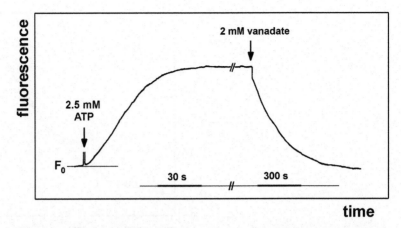

Fig. 3 Oxonol VI experiment with Na,K-ATPase-containing liposomes showing the discharge of the membrane capacitor upon inactivation of the ion pumps by o-vanadate as indicated by the decreasing fluorescence (i.e., membrane potential). The temperature was 12 °C

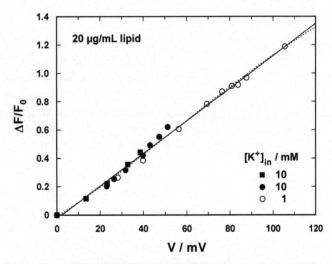

Fig. 4 Fluorescence–voltage calibration of Na,K-ATPase containing liposomes with Oxonol VI. Under the condition of 30 nM Oxonol VI and 20 µg/mL lipid, a linear calibration line is obtained which allows a simple conversion of fluorescence into membrane voltage. The *solid line* represents a linear regression with a slope of 0.0115 mV^{-1} and an ordinate intercept of −0.02. For comparison, a cubic regression line is added as *dotted line* to show that deviations from the linearity are not significant. The results of three experiments with different intralamellar K$^+$ concentrations are combined

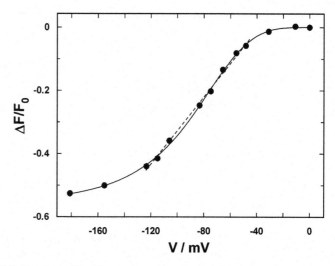

Fig. 5 Calibration of the fluorescence responses of DiSC$_3$(5) dye. Lipid vesicles without reconstituted KdpFABC pumps loaded with 140 mM K$^+$ were added to buffer containing 0.14 mM K$^+$. After equilibration, 1 μM valinomycin and subsequently aliquots of K$_2$SO$_4$ were added. The equilibrium potential, V, was calculated at each K$^+$ concentration according to Eq. 1 from both K$^+$ concentrations, inside and outside, and the fluorescence levels were plotted against the respective value of V. The *dashed regression line* indicates the range of a linear voltage–fluorescence relation

pump activity has been started by ATP and the steady-state membrane potential is obtained. In this state, the ion pumps are completely inhibited by addition of 2 mM *o*-vanadate. Under this condition only passive ion transport, namely the leak current, preferentially of K$^+$ ions, is observed which will degrade the membrane potential eventually to zero.

The fluorescence amplitude, which is proportional to the membrane potential, decays exponentially and can be fitted by the function, $F(t) = \Delta F_{max} \cdot \exp(-t/\tau)$. The characteristic time constant, τ, carries information on the specific leak conductance of the membrane. (The fact that the fluorescence amplitude drops below the starting value may be the result of a change of the interfacial potential of the membrane surface by *o*-vanadate binding to the lipid head groups, as reflected by the fast fluorescence step upon *o*-vanadate addition.) The calculation of the specific leak conductance is based on following considerations. When the pumps are inhibited, the liposome membrane can be treated as a parallel connection of the membrane capacitance, C_m, and the leak conductance, G_m. Since the membrane potential, V_m, is different from zero, a discharge of the membrane capacitor is observed, which is described, according to physics textbooks, as:

$$V_\mathrm{m}(t) = V_\mathrm{max} \cdot \exp\left(-\frac{t}{\tau}\right); \ \tau = \frac{C_\mathrm{m}}{G_\mathrm{m}} \quad (5)$$

Since the specific membrane capacitance, C_m, is a constant, commonly ~1 µF/cm^2, without significant changes for all lipid membranes, the specific leak conductance can be determined as $G_\mathrm{m} = C_\mathrm{m}/\tau$. With the characteristic time constant of ~200 s in the experiment shown in Fig. 3, G_m is 1 (µF/cm^2)/200 s = 5 nS/cm^2.

4. Initial slope analysis: When the fluorescence signals are recorded with a time resolution of 0.1 s (or faster), it is possible to obtain detailed information on the pump currents, I_P, generated by the ATPases in the liposome membrane. The following prescription is given for the Na,K-ATPase, however, it can be applied also to the KdpFABC complex in liposomes [28] or other reconstituted electrogenic ATPases. When ATP is added, the ATPases start to pump simultaneously, and in the beginning, i.e., before a considerable membrane potential is build up and the K$^+$ concentration in the liposomes becomes altered significantly, the pump is controlled by the rate-limiting reaction step of the pump cycle, which is Na$^+$ binding at low Na$^+$ concentrations, and the conformation transition, (Na$_3$)E$_1$-P → P-E$_2$Na$_3$ at saturating Na$^+$ concentrations.

To study the substrate dependence of the pump current, the following system properties have to be exploited. The equivalent circuit diagram of the liposome membrane can be represented as shown in Scheme 1. The lipid bilayer has the capacitance, C_m, and the leak conductance, G_m. The ion pumps are represented as current generator, I_P.

According to basic physics, these elements may be linked up in the following way:

$$Q = C_\mathrm{m} \cdot V; \ I = \frac{dQ}{dt} = C_\mathrm{m} \cdot \frac{dV}{dt}, \quad (6)$$

where Q is the electric charge on the capacitor, V the electric potential across the membrane, and I the current flowing across the membrane. The net current, I, is the sum of two components, the pump current, I_P, and the leak current, I_L, both flowing in opposing directions,

$$I = I_\mathrm{P} - I_\mathrm{L} = n_\mathrm{P} \cdot v_\mathrm{P} \cdot z_\mathrm{P} \cdot e_0 - G_\mathrm{m} \cdot V \quad (7)$$

where n_P is the number of Na,K-ATPases in the liposome membrane, v_P is the pump rate, z_P is the number of charges transferred per pump cycle (= 1 in the case of the Na,K-ATPase), and e_0 is the elementary charge. G_m is the leak conductance of

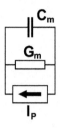

Scheme 1 Equivalent circuit of the liposome membrane. The membrane can be represented by a parallel connection of the membrane capacitance, C_m, the membrane conductance, G_m, and the current generator (ion pumps) that generates the current, I_p.

the liposome. Combining Eqs. 6 and 7 leads to an inhomogeneous differential equation,

$$\frac{dV}{dt} = \frac{I_P}{C_m} - \frac{G_m}{C_m} \cdot V \tag{8}$$

It is solved with the boundary condition that at long times, $t \to \infty$, $I = 0$. This leads to:

$$I = I_P - I_L = 0 \text{ and} \tag{9}$$

$$I_P^\infty = G_m \cdot V_\infty \tag{10}$$

where V_∞ is the stationary voltage level at which the pump current, I_P, is compensated by the leak current. The solution of the differential equation is then:

$$V(t) = V_\infty \cdot (1 - \exp(-\frac{t}{\tau})); \quad \tau = \frac{C_m}{G_m}; \quad V_\infty = \frac{I_P^\infty}{G_m} \tag{11}$$

It has been shown that the (normalized) fluorescence of Oxonol VI in liposome experiments, $\Delta F/F_0$, is proportional to the membrane potential built up by the Na,K-ATPase [17]. Therefore, the function in Eq. 11 may be used to fit the time course of the fluorescence signal upon addition of ATP.

During the initial phase of the experiments the substrate concentrations do not change significantly and the electric potential across the membrane is still small so that v_P may be assumed to be constant, and the pump current, I_P, is then constant too. Therefore, the initial slope of the fluorescence signal contains information on the pump current,

$$(\frac{dV}{dt})_{t=0} = V_\infty \cdot \frac{1}{\tau} \cdot \exp(-\frac{t}{\tau})|_{t=0} = \frac{V_\infty}{\tau} = \frac{1}{C_m} \cdot I_{P,t=0} \tag{12}$$

$$\left(\frac{dF_{norm}}{dt}\right)_{t=0} = K^* \cdot \left(\frac{dV}{dt}\right)_{t=0} = \frac{K^*}{C_m} \cdot I_{P,t=0} \qquad (13)$$

Both equations may be combined to obtain a conditional equation for I_P:

$$I_{P,t=0} = \frac{C_m}{K^*} \cdot \frac{F_{norm,\infty}}{\tau} = K^{\#} \cdot \frac{F_{norm,\infty}}{\tau} \qquad (14)$$

The initial pump current, $I_{P,t=0}$, is therefore directly proportional to the initial slope of the fluorescence signal, $(dF_{norm}/dt)_{t=0}$. The proportionality constant, $K^{\#}$, in Eq. 14 depends on the physical and chemical properties of the specific liposome preparation and it is invariable within the same preparation. Therefore, the results obtained from the same liposome preparation can be directly compared, at least within a time period in which no aging processes modify the enzyme activity. A fit of Eq. 11 to the ATP-induced fluorescence change of an experiment similar to Fig. 1 is shown in Fig. 6. An application of this method of analysis can be found in refs. 28, 29.

5. Fluorescence–voltage transformation: This step of analysis is exemplified for an experiment with Na,K-ATPase containing liposomes with the Oxonol VI approach. The time course of the fluorescence signal was recorded under the experimental condition of 20 µg/mL lipid and 30 nM Oxonol VI after the addition of 80 µM/mL Na$_2$-ATP. The fluorescence signal was normalized according to Eq. 3 and plotted against time (Fig. 7).

The experimental condition of Fig. 7 allowed the use of the fluorescence–voltage relation shown in Fig. 4 and quantified in Eq. 4 to calculate the membrane potential built up by the ion pump across the liposome membrane. In this liposome preparation, the maximal potential, i.e., when pump and leak current balance each other, was ~120 mV. During the first 2–3 s, a delayed increase of the fluorescence can be observed which is caused by the mixing process of the 4 µL of 20 mM ATP added to the 1 mL solution in the fluorescence cuvette stirred by a magnetic stirrer.

6. Determination of the current–voltage curves: Current–voltage (or *I–V*) curves of the ion pumps contain important information on the mechanism of ion transporters. Those *I–V* curves are typically measured by electrophysiological experiments with whole cells. In native membranes, in which numerous ion transporters are present, *I–V* curves have to be determined by subtracting the current contribution of all other transporters. Typically, this is obtained by performing the experiment without and with a specific pump inhibitor. The pump-specific action is calculated as the difference between both signals.

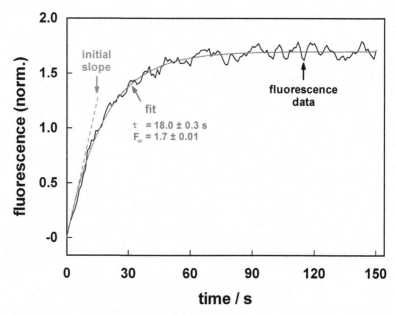

Fig. 6 Analysis of the ATP-induced Oxonol VI fluorescence increase upon addition of ATP to the Na,K-ATPase-containing liposome suspension. The fluorescence trace has been fitted with the function of Eq. 11, shown as *solid curve*. The initial slope (*dashed line*) is indicated in addition

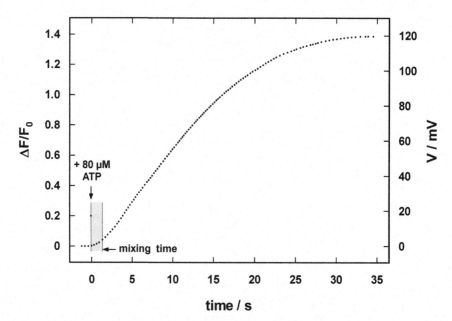

Fig. 7 Fluorescence–voltage transformation of Oxonol VI fluorescence. The digitized fluorescence trace is plotted against time. The *left ordinate*, $\Delta F/F_0$, represents the normalized fluorescence. The *right ordinate* represents the respective membrane potential calculated according to Eq. 4

An experiment as shown in Fig. 7 can be used to determine directly the I–V curve of the ion pump in the voltage range up to the maximal voltage without having to deal with other ion transporters. As has been discussed in detail in the literature, the number of pumps reconstituted in a single liposome varies around a mean value, and this value has to be determined from electron microscopic images [25]. Since, however, the number of ion pumps per liposome and the membrane area of the liposome, A, are voltage independent, a "mean" I–V curve can easily be calculated without knowing the number of active pumps reconstituted per liposome, n_P. Based on Eqs. 6 and 7, one obtains an equation that expresses the voltage dependence of the pump rate, $v_p(V)$, as function of parameters provided by the experiment:

$$n_p \cdot v_p(V) = \frac{A}{z_p \cdot e_0} \cdot \left(C_m \cdot \frac{dV}{dt} + G_m \cdot V \right) \tag{15}$$

Since the number of pumps, n_p, is constant (although unknown), the voltage dependence of the product $n_p \cdot v_p$ is studied. A is the area of the liposome membrane. For Na,K-ATPase-containing liposomes produced by dialysis a typical radius is $R = 45$ nm [7], therefore $A = 4\pi \cdot R^2 = 2.54 \times 10^{-10}$ cm^2. The specific capacitance is 1 µF/cm^2, and the leak conductance $G_m = 5$ nS/cm^2 (*see* **step 3** in Subheading 3.9). The number of elementary charges, e_0, translocated per pump cycle is $z_p = 1$. At each given data point, i, in Fig. 7 the voltage Vi is known after the fluorescence–voltage transformation, and the slope of the voltage, dV/dt, can be calculated. To reduce the noise of the resulting data, the mean of the slope before and after data point i is taken

$$\frac{dV_i}{dt} = \frac{1}{2}\left[\frac{V_i - V_{i-1}}{t_i - t_{i-1}} + \frac{V_{i+1} - V_i}{t_{i+1} - t_i} \right] \tag{16}$$

Using the numbers of V_i and (dV_i/dt), and the constants given above, $n_p \cdot v_p$ can be calculated for each V_i according to Eq. 16 and plotted against V_i (Fig. 8).

Due to the mixing period of ~3 s at the beginning of the fluorescence experiment (Fig. 7), the I/V curve cannot be determined for small voltages. However, in this voltage range no significant voltage dependence is expected, and the value of $n_p \cdot v_p$ can be extrapolated to $V = 0$.

4 Notes

1. Example: When you work with membrane fragments of 4 mg/mL protein concentration, you have to take 75 µL to get 300 µg protein. After centrifugation you remove with a micropipette

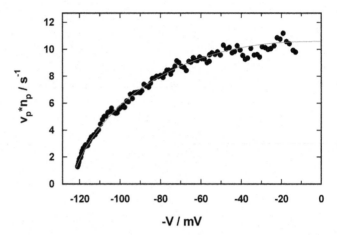

Fig. 8 Current–voltage dependence of the Na,K-ATPase determined from the fluorescence signal shown in Fig. 7. The *grey line* drawn through the data points is included to guide the eye

say 70 µL not to disturb the pellet. To obtain a final protein concentration of 2 mg/mL, the 300 µg protein have to be dissolved in 150 µL cholate buffer. Since the pellet has already 5 µL, one has to add 145 µL cholate buffer.

2. The dialysis tubing has a nominal dry wall thickness of 50 µm and a pore diameter of approximately 25 Å. To prepare the delivered dry and glycerol-treated tubing, a piece of ~1 m length is boiled for 30 min in 10 mM EDTA at pH 7.5. Then it is stored in 1 mM EDTA, pH 7.5, at 4 °C in a closed glass flask. For a dialysis, pieces of appropriate lengths are cut off. To avoid contamination of the remaining tubing, it is always touched only with gloves.

3. To determine the protein concentration of the Na,K-ATPase reconstituted in liposomes the Markwell method has to be used [30], which is a modified Lowry assay with sodium dodecyl sulfate (SDS).

4. The final activity depends significantly on the lipid composition of the membrane [31]. In addition, it has to be taken into account that only half of the reconstituted pumps are oriented cytoplasmic side out and the other half in opposite orientation. The latter is not accessible by ATP, and therefore, does not contribute to the measured enzyme activity.

5. The enzyme activity in this preparation cannot be determined accurately because the pumps are unable to perform the Na,K-mode due to the low internal K^+ concentration.

6. The lipid/protein ratio affects the size of the formed proteoliposomes and the actual protein reconstitution, and it has to be individually optimized. For the purposes of experiments performed with reconstituted KdpFABC in our lab, the most

suitable protein/lipid ratio was found to be 1:5 (w/w), which results in high fluorescence responses and a stable enzyme activity, lasting for about 1 day.

7. To pre-equilibrate Bio-Beads add 600 µL Dialysis buffer 2 to 200 mg of Bio-Beads, incubate for 30 min at room temperature. The buffer above the sedimented beads is removed before use by aspiration with a pipette and capillary tips.

8. Dynamic light scattering measurements are performed at room temperature, using a DLS Viscotek 802 spectrometer. Liposomes are diluted to achieve a final concentration of approximately 0.1 mg of lipid/mL of buffer, filtered with a 0.45 µm PVDF centrifugal filter (Millipore), and transferred to a 12 µL square cuvette (Viscotek, 802DLS quartz cell). Distilled water and Dialysis buffer 2 are filtered with 0.02 µm inorganic membrane filters (Whatman). Intensity distributions of the scattered light are recorded and analyzed with the program Omnisize (V. 2.0) to obtain the corresponding liposome diameters distribution. Typical liposome diameters are 117 ± 21 nm (SD) under the condition described.

9. To obtain the 30 µM working solution, 3 µL of the oxonol VI stock solution are pipetted into a black 500 µL Eppendorf vessel to which 97 µL ethanol (spectroscopic quality) are added. The effective concentration can be checked spectroscopically. The molar extinction coefficient is 136,000 cm^{-1} M^{-1}. The vessel is kept on ice during the experimental session. It is discarded at the end of the day. If no black Eppendorf vessel is available, the used vessel has to be wrapped in aluminum foil to prevent bleaching effects of the light in the lab.

10. A concentration of 250 µM in the cuvette is sufficient to make sure that ATP does not become limiting even for experiments lasting more than 30 min.

11. Both absorption and emission spectra of DiSC$_3$(5) depend on solvent, ionic strength, and concentration of the dye. Due to the mechanism of the potential detection, the DiSC$_3$(5) dye should be added to the liposome suspension (and not vice versa). In this manner, a constant fluorescence signal is reached already after 50 s, and a slow redistribution of the dye between non-fluorescent dye aggregates in the aqueous phase and the membranes is avoided. To minimize dye aggregation, it is essential to use the lowest amount of the dye that produces an acceptably large and stable signal. This should be tested in the used setup by checking various dye concentrations.

12. A limitation of the application of DiSC$_3$(5) consists in the interaction with a number of compounds commonly used as ionophores, including CCCP and dinitrophenol, which tend to produce non-fluorescent complexes with the dye [32, 33]. Valinomycin produces only minor artifacts.

13. When the condition of very low K^+ concentration in the liposomes is chosen, it has to be taken into account that the liposome volume is in the order of 4×10^{-18} L, and therefore, the number of K^+ ions is only in the order of a thousand or less. When the K^+ equilibrium potential is generated, the uptake of K^+ may increase significantly the K^+ concentration in the liposomes and has to be considered. The prescription for a correct calculation of very low initial K^+ concentrations at the Nernst equilibrium is provided in ref. 17.

14. According to Eq. 1, the Nernst potential can be calculated from the initial K^+ concentration in the liposomes and the actual K^+ concentration in the external solution. The following table provides some examples of K_2SO_4 additions and the calculated membrane potential. It is based on the initial condition of 1 mL buffer with 1 mM K^+ in the cuvette and liposomes containing 1 mM K^+.

Addition of K_2SO_4	$[K^+]_{out}$ / mM	V_m / mV
1 µL 0.3 M	1.6	11.9
2 µL 0.3 M	2.2	19.9
3 µL 0.3 M	2.8	26.0
4 µL 0.3 M	3.4	30.9
5 µL 0.3 M	4.0	35.0
1 µL 3 M	7.0	49.1
2 µL 3 M	13.0	64.7
5 µL 3 M	31.0	86.7
10 µL 3 M	61.0	103.7

15. This fluorescence level corresponds to the maximum inside-negative membrane potential. According to Eq. 1, under the initial condition of $[K^+]_{in}$ = 140 mM and $[K^+]_{out}$ = 0.14 mM, at 18 °C a Nernst potential, E_K, of –173 mV is obtained (Eq. 1).

16. To restrict the volume of K_2SO_4 additions and avoid significant dilution effects that would have to be corrected numerically, the experiments can be started also in buffers containing initially, e.g., $[K^+]_{out}$ = 1.4 or 14 mM.

Acknowledgement

This work was supported by the Konstanz Research School Chemical Biology, University of Konstanz, Germany, and the University of Konstanz (AFF 4/68).

References

1. Goldin SM, Tong SW (1974) Reconstitution of active transport catalyzed by the purified sodium and potassium ion-stimulated adenosine triphosphatase from canine renal medulla. J Biol Chem 249:5907–5915
2. Anner BM, Lane LK, Schwartz A, Pitts BJR (1977) A reconstituted Na$^+$+K$^+$ pump in liposomes containing purified (Na$^+$+K$^+$)-ATPase from kidney medulla. Biochim Biophys Acta 467:340–345
3. Karlish SJD, Pick U (1981) Sidedness of the effects of sodium and potassium ions on the conformational state of the sodium-potassium pump. J Physiol 312:505–529
4. Brotherus JR, Jacobsen L, Jørgensen PL (1983) Soluble and enzymatically stable (Na$^+$ + K$^+$)-ATPase from mammalian kidney consisting predominantly of protomer $\alpha\beta$-units. Preparation, assay and reconstitution of active Na$^+$, K$^+$ transport. Biochim Biophys Acta 731:290–303
5. Cornelius F, Skou JC (1984) Reconstitution of (Na$^+$+K$^+$)-ATPase into phospholipid vesicles with full recovery of its specific activity. Biochim Biophys Acta 772:357–373
6. Forbush B III (1984) An apparatus for rapid kinetic analysis of isotopic efflux from membrane vesicles and of ligand dissociation from membrane proteins. Anal Biochem 140:495–505
7. Apell H-J, Marcus MM, Anner BM, Oetliker H, Läuger P (1985) Optical study of active ion transport in lipid vesicles containing reconstituted Na K-ATPase. J Membr Biol 85:49–63
8. Fürst P, Solioz M (1986) The vanadate-sensitive ATPase of *Streptococcus faecalis* pumps potassium in a reconstituted system. J Biol Chem 261:4302–4308
9. Fendler K, Drose S, Altendorf K, Bamberg E (1996) Electrogenic K+ transport by the Kdp-ATPase of Escherichia coli. Biochemistry 35:8009–8017
10. Anner BM, Moosmayer M (1981) Preparation of Na, K-ATPase-containing liposomes with predictable transport properties by a procedure relating the Na, K-transport capacity to the ATPase activity. J Biochem Biophys Meth 5:299–306
11. Forbush B III (1984) Na$^+$ movement in a single turnover of the Na pump. Proc Natl Acad Sci U S A 81:5310–5314
12. Anner BM, Marcus MM, Moosmayer M (1984) Reconstitution of Na, K-ATPase. In: Azzi A, Brodbeck U, Zahler P (eds) Enzymes, receptors and carriers of biological membranes – a laboratory manual. Springer, Heidelberg, pp 81–96
13. Goldshlegger R, Karlish SJ, Rephaeli A, Stein WD (1987) The effect of membrane potential on the mammalian sodium-potassium pump reconstituted into phospholipid vesicles. J Physiol 387:331–355
14. Clarke RJ, Apell H-J (1989) A stopped-flow kinetic study of the interaction of potential-sensitive oxonol dyes with lipid vesicles. Biophys Chem 34:225–237
15. Damnjanovic B, Weber A, Potschies M, Greie JC, Apell H-J (2013) Mechanistic analysis of the pump cycle of the KdpFABC P-type ATPase. Biochemistry 52:5563–5576
16. Bashford CL, Chance B, Smith JC, Yoshida T (1979) The behavior of oxonol dyes in phospholipid dispersions. Biophys J 25:63–85
17. Apell H-J, Bersch B (1987) Oxonol VI as an optical indicator for membrane potentials in lipid vesicles. Biochim Biophys Acta 903:480–494
18. Clarke RJ, Apell H-J, Läuger P (1989) Pump current and Na$^+$/K$^+$ coupling ratio of Na$^+$/K$^+$-ATPase in reconstituted lipid vesicles. Biochim Biophys Acta 981:326–336
19. Jørgensen PL (1974) Isolation of (Na$^+$+K$^+$)-ATPase. Meth Enzymol 32:277–290
20. Jørgensen PL (1982) Mechanism of the Na$^+$, K$^+$ pump. Protein structure and conformations of the pure (Na$^+$+K$^+$)-ATPase. Biochim Biophys Acta 694:27–68
21. Schwartz AK, Nagano M, Nakao M, Lindenmayer GE, Allen JC (1971) The sodium- and potassium-activated adenosinetriphosphatase system. Meth Pharmacol 1:361–388
22. Vagin O, Denevich S, Munson K, Sachs G (2002) SCH28080, a K$^+$-competitive inhibitor of the gastric H, K-ATPase, binds near the M5-6 luminal loop, preventing K$^+$ access to the ion binding domain. Biochemistry 41:12755–12762
23. Lowry OH, Rosebrough AL, Farr AL, Randall RJ (1951) Protein measurement with the Folin phenol reagent. J Biol Chem 193:265–275
24. Apell H-J, Marcus MM (1985) Effects of vesicle inhomogeneity on the interpretation of flux data obtained with reconstituted Na, K-ATPase. In: Glynn J, Ellory C (eds) The sodium pump. The Company of Biologists Ltd., Cambridge, pp 475–480
25. Apell H-J, Läuger P (1986) Quantitative analysis of pump-mediated fluxes in reconstituted lipid vesicles. Biochim Biophys Acta 861:302–310

26. Apell H-J, Häring V, Roudna M (1990) Na, K-ATPase in artificial lipid vesicles. Comparison of Na, K and Na-only pumping mode. Biochim Biophys Acta 1023:81–90
27. Hoffman JF, Laris PC (1974) Determination of membrane potentials in human and Amphiuma red blood cells by means of fluorescent probe. J Physiol 239:519–552
28. Damnjanovic B, Apell H-J (2014) KdpFABC reconstituted in E. coli lipid vesicles: substrate dependence of the transport rate. Biochemistry 53(35):5674–5682
29. Cirri E, Kirchner C, Becker S, Katz A, Karlish SJ, Apell H-J (2013) Surface charges of the membrane crucially affect regulation of Na, K-ATPase by phospholemman (FXYD1). J Membr Biol 246:967–979
30. Markwell MA, Haas SM, Bieber LL, Tolbert NE (1978) A modification of the Lowry procedure to simplify protein determination in membrane and lipoprotein samples. Anal Biochem 87:206–210
31. Marcus MM, Apell H-J, Roudna M, Schwendener RA, Weder HG, Läuger P (1986) (Na^+ + K^+)-ATPase in artificial lipid vesicles: influence of lipid structure on pumping rate. Biochim Biophys Acta 854:270–278
32. Waggoner AS (1979) Dye indicators of membrane potential. Annu Rev Biophys Bioeng 8:47–68
33. Bashford CL (1981) The measurement of membrane potential using optical indicators. Biosci Rep 1:183–196

Chapter 15

Coupling Ratio for Ca²⁺ Transport by Calcium Oxalate Precipitation

Pankaj Sehgal, Claus Olesen, and Jesper V. Møller

Abstract

The SERCA isoform 1a is constructed to transport 2 Ca^{2+} ions across the sarcoplasmic reticulum membrane coupled to the hydrolysis of one molecule of MgATP. However, observed coupling ratios for Ca^{2+} transported/ATP hydrolzyed are usually less than 2:1, since part of the Ca^{2+} accumulated at high intravesicular concentrations by the active transport of Ca^{2+} leaks out of the vesicles because of Ca^{2+}-induced Ca^{2+} exchange. However, in the presence of a high concentration of oxalate (5 mM) Ca^{2+} will precipitate as Ca-oxalate inside the vesicles and thereby be prevented from leaking out and, in addition, this treatment will reduce the intravesicular free concentration of Ca^{2+} to a level where optimal coupling ratios of 2:1 can be achieved.

Key words Calcium transport, Oxalate, Sarcoplasmic reticulum vesicles, Calcium accumulation, Calcium ATPase, SERCA

1 Introduction

The rate of Ca^{2+} transport into sarcoplasmic reticulum vesicles is a highly regulated process, with major determinants not only by cytosolic Ca^{2+} and ATP participants, but also by the intravesicular Ca^{2+} accumulation which exerts important effects on transport by both inhibiting enzyme activity and increasing Ca^{2+}/Ca^{2+} exchange [1, 2]. To be able to follow with sufficient accuracy the correlation between Ca^{2+} transport and ATP hydrolysis under these conditions it is necessary to create experimental situations that allow the vesicles unhindered to continue their Ca^{2+} uptake beyond the normal level while at the same time maintaining a relatively low level of Ca^{2+} accumulation inside the vesicles. This feat can be achieved by the addition of oxalate that, as shown by the classical studies of Hasselbach and Makinose [3], forms a fairly insoluble precipitate with Ca^{2+} when the solubility product for calcium oxalate formation is exceeded by Ca^{2+} transport into the vesicles. Under these conditions a steady-state situation is reached where Ca^{2+} transport is continued, while the intravesicular concentration of free Ca^{2+} inside the vesicles remains low and virtually constant, and where Ca^{2+} uptake only reflects active Ca^{2+} influx, not complicated by the inhibition and Ca^{2+}/Ca^{2+} exchange taking place at high levels of Ca^{2+} accumulation.

2 Materials

1. Filtration apparatus.
2. Water bath at 23 °C.
3. Liquid scintillation counter.
4. Ca^{2+}-uptake medium: SR vesicles at a protein concentration of 0.05 mg/mL is suspended in 100 mM K-oxalate and 20 mM MOPS, adjusted to pH 7.0 with KOH, 1 mM EGTA, 1.05 mM $CaCl_2$, $^{45}CaCl_2$ (added at a specific radioactivity (SRA) of around 150 CPM/nmol), 80 mM KCl, and 5 mM $MgCl_2$.
5. Rinsing buffer (ice cold): 20 mM MOPS, pH 7.0, 80 mM KCl, and 5 mM $MgCl_2$.
6. 100 mM MgATP, adjusted to pH 7.0 with KOH.
7. 10 N H_2SO_4: Add slowly concentrated sulfuric acid to demineralized water to a final volume of 1 L (see **Note 1**).
8. 5 % ammonium molybdate (w/v) solubilized in demineralized water.
9. ANSA reagent: 28.5 g Na-metabisulfite ($Na_2S_2O_5$), 1 g sodium sulfite (Na_2SO_3), 0.5 g ANSA (1-amino-2-naphthol-4-sulfonic acid): Add H_2O to a final volume of 200 mL. Warm slightly with magnetic stirrer overnight, then cool on ice bath and filter. Keep in light tight bottle at an ambient temperature.
10. 0.1 and 0.2 mM KH_2PO_4 as standards.

3 Methods

1. Start the reaction by addition of MgATP stock to 50 mL Ca^{2+} uptake medium to a final MgATP concentration of 1 mM.
2. Zero time blank: Immediately after the addition of MgATP withdraw two 1-mL aliquots from the reaction medium, quench with 4 mL ice-cold rinsing buffer, and filter the samples on the Millipore equipment. Complete the filtration with two further 3 mL to rinse the filter.
3. Repeat the same procedure after the following time intervals: 2, 4, 7, 10, 15, 20, and 30 min.
4. Count the radioactive content deposited on the filters after drying and addition of 5 mL scintillation solvent. Measure the

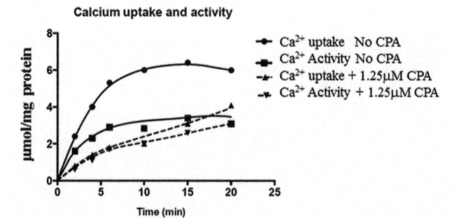

Fig. 1 An illustrative example how cyclopiazonic acid (CPA) affects the Ca^{2+}/ATP coupling ratio. Note that in the absence of CPA (the *unbroken curves*) uptake of Ca^{2+} is higher than the ATP which is hydrolyzed, signifying a coupling ratio >1, while in the presence of CPA Ca^{2+} uptake and ATP hydrolysis rates are almost equal, signifying a decreased coupling ratio of about 1:1. Furthermore, in the presence of CPA both Ca^{2+}-uptake and ATP hydrolysis are still rising after 20 min because equilibrium has not been reached due to the inhibitory effect of CPA

$^{45}Ca^{2+}$ counts present in the Ca^{2+} uptake medium after depositing 20 μL sample aliquots on separate dry filters. From these data calculate the Ca^{2+} content of the vesicles in terms of micromoles/milligram protein after subtraction of the zero time blank.

5. For determination of hydrolyzed ATP take out 2×1.2 mL reaction mixture, precipitate the protein and membranes with 70 μL 10 N H_2SO_4, and centrifuge for 5 min at $1000 \times g$. Include a zero time sample in the procedure as a blank.

6. 1 mL of the supernatants are pipetted off and color is developed by addition of 50 μL 5 % ammonium molybdate and 50 μL ANSA reagent. Include in this step color development of the phosphate standards (*see* **Note 2**). Calculate from the data ATP hydrolysis as a function of time, with correction for ATP and reagent blanks. Compare the ATP hydrolysis with the accumulation of Ca^{2+}. Coupling ratios are often best estimated after 5–15 min (*see* Fig. 1).

4 Notes

1. To avoid excessive heating add first *slowly* 250 mL concentrated sulfuric acid to around 650 mL demineralized water and then make up to a complete 1 L volume by addition of demineralized water.

2. There should be absolute proportionality between the spectrophotometric absorption at 700 nm and inorganic phosphate concentration.

References

1. Gerdes U, Møller JV (1983) Biochim Biophys Acta 734:191–200
2. Møller JV, Olesen C, Winther A-ML, Nissen P (2010) Q Rev Biophys 43:501–566
3. Hasselbach W, Makinose M (1961) Biochem Z 333:518–528

Chapter 16

Calcium Uptake in Crude Tissue Preparation

Philip A. Bidwell and Evangelia G. Kranias

Abstract

The various isoforms of the sarco/endoplasmic reticulum Ca^{2+} ATPase (SERCA) are responsible for the Ca^{2+} uptake from the cytosol into the endoplasmic or sarcoplasmic reticulum (ER/SR). In some tissues, the activity of SERCA can be modulated by binding partners, such as phospholamban and sarcolipin. The activity of SERCA can be characterized by its apparent affinity for Ca^{2+} as well as maximal enzymatic velocity. Both parameters can be effectively determined by the protocol described here. Specifically, we describe the measurement of the rate of oxalate-facilitated ^{45}Ca uptake into the SR of crude mouse ventricular homogenates. This protocol can easily be adapted for different tissues and animal models as well as cultured cells.

Key words Calcium uptake, Calcium affinity, SERCA, Phospholamban, Sarcolipin, ^{45}Ca, K_m, V_{max}, Oxalate

1 Introduction

The ability to mobilize Ca^{2+} has been, for decades, a well-established characteristic of ER/SR microsomes [1–3]. This observation was of particular importance in muscle tissue where Ca^{2+} is the trigger for contraction [4, 5]. This Ca^{2+} uptake activity was later determined to be produced by an integral membrane ATPase protein, SERCA [6–8]. Further research identified phospholamban [9, 10], and later sarcolipin [11, 12], as inhibitors of SERCA activity in cardiac and skeletal muscles, respectively.

SERCA activity follows traditional Michaelis-Menten enzyme kinetics [13] described by the equation

$$v = \frac{V_{max}[S]}{K_m + [S]}$$

Here v is the observed rate and $[S]$ is the concentration of rate-limiting substrate, which is Ca^{2+} in this case. In other words, the rate of Ca^{2+} binding to SERCA is the determinant of SERCA activity. (ATP is also a critical substrate for SERCA, but generally is well above saturating concentrations and therefore not rate limiting and not included in basic descriptions of SERCA activity). V_{max} is the maximal rate of the enzyme and K_m is the Michaelis constant which corresponds to the substrate concentration at the half maximal rate. K_m is an inverse measure of substrate affinity, meaning that a low value corresponds to a high affinity and vice versa. In simpler terms — at low Ca^{2+} concentrations, where little is available to be pumped across the membrane, SERCA activity is low. Elevating Ca^{2+} increases SERCA activity up to a maximal level.

The Ca^{2+} uptake protocol described here produces the Michaelis-Menten response curve of SERCA over the physiologically relevant Ca^{2+} concentrations (10 nM to 10 μM), which also corresponds to the full dynamic range of the enzyme [10] (Fig. 1). For easy visualization and discussion, Ca^{2+} concentrations are represented with pCa values, which correspond to the inverse log of the molar concentration.

Changes in PLN or SLN inhibition will generate a leftward or rightward shift in this activity curve signifying an altered Ca^{2+} affinity. V_{max} typically varies depending on SERCA expression level. Variations in K_m and V_{max} will also vary between species, tissue type, and SERCA isoform.

The protocol presented here is a detailed description of our standard laboratory procedure [14–20] and is adapted from the Millipore filtration technique [21]. In principle, this assay measures the amount of ^{45}Ca retained in homogenate microsomes over time after being transported by SERCA. These microsomes

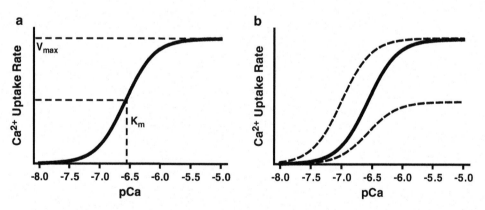

Fig. 1 *Representative SERCA activity curves following Michaelis-Menten enzyme kinetics*. (**a**) V_{max} describes the maximal uptake rate of the sample and K_m indicates the Ca^{2+} concentration at half the maximal rate. (**b**) The *dashed lines* showing downward and leftward shifts from the *solid black line* correspond to a decrease maximal rate (V_{max}) and an increase in Ca^{2+} affinity (K_m), respectively. For cardiac preparations, a downward shift would correspond to the diminished SERCA protein expression as observed in heart failure. The leftward shift would indicate relief of PLN inhibition either through phosphorylation or ablation

are collected by a nitrocellulose membrane and subsequently washed to allow excess Ca^{2+} that is not sequestered by the microsomes to pass through. Ruthenium Red blocks extrusion of Ca^{2+} out of the microsomes through ion channels [22] and prevents uptake into the mitochondria [23]. Ca^{2+} precipitates with oxalate inside ER/SR microsomes [24–26], which serves multiple purposes in this assay. First, this precipitation lowers the free Ca^{2+} inside the microsomes, which eliminates the generation of a concentration gradient that would slow SERCA activity over time, thereby allowing consistent Ca^{2+} transport for the duration of the assay [27, 28]. Secondly, it further prevents Ca^{2+} extrusion out of the microsomes. Oxalate also preferentially accumulates in ER/SR microsomes via a nonspecific anion transporter [24–26, 29]. Therefore, the oxalate-trapped Ca^{2+} resides in only ER/SR microsomes which eliminates the need for ER/SR purification that may introduce significant variability between samples.

It is important to note that this assay describes the initial rates of steady-state activity of SERCA [27], although the cytosolic environment is not at steady-state. Increased SERCA activity decreases cytosolic Ca^{2+}, thereby decreasing its own enzymatic activity.

2 Materials

2.1 Solutions

Prepare all stock solutions using ultrapure water and analytical grade reagents and store at 4 °C unless otherwise noted.

1. Homogenization Buffer: Prepare on the day of the experiment according to Table 1 and keep on ice until use.

2. Reaction Mixture: Prepare on the day of the experiment according to Table 2 and keep on ice until use.

3. 0.1 M ATP: For 75 mL, dissolve 4.27 g ATP (MW 569.1 g/mol) in 40 mL of H_2O and adjust the pH to 7.0 using 1 N NaOH. Keep on ice. Bring the volume to 70 mL. Calculate the true concentration by measuring and averaging the absorbance at 259 nm of multiple dilutions (1:1000–1:4000). M = Abs at 259 nm/15.4×10^3. Dilute the sample to exactly 0.1 M, aliquot, and store at –80 °C.

4. 1.14×10^{-4} M Ruthenium Red: The day of the experiment, dilute approximately 0.1 mg in 1 mL of water. Calculate the true concentration by measuring the absorbance at 533 nm at multiple dilutions (1:100–1:300). M = Abs at 533 nm/6.4×10^4. Dilute the sample to 1.14×10^{-4} M. 400 µL are needed for each assay in duplicate.

Table 1
Homogenization buffer

Stock	[Final]	Amount for 5 mL
0.5 M phosphate buffer, pH 7.0	50 mM	0.5 mL
0.1 M NaF	10 mM	0.5 mL
0.5 M EDTA	1 mM	0.01 mL
Sucrose	0.3 M	0.52 g
0.02 M PMSF (−20 °C)	0.3 mM	0.075 mL
0.1 M DTT (−20 °C)	0.5 mM	0.0125 mL

Table 2
Reaction mixture

Stock	[Final] (mM)	Amount for 30 reactions (mL)
0.08 M Imidazol, pH 7.0	40	22.5
1 M KCl	95	4.275
0.1 M NaN$_3$	5	2.25
0.1 M MgCl$_2$	5	2.25
0.5 M EGTA	0.5	0.045
0.1 M K$^+$ oxalate	5	2.25

5. ^{45}Ca: Prepare an initial stock of ^{45}Ca to a concentration of 2.5 μCi/μL in H$_2$O. For each assay in duplicate, 900 μL of 40 μCi/mL (36 μCi) is needed. 36 μCi corresponds to 14.4 μL of a 2.5 μCi/μL stock. To account for decay, divide 14.4 μL by the decay factor obtained from a ^{45}Ca decay chart. Add H$_2$O to bring final volume to 900 μL.

6. 10.00 mM CaCl$_2$: Either purchase an analytical grade calcium solution or have the concentration of a prepared stock analytically verified.

7. Wash Buffer: 20 mM Tris–HCl, 2 mM EGTA, pH 7.0.

8. Tissue of Interest: This assay is optimized for whole mouse ventricular cardiac tissue (approximately 20 mg) and can be adapted

for other tissue types or cultured cell lines. The abundance of SERCA protein, which is high in muscle, should be taken into account when adapting to non-muscle tissue or cells.

2.2 Equipment

1. Vacuum filtration system.
2. 0.45 μm nitrocellulose Millipore filters.
3. Water bath set to 37 °C.
4. Reaction tubes: 15 × 85 mm borosilicate glass culture tubes.
5. 20 mL scintillation vials.
6. Scintillation counter.
7. Tissue homogenizer.
8. Vortex.

3 Methods

3.1 Uptake Reaction

The key to this assay is consistent pipetting and great care should be taken to yield accurate and precise results. To further enhance accuracy, we recommend performing the entire assay in duplicate. Also, start at the lowest Ca^{2+} concentration and move to higher ones. On the duplicate set of reactions, start at the highest Ca^{2+} concentration and move to lower ones.

1. Set up the 13 reaction tubes in duplicate (26 total).
2. Set up 5 scintillation vials for each reaction: 1 blank, 1 standard, and 1 for each of the 3 time points (130 total).
3. Prepare reaction tubes according to Table 3.
4. Homogenize tissue in homogenization buffer and keep on ice until used for each reaction. 75 μL is necessary for each of the 26 reactions (1.950 mL total). Approximate protein concentration should be between 0.5 and 5.0 mg/mL. Final rates will be normalized to quantified concentration.
5. Place four 0.45 μm Millipore filters on filtration manifold for the first reaction tube.
6. Wash each filter two times with 2.5 mL Wash Buffer.
7. Add 13.2 μL of 0.114 mM Ruthenium Red (1 μM final) to the first reaction tube.
8. Add 75 μL of homogenates to the first reaction tube and slightly vortex (avoiding air bubble formation) to thoroughly mix.
9. Place the reaction tube in a 37 °C water bath.
10. After 30 s remove an aliquot of 290 μL for nonspecific binding (also called blank). Pass the aliquot through a 0.45 μm Millipore membrane.

Table 3
Calcium dilutions

Buffer #	pCa	Volume to add (μL)			Reaction mixture
		10.00 mM CaCl$_2$	dH$_2$O	^{45}Ca (40 μCi/mL)	
1	8	2.6	228.3	1.9	1119
2	7.5	7.7	219.4	5.7	1119
3	7.2	13.9	208.7	10.2	1119
4	7	19.9	198.3	14.6	1119
5	6.8	27.4	185.2	20.2	1119
6	6.6	36.0	170.3	26.5	1119
7	6.4	44.8	155.0	33.0	1119
8	6.3	49.1	147.6	36.1	1119
9	6.2	53.2	140.5	39.1	1119
10	6	60.5	127.8	44.5	1119
11	5.8	66.8	116.9	49.1	1119
12	5.5	74.7	103.1	55.0	1119
13	5	89.9	76.8	66.1	1119

11. Initiate the reaction by adding 60 μL 0.1 M ATP, [Final] = 5 mM. Mix thoroughly with the pipet.
12. Take out 300 μL at 30, 60, and 90 s time points and pass each aliquot through a different membrane.
13. Wash each membrane two times with 2.5 mL wash buffer.
14. Place the filters in the scintillation vials.
15. Repeat **steps 5–14** for the remaining reaction tubes.
16. When finished, add 60 μL from the remaining reaction mixture to the standard scintillation vial for each reaction.
17. Add 10 mL of scintillation fluid to each scintillation vial.
18. Measure the samples with a scintillation counter.

3.2 Data Analysis

1. Quantify homogenate protein concentration using the Bradford method.
2. The scintillation counter yields values in counts per minute (cpm) which can be converted to moles of Ca^{2+} using the standard sample from **step 16** in Subheading 3.1. This sample has not been filtered and therefore is representative of the total

Ca^{2+} in each reaction mixture. The amount of microsome sequestered Ca^{2+} is represented by the following equation:

$$\text{Ca Uptake (nmol)} = \frac{\text{sample cpm} - \text{blank cpm}}{\text{standard cpm} \times 5} \times \text{Ca (nmol)}$$

Here Ca (nmol) represents the amount of cold Ca^{2+} present in each aliquot (300 μL) of each individual reaction mixture (Table 3: 10.00 mM $CaCl_2$ column) which is one-fifth of the amount in the total reaction mixture (1.5 mL).

3. After all Ca-uptake values are calculated, perform a linear regression from the three time points (30, 60, and 90 s) for each pCa. The slope of this line corresponds to the rate of Ca^{2+} uptake in nanomole/minute. These values should then be normalized to the amount of protein giving a final normalized rate in nmol/min/mg (Fig. 2a).

4. Plot the average of the duplicate calculations against the corresponding Ca^{2+} concentration. This plot can then be fit with the Hill equation to determine the V_{max} and K_m parameters (Fig. 2b).

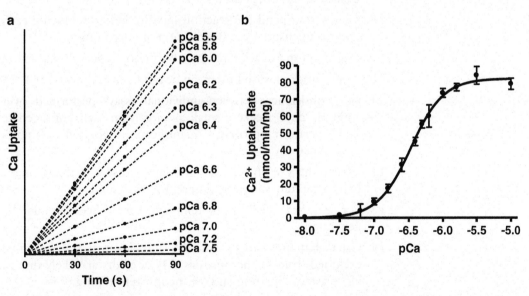

Fig. 2 Examples plots of the initial rates of SERCA activity from C57/BL6 mouse ventricular homogenates. (**a**) The rate at each Ca^{2+} concentration is determined by calculating the slope from the Ca-uptake values from the 3 time points. These rates can then be plotted against the Ca^{2+} concentration (**b**) and fit with the Hill equation to determine the V_{max} and K_m parameters. For this example plot, $V_{max} = 80.2 \pm 0.5$ and $K_m = $ pCa 6.468 ± 0.007 (340 nM)

4 Notes

1. Be sure to follow the specific radiation safety guidelines at your facility for use and disposal of ^{45}Ca.
2. For consistency between assays, use the same batch of stock solutions for all experiments if possible.
3. ATP and EGTA are critical for buffering Mg^{2+} and Ca^{2+} as changes in ATP concentration will cause variations in the free Ca^{2+} concentration.
4. The stock $CaCl_2$ solution is perhaps the most critical element in this assay and is essential for accurate K_m measurements.
5. Reaction tubes and scintillation vials can be organized and labeled the day prior to the assay.
6. To limit the time spent in the radioactive lab and overall exposure, ^{45}Ca should be added to the reaction mixture just prior to initiating the reaction.
7. The 30, 60, and 90 s time points correspond to when the sample is passed through the membrane. To ensure proper timing, start a timer when adding ATP. Remove an aliquot from the reaction tube in water bath 5–6 s before each time point for transport to filtration system. Then pipet the sample exactly at 30, 60, and 90 s.
8. Gently resuspend the reaction mixture before removing each aliquot to ensure even distribution of microsomes.
9. Washing can be facilitated by a bottle top repeat dispenser.
10. A 50× stock of wash buffer can be prepared and stored at 4 °C.
11. For consistency between individual assays performed over multiple days, freeze a small aliquot from each homogenate and perform protein quantification for all samples simultaneously.
12. Gently shake each scintillation vial to ensure that the membranes are fully submerged in scintillation fluid.
13. For ease of analysis, a template spreadsheet should be generated.
14. Calculated rates (slopes) with R^2 values less than 0.9 should be excluded from further analysis. These errors are likely due to inconsistent pipetting and/or inconsistent timing.
15. K_m values are generally consistent and may only require a few repetitions to generate statistical significance. V_{max} values are more inconsistent, though reliable values can be obtained by additional repetitions. If necessary, data can be normalized to V_{max} values.

Acknowledgement

This work was supported by NIH grants HL-26057 and HL-64018 to E. G. K. and AHA grant 13POST13860006 to P. A. B.

References

1. Ebashi S, Lipmann F (1962) Adenosine triphosphate-linked concentration of calcium ions in a particulate fraction of rabbit muscle. J Cell Biol 14:389–400
2. Hasselbach W, Makinose M (1962) The calcium pump of the "relaxing granules" of muscle and its dependence on ATP-splitting. Biochem Z 333:518–528
3. Hasselbach W, Makinose M (1963) on the Mechanism of Calcium Transport Across the Membrane of the Sarcoplasmic Reticulum. Biochem Z 339:94–111
4. Gordon AM, Homsher E, Regnier M (2000) Regulation of contraction in striated muscle. Physiol Rev 80:853–924
5. Bers DM (2008) Calcium cycling and signaling in cardiac myocytes. Annu Rev Physiol 70:23–49
6. MacLennan DH, Brandl CJ, Korczak B, Green NM (1985) Amino-acid sequence of a Ca2+ + Mg2+-dependent ATPase from rabbit muscle sarcoplasmic reticulum, deduced from its complementary DNA sequence. Nature 316:696–700
7. Brandl CJ, Green NM, Korczak B, MacLennan DH (1986) Two Ca2+ ATPase genes: homologies and mechanistic implications of deduced amino acid sequences. Cell 44:597–607
8. MacLennan DH (1970) Purification and properties of an adenosine triphosphatase from sarcoplasmic reticulum. J Biol Chem 245:4508–4518
9. Tada M, Kirchberger MA, Katz AM (1975) Phosphorylation of a 22,000-dalton component of the cardiac sarcoplasmic reticulum by adenosine 3′:5′-monophosphate-dependent protein kinase. J Biol Chem 250:2640–2647
10. MacLennan DH, Kranias EG (2003) Phospholamban: a crucial regulator of cardiac contractility. Nat Rev Mol Cell Biol 4:566–577
11. Odermatt A, Becker S, Khanna VK et al (1998) Sarcolipin regulates the activity of SERCA1, the fast-twitch skeletal muscle sarcoplasmic reticulum Ca2+-ATPase. J Biol Chem 273:12360–12369
12. MacLennan DH, Asahi M, Tupling AR (2003) The regulation of SERCA-type pumps by phospholamban and sarcolipin. Ann N Y Acad Sci 986:472–480
13. Inesi G, Kurzmack M, Coan CR, Lewis DE (1980) Cooperative calcium binding and ATPase activation in sarcoplasmic reticulum vesicles. J Biol Chem 255:3025–3031
14. Haghighi K, Pritchard T, Bossuyt J et al (2012) The human phospholamban Arg14-deletion mutant localizes to plasma membrane and interacts with the Na/K-ATPase. J Mol Cell Cardiol 52:773–782
15. Schmidt AG, Zhai J, Carr AN et al (2002) Structural and functional implications of the phospholamban hinge domain: impaired SR Ca2+ uptake as a primary cause of heart failure. Cardiovasc Res 56:248–259
16. Frank K, Tilgmann C, Shannon TR et al (2000) Regulatory role of phospholamban in the efficiency of cardiac sarcoplasmic reticulum Ca2+ transport. Biochemistry 39:14176–14182
17. Zhai J, Schmidt AG, Hoit BD et al (2000) Cardiac-specific overexpression of a superinhibitory pentameric phospholamban mutant enhances inhibition of cardiac function in vivo. J Biol Chem 275:10538–10544
18. Harrer JM, Haghighi K, Kim HW et al (1997) Coordinate regulation of SR Ca(2+)-ATPase and phospholamban expression in developing murine heart. Am J Physiol 272:H57–H66
19. Szymanska G, Grupp IL, Slack JP et al (1995) Alterations in sarcoplasmic reticulum calcium uptake, relaxation parameters and their responses to beta-adrenergic agonists in the developing rabbit heart. J Mol Cell Cardiol 27:1819–1829
20. Zhao W, Waggoner JR, Zhang Z-G et al (2009) The anti-apoptotic protein HAX-1 is a regulator of cardiac function. Proc Natl Acad Sci U S A 106:20776–20781
21. Solaro RJ, Briggs FN (1974) Estimating the functional capabilities of sarcoplasmic reticulum in cardiac muscle. Circ Res 34:531–540
22. Chamberlain BK, Volpe P, Fleischer S (1984) Inhibition of calcium-induced calcium release from purified cardiac sarcoplasmic reticulum vesicles. J Biol Chem 259:7547–7553

23. Moore CL (1971) Specific inhibition of mitochondrial Ca++ transport by ruthenium red. Biochem Biophys Res Commun 42:298–305
24. Beil FU, von Chak D, Hasselbach W, Weber HH (1977) Competition between oxalate and phosphate during active calcium accumulation by sarcoplasmic vesicles. Z Naturforsch C 32:281–287
25. De Meis L, Hasselbach W, Machado RD (1974) Characterization of calcium oxalate and calcium phosphate deposits in sarcoplasmic reticulum vesicles. J Cell Biol 62:505–509
26. Feher JJ, Lipford GB (1985) Calcium oxalate and calcium phosphate capacities of cardiac sarcoplasmic reticulum. Biochim Biophys Acta 818:373–385
27. Madeira VM (1984) State of translocated Ca2+ by sarcoplasmic reticulum inferred from kinetic analysis of calcium oxalate precipitation. Biochim Biophys Acta 769:284–290
28. Inesi G, de Meis L (1989) Regulation of steady state filling in sarcoplasmic reticulum. Roles of back-inhibition, leakage, and slippage of the calcium pump. J Biol Chem 264:5929–5936
29. MacLennan DH, Reithmeier RA, Shoshan V et al (1980) Ion pathways in proteins of the sarcoplasmic reticulum. Ann N Y Acad Sci 358:138–148

Chapter 17

Measuring H⁺ Pumping and Membrane Potential Formation in Sealed Membrane Vesicle Systems

Alex Green Wielandt, Michael G. Palmgren, Anja Thoe Fuglsang, Thomas Günther-Pomorski, and Bo Højen Justesen

Abstract

The activity of enzymes involved in active transport of matter across lipid bilayers can conveniently be assayed by measuring their consumption of energy, such as ATP hydrolysis, while it is more challenging to directly measure their transport activities as the transported substrate is not converted into a product and only moves a few nanometers in space. Here, we describe two methods for the measurement of active proton pumping across lipid bilayers and the concomitant formation of a membrane potential, applying the dyes 9-amino-6-chloro-2-methoxyacridine (ACMA) and oxonol VI. The methods are exemplified by assaying transport of the *Arabidopsis thaliana* plasma membrane H⁺-ATPase (proton pump), which after heterologous expression in *Saccharomyces cerevisiae* and subsequent purification has been reconstituted in proteoliposomes.

Key words pH-sensitive dye, Proteoliposome, Proton pumping, Membrane potential

1 Introduction

Cells and organelles depend on electrochemical gradients across their membranes. Membrane-embedded biological pumps are key players in maintaining these gradients through primary, active transport of ions. Among these pumps, plasma membrane H⁺-ATPases form a subfamily of P-type ATPases that use the energy derived from hydrolysis of ATP to pump protons out of cells. In protists, fungi, and plants, they are essential for establishing and maintaining the crucial transmembrane electrochemical gradients of protons which provides the driving force for the uptake of nutrients and other cell constituents through an array of secondary transport systems [1–4].

This work is dedicated to the memory of Alex Green Wielandt, who set the basis of this study until his untimely death.

The activity of the plasma membrane H⁺-ATPases can be conveniently studied when the enzyme resides in the membrane of a closed vesicle. The closed vesicle may be the native membrane of the protein (see, e.g., [5]) or a liposome prepared from a known phospholipid composition. The latter offers a well-defined experimental system where no unknown or unwanted proteins are present. In both cases, the membrane itself must be proton impermeable on the time scale of the experiment. Proton fluxes in this system are typically followed indirectly by use of ΔpH-sensitive dyes that change absorbance or fluorescence properties in response to the formation of pH gradients across the membrane. Acridine orange is a classical ΔpH-sensitive dye, and a number of derivatives of this probe are used for measurement of ΔpH. The spectral properties of acridine orange are relatively insensitive to medium pH at physiological values [6]. However, acridine orange is a monomer in dilute solutions and a dimer at high probe concentrations, and since the monomer and the dimer have different absorbance as well as fluorescence spectra, dimer formation can be followed by assaying the decrease in monomer absorbance or fluorescence [6]. In vesicles with a higher H⁺ concentration than the surrounding medium, monomeric acridine orange is protonated and trapped, and dimerizes as the concentration inside increases, leading to quenching of the fluorescence [6, 7]. As the degree of dimerization of acridine orange depends strongly on temperature and the anion composition of the vesicle lumen, great care should be taken when using this probe to compare proton pumping activities as a function of temperature and anions in the medium [6].

Here we describe the use of 9-amino-6-chloro-2-methoxyacridine (ACMA), a derivative of the probe acridine orange. This probe is yellow in solution and can cross bilayers in its unprotonated form. Because of its derived nature ACMA can exist in four states, namely as an anion, as a neutral molecule, as a monocation, and as a dication [8]. The monocation is protonated at the acridinic nitrogen with a pK_a ranging between 8.3 [9] and 8.6 [8] implying that at neutral pH the monocation is already the predominant form. The dication is further protonated at the amino group with a pK_a ranging between 4.7 [9] and 5.7 [8]. The mechanism by which ACMA reports ΔpH gradients is not clear, as at least the monocation does not dimerize in aqueous solutions even as high as 200 μM [8]. It has been suggested that decrease of fluorescence of ACMA is the result of adsorption of the probe on the membrane, which results in stacking and aggregation at this interface [10]. The related ΔpH probe 9-aminoacridine has been suggested to form a similar dimer-exited state at the membrane [11]. Because of the structural similarities with acridine orange, such stacking would presumably also be sensitive to temperature and anions present. Alternatively, as the mono- and dications have very different fluorescence spectra with the monocation having a fluorescence peak at 480 nm and the dication exhibiting almost no fluorescence at this wavelength [8], it is possible that what is assayed when luminal vesicular pH decreases is simply disappearance of the monocation.

While acridine orange and derivatives measure formation of a ΔpH, oxonol V and VI can be used to measure differences in charge balance across a liposomal membrane [12–15]; the latter is not limited to proton transport, but is highly applicable to other ions as well [12]. The oxonol dyes are blue in solution, negatively charged (pK_a ~4.2, [12]), and show a large increase in fluorescence upon insertion into a hydrophobic environment. Generation of an inside positive potential causes the dye to preferentially accumulate inside the liposome, and leads to an increased interaction with the liposomal membrane. Oxonol VI does not interact equally well with all membranes, and particularly high phosphatidylethanolamine concentrations can interfere with the probe [16]. The assay presented here is intended for obtaining relative levels of the gradient achieved in relation to a reference standard (e.g., for comparing variants of the H^+-ATPase); if a quantification of the actual membrane potential is desired, the system can be calibrated by the method of Apell and Bersch [12]. Reduction of the temperature to 10 °C can help to resolve the steady state fluorescence level.

The assays described here are based on proteoliposomes prepared from protein heterologously expressed in *S. cerevisiae*, and purified by affinity chromatography [17]. Measurement of H^+ transport is also directly applicable to sealed vesicles from plant plasma membrane preparations [18], as well as measurement of both H^+-transport and membrane potential in proteoliposomes prepared from crude membranes from *S. cerevisiae* [19].

2 Materials

2.1 Instrumentation

1. A fluorometer with excitation and emission filters in the 350–700 nm range. It is preferable to have a temperature control unit and a magnetic stirrer mounted.

2. Fluorometer cuvettes, preferably suitable for magnetic stirring. This protocol is scaled for a 3 mL cuvette.

2.2 Stock Solutions

1. ACMA solution: Prepare 10 mM ACMA in DMSO. Weigh 2.59 mg of ACMA and dissolve in 1 mL DMSO. Dilute to 1 mM in DMSO and store in 100 μL aliquots at –20 °C in the dark.

2. Oxonol VI solution: Prepare 2.5 mM solution in 50 % ethanol. Weigh 0.79 mg of oxonol VI and dissolve in 1 mL 50 % ethanol. Dilute to 50 μM in 50 % ethanol and store at –20 °C in the dark.

3. Valinomycin solution: Prepare 1.25 mM valinomycin in 96 % ethanol. Weigh 1.39 mg of valinomycin and dissolve in 1 mL

96 % ethanol. Dilute to 125 µM in 96 % ethanol and store in aliquots of 100 µL at −20 °C.

4. Vanadate solution: Prepare 0.1 M Sodium orthovanadate in ultrapure water. Weigh 183.9 mg of Na_3VO_4 and dissolve in 10 mL ultrapure water. Adjust to pH 7 using HCl.

5. ATP-KOH, 0.5 M: Dissolve 5.51 g $Na_2ATP \cdot H_2O$ in 10 mL ultrapure water. Adjust pH to 7.0 with KOH, and dilute to 20 mL. Store in aliquots of 1 mL at −20 °C.

6. Carbonyl cyanide *m*-chlorophenylhydrazone (CCCP) solution: Prepare 5 mM in 70 % ethanol. Weigh 1.023 mg of CCCP and dissolve in 1 mL 70 % ethanol. Store in aliquots of 100 µL at −20 °C.

7. 0.5 M MOPS-KOH, pH 7.0: Weigh 52.32 g MOPS and dissolve in 300 mL ultrapure water. Adjust pH to 7.0 using KOH, and add ultrapure water to a final volume of 0.5 L.

8. 0.5 M K_2SO_4: Weigh 4.36 g K_2SO_4 and dissolve in 50 mL ultrapure water.

9. 0.5 M $MgSO_4$: Weigh 616.25 mg $MgSO_4 \cdot 7H_2O$ and dissolve in 5 mL ultrapure water.

2.3 Assay Buffers

1. Proton pumping buffer: Typically 20 mL is prepared, containing 20 mM MOPS-KOH pH 7.0, 50 mM K_2SO_4, 1 µM ACMA, 1 mM ATP, and 60 nM Valinomycin. Store at 4 °C (*see* **Note 1**).

2. Membrane potential buffer: Typically 20 mL is prepared, containing 10 mM MOPS-KOH pH 7.0, 50 mM K_2SO_4, 50 nM oxonol VI, 1 mM ATP. Store at 4 °C (*see* **Note 2**).

2.4 Biological Materials

1. *Arabidopsis thaliana* plasma membrane H^+-ATPase isoform 2, purified and reconstituted into preformed asolectin or dioleoylphosphatidylcholine vesicles as described [20]: 12 µg purified protein per 220 µL reconstitution mixture, containing MOPS-KOH buffer (pH 7), 50 mM octylglucoside, and 10.6 mg/mL asolectin (*see* **Note 3**).

3 Methods

3.1 Measurement of Proton Transport

1. Turn on the fluorometer and set up the parameters as follows: excitation wavelength 412 nm, emission wavelength 480 nm, and measurement duration approximately 10 min with 1 s resolution. Adjust slits as necessary; a bandpass of 2 nm is usually sufficient.

2. Fill the fluorometer cuvette with 1 mL of proton pumping assay buffer. Add 10 µL proteoliposomes from the reconstitution

mixture, corresponding to about 0.5 μg protein, and homogenize gently with a pipette.

3. Introduce the cuvette into the fluorometer (remember to include the stirring bar), and start the monitoring of the emission intensity (*see* **Note 4**).

4. Wait until fluorescence is stable (60–120 s).

5. Add MgSO$_4$ to start the assay (12 μL 0.5 M) (*see* **Note 5**).

6. Let the experiment run for a couple of minutes (*see* **Note 6**).

7. Dissipate the membrane potential by addition of 1 μL of 5 mM CCCP when the fluorescence has stabilized.

8. Optional: Repeat the measurements on a sample supplemented with 3 μL of 0.1 M vanadate as the background control (*see* **Note 7**).

An example of H$^+$ transport measurement by the ACMA assay is shown in Fig. 1.

3.2 Measurement of Membrane Potential

1. Turn on the fluorometer, and set up the parameters as follows: excitation wavelength 610 nm, emission wavelength 640 nm, and measurement duration up to approximately 30 min with 1 s resolution. Adjust slits as necessary; a bandpass of 5 nm is usually sufficient.

2. Fill the fluorometer cuvette with 3 mL membrane potential assay buffer, and 10 μL of proteoliposomes containing approximately 0.5 μg of purified protein.

3. Insert the cuvette into the fluorometer, and allow for the temperature to stabilize to the desired level (usually room temperature or 10 °C), i.e., 5 min (*see* **Note 8**).

4. Start monitoring the fluorescence emission.

5. Wait for the fluorescence to stabilize (60–120 s).

6. Start the reaction by the addition of MgSO$_4$ (12 μL of 0.5 M) (*see* **Note 5**).

7. Wait for the fluorescence to stabilize. After an initial stable level, the fluorescence may decrease as the vesicles can become leaky to H$^+$.

8. Dissipate the membrane potential with valinomycin (3 μL of 125 μM).

9. Wait for the fluorescence level to stabilize.

10. The initial rate of transport can be determined from a linear fit of the initial change in fluorescence; this can be correlated to rates derived from measurements with ACMA. The maximal membrane potential achieved can be determined as the differ-

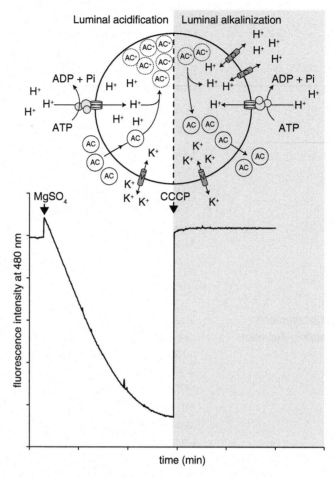

Fig. 1 Example of a typical measurement of H+ transport into phospholipid vesicles using the ACMA assay. Proteoliposomes with reconstituted AHA2 were mixed with proton pumping buffer containing the fluorophore ACMA (AC in *closed ring*), valinomycin (denoted V), and ATP, and the fluorescence intensity at 480 nm was monitored. After a stable fluorescence emission was obtained, proton pumping was initiated by the addition of MgSO$_4$. The ionophore valinomycin ensures that K+ ions are free to diffuse out of the proteoliposomes, thereby eliminating any contribution from the electrical potential against proton pumping. Upon proton accumulation and luminal acidification, ACMA is protonated and forms dimeric or higher order complexes within the proteoliposomes (AC in *dashed ring*), resulting in a shift in its fluorescence maxima, measured as a drop in fluorescence intensity at 480 nm. After maximum proton influx is achieved, the proton gradient is disrupted by the addition of CCCP (denoted C). This leads to a luminal alkalization followed by deprotonation and dissociation of the ACMA oligomers, resulting in restored fluorescence intensity at 480 nm. Figure is modified from [21]

ence in fluorescence between the peak fluorescence under pumping, and the level obtained after addition of valinomycin.

An example of the fluorescence response of oxonol VI to energization of proteoliposomes containing incorporated plasma membrane H^+-ATPase is shown in Fig. 2.

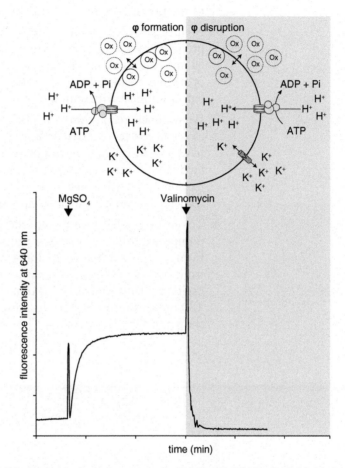

Fig. 2 Example of a typical measurement of the electrical potential (φ) using the assay based on the oxonol VI dye. Proteoliposomes with reconstituted AHA2 were mixed with membrane potential buffer containing oxonol VI (Ox in *dashed ring*) and ATP, and the fluorescence intensity at 640 nm was monitored. After a stable fluorescence emission was obtained, proton pumping was initiated by the addition of $MgSO_4$. Upon proton accumulation and buildup of an electrical potential, oxonol VI preferentially accumulates inside the proteoliposomes, leading to an increased interaction with the liposomal membrane (Ox in *closed ring*), which is monitored by an increase in fluorescence at 640 nm. After maximum electrical potential was achieved it was disrupted by the addition of the ionophore valinomycin (denoted V), which allows K^+ to diffuse out of the proteoliposomes. This results in dissociation of oxonol VI from the membrane, and a decrease in fluorescence intensity at 640 nm. This assay was performed at 10 °C. Note that the *large spikes* at the addition of $MgSO_4$ and valinomycin are due to opening of the fluorometer chambers. Figure is modified from [21]

4 Notes

1. Proton pumping buffer without ACMA, ATP, and valinomycin can be prepared in bulk and stored up to 6 months. Buffer containing all ingredients should be prepared immediately before use and protected from light by wrapping the container of the prepared solution with aluminum foil.

2. Membranepotential buffer without oxonol VI and ATP can be prepared in bulk and stored up to 6 months. Buffer containing all ingredients should be prepared immediately before use and protected from light by wrapping the container of the prepared solution with aluminum foil.

3. We routinely reconstitute the plasma membrane H^+-ATPase into asolectin vesicles of mixed phospholipid composition, although we have also successfully reconstituted using vesicles with a defined lipid composition, e.g., dioleoylphosphatidylcholine, palmitoyloleoylphosphatidylcholine, or mixtures of palmitoyloleoylphosphatidylcholine and palmitoyloleoylphosphatidylglycerol.

4. To ensure your setting and the ACMA containing proton pumping buffer are working properly, run an excitation scan from 350 to 450 nm (emission 480 nm) and an emission scan from 450 to 500 nm (excitation 412 nm). For ACMA, the excitation and emission maxima should be at 412 nm and 480 nm, respectively. The fluorescence spectrum for ACMA is shown in Fig. 3.

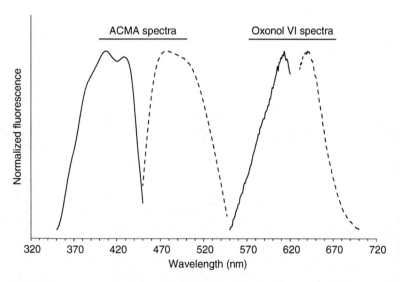

Fig. 3 Fluorescence spectrum of ACMA and oxonol VI, showing excitation (*solid line*) and emission (*dashed line*) maxima. For details of the spectra, *see* **Notes 4** and **8**. These measurements were performed at room temperature

5. We typically initiate the reaction by the addition of Mg^{2+} as the true substrate of the plasma membrane H^+-ATPase is MgATP. ATP is already present in the assay buffer.

6. The initial rate of H^+ flux is used when calculating the relative rate of H^+ transport. After 5–10 min the vesicles might become leaky to protons and you will observe a slight increase in fluorescence. The maximal degree of fluorescence quenching is reached when proton influx into vesicles matches efflux and depends on both proton pumping activity and leakiness of the vesicles. Proton fluxes cannot be quantified in terms of numbers of H^+ transported due to the non-ideal behavior of the ΔpH probe.

7. Orthovanadate inhibits many NTPases, including P-type ATPases, and will therefore block all proton pumping that is dependent on the plasma membrane H^+-ATPase. Solutions of orthovanadate are prone to polymerization when stored for prolonged periods of time, and these polyvanadate ions are unable to properly inhibit NTPases. To ensure that the orthovanadate solution is properly monomeric before use, the solution should be boiled for 5 min and placed on ice.

8. To ensure your settings and the oxonol VI containing membrane potential buffer are working properly, run an excitation scan from 550 to 630 nm (emission 640 nm) and an emission scan from 620 to 700 nm (excitation 610 nm). The properties of oxonol VI are as follows: Excitation and emission maxima at 610 nm and 640 nm, respectively. It is important to measure the spectra in the presence of liposomes since these both shift and greatly increase the fluorescence yield of oxonol VI. The fluorescence spectrum for oxonol VI is shown in Fig. 3.

Acknowledgement

This work was supported by the UNIK research initiative of the Danish Ministry of Science, Technology and Innovation through the "Center for Synthetic Biology" at the University of Copenhagen and the Danish National Research Foundation through the PUMPKIN Center of Excellence (DNRF85).

References

1. Meade JC, Li C, Stiles JK et al (2000) The Trypanosoma cruzi genome contains ion motive ATPase genes which closely resemble Leishmania proton pumps. Parasitol Int 49:309–320
2. Nakamoto RK, Slayman CW (1989) Molecular properties of the fungal plasma-membrane [H+]-ATPase. J Bioenerg Biomembr 21:621–632
3. Slayman CL, Long WS, Lu CY (1973) The relationship between ATP and an electrogenic pump in the plasma membrane of Neurospora crassa. J Membr Biol 14:305–338
4. Palmgren MG (2001) Plant plasma membrane H+-ATPases: powerhouses for nutrient uptake. Annu Rev Plant Physiol Plant Mol Biol 52:817–845
5. Palmgren MG (1990) An H+-ATPase assay: proton pumping and ATPase activity determined simultaneously in the same sample. Plant Physiol 94:882–886
6. Palmgren MG (1991) Acridine orange as a probe for measuring pH gradients across membranes: Mechanism and limitations. Anal Biochem 192:316–321
7. Clerc S, Barenholz Y (1998) A quantitative model for using acridine orange as a transmembrane pH gradient probe. Anal Biochem 259:104–111
8. Marty A, Bourdeaux M, Dell'Amico M et al (1986) 9-amino-2-methoxy-6-chloroacridine monocation fluorescence analysis by phase-modulation fluorometry. Eur Biophys J 13:251–257
9. Capomacchia AC, Schulman SG (1975) Electronic absorption and fluorescence spectrophotometry of quinacrine. Anal Chim Acta 77:79–85
10. Casadio R (1991) Measurements of transmembrane pH differences of low extents in bacterial chromatophores. Eur Biophys J 19:189–201
11. Grzesiek S, Otto H, Dencher NA (1989) delta pH-induced fluorescence quenching of 9-aminoacridine in lipid vesicles is due to excimer formation at the membrane. Biophys J 55:1101–1109
12. Apell HJ, Bersch B (1987) Oxonol VI as an optical indicator for membrane potentials in lipid vesicles. Biochim Biophys Acta 903:480–494
13. Kiehl R, Hanstein WG (1984) ATP-dependent spectral response of oxonol VI in an ATP-Pi exchange complex. Biochim Biophys Acta 766:375–385
14. Cooper CE, Bruce D, Nicholls P (1990) Use of oxonol V as a probe of membrane potential in proteoliposomes containing cytochrome oxidase in the submitochondrial orientation. Biochemistry 29:3859–3865
15. Holoubek A, Večeř J, Opekarová M et al (2003) Ratiometric fluorescence measurements of membrane potential generated by yeast plasma membrane H+-ATPase reconstituted into vesicles. Biochim Biophys Acta Biomembr 1609:71–79
16. Gibrat R, Grignon C (2003) Liposomes with multiple fluorophores for measurement of ionic fluxes, selectivity, and membrane potential. Methods Enzymol 372:166–186
17. Ekberg K, Wielandt AG, Buch-Pedersen MJ et al (2013) A conserved asparagine in a P-type proton pump is required for efficient gating of protons. J Biol Chem 288:9610–9618
18. Lund A, Fuglsang AT (2012) Purification of plant plasma membranes by two-phase partitioning and measurement of H+ pumping. Methods Mol Biol 913:217–223
19. Venema K, Palmgren MG (1995) Metabolic modulation of transport coupling ratio in yeast plasma membrane H(+)-ATPase. J Biol Chem 270:19659–19667
20. Lanfermeijer FC, Venema K, Palmgren MG (1998) Purification of a histidine-tagged plant plasma membrane H+-ATPase expressed in yeast. Protein Expr Purif 12:29–37
21. Rodriguez A, Benito B, Cagnac O (2012) Using heterologous expression systems to characterize potassium and sodium transport activities. Methods Mol Biol 913:371–386

Chapter 18

Assay of Flippase Activity in Proteoliposomes Using Fluorescent Lipid Derivatives

Magdalena Marek and Thomas Günther-Pomorski

Abstract

Specific membrane proteins, termed lipid flippases, play a central role in facilitating the movement of lipids across cellular membranes. In this protocol, we describe the reconstitution of ATP-driven lipid flippases in liposomes and the analysis of their in vitro flippase activity based on the use of fluorescent lipid derivatives. Working with purified and reconstituted systems provides a well-defined experimental setup and allows to directly characterize these membrane proteins at the molecular level.

Key words ABC transporter, Dithionite, Fluorescence, NBD-lipid, P-type ATPase, Reconstitution

1 Introduction

The distribution of lipids across cellular membranes is regulated by specific membrane proteins termed lipid flippases that control the movement of lipids across cellular membranes [1–4]. While part of these membrane proteins functions without any direct metabolic energy input, others are ATPases that function by coupling lipid translocation to ATP hydrolysis. Among the ATP-driven transporters, type IV P-type ATPases (P4-ATPases) in complex with their Cdc50 subunits have been characterized as lipid flippases that catalyze the translocation of (amino)phospholipids from the exoplasmic to the cytoplasmic leaflet of membranes in eukaryotes [1, 2]. This process enables eukaryotic cells to create and maintain nonrandom lipid distributions between the two leaflets of membrane-bound compartments along the late secretory and endocytic secretory pathways, such that the aminophospholipids phosphatidylserine and phosphatidylethanolamine are largely confined to the cytosolic leaflet and sphingolipids (i.e., sphingomyelin and glycosphingolipids) are enriched in the exoplasmic leaflet [5, 6].

Lipid flippase activities have been mostly demonstrated by studies performed on intact cells or isolated organelles. Only recently, advances in purification and reconstitution techniques have allowed demonstrating the capacity of P4-ATPases to directly participate in lipid translocation [7–9]. Compared to in vivo experiments, working with purified and reconstituted systems is challenging given the difficulties of purifying and reconstituting membrane proteins. Another problem associated with flippase experiments in artificial vesicles is that the net transfer of lipid to one leaflet can result in an increased surface tension, which eventually inhibits the flippase activity [10]. This is supported by the observation that extent of lipid transport observed so far in reconstituted systems made of large unilamellar vesicles (diameter of about 100–250 nm) containing purified lipid flippases has been very limited [8, 11–15]. Nevertheless, such studies are key to directly analyze lipid flippases at the molecular level under well-defined conditions in which parameters can be varied systematically.

This chapter deals with recently described protocols to reconstitute ATP-driven lipid flippases into large unilamellar vesicles and to measure phospholipid flip-flop in the reconstituted vesicles. The procedure described here starts with the preparation of large unilamellar vesicles composed of dioleoyl-phosphatidylcholine (DOPC) and trace amounts of a labeled reporter lipid (Fig. 1). For membrane proteins that depend on a specific lipid composition of the membrane, liposomes with a more complex lipid mixture need to be prepared. The preformed and presized lipid vesicles are destabilized with mild detergent such as n-dodecyl-β-maltoside (DDM) or Triton X-100 to promote protein insertion. The optimal detergent concentration for the destabilization depends on the vesicle lipid composition as well as the type of detergent and should be determined by titration before reconstitution (Fig. 2). Subsequently, the detergent-solubilized membrane protein is mixed with the detergent-destabilized liposomes, and slow removal of detergent (using polystyrene beads) leads to formation of tightly sealed proteoliposomes with diameters in the range of 150–300 nm. Alternatively, reconstitution procedures based on fully solubilized phospholipid can be used [8, 13–15]. Upon formation, proteoliposomes are assayed for their ability to flip the labeled reporter phospholipid across the bilayer. Additionally, the proteoliposomes can be used for other types of assays, e.g., ATPase assays, or serve as basis for the formation of giant unilamellar vesicles (GUVs) that allow characterization of flippase-mediated transport of unlabeled phospholipids [16].

The assay to measure lipid flippase activities in proteoliposomes makes use of fluorescent 7-nitro-2,1,3-benzoxadiazol (NBD) acyl-labeled phospholipids as a transport reporter (Fig. 1). NBD-lipids have been extensively used as reporters of phospholipid flip-flop

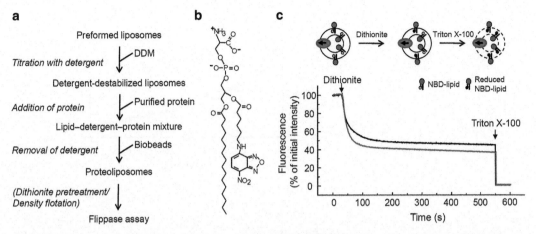

Fig. 1 Outline of the described procedures for assaying of lipid flippase activity in proteoliposomes. (a) Membrane reconstitution: liposomes containing trace amounts of NBD-lipids are detergent-destabilized and mixed with the detergent-solubilized pump complex. Subsequent removal of the detergent by Bio-Beads results in the formation of sealed proteoliposomes. Optional NBD-lipids in the outer leaflet can be prequenched with dithionite and proteoliposomes can be purified by floatation in a density gradient. (b) NBD-lipids with a C6 or C12 acyl chain in the *sn*-2 position such as C6-NBD-PS are used to analyze lipid flippase activities. (c) Illustration of the dithionite quenching assay on proteoliposomes. Dithionite reduces NBD to 7-amino-2,1,3-benzoxadiazol and thereby irreversibly eliminates the fluorescence of NBD-phospholipids situated in the outer leaflet of the vesicles. Thus, proteoliposomes with lipid flippases that promote an energy-dependent transport of the NBD-lipid from the extracellular/luminal leaflet to the cytosolic leaflet of cells show increased bleaching after incubation with Mg-ATP (*red trace*) relative to control vesicles incubated with Na-ATP (*black trace*). Triton X-100 is added to disrupt all vesicles and quench the remaining inner leaflet NBD-lipids

processes and the data obtained are qualitatively similar to that for other phospholipid analogs (e.g., radiolabeled dibutyryl phospholipids or acyl-spin-labeled phospholipids) [17, 18] and natural phospholipids [19, 20]. Their transmembrane transport is quantified by either the back-exchange or dithionite assay. In the back-exchange assay, fatty acid-free bovine serum albumin is used to deplete the outer membrane leaflet from short-chain NBD-lipid analogs, whereas, in the dithionite assay, the NBD moiety of NBD-lipid analogs present in the outer membrane leaflet is reduced and its fluorescence quenched by treatment with the membrane-impermeant dianion dithionite ($S_2O_4^{2-}$). Both methods have been widely used to monitor lipid flippase activities in proteoliposomes (e.g., refs. 14, 21, 22), isolated cellular organelles (e.g., refs. 21, 23), and intact cells (e.g., refs. 24–27). Here, we describe the use of dithionite quenching to assess ATP-dependent, lipid flippase-mediated changes in the distribution of the NBD-lipid over time.

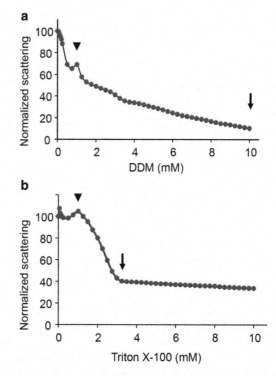

Fig. 2 Titration of preformed liposomes with detergent. Liposomes (3 μmol/mL) composed of DOPC lipids were extruded through a 200-nm polycarbonate filter and exposed to increasing concentrations of DDM (**a**) or Triton X-100 (**b**). Destabilization of the vesicles was followed by light scattering measurements at 600 nm on a fluorescence spectrometer. Increasing concentrations of DDM added to the vesicles caused a decrease in absorbance until full lysis of the vesicles. By contrast, addition of Triton X-100 caused an initial increase in absorbance until the membrane is saturated with detergent followed by a decrease in absorbance until full lysis of the vesicles. The best stage for protein insertion is between the onset of solubilization (indicated by *arrow heads*) and the complete solubilization of liposomes (indicated by *arrows*)

2 Materials

2.1 Instrumentation

1. A rotary evaporator equipped with vacuum pump set at 200 mbar and a water bath.

2. A Mini-extruder from Avanti Polar Lipids or similar model assembled with 200-nm polycarbonate membranes as specified by the manufacturer and prewashed with reconstitution buffer before use.

3. A fluorometer with excitation and emission filters in the 350–700 nm range is required. It is preferable to have a temperature

control unit and a magnetic stirrer mounted, and to use fluorimeter cuvettes suitable for magnetic stirring. This protocol is scaled for a 3 mL cuvette.

2.2 Lipids

1. Unlabeled phospholipids and fluorescent NBD-lipids with a C6 or C12 acyl chain in the *sn*-2 position are available from Avanti Polar Lipids (Alabaster, AL). In addition, synthesis of several NBD-lipid derivatives has been previously described [28–30] and will not be reviewed here. Lipid stocks are stored at −20 °C in chloroform using 1 mL glass vials with Teflon-lined screw caps and periodically monitored for purity by thin-layer chromatography (*see* **Note 1**).

2.3 Stock Solutions

1. Reconstitution buffer: 20 mM HEPES pH 7.4, 150 mM NaCl.
2. DDM prepared as 10 % (w/v) stock solution in water and stored at −20 °C.
3. ATP prepared as 0.5 mM sock solution. Dissolve 5.51 g $Na_2ATP \cdot H_2O$ in 10 mL water. Adjust pH to 7.0 with NaOH, and dilute to 20 mL. Store in aliquots of 1 mL at −20 °C or lower.
4. $MgCl_2$ prepared from $MgCl_2 \cdot 6H_2O$ as 0.5 M stock solution in water.
5. Sodium dithionite freshly prepared as 1 M stock solution in unbuffered 0.5 M Tris-HCl and kept on ice. The stock has a short shelf-life and should be used within 1 h.
6. Triton X-100 prepared as 20 % (w/v) stock solution in water and stored at 4 °C.

2.4 Biological Materials

Purified P4-ATPase in complex with the Cdc50 subunit at about 1 mg/mL of protein in reconstitution buffer plus 0.05 % (w/v) DDM; examples reported are the yeast P4-ATPase Drs2p [7] and the mammalian P4-ATPase ATP8A2 [8]. For purification of the P4-ATPase complex, it is recommended to co-express the P4-ATPase and the Cdc50 subunit [9, 31].

2.5 Miscellaneous

1. Glass beads (4–5 mm).
2. BioBeads SM-2. Before use wash the beads twice with 10 volumes of methanol and 10 times with 10 volumes of water. Remove beads that are floating on the surface. BioBeads can be stored in water for 6 months at 4 °C. Before use wash the beads twice with 10 volumes of reconstitution buffer.
3. Nycodenz (or OptiPrep) gradient medium.
4. Ultraclear centrifuge tubes (11×60 mm).

3 Method

3.1 Preparation of Unilamellar Vesicles

The procedure described is for the reconstitution of vesicles from a 1-mL sample and is adapted from [32].

1. Mix the desired amount of unlabeled (e.g., 3 µmol) and labeled (30 nmol or 1 mol % of total phospholipid) lipids dissolved in chloroform and transfer to round bottom glass flask (*see* **Note 2**). Use Hamilton syringes specifically set aside for this purpose—avoid plastic-ware.

2. Dry by rotary evaporation under vacuum to form thin lipid film on the wall of the flask.

3. After drying add 1 mL of reconstitution buffer and four glass beads to the glass vial. Mix gently until the lipids are dissolved in buffer (*see* **Note 3**). Hydration of the lipid film results in formation of multivesicular liposomes.

4. Extrude the lipid suspension 31 times through a 200-nm polycarbonate filter to form large unilamellar vesicles. The liposome suspension starts off as milky and becomes more transparent after the first extrusion pass.

5. Keep the resulting liposomes at 4 °C and use within the day of preparation.

3.2 Detergent Titration Assay of Preformed Liposomes

1. Turn on the fluorometer and set up the parameters as follows: excitation wavelength 600 nm, emission wavelength 600 nm (*see* **Note 4**).

2. Fill a semi-micro fluorimeter cuvette with 1 mL of liposome suspension (3 µmol/mL lipid).

3. Introduce the cuvette in the fluorimeter (remember to include the stirring bar), and start monitoring of the emission intensity.

4. Titrate with increasing amount of detergent solution (i.e., 10 % (w/v) DDM stock solution), allow the detergent to equilibrate for at least 1 min before each addition, and monitor of the emission intensity (*see* **Note 5**).

5. Based on the data, estimate the detergent concentrations for the onset of solubilization and for total solubilization (*see* Fig. 2).

3.3 Reconstitution of Proteoliposomes

1. Destabilize 1 mL of liposome suspension (2.5 mg/mL lipid) by addition of detergent to a final concentration determined in the titration assay (*see* Subheading 3.2 and **Note 6**).

2. Add the purified protein to the detergent destabilized liposomes to obtain a molar protein:lipid ratio between 1:6000 and 1:60,000 (*see* **Note 7**). Volume of added protein should not exceed 10 % of total mixture volume.

3. Incubate for 15 min at room temperature with gentle mixing.

4. Add 0.3 g of wet washed BioBeads per 1 mL of reconstitution mixture and incubate for 16 h at 4 °C with rotation (*see* **Note 8**).

5. Allow the BioBeads to settle for about 2 min and collect the proteoliposome solution by pipetting (*see* **Note 9**).

6. Collect the proteoliposomes by ultracentrifugation at $100,000 \times g$ for 1 h at 4 °C and gently resuspend in 250 µL of reconstitution buffer.

3.4 Density Flotation (Optional)

Flotation of liposomes on a gradient allows for separation of protein reconstituted into liposomal membrane (top gradient fraction) from free protein (bottom fraction of the gradient). This approach can also be used to recover proteoliposomes after pretreatment with dithionite to obtain vesicles exclusively labeled on the luminal side which allow for a higher sensitivity for detecting lipid flip toward the outer leaflet by P4-ATPases [15]. The flotation procedure described here is adapted from [33].

1. Mix proteoliposomes from Subheading 3.3, **step 6** with an equal volume of 80 % Nycodenz solution and transfer to a 11 × 60 mm centrifuge tube. This will result in 500 µL of 40 % Nycodenz solution containing proteoliposomes.

2. Overlay the proteoliposomes with 0.5 mL 30 %, 0.5 mL 20 %, 0.5 mL 10 %, 0.5 mL 5 %, 0.5 mL 2.5 % (w/v) Nycodenz, and 0.5 mL reconstitution buffer.

3. Centrifuge at $130,000 \times g$ for 4 h at 4 °C.

4. Collect 250 µL fractions from the top of the gradient. Proteoliposomes are recovered in the first top gradient fractions (*see*, e.g., ref. 34).

3.5 Flippase Assay

1. Prepare the assay buffer by supplementing the reconstitution buffer with ATP (final nucleotide concentration 1 mM). Store at 4 °C.

2. Turn on the fluorometer and set up the parameters as follows: excitation wavelength 470 nm, emission wavelength 540 nm, and measurement duration approximately 10 min with 1 s resolution.

3. Start the flippase assays by mixing a 100 µL aliquot of the proteoliposomes with 1.9 mL assay buffer containing 5 mM $MgCl_2$ and incubate at 37 °C (*see* **Note 10**). A sample without $MgCl_2$ supplementation serves as control (*see* **Note 11**).

4. At a given incubation period (e.g., 60 min, *see* **Note 12**), transfer the mixture to a fluorimeter cuvette (remember to include the stirring bar) and start recording the emission intensity.

5. After 30 s, add dithionite to the final concentration of 10 mM to quench the fluorescent probes in the outer leaflet of liposomes and record fluorescence until a stable line is obtained (*see* **Note 13**).

6. Add Triton X-100 to a final concentration of 0.5 % for complete reduction of all analogs by dithionite and record the background fluorescence for another 30 s.

7. Calculate the accessible pool of fluorescent lipid P_{ext} as $(F_{buffer} - F_{dithionite})/(F_{buffer} - F_0) \times 100$, where F_{buffer} is the initial fluorescence of vesicles in buffer before addition of dithionite, $F_{dithionite}$ is the plateau value of fluorescence of vesicles after incubation with dithionite, and F_0 is the final background fluorescence after addition of Triton X-100. F_{buffer}, $F_{dithionite}$, and F_0 are determined by averaging the last 10–15 data points (covering 10–15 s) of the recorded traces.

8. Calculate the extent of ATP-dependent NBD-lipid transport as $(P_{ext, ATP} - P_{ext, control})$, where ATP and control represent proteoliposomes incubated with $ATP + MgCl_2$ and ATP alone (or $AMP\text{-}PNP + MgCl_2$), respectively.

4 Notes

1. If lipids should be stored for an extended period of time, purchase the powder form for storage and then dissolve the powder using chloroform when ready to use. It is highly recommended to use ethanol-stabilized chloroform with a certificate for absence of phosgene and HCl.

2. Fluorescent lipids should not be incorporated over 2 mol % since the fluorescent groups start to self-quench at higher incorporation levels.

3. When lipid mixtures are used, the temperature of the hydrating medium should be above the gel-liquid crystal transition temperature (T_c or T_m) of the lipid with the highest T_c.

4. Destabilization of the liposomes can also be monitored by measuring the optical density of the suspension at 540 nm.

5. The equilibration time depends on the detergent; while equilibration with Triton X-100 occurs within seconds, DDM requires longer times for full equilibration (> 30 min).

6. DOPC liposomes are typically destabilized in a 1-mL reaction containing 1.7–3.5 mM DDM in a 2-mL Eppendorf tube.

7. At a protein:phospholipid ratio of 1:6000, an average of 90 lipid flippases per vesicle can be expected based on an average vesicle size of 250 nm (determined by dynamic light scattering or other means) and on assumptions concerning the thickness of the membrane bilayer (~5 nm), the cross-sectional area of a

phospholipid (~ 0.7 nm^2), and the number of phospholipid molecules per vesicle (~550,000).

8. It is essential to remove all the detergent. The kinetics and efficiency of detergent removal by Bio-beads depends on the type of detergent, the amount of beads, and temperature [35].

9. Alternatively, punch a small hole in the bottom of the Eppendorf tube with a needle and collect the proteoliposome solution in a fresh Eppendorf tube by brief centrifugation ($3000 \times g$).

10. The optimal incubation temperature depends on the membrane protein and should be selected carefully.

11. The reaction requires both ATP and Mg^{2+} as the true substrate of P-type ATPases and ABC transporters is Mg-ATP. Alternatively, a non-hydrolysable analog of ATP such as 5-Adenylylimidodiphosphate (AMP-PNP) in the presence of MgCl$_2$ can be used as control sample.

12. During prolonged incubation (> 60 min at 37 °C), the buffer might get depleted of ATP and the use of an ATP regenerating system is recommended.

13. Membranes might be relatively permeable to dithionite and even NBD-lipids positioned in the luminal vesicle layer are then destroyed. In this case, measurements can be performed at a lower temperature, e.g., 10 °C or a lower dithionite concentration is used.

Acknowledgements

This work was supported by the Danish National Research Foundation through the PUMPKIN Center of Excellence (DNRF85), the Danish Council for Independent Research | Natural Sciences (FNU, project number 10-083406), and the Lundbeck Foundation.

References

1. Lopez-Marques RL, Poulsen LR, Bailly A, Geisler M, Pomorski TG, Palmgren MG (2014) Structure and mechanism of ATP-dependent phospholipid transporters. Biochim Biophys Acta 1850:461–475

2. Coleman JA, Quazi F, Molday RS (2013) Mammalian P4-ATPases and ABC transporters and their role in phospholipid transport. Biochim Biophys Acta 1831:555–574

3. Daleke DL (2007) Phospholipid flippases. J Biol Chem 282:821–825

4. Sharom FJ (2011) Flipping and flopping--lipids on the move. IUBMB Life 63:736–746

5. Leventis PA, Grinstein S (2010) The distribution and function of phosphatidylserine in cellular membranes. Annu Rev Biophys 39:407–427

6. van Meer G, Voelker DR, Feigenson GW (2008) Membrane lipids: where they are and how they behave. Nat Rev Mol Cell Biol 9:112–124
7. Zhou X, Sebastian TT, Graham TR (2013) Auto-inhibition of Drs2p, a yeast phospholipid flippase, by its carboxyl-terminal tail. J Biol Chem 288:31807–31815
8. Coleman JA, Kwok MC, Molday RS (2009) Localization, purification, and functional reconstitution of the P4-ATPase Atp8a2, a phosphatidylserine flippase in photoreceptor disc membranes. J Biol Chem 284:32670–32679
9. Coleman JA, Molday RS (2011) Critical role of the beta-subunit CDC50A in the stable expression, assembly, subcellular localization, and lipid transport activity of the P4-ATPase ATP8A2. J Biol Chem 286:17205–17216
10. Traikia M, Warschawski DE, Lambert O, Rigaud JL, Devaux PF (2002) Asymmetrical membranes and surface tension. Biophys J 83:1443–1454
11. Rothnie A, Theron D, Soceneantu L, Martin C, Traikia M, Berridge G et al (2001) The importance of cholesterol in maintenance of P-glycoprotein activity and its membrane perturbing influence. Eur Biophys J 30:430–442
12. Auland ME, Roufogalis BD, Devaux PF, Zachowski A (1994) Reconstitution of ATP-dependent aminophospholipid translocation in proteoliposomes. Proc Natl Acad Sci U S A 91:10938–10942
13. Romsicki Y, Sharom FJ (2001) Phospholipid flippase activity of the reconstituted P-glycoprotein multidrug transporter. Biochemistry 40:6937–6947
14. Eckford PD, Sharom FJ (2005) The reconstituted P-glycoprotein multidrug transporter is a flippase for glucosylceramide and other simple glycosphingolipids. Biochem J 389:517–526
15. Zhou X, Graham TR (2009) Reconstitution of phospholipid translocase activity with purified Drs2p, a type-IV P-type ATPase from budding yeast. Proc Natl Acad Sci U S A 106:16586–16591
16. Papadopulos A, Vehring S, Lopez-Montero I, Kutschenko L, Stöckl M, Devaux PF et al (2007) Flippase activity detected with unlabeled lipids by shape changes of giant unilamellar vesicles. J Biol Chem 282:15559–15568
17. Menon AK, Watkins WE 3rd, Hrafnsdottir S (2000) Specific proteins are required to translocate phosphatidylcholine bidirectionally across the endoplasmic reticulum. Curr Biol 10:241–252
18. Devaux PF, Fellmann P, Herve P (2002) Investigation on lipid asymmetry using lipid probes: comparison between spin-labeled lipids and fluorescent lipids. Chem Phys Lipids 116:115–134
19. Gummadi SN, Menon AK (2002) Transbilayer movement of dipalmitoylphosphatidylcholine in proteoliposomes reconstituted from detergent extracts of endoplasmic reticulum. Kinetics of transbilayer transport mediated by a single flippase and identification of protein fractions enriched in flippase activity. J Biol Chem 277:25337–25343
20. Chang QL, Gummadi SN, Menon AK (2004) Chemical modification identifies two populations of glycerophospholipid flippase in rat liver ER. Biochemistry 43:10710–10718
21. Vehring S, Pakkiri L, Schroer A, Alder-Baerens N, Herrmann A, Menon AK et al (2007) Flip-flop of fluorescently labeled phospholipids in proteoliposomes reconstituted with Saccharomyces cerevisiae microsomal proteins. Eukaryot Cell 6:1625–1634
22. Malvezzi M, Chalat M, Janjusevic R, Picollo A, Terashima H, Menon AK et al (2013) Ca2+-dependent phospholipid scrambling by a reconstituted TMEM16 ion channel. Nat Commun 4:2367
23. Alder-Baerens N, Lisman Q, Luong L, Pomorski T, Holthuis JC (2006) Loss of P4 ATPases Drs2p and Dnf3p disrupts aminophospholipid transport and asymmetry in yeast post-Golgi secretory vesicles. Mol Biol Cell 17:1632–1642
24. Pomorski T, Muller P, Zimmermann B, Burger K, Devaux PF, Herrmann A (1996) Transbilayer movement of fluorescent and spin-labeled phospholipids in the plasma membrane of human fibroblasts: a quantitative approach. J Cell Sci 109:687–698
25. Araujo-Santos JM, Gamarro F, Castanys S, Herrmann A, Pomorski T (2003) Rapid transport of phospholipids across the plasma membrane of Leishmania infantum. Biochem Biophys Res Commun 306:250–255
26. Buton X, Herve P, Kubelt J, Tannert A, Burger KN, Fellmann P et al (2002) Transbilayer movement of monohexosylsphingolipids in endoplasmic reticulum and Golgi membranes. Biochemistry 41:13106–13115
27. Pomorski T, Herrmann A, Zachowski A, Devaux PF, Müller P (1994) Rapid determination of the transbilayer distribution of NBD-phospholipids in erythrocyte membranes with dithionite. Mol Membr Biol 11:39–44
28. Colleau M, Herve P, Fellmann P, Devaux PF (1991) Transmembrane diffusion of fluorescent phospholipids in human erythrocytes. Chem Phys Lipids 57:29–37

29. Fellmann P, Herve P, Pomorski T, Müller P, Geldwerth D, Herrmann A et al (2000) Transmembrane movement of diether phospholipids in human erythrocytes and human fibroblasts. Biochemistry 39:4994–5003
30. Sleight RG (1994) Fluorescent glycerolipid probes. Synthesis and use for examining intracellular lipid trafficking. Methods Mol Biol 27:143–160
31. Jacquot A, Montigny C, Hennrich H, Barry R, le Maire M, Jaxel C et al (2012) Phosphatidylserine stimulation of Drs2p.Cdc50p lipid translocase dephosphorylation is controlled by phosphatidylinositol-4-phosphate. J Biol Chem 287:13249–13261
32. Geertsma ER, Nik Mahmood NA, Schuurman-Wolters GK, Poolman B (2008) Membrane reconstitution of ABC transporters and assays of translocator function. Nat Protoc 3:256–266
33. Chen X, Arac D, Wang TM, Gilpin CJ, Zimmerberg J, Rizo J (2006) SNARE-mediated lipid mixing depends on the physical state of the vesicles. Biophys J 90:2062–2074
34. Marek M, Milles S, Schreiber G, Daleke DL, Dittmar G, Herrmann A et al (2011) The yeast plasma membrane ATP binding cassette (ABC) transporter Aus1: purification, characterization, and the effect of lipids on its activity. J Biol Chem 286:21835–21843
35. Rigaud JL, Mosser G, Lacapere JJ, Olofsson A, Levy D, Ranck JL (1997) Bio-Beads: an efficient strategy for two-dimensional crystallization of membrane proteins. J Struct Biol 118:226–235

Part III

In Vitro Functional Studies

Chapter 19

The Use of Metal Fluoride Compounds as Phosphate Analogs for Understanding the Structural Mechanism in P-type ATPases

Stefania J. Danko and Hiroshi Suzuki

Abstract

The membrane-bound protein family, P-type ATPases, couples ATP hydrolysis with substrate transport across the membrane and forms an obligatory auto-phosphorylated intermediate in the transport cycle. The metal fluoride compounds, BeF_x, AlF_x, and MgF_x, as phosphate analogs stabilize different enzyme structural states in the phosphoryl transfer/hydrolysis reactions, thereby fixing otherwise short-lived intermediate and transient structural states and enabling their biochemical and atomic-level crystallographic studies. The compounds thus make an essential contribution for understanding of the ATP-driven transport mechanism. Here, with a representative member of P-type ATPase, sarco(endo)plasmic reticulum Ca^{2+}-ATPase (SERCA), we describe the method for their binding and for structural and functional characterization of the bound states, and their assignments to states occurring in the transport cycle.

Key words Metal fluoride, Phosphate analog, Beryllium fluoride (BeF_x), Magnesium fluoride (MgF_x), Aluminum fluoride (AlF_x), P-type ATPase, SERCA (sarco(endo)plasmic reticulum Calcium ATPase), Ca^{2+} pump, Phosphoryl transfer, Phosphorylated intermediate

1 Introduction

Phosphoryl transfer from a donor molecule to an acceptor is a very basic chemical reaction occurring in many biochemical processes. Because of the inherent characteristics of phosphorous being able to form trivalent, tetravalent, or pentavalent compounds, the phosphoryl transfer can proceed by either of two general types of mechanism; dissociative or associative depending on which takes place first, P-O bond breaking or forming [1, 2]. In the in-line associative mechanism, which is employed in the P-type ATPases [3–5], the attacking nucleophile enters opposite the leaving group producing a pentacovalent transition state with bipiramidal geometry in the transition state and then releasing a product with again a tetrahedral geometry but with an inverted stereochemical configu-

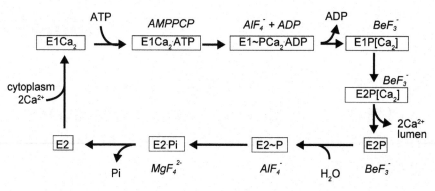

Scheme 1 Ca^{2+}-ATPase Reaction cycle. $E1P$, ADP-sensitive phosphorylated intermediate (EP) that can produce ATP in the reverse reaction with ADP. $E2P$, ADP-insensitive EP. The bound Ca^{2+} ions become occluded at the transport sites in $E1P$ ($E1P[Ca_2]$). Mg^{2+} binding at the catalytic site (*see* Fig. 1) is required for the EP formation and hydrolysis, but not depicted for simplicity. In $E2P$ after Ca^{2+} release, protons bind to the Ca^{2+} ligands from the lumenal side and are released to the cytoplasmic side in $E2 \rightarrow E1Ca_2$, thus the Ca^{2+} pump is actually a Ca^{2+}/H^+ pump (also not depicted for simplicity). The stable analogs for the transient and intermediate structural states are produced with metal fluoride compounds and AMPPCP (a non-hydrolyzable ATP analog) as indicated

ration (imagine, your umbrella turns inside out by wind). Mg^{2+} bound at the catalytic site, as in the case of P-type ATPases, generally functions as a catalytic cofactor for the proper orientation and charge-shielding for the nucleophilic attack, i.e., the Mg^{2+} makes the phosphorous more electrophilic.

The well-studied representative member of this group, sarco(endo)plasmic reticulum Ca^{2+}-ATPase (SERCA), couples the energy gained by one ATP molecule hydrolysis with transport of two Ca^{2+} from the cytoplasm into the endoplasmic reticulum lumen against its concentration gradient (Scheme 1) (for recent review, *see* refs. 6–8).

Thus the Ca^{2+}-ATPase is a "Ca^{2+} pump." In the transport cycle, the ATP γ-phosphate is first transferred to the conserved catalytic aspartate, Asp351 in the Ca^{2+}-ATPase, by the nucleophilic attack of the aspartyl oxygen forming an auto-phosphorylated intermediate, then later the aspartylphosphate is hydrolyzed by a nucleophilic attack with a specific water molecule in the catalytic site releasing an inorganic phosphate; thus ATP is hydrolyzed. Both the phosphorylation and hydrolysis proceed via an in-line associative mechanism, i.e., backside-in-line nucleophilic substitution with inversion of configuration at the phosphorus. Kinetically, it has been well characterized that the auto-phosphorylated intermediate $E1PCa_2$ (formed from the Ca^{2+}-bound activated enzyme with ATP) proceeds its large structural change, EP isomerization to the $E2P$ ground state, releasing Ca^{2+} into the lumen and producing the catalytic site with an ability for the hydrolysis, thereby the ATP-driven transport will be accomplished (ATP hydrolysis occurs with Ca^{2+} transport). The question is how the mutual structural communication between the catalytic site and transport sites takes place in each of processes: EP formation, isomerization, and hydrolysis in the transport cycle. Note that the pump cycles maximally ~50 times per second, very rapidly. Here the phosphate analogs, BeF_x, AlF_x, and MgF_x, in which the phosphorous and oxygen in phosphate are replaced by the metal and fluoride respectively, made an essential contribution as they "freeze" and "mimic" the short-lived phosphorylated intermediates and transient states (*see* review [9] and original works [10–14]). Assignment of the ATPase with a tightly bound metal fluoride compound to a specific intermediate/transient state was done biochemically by the detailed analyses of its structural and functional properties and by comparing with the known nature of the enzyme states occurring in transport cycle [11–14], thereby ensured which state is represented by the ATPase complex and by its crystal structure [15–21]. Furthermore, the biochemical analyses of the stabilized structural states revealed unknown properties of the intermediate/transient states otherwise not feasible to be explored [13, 14].

For using BeF_x, AlF_x, and MgF_x compounds, it is critical to note the coordination chemistry; i.e., the size, charge, and coordination number and angle of the central cations differ, therefore, each compound has its specific distance between the metal and fluoride, charge distribution, and geometry. Thereby as described below, each mimics a specific phosphate state with characteristic and well defined features, which change slightly but distinctly in the phosphoryl transfer reactions. Thus, by the three different metal fluoride compounds, all the possible structural states occurring in the phosphoryl transfer reactions can be mimicked. Also, in combination with the strong electronegativity of fluoride ions, the metal fluoride compounds bind very tightly to the catalytic site and fix the enzyme structural states enabling even the crystallization. In Fig. 1, the crystal structures around the catalytic Asp351 with bound metal fluorides are depicted very nicely by Toyoshima throughout the catalytic cycle (*see* also ref. 6).

1.1 BeF$_x$

Be^{2+} with its high charge density due to its small size is able to attract the oxygen atom of the catalytic aspartate to coordinate. Actually in the Ca^{2+}-ATPase, the stable analogs of phosphorylated intermediates $E1PCa_2$, $E2PCa_2$, and the $E2P$ ground state were successfully developed by the BeF_3^- ligation at Asp351, as $E1Ca_2 \cdot BeF_3^-$, $E2Ca_2 \cdot BeF_3^-$, and $E2 \cdot BeF_3^-$, respectively [13, 14], and the crystal structure of $E2 \cdot BeF_3^-$ was solved (Fig. 1) [20, 21]. In the crystal structure, the BeF_x bound at the catalytic site is actually the BeF_3^- compound that adopts a tetrahedral geometry with the Be-F 1.55-Å bond length and with the fourth coordination from the aspartyl oxygen, thereby making them strictly isomorphous to the tetrahedral phosphate group and superimposable with the geometry of covalently bound phosphate at the catalytic

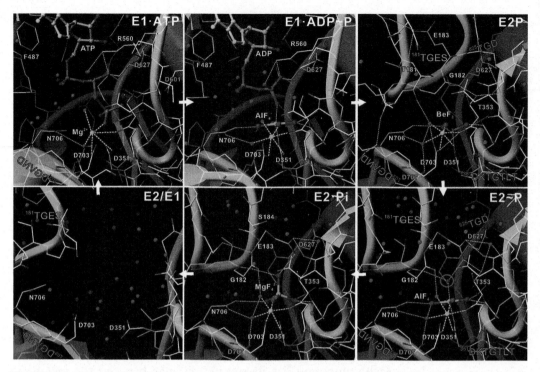

Fig. 1 Structural events that occur around the phosphorylation site (Kindly provided by Prof. Chikashi Toyoshima, University of Tokyo, *see* also review article by Toyoshima ref. 6). Crystal structures of $E1Ca_2 \cdot AMPPCP$, $E1Ca_2 \cdot ATP$ analog; $E1Ca_2 \cdot AlF_4^- \cdot ADP$, $E1 \sim PCa_2 \cdot ADP$ transition state analog; $E2 \cdot BeF_3^-$ (with bound TG), E2P ground state analog; $E2 \cdot AlF_4^-$ (with bound TG), $E2 \sim P^\ddagger$ transition state analog; $E2 \cdot MgF_4^{2-}$ (with bound TG), $E2 \cdot P_i$ product state analog; E2 (with bound TG and BHQ). In the figure, $E1Ca_2$ is denoted as E1 for simplicity. *Broken lines* in *pink* show likely hydrogen bonds and those in light green Mg^{2+} coordination. *Small red spheres* represent water molecules. Conserved sequence motifs are labeled. *TG* thapsigargin, *BHQ* 2,5-di-tert-butyl-1,4-dihydroxybenzene

aspartate [9, 20, 21]. The structures of the other two, $E1Ca_2 \cdot BeF_3^-$ and $E2Ca_2 \cdot BeF_3^-$, are not yet solved; nevertheless, their biochemical structural analyses demonstrated the dramatic and critical changes in the catalytic site structure, the cytoplasmic domains organization, and the transmembrane helices arrangement (i.e., the transport sites structure) during the phosphoryl transfer, *EP* isomerization, and Ca^{2+} release processes $E1 \sim PCa_2 \cdot ADP \rightarrow E1PCa_2 \rightarrow E2PCa_2 \rightarrow E2P + 2Ca^{2+}$ [13, 14].

1.2 AlF$_x$

Al^{3+} is able to make five- or six-fold coordinations, and AlF_x bound to the enzymes was shown to be either the AlF_3 or AlF_4^- compound, and their formation also depends on pH in the incubation medium [22]. Importantly, the bound AlF_3 and AlF_4^- both possess a planar (trigonal (AlF_3) and square (AlF_4^-)) geometry, in which two oxygen atoms coordinate to the aluminum at apical positions, e.g., from the ADP β-phosphate or the hydrolytic water molecule and from the specific aspartate to produce the state superimposable (AlF_3) or analogous (AlF_4^-) to the trigonal bipyramidal structure of the penta-coordinated phosphorus in the transition state of the in-line associative mechanism. In the Ca^{2+}-ATPase, the transition states, $E1 \sim PCa_2 \cdot ADP^\ddagger$ in the phosphoryl transfer from ATP and $E2 \sim P^\ddagger$ in the aspartylphosphate hydrolysis, were successfully mimicked and fixed as $E1Ca_2 \cdot AlF_4^- \cdot ADP$ and $E2 \cdot AlF_4^-$, respectively [13, 14] and their crystal structures were solved (Fig. 1) [18–21].

1.3 MgF_x

MgF_x bound at the catalytic site of the Ca^{2+}-ATPase was shown to be MgF_4^{2-} [18, 23] of which tetrahedral geometry is superimposable to the non-covalently bound P_i and therefore represents the product state of the $E2P$ hydrolysis ($E2 \cdot P_i$ complex). Notably here, although both the bound MgF_4^{2-} and BeF_3^- (BeF_3O adduct with the fourth coordination from the aspartate) produce the tetrahedral geometry, Be^{2+} with its high charge density (due to its small size) is able to attract the Asp351 oxygen atom to coordinate but it is not the case in the bound MgF_4^{2-} (the oxygen atom cannot substitute the fluoride). Thus MgF_4^{2-} in the $E2 \cdot MgF_4^{2-}$ complex formed in the absence of Ca^{2+} mimics the non-covalently bound P_i in the hydrolysis product complex $E2 \cdot P_i$, in contrast to BeF_3^- [14] (Fig. 1). It is also noteworthy that the ATP-induced phosphorylation in the Ca^{2+}-ATPase does not involve a possible P_i bound state (i.e., due to the in-line associative mechanism), and in agreement, the ATPase with bound Ca^{2+}, $E1Ca_2$, does not form a possible complex $E1Ca_2 \cdot MgF_4^{2-}$ [14].

Thus, due to the inherent charge and geometrical arrangements in BeF_x, AlF_x, and MgF_x, each compound binds tightly at the catalytic site and mimics and fixes a specific enzyme structural state with the bound phosphate during the phosphoryl transfer reactions. Also these enzyme structural analogs are remarkably stable and perfectly suitable for their biochemical characterizations and crystallographic studies (Fig. 1).

2 Materials

2.1 Protein Samples

1. Sarcoplasmic reticulum vesicles (SR vesicles) prepared from rabbit muscle, e.g., from procedures according to Nakamura et al. [24].

2. SERCA protein expressed in COS-1 cells, and preparations of microsomes containing expressed SERCA protein, e.g., from procedures according to Maruyama and MacLennan [25].

3. Optional: microsomes expressing the mutant **4Gi-46/47** (4 glycine insertion between amino acid 46 and 47 on the linker connecting the A domain and the first transmembrane helix (M1)), a mutant Ca^{2+}-ATPase specially designed for trapping the transient $E2PCa_2$ state [26].

2.2 Stock Solutions for Metal Fluoride Binding

Use analytical grade reagents. Prepare solutions with ultrapure water (prepared by purifying deionized water to attain a sensitivity of 18 MΩ cm at 25 °C) and keep stock solutions at 4 °C. Wear mask and vinyl gloves when measuring materials hazardous for health!

1. 0.5 M KF: Dissolve 0.29 g KF in 10 mL of water (*see* **Note 1**).
2. 0.1 M BeSO$_4$: Dissolve 0.177 g BeSO$_4$ tetrahydrate in 10 mL of water (*see* **Note 2**).
3. 0.1 M AlCl$_3$: Dissolve 0.241 g AlCl$_3$ hexahydrate in 10 mL of water.
4. 1 M MgCl$_2$: Dissolve 20.3 g MgCl$_2$ hexahydrate in 100 mL of water.

2.3 Buffers

1. 1 M Tris solution for pH adjustment (100 mL): Dissolve 12.14 g of Tris (hydroxymethyl) aminomethane in water and fill up to 100 mL.
2. 0.5 M MOPS/Tris buffer, pH 7.0 at 25 °C (500 mL): Measure 52.32 g MOPS into a beaker and dissolve in about 450 mL of water. Equilibrate the solution in a water bath to 25 °C, and under continuous stirring adjust the pH with the 1 M Tris solution to pH 7.0. Adjust the volume of the buffer solution to 500 mL and check the pH once more and assure the desired pH.
3. 0.5 M MES/Tris buffer, pH 6.0 at 25 °C (500 mL): Measure 53.31 g MES monohydrate into a beaker and dissolve in about 450 mL of water. Equilibrate the solution in a water bath to 25 °C and under continuous stirring adjust the pH with the 1 M Tris solution to pH 6.0. Adjust the volume of the buffer solution to 500 mL and check the pH once more and assure the desired pH.

2.4 Other Solutions

1. 3 M KCl (100 mL): Dissolve 22.37 g KCl in 100 mL of water.
2. 3 M LiCl (100 mL): Dissolve 12.72 g LiCl in 100 mL of water.
3. 0.5 M CaCl$_2$ (100 mL): Dissolve 7.35 g CaCl$_2$ dihydrate in 100 mL of water.
4. 0.2 M EGTA at pH 7.0 or 6.0 at 25 °C (100 mL): Measure 7.607 g EGTA into a beaker and dissolve in about 80 mL of water. Similarly to the buffer preparation, adjust the pH to the required value with the 1 M Tris solution. Adjust the volume of the solution to 100 mL and check the pH once more and assure the desired pH (*see* **Note 3**).
5. 100 mM ADP, pH 7.0 (10 mL): Dissolve 0.4272 g adenosine 5′-diphosphate sodium salt in about 10 mL of water and adjust

pH to 7.0 at 25 °C with the 1 M Tris solution. Measure the optical density of the solution at 259 nm and calculate the exact concentration by using the molar extinction coefficient of 15,400 M^{-1} cm^{-1}. Keep the ADP solution frozen at −20 °C in small aliquots (*see* **Note 4**).

6. 3 mM A23187: Dissolve 2 mg calcium ionophore A23187 in 1.27 mL ethanol. Keep in small aliquots at −20 °C and protect from light (*see* **Note 5**).
7. Reaction medium 1: 30 mM MOPS/Tris pH 7.0, 0.1 M KCl, 15 mM $MgCl_2$, 0.1 mM $CaCl_2$ (or $CaCl_2$/EGTA solution with the desired free Ca^{2+} concentration).
8. Reaction medium 2: 50 mM MOPS/Tris pH 7.0, 0.1 M KCl, 15 mM $MgCl_2$, 0.01 mM $CaCl_2$.
9. Reaction medium 3: 50 mM MOPS/Tris pH 7.0, 50 mM LiCl, 7 mM $MgCl_2$, 1 mM EGTA.
10. Reaction medium 4: 50 mM MOPS/Tris pH 7.0, 50 mM LiCl, 7 mM $MgCl_2$, 5 μM Ca^{2+} ionophore A23187 and 1 mM $CaCl_2$.
11. Reaction medium 5: 50 mM MOPS/Tris pH 7.0, 50 mM LiCl, 5 mM $MgCl_2$, 2 mM EGTA.
12. Reaction medium 6: 50 mM MES/Tris pH 6.0, 100 mM KCl, 0.1 mM $MgCl_2$, 2 mM EGTA.
13. Reaction medium 7: 50 mM Tris/HCl, pH 7.5, 50 mM LiCl, 10 mM $MgCl_2$, 2 mM EGTA, 20 % (v/v) glycerol, 20 % (v/v) Me_2SO.
14. Homogenization buffer: 5 mM MOPS/Tris pH 7.0, 50 mM LiCl, 0.5 mM EGTA, and 0.3 M sucrose.

3 Methods

Important: The conditions to produce $E1$ and $E2$ states depend on each member of P-type ATPases, and those described here apply to SERCA! Also the samples described are representative ones, SR vesicles and SERCA expressed in microsomes, but the protocols for the use of metal fluoride compounds generally apply to all the P-type ATPase members and the ATPases (wild type and mutants) in microsomes or purified (for more general information, *see* **Note 6**).

3.1 E1PCa$_2$ Analogs: AlF$_x$ and BeF$_x$ Binding to the Ca-Bound State E1Ca$_2$

This section describes how to generate $E1Ca_2 \cdot AlF_4^- \cdot ADP$ and $E1Ca_2 \cdot BeF_3^-$ forms as analogs for $E1 \sim PCa_2 \cdot ADP^{\ddagger}$ (the transition state in the ATP-induced phosphorylation) and $E1PCa_2$ (the covalently phosphorylated E1P ground state) [12, 14]

1. Pipette SR vesicles into reaction medium 1 to give 0.05–1 mg protein/mL and incubate for a few minutes at 25 °C (*see* **Note 7**).
2. Add the ligands to give 2 mM KF, 0.1 mM AlCl$_3$, and 50 μM ADP (for $E1\text{Ca}_2 \cdot \text{AlF}_4^- \cdot \text{ADP}$ (*see* **Note 8**)), or 2 mM KF and 0.1 mM BeSO$_4$ (for $E1\text{Ca}_2 \cdot \text{BeF}_3^-$, *see* **Note 9**) and incubate for 30 min at 25 °C for the metal fluoride binding (*see* **Note 10**). The sample is ready for analyses. $E1\text{Ca}_2 \cdot \text{AlF}_4^-$ as an analog for possible $E1 \sim \text{PCa}_2^\ddagger$ without ADP can be formed by omitting ADP otherwise as above for $E1\text{Ca}_2 \cdot \text{AlF}_4^- \cdot \text{ADP}$.

3.2 E2PCa$_2$ Analog: E2Ca$_2 \cdot$BeF$_3^-$

This sections describes BeF$_x$ binding to a mutant Ca-ATPase **4Gi-46/47**, specially designed for trapping the transient E2PCa$_2$ state [26, 27] (*see* **Note 11**).

3.2.1 Obtaining E2Ca$_2 \cdot$BeF$_3^-$ from E1Ca$_2$ State by BeF$_x$ Binding

1. Treat 0.35 mg protein/mL of the microsomes expressing the mutant **4Gi-46/47** with 3 mM KF and 0.05 mM BeSO$_4$ at 25 °C for 30 min in reaction medium 2.

3.2.2 Obtaining E2Ca$_2 \cdot$BeF$_3^-$ from E2$\cdot$BeF$_3^-$ by Ca^{2+} Binding to Lumenally Oriented Low-Affinity Transport Sites

1. Treat 1 mg protein/mL of the microsomes expressing the mutant **4Gi-46/47** with 1 mM KF plus 0.02 mM BeSO$_4$ in reaction medium 3 at 25 °C for 30 min to produce first $E2 \cdot \text{BeF}_3^-$.
2. Dilute the mixture 2.5-fold with reaction medium 4, and incubate at 25 °C for 1 min (for the Ca^{2+}-dependence curve, *see* Fig. 8 in ref. 27).

3.3 E2P Analogs: BeF$_x$, AlF$_x$, and MgF$_x$ Binding to Ca^{2+}-Free E2 State (See Note 12)

3.3.1 E2$\cdot$BeF$_3^-$, an Analog for E2P Ground State [13, 28]

1. Treat 1 mg/mL SR vesicles with 0.5 mM KF, and 40 μM BeSO$_4$ in reaction medium 5 for 3 h at 25 °C.

3.3.2 E2$\cdot$AlF$_4^-$, an Analog for the Transition State E2 $\sim$ P‡ in E2P Hydrolysis [13]

1. Treat 1 mg/mL SR vesicles with 3 mM KF and 50 μM AlCl$_3$ in reaction medium 6 for 1 h at 25 °C (*see* **Note 13**).

3.3.3 E2$\cdot$MgF$_4^{2-}$, an Analog for the Product Complex E2$\cdot$P$_i$ in E2P Hydrolysis [11, 13, 23]

1. Treat 2 mg/mL SR vesicles with 1 mM KF in reaction medium 7 for 3 h at 25 °C (*see* **Note 14**).

3.4 Removal of Unbound F^-, Be^{2+}, Al^{3+}, and Mg^{2+} and Storage [13, 23]

In the complexes $E2 \cdot BeF_3^-$, $E2 \cdot AlF_4^-$, and $E2 \cdot MgF_4^{2-}$, the metal fluoride compounds are tightly bound and Mg^{2+} at the catalytic site is tightly occluded, therefore the unbound ligands can be removed by washing without decomposition of the ATPase complexes as follows:

1. Cool down to 4 °C after their formation, centrifuge for 10 min at 4 °C and $541,000 \times g$.
2. Homogenize the pellet gently in homogenization buffer, and repeat once more the centrifugation/homogenization.
3. You may store the sample in tubes at −80 °C after flash freezing by liquid nitrogen (*see* **Note 15**).

3.5 Characterization of Structural Analogs of Ca^{2+}-ATPase with Bound Metal Fluoride

The biochemical analyses of structural and functional properties of the analogs and their systematic comparison with the states occurring in the transport cycle enabled to assign each of these analogs to a particular structural state in the cycle. The results uncovered the structural events occurring in the catalytic site, cytoplasmic domains organization, and transport sites during the phosphorylation, Ca^{2+}-occlusion, EP isomerization, Ca^{2+}-deocclusion/release, and $E2P$ hydrolysis with the gate-closure, thereby provided essential information for understanding the mechanism of the Ca^{2+}-ATPase via mutual structural communication between the catalytic site in the cytoplasmic domains and the transport sites in the transmembrane region. We describe below briefly the series of analyses and findings to emphasize critical importance of such biochemical analyses as well as of the crystallographic structural studies, which can be applied to other P-type ATPases.

3.5.1 ATPase Activities

The ATPase activity and EP formation from ATP and from P_i are all completely inhibited by the tight binding of the metal fluoride to the catalytic site. To ensure that the inactivation is not due to an irreversible denaturation, the reactivation in the $E2P$ analogs was performed [13, 23] by incubating them with a very high concentration of Ca^{2+} (20 mM) at 25 °C for 50 min in 0.1 M KCl, 7 mM $MgCl_2$, and 50 mM MOPS/Tris pH 7.0 in the presence of Ca^{2+} ionophore A23187. Upon the Ca^{2+} binding to the transport sites from lumenal side, the complexes decompose and release the bound metal fluoride; thereby the ATPase activity and the EP formation ability are entirely restored. Furthermore, the Ca^{2+} concentration dependence of this lumenal Ca^{2+}-induced reactivation revealed a marked difference in luminal Ca^{2+} accessibility between $E2 \cdot BeF_3^-$ and $E2 \cdot AlF_4^-/E2 \cdot MgF_4^{2-}$; namely $E2 \cdot AlF_4^-/E2 \cdot MgF_4^{2-}$ are much more resistant to the lumenal Ca^{2+} than $E2 \cdot BeF_3^-$ showing that the Ca^{2+} release gate is opened in the $E2P$ ground state as the Ca^{2+}-releasing structure, but becomes closed in the transition state ($E2 \sim P^{\ddagger}$) thereby preventing a possible Ca^{2+} leakage [13]. Thus in $E2P \rightarrow E2 \sim P^{\ddagger}$, the lumenal gate closure takes place and is tightly coupled with the structural change in the catalytic site ([13, 27], see also the following section "Nature of the Catalytic Site"). The decomposition of the analogs with bound Ca^{2+}, i.e., $E1Ca_2 \cdot AlF_4^- \cdot ADP$, $E1Ca_2 \cdot AlF_4^-$, and $E1Ca_2 \cdot BeF_3^-$ occurs during incubation after removal of free metal fluoride, and thus the reactivation occurs.

3.5.2 Ca^{2+} handling

Ca^{2+} binding and occlusion can be assessed by a membrane filtration method using $^{45}Ca^{2+}$ radioisotope (see also Chap. 23 and Chap. 20). The analyses revealed the stoichiometric binding and occlusion of Ca^{2+} in $E1Ca_2 \cdot AMPPCP$ ($E1Ca_2 \cdot ATP$ complex mimicked with the non-hydrolyzable ATP analog), $E1Ca_2 \cdot AlF_4^- \cdot ADP$, $E1Ca_2 \cdot BeF_3^-$, and $E2Ca_2 \cdot BeF_3^-$, the structural mechanism of the Ca^{2+} handling during the EP formation, isomerization, and hydrolysis for the Ca^{2+} transport [14, 27].

3.5.3 Cytoplasmic Domains Organization

Limited proteolysis was found to be essential to probe the changes in the cytoplasmic domains organization during the transport cycle. Specific cleavage sites for proteinase K, trypsin, and V8 protease are located on the A domain, top part of the second transmembrane helix (M2), and A-domain/M3-linker. When the A and P domain and these M2 and A/M3-linker regions gather and associate with each other, the sites become sterically hindered and therefore completely resistant to the proteases, [11–14]. In contrast, the sites are rapidly cleaved in "open" states. The analysis on the stable structural analogs with metal fluoride and on the non-phosphorylated $E1Ca_2$ and $E2$ states revealed how the cytoplasmic domains move and change their organization state in each step of the transport cycle, thereby predicted the critical importance of these changes in the structural communication between the cytoplasmic domains and transmembrane region to achieve the transport (see **Note 16**).

3.5.4 Nature of the Catalytic Site

TNP-AMP (2′(3′)-O-(trinitrophenyl)-AMP) binds to the ATP binding site with an extremely high affinity, and its fluorescence monitors a change in the hydrophobicity; i.e., it develops "superfluorescence," an extremely high fluorescence, in the hydrophobic atmosphere around the TNP-moiety (see also Chap. 22). The fluorescence analysis of the structural analogs with bound metal fluoride compounds revealed that the catalytic site in $E1PCa_2$ and in the $E2P$ ground state is strongly hydrophobic, but hydrophilic in other structural states. The results further showed that in $E2P \rightarrow E2 \sim P^{\ddagger}$ of the $E2P$ hydrolysis, the catalytic site becomes hydrophilic and thus opened, and this structural rearrangements couples the luminal gate closure [13, 14, 18, 26, 27].

3.5.5 Structural Change in Transmembrane Region	Intrinsic tryptophan fluorescence of the Ca^{2+}-ATPase reflects conformational changes near the Ca^{2+} binding sites in the transmembrane region, as most of the tryptophan residues, 12 among 13 are located at or near this region. The fluorescence actually changes upon Ca^{2+} binding/dissociation in the unphosphorylated state $E2/E1Ca_2$ and during the EP formation, isomerization, and hydrolysis (see also Chap. 21). The fluorescence analysis with the stabilized structural analogs assigned the reaction processes in which the fluorescence changes, thereby uncovered the structural communication between the catalytic site and the transport sites for the Ca^{2+} handling, i.e., the cytoplasmic gate closure for Ca^{2+} occlusion at the transport sites, the rearrangements of the Ca^{2+} ligands at the transport sites for affinity-reduction/deocclusion to release Ca^{2+}, and the lumenal gate closure to prevent Ca^{2+} leak [13, 18].

4 Notes

1. Keep fluoride powder in a desiccator and dispose the fluoride solutions in accordance with local regulations. In the incubation medium for binding, the K^+ and F^- concentrations are in the mM range, and if you should avoid the presence of mM K^+, use NaF or LiF instead of KF.

2. Beryllium solution can be prepared from $BeSO_4$ tetrahydrate powder dissolved in deionized water [13, 29, 30] or from $BeCl_2$ dissolved in 1 % HCl [20, 28]. The $BeCl_2$ solution might emit chloride fumes, therefore be extremely careful. The pH of the $BeCl_2$ solution is acidic, and pH can be adjusted by NaOH. Because beryllium is very toxic, keep properly and handle with great care; follow instructions for disposal.

3. EGTA (ethylene glycol-bis(2-aminoethylether)-N,N,N',N'-tetraacetic acid) chelates Ca^{2+} with a high affinity at a ratio of 1:1. Its solubility increases with increasing pH. The free Ca^{2+} concentration can be adjusted precisely by a Ca^{2+}/EGTA buffer (appropriate concentrations of Ca^{2+} and EGTA). For calculation of free Ca^{2+} concentration, use programs such as the Calcon program.

4. When the presence of sodium is undesired in your experiment, use adenosine 5'-diphosphate monopotassium salt.

5. Calcium ionophore A23187 is a divalent cation ionophore but highly selective for Ca^{2+}. Because this strongly hydrophobic reagent is dissolved in ethanol which is a protein-denaturant, do not exceed an ethanol concentration of 1 % (v/v), preferably 0.5 % or less. The ionophore should be added to a reaction mixture that contains the membrane protein, so as to

incorporate it into the membrane phase, otherwise it will be precipitated in an aqueous solution.

6. General: The metal fluoride binding can be done in any volume and protein concentrations, in Eppendorf tubes, plastic tubes, or beakers, and at any temperature (4–37 °C) depending on the type of the experiments you design. Glass-wear contains aluminum and may elute a trace amount of Al^{3+}, therefore use an aluminum chelator if even its very low concentration might possibly disturb your system.

7. The ATP-induced and P_i-induced phosphorylation of Ca^{2+}-ATPase and the resulting $E1PCa_2$ and $E2P$ require a divalent cation bound at the catalytic Mg^{2+} site (*see* Subheading 1). Mg^{2+} is the physiological catalytic cofactor, but Mn^{2+} or in some cases Ca^{2+} can substitute for Mg^{2+}. For the metal fluoride binding and formation of the structural analogs too, the divalent cation binding is obligatory; Mg^{2+}, Mn^{2+}, or Ca^{2+} is the choice.

8. $E1Ca_2 \cdot AlF_4^- \cdot ADP$ can be formed with the Ca^{2+} binding at the catalytic Mg^{2+} site in the absence of Mg^{2+} and presence of Ca^{2+} as low as 0.1 mM. Also, the $E1Ca_2 \cdot AMPPCP$ complex ($E1Ca_2 \cdot ATP$ analog) can be formed with bound Ca^{2+} at the Mg^{2+} site, and this complex occludes more strongly Ca^{2+} at the transport sites than that with the bound Mg^{2+} [31]. The ligation of Ca^{2+} at the nucleotide-bound catalytic site causes a marked stabilization of these ATPase structures.

9. The $E1Ca_2 \cdot BeF_3^-$ complex loses its bound Ca^{2+} by spontaneously isomerizing to the Ca^{2+}-released $E2 \cdot BeF_3^-$ structure in a time- and Ca^{2+} concentration-dependent manner, e.g., in ~10 h at 25 °C in 0–50 μM Ca^{2+} [14]. Also, the complex is decomposed to the $E1Ca_2$ state by the Ca^{2+} substitution for Mg^{2+} at the catalytic site in the presence of high concentration of Ca^{2+}; namely the Ca^{2+} binding at the catalytic site is unfavorable for the $E1Ca_2 \cdot BeF_3^-$ structure probably due to a disruption of the precise geometry of the catalytic site because of the difference in the coordination distance and number between Mg^{2+} and Ca^{2+} [14]. Therefore, for the stabilization of this complex, the balance between the concentrations of Mg^{2+} and Ca^{2+} is critical. We found that the presence of 0.7 mM Ca^{2+} and 15 mM Mg^{2+} in the ionophore A23187 stabilizes perfectly this $E1Ca_2 \cdot BeF_3^-$ complex; no decomposition even after 300 h at 25 °C [14].

10. Al^{3+} or Be^{2+} at high concentrations may cause protein precipitation; therefore keep their concentrations low, below 100 μM. Also the sequence of addition is important; because the presence of Mg^{2+} (or its substitute) bound at the catalytic site is obligatory for the formation of the structural analog by the AlF_x or BeF_x binding, the F^- addition in the presence of Mg^{2+} should be followed by an immediate addition of Be^{2+} or Al^{3+} to prevent the MgF_x formation and binding.

11. The transient state before the Ca^{2+} release $E2P$ with bound Ca^{2+} ($E2PCa_2$) has been postulated but never been identified, due to the rate-limiting slow $E1PCa_2$ to $E2PCa_2$ isomerization followed by a very rapid Ca^{2+} release. By elongating the linker between the A domain and first transmembrane helix (A/M1-linker, a.a. 40–48) with four glycine insertions, we successfully trapped and identified this transient $E2P$ with occluded Ca^{2+}, $E2PCa_2$ [26]. The finding revealed the structural mechanism with critical function of this linker for deocclusion and release of Ca^{2+} in $E2PCa_2 \rightarrow E2P + 2Ca^{2+}$. Importantly, we further found [27] that the $E2PCa_2$ analog is formed only by the tetrahedral BeF_x (BeF_3^-), but not by AlF_x or MgF_x, from $E1Ca_2 \cdot BeF_3^-$ (mimicking $E1PCa_2 \rightarrow E2PCa_2$ forward reaction) and from $E2 \cdot BeF_3^-$ by binding of Ca^{2+} at the lumenally oriented low-affinity transport sites (mimicking the lumenal Ca^{2+}-induced reverse reaction $E2P + 2Ca^{2+} \rightarrow E2PCa_2$).

12. The methods were described basically according to the original reports with Ca^{2+}-ATPase, but the conditions to form the complexes $E2 \cdot BeF_3^-$, $E2 \cdot AlF_4^-$, and $E2 \cdot MgF_4^{2-}$ can fairly be modified, i.e., we found their formation at any pH in 6–7.5 and without Me_2SO and glycerol (described for $E2 \cdot MgF_4^{2-}$). These complexes can be produced with bound TG (thapsigargin), a specific inhibitor for SERCA that binds very tightly to SERCA and stabilizes strongly the $E2$ state. Note here that the conditions to produce the $E2$ state in the $E1/E2$ equilibrium depend totally on each P-type ATPase member; you should find out the optimal conditions by varying different parameters.

13. Here, we modified slightly the conditions originally described by Troullier et al. [32]; we employ the low concentration of Mg^{2+} to avoid a possible MgF_x binding but to attain full AlF_x binding.

14. The treatment is done according to the conditions by Daiho et al. [23], which include Me_2SO and glycerol. Be careful that the addition of Me_2SO to an aqueous solution produces heat; therefore, make first the medium containing Me_2SO and cool down to the desired temperature, then add protein to the reaction medium.

15. Sucrose and monovalent cations (K^+, Li^+, or Na^+ choice according to the needs in the assays) stabilize the Ca^{2+}-ATPase protein in the storage, and EGTA avoids the possible Ca^{2+}-induced decomposition of the complexes. The low buffer concentration (5 mM) makes pH adjustment (by a different buffer) easier for subsequent biochemical assays.

16. Proteinase K, trypsin, and protease V8 are the choice for Ca^{2+}-ATPase and Na^+/K^+-ATPase, but other proteases can be more informative for other P-type ATPases (for example, papain is used for the copper ATPase (CopA)) [33].

Acknowledgement

This work was supported by JSPS KAKENHI Grant Number 23370058.

References

1. Lassila JK, Zalatan JG, Herschlag D (2011) Biological phosphoryl-transfer reactions: understanding mechanism and catalysis. Annu Rev Biochem 80:669–702
2. Fersht AR (1999) Structure and mechanism in protein science. W. H. Freeman and Co., New York, NY
3. Allen KN, Dunaway-Mariano D (2004) Phosphoryl group transfer: evolution of a catalytic scaffold. Trends Biochem Sci 29:495–503
4. Webb MR, Trentham DR (1981) The stereochemical course of phosphoric residue transfer catalyzed by sarcoplasmic reticulum ATPase. J Biol Chem 256:4884–4887
5. Bublitz M, Poulsen H, Morth JP, Nissen P (2010) In and out of the cation pumps: P-type ATPase structure revisited. Curr Opin Struct Biol 20:431–439
6. Toyoshima C (2008) Structural aspects of ion pumping by Ca^{2+}-ATPase of sarcoplasmic reticulum. Arch Biochem Biophys 476:3–11
7. Toyoshima C (2009) How Ca^{2+}-ATPase pumps ions across the sarcoplasmic reticulum membrane. Biochim Biophys Acta 1793:941–946
8. Møller JV, Olesen C, Winther A-ML, Nissen P (2010) The sarcoplasmic Ca^{2+}-ATPase: design of a perfect chemi-osmotic pump. Q Rev Biophys 43:501–566
9. Goličnik M (2010) Metallic fluoride complexes as phosphate analogues for structural and mechanistic studies of phosphoryl group transfer enzymes. Acta Chim Slov 57:272–287
10. Wang W, Cho HS, Kim R, Jancarik J, Yokota H, Nguyen HH, Grigoriev IV, Wemmer DE, Kim S-H (2002) Structural characterization of the reaction pathway in phosphoserine phosphatase: "crystallographic snapshots" of intermediate states. J Mol Biol 319:421–431
11. Danko S, Daiho T, Yamasaki K, Kamidochi M, Suzuki H, Toyoshima C (2001) ADP-insensitive phosphoenzyme intermediate of sarcoplasmic reticulum Ca^{2+}-ATPase has a compact conformation resistant to proteinase K, V8 protease and trypsin. FEBS Lett 489:277–282
12. Danko S, Yamasaki K, Daiho T, Suzuki H, Toyoshima C (2001) Organization of cytoplasmic domains of sarcoplasmic reticulum Ca^{2+}-ATPase in E1P and E1ATP states: a limited proteolysis study. FEBS Lett 505:129–135
13. Danko S, Yamasaki K, Daiho T, Suzuki H (2004) Distinct natures of beryllium fluoride-bound, aluminum fluoride-bound, and magnesium fluoride-bound stable analogues of an ADP-insensitive phosphoenzyme intermediate of sarcoplasmic reticulum Ca^{2+}-ATPase. J Biol Chem 279:14991–14998
14. Danko S, Daiho T, Yamasaki K, Liu X, Suzuki H (2009) Formation of the stable structural analog of ADP-sensitive phosphoenzyme of Ca^{2+}-ATPase with occluded Ca^{2+} by beryllium fluoride: structural changes during phosphorylation and isomerization. J Biol Chem 284:22722–22735
15. Toyoshima C, Nakasako M, Nomura H, Ogawa H (2000) Crystal structure of the calcium pump of sarcoplasmic reticulum at 2.6 Å resolution. Nature 405:647–655
16. Toyoshima C, Nomura H (2002) Structural changes in the calcium pump accompanying the dissociation of calcium. Nature 418:605–611
17. Toyoshima C, Mizutani T (2004) Crystal structure of the calcium pump with a bound ATP analogue. Nature 430:529–535
18. Toyoshima C, Nomura H, Tsuda T (2004) Lumenal gating mechanism revealed in calcium pump crystal structures with phosphate analogues. Nature 432:361–368
19. Sørensen TL-M, Møller JV, Nissen P (2004) Phosphoryl transfer and calcium ion occlusion in the calcium pump. Science 304:1672–1675
20. Toyoshima C, Norimatsu Y, Iwasawa S, Tsuda T, Ogawa H (2007) How processing of aspartylphosphate is coupled to lumenal gating of the ion pathway in the calcium pump. Proc Natl Acad Sci U S A 104:19831–19836
21. Olesen C, Picard M, Winther AM, Gyrup C, Morth JP, Oxvig C, Møller JV, Nissen P (2007) The structural basis of calcium transport by the calcium pump. Nature 450:1036–1042
22. Schlichting I, Reinstein J (1999) pH influences fluoride coordination number of the AlF_x phosphoryl transfer transition state analog. Nat Struct Biol 6:721–723
23. Daiho T, Kubota T, Kanazawa T (1993) Stoichiometry of tight binding of magnesium and fluoride to phosphorylation and high-

affinity binding of ATP, vanadate, and calcium in the sarcoplasmic reticulum Ca^{2+}-ATPase. Biochemistry 32:10021–10026
24. Nakamura S, Suzuki H, Kanazawa T (1994) The ATP-induced change of tryptophan fluorescence reflects a conformational change upon formation of ADP-sensitive phosphoenzyme in the sarcoplasmic reticulum Ca^{2+}-ATPase. J Biol Chem 269:16015–16019
25. Maruyama K, MacLennan DH (1988) Mutation of aspartic acid-351, lysine-352, and lysine-515 alters the Ca^{2+} transport activity of the Ca^{2+}-ATPase expressed in COS-1 cells. Proc Natl Acad Sci U S A 85:3314–3318
26. Daiho T, Yamasaki K, Danko S, Suzuki H (2007) Critical role of Glu40-Ser48 loop linking actuator domain and first transmembrane helix of Ca^{2+}-ATPase in Ca^{2+} deocclusion and release from ADP-insensitive phosphoenzyme. J Biol Chem 282:34429–34447
27. Daiho T, Danko S, Yamasaki K, Suzuki H (2010) Stable structural analog of Ca^{2+}-ATPase ADP-insensitive phosphoenzyme with occluded Ca^{2+} formed by elongation of A-domain/M1′-linker and beryllium fluoride binding. J Biol Chem 285:24538–24547
28. Murphy AJ, Coll RJ (1993) Formation of a stable inactive complex of the sarcoplasmic reticulum calcium ATPase with magnesium, beryllium, and fluoride. J Biol Chem 268:23307–23310
29. Abe K, Tani K, Fujiyoshi Y (2010) Structural and functional characterization of H^+, K^+-ATPase with bound fluorinated phosphate analogs. J Struct Biol 170:60–68
30. Cornelius F, Mahmmoud YA, Toyoshima C (2011) Metal fluoride complexes of Na, K-ATPase: characterization of fluoride-stabilized phosphoenzyme analogues and their interaction with cardiotonic steroids. J Biol Chem 286:29882–29892
31. Picard M, Jensen AM, Sørensen T, Champeil L, Møller JV, Nissen P (2007) Ca^{2+} versus Mg^{2+} coordination at the nucleotide-binding site of the sarcoplasmic reticulum Ca^{2+}-ATPase. J Mol Biol 368:1–7
32. Troullier A, Girardet J-L, Dupont Y (1992) Fluoroaluminate complexes are bifunctional analogues of phosphate in sarcoplasmic reticulum Ca^{2+}-ATPase. J Biol Chem 267:22821–22829
33. Hatori Y, Lewis D, Toyoshima C, Inesi G (2009) Reaction cycle of Thermotoga maritima copper ATPase and conformational characterization of catalytically deficient mutants. Biochemistry 48:4871–4880

Chapter 20

Phosphorylation/Dephosphorylation Assays

Hiroshi Suzuki

Abstract

The P-type ATPases form an autophosphorylated intermediate with ATP, and its isomeric transition and hydrolysis are obligatory events in the ATP-driven pump and thus for the energy coupling. The analyses of these reactions are therefore crucial for understanding the mechanism of the pump function and diseases caused by its defects. Here we describe the methods to analyze these processes in the transport cycle with a representative member of P-type ATPase family, SERCA1a, sarco(endo)plasmic reticulum Ca^{2+}-ATPase.

Key words P-type ATPases, ATP-driven pumps, Phosphorylation, Phosphorylated intermediates, Aspartyl phosphate, ATP, Inorganic phosphate, SERCA, SDS-PAGE, Autoradiography

1 Introduction

The P-type ATPase functions for a substrate transport across the membrane. The activated ATPase (denoted as E1, produced with, e.g., the binding of substrate cation) forms an obligatory autophosphorylated intermediate (EP) by transferring the γ-phosphate of ATP bound at the N (*N*ucleotide binding) domain to a conserved essential aspartate in the catalytic site on the P (*P*hosphorylation) domain in the presence of the catalytic cofactor Mg^{2+}, thereby producing the covalently bound aspartyl phosphate (*see* recent reviews, refs. 1–3). The EP formed is able to reproduce ATP rapidly with ADP in the reverse reaction, thus denoted as ADP-sensitive EP (E1P). Subsequently, E1P proceeds to its isomeric transition to an ADP-insensitive form (E2P), i.e., chemically the process for the loss of the ADP-reactivity of the aspartyl phosphate. The isomerization involves an extensive structural change of the whole ATPase molecule; large motions of these cytoplasmic domains, a large rotation of the cytoplasmic A (*A*ctuator) domain, and resulting tight association of the A and P domains and rearrangement of transmembrane helices where the substrate transport sites are situated. Thereby in E2P, the catalytic site is now composed of the associated A and P domains and the A domain sterically prevents the ADP access to the aspartyl phosphate in the P domain. Also due to the A domain association in E2P, the catalytic site gains the ability for the aspartyl phosphate hydrolysis, and a specific water molecule situated in the catalytic site is able to attack the aspartyl phosphate, releasing P_i. E2P can be formed with P_i in the reverse reaction of the hydrolysis [4, 5].

These ATP hydrolysis processes couple the substrate transport, as shown in Scheme 1 with a representative example; the Ca^{2+} transport from the cytoplasm to endoplasmic reticulum lumen by SERCA1a (sarcoplasmic reticulum (SR) Ca^{2+}-ATPase) [1–3]. This ATPase is first activated by binding of two cytoplasmic Ca^{2+} ions to the high affinity transport sites composed of the transmembrane helices, then forms E1P with the occlusion of the Ca^{2+} (E1Ca$_2$ to E1P[Ca$_2$]). Subsequently E1P[Ca$_2$] proceeds to E2P with a large reduction of Ca^{2+}-affinity, opening of the lumenal gate, and thus Ca^{2+}-release into the lumen. E2P is then hydrolyzed to the Ca^{2+}-free inactive form (E2).

Setting conditions to produce the E1 and E2 states, EP formation, isomeric transition, and hydrolysis totally depend on each member of the P-type ATPases, nevertheless the autophosphorylated intermediates can be detected and quantified by using radioactive [γ-^{32}P]ATP and ^{32}P$_i$. In other words, searching crucial factors and examining variables in experimental conditions, such as specific substrates, ions, pH, and lipids, and extensive studies of the effects of these as well as of mutations on the processes in the ATP hydrolysis and coupled substrate-transport will open the way for understanding of the pump mechanism, structure/function relationship, and defects and disease mechanism. The steady-state distribution of E1P and E2P and the E1P $\rightarrow$ E2P isomeric transition can be studied by an ADP-addition followed by an acid-quenching, because E1P is very rapidly dephosphorylated to the E1 state with the added ADP, reproducing ATP; on the other hand, E2P is slowly hydrolyzed. E2P can be formed with ^{32}P$_i$ in the reverse reaction of the hydrolysis, and the hydrolysis of E2P thus formed can be analyzed independently from other processes in the transport cycle. Furthermore, the reverse isomeric transition E2P $\rightarrow$ E1P and subsequent dephosphorylation of E1P to E1 by ADP can also be studied by setting conditions appropriate and specific for each of the P-type ATPases you study.

The aspartyl phosphate is stable in acidic conditions but rapidly hydrolyzed in alkaline conditions [6–8]; therefore the ^{32}P-labeled EP species (i.e., possessing the aspartyl-^{32}phosphate) can be quantified after removal of free [γ-^{32}P]ATP and ^{32}P$_i$ by extensive washing in acidic conditions. Here with SR Ca^{2+}-ATPase (SERCA1a) as an example for the P-type ATPases, we describe a simple and convenient method using SDS-PAGE at acidic conditions to separate the ATPase polypeptide chain followed by

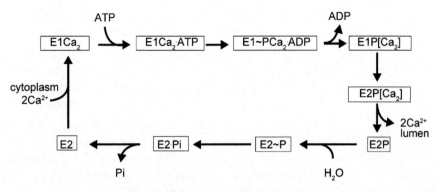

Scheme 1 Ca^{2+}-ATPase reaction cycle. *E*1P, ADP-sensitive phosphorylated intermediate (*E*P) that can produce ATP in the reverse reaction with ADP. *E*2P, ADP-insensitive *E*P. The bound Ca^{2+} ions become occluded at the transport sites in *E*1P (*E*1P[Ca_2]). Mg^{2+} binding at the catalytic site is required for the *E*P formation and hydrolysis as the cofactor, but not depicted for simplicity. In *E*2P after Ca^{2+} release, protons bind to the Ca^{2+}-ligands from the lumenal side and are released to the cytoplasmic side; thus the Ca^{2+}-pump is actually a Ca^{2+}/H^+ pump (also not depicted for simplicity)

an autoradiography for the aspartyl-^{32}phosphate determination. The protocol is applicable to all the P-type ATPases even in a very small quantity of the wild type or mutant obtained in a protein expression system.

2 Materials

Prepare all solutions using ultrapure water (obtained by purifying deionized water to attain a sensitivity of 18 MΩ cm at 25 °C) and analytical grade reagents. Prepare reagents at room temperature and store at 4 °C, unless otherwise indicated. Diligently follow all waste disposal regulations when disposing waste materials.

2.1 Preparation of Protein Samples

1. Sarcoplasmic reticulum vesicles (SR vesicles) can be prepared from rabbit muscle, e.g. according to Nakamura et al. [9].
2. Expression of SERCA1a protein in COS-1 cells and preparation of microsomes containing expressed SERCA1a protein can be done, e.g., according to Maruyama and MacLennan [10]. The expression level of wild-type SERCA1a is 2–3 % of the total microsomal proteins in weight.

2.2 ATP, [γ-^{32}P]ATP, P_i, $^{32}P_i$, ADP Solutions (See Note 1)

1. ATP, pH 7.0 (100 mM, 10 mL): Dissolve 0.551 g adenosine 5′-triphosphate (ATP) disodium salt hydrate in about 8 mL of water, adjust pH to 7.0 at 0 °C by adding Tris solution (*see* section "Solutions for *E*P Assay") and subsequently the final volume to 10 mL. Measure the optical density of the solution at 259 nm after an appropriate dilution by water, and calculate

the exact concentration using the molar extinction coefficient of 15,400 M^{-1} cm^{-1}. Store at −20 °C in small aliquots.

2. [γ-^{32}P]ATP solution can be purchased (e.g., from Perkin Elmer, 9.25 MBq). For the EP formation experiments with [γ-^{32}P]ATP, adjust the specific radioactivity to 10^7–10^8 Bq/μmol by diluting with nonradioactive ATP (see **Note 2**).

3. KH_2PO_4 (0.1 M, 500 mL): Dissolve 6.8 g potassium phosphate monobasic in 500 mL water. The solution is for the "cold-P_i chase" of ^{32}P-labeled $E2P$ in section "Dephosphorylation (Hydrolysis) of $E2P$", **step 2** (see **Note 3**), and for $^{32}P_i$ purification in section "$^{32}P_i$ Purification".

4. Carrier-free $^{32}P_i$ can be purchased (e.g., from Perkin Elmer, 37 MBq) and should be purified according to the method described in section "$^{32}P_i$ Purification". For the EP formation experiments with $^{32}P_i$, adjust the specific radioactivity to 10^7–10^8 Bq/μmol by diluting with nonradioactive P_i (see **Note 2**).

5. ADP, pH 7.0 (100 mM, 10 mL): Dissolve 0.427 g adenosine 5′-diphosphate sodium salt in about 8 mL of water, adjust pH to 7.0 at 0 °C by adding Tris solution (see section "Solutions for EP Assay") and subsequently the final volume to 10 mL. Measure the optical density of the solution at 259 nm after an appropriate dilution with water, and calculate the exact concentration using the molar extinction coefficient of 15,400 M^{-1} cm^{-1}. Store at −20 °C in small aliquots.

2.3 Solutions for $^{32}P_i$ Purification (See Note 4)

1. Hydrochloric acid (HCl) solutions.
   Dilute commercial 12 N HCl to 6 and 1 N. You will need:

   12 N HCl, 1 mL.

   6 N HCl, 1 mL.

   1 N HCl, 5 mL.

2. Ammonium hydroxide (NH_4OH) solutions.
   Dilute commercial 15 N NH_4OH (28 % NH_3 in H_2O) to 7.5 and 5 N. You will need:

   7.5 N NH_4OH, 3 mL.

   5 N NH_4OH, 3 mL.

3. KP_i solutions.
   Dilute 0.1 M KH_2PO_4 (see **item 3** in section "ATP, [γ-^{32}P]ATP, P_i, $^{32}P_i$, ADP Solutions") to 1 mM. You will need:

   100 mM KH_2PO_4, 1 mL.

   1 mM KH_2PO_4, 1 mL.

4. Aqueous Ammonium molybdate (50 mM, 10 mL).
   Dissolve 0.618 g ammonium molybdate tetrahydrate $(NH_4)_6Mo_7O_{24} \cdot 4\ H_2O$ in 10 mL of water. Use a freshly prepared solution.

5. Mg^{2+}/NH_4^+ mixture (10 mL).
   Dissolve 0.51 g $MgCl_2 \cdot 6\ H_2O$, 1 g NH_4Cl, and 1 mL of 15 N NH_4OH in water at a final volume of 10 mL. Use a freshly prepared solution.

6. Isobutyl alcohol/benzene mixture (IBB) (15 mL) (*see* **Note 5**).
   Mix isobutyl alcohol and benzene in a 1:1 (v/v) ratio, and shake well.

2.4 Solutions for EP Assay

1. Tris solution for pH adjustment (1.0 M, 100 mL).
   Dissolve 12.14 g of Tris (Tris(hydroxymethyl)aminomethane) in water at the final volume of 100 mL. For the final precise pH adjustment, use a diluted Tris solution.

2. MOPS/Tris buffer, pH 7.0 at 25 or 0 °C (0.5 M, 500 mL).
   Weigh 52.32 g MOPS (3-morpholinopropanesulfonic acid) into a beaker and dissolve in about 450 mL of water. Equilibrate the solution in a water bath (or ice-water bath) to the desired temperature, and under continuous stirring adjust the pH by adding the 1 M Tris solution to 7.0. Bring the volume of the buffer solution to 500 mL and check pH once more to assure the desired pH.

3. MES/Tris buffer, pH 6.0 at 25 or 0 °C (0.5 M, 500 mL).
   Weigh 53.31 g MES monohydrate (2-morpholinoethanesulfonic acid monohydrate) into a beaker and dissolve in about 450 mL of water. Equilibrate the solution in a water bath (or ice-water bath) to the desired temperature, and under continuous stirring adjust the pH by adding the 1 M Tris solution to pH 6.0. Bring the volume of the buffer solution to 500 mL and check pH once more to assure the desired pH.

4. $MgCl_2$ (1 M, 100 mL).
   Dissolve 20.33 g $MgCl_2$ hexahydrate in water at the final volume of 100 mL.

5. KCl (3 M, 100 mL).
   Dissolve 22.37 g KCl in water at the final volume of 100 mL.

6. $CaCl_2$ (0.5 M, 100 mL).
   Dissolve 7.35 g $CaCl_2$ dihydrate in water at the final volume of 100 mL.

7. EGTA at pH 7.0 or 6.0 at 25 °C (0.2 M, 100 mL).
   Weigh 7.607 g EGTA into a beaker and dissolve in about 80 mL of water. Similarly to the buffer preparation, adjust the pH to the desired value by Tris solution. Add water and adjust the final volume to 100 mL and check the pH once more to assure the desired pH (*see* **Note 6**).

8. A23187 (3 mM, 1.27 mL ethanol).
   Dissolve 2 mg calcium ionophore A23187 in 1.27 mL ethanol. Keep in small aliquots at −20 °C and protect from light (*see* **Note 7**).

9. Trichloroacetic acid (20 % (w/v)) containing 0.1 M P_i (100 mL).
   Dissolve 20 g trichloroacetic acid and 1.56 g $NaH_2PO_4 \cdot 2\ H_2O$ in water at the final volume of 100 mL (*see* **Note 8**).

10. Reaction mixture 1.
    50 mM MOPS/Tris (pH 7.0 at 0 °C), 0.1 M KCl, 5 mM $MgCl_2$, 0.05 mM $CaCl_2$.

11. Reaction mixture 2.
    50 mM MOPS/Tris (pH 7.0 at 25 °C), 30 % (v/v) Me_2SO, 10 mM $MgCl_2$, 2 mM EGTA.

12. Chasing solution.
    50 mM MOPS/Tris (pH 7.0 at 0 °C), 105 mM KCl (or LiCl in place of KCl), 10 mM $MgCl_2$, 0.5 mM EGTA (pH 7.0), 2 mM nonradioactive P_i (prepared in **item 3** of section "ATP, [γ-^{32}P]ATP, P_i, $^{32}P_i$, ADP Solutions").

2.5 Solutions for SDS-PAGE (Sodium Dodecyl Sulfate-Polyacrylamide Gel Electrophoresis) at pH 6.0 According to Weber and Osborn [11]

1. Acrylamide 30 % (w/v) containing 0.8 % (w/v) N,N'-methylenebisacrylamide (200 mL).
   Dissolve 60 g acrylamide and 1.6 g N,N'-methylenebisacrylamide in water at the final volume of 200 mL. Store in a dark bottle at 4 °C (*see* **Note 9**).

2. Ammonium persulfate (10 % (w/v), 2 mL).
   Dissolve 0.2 g ammonium persulfate in 2 mL of water. Use a freshly prepared solution.

3. TEMED (N,N,N',N'-tetramethyl-ethylenediamine).
   Store at 4 °C.

4. SDS (10 % (w/v), 10 mL).
   Dissolve 1 g sodium lauryl sulfate (sodium dodecyl sulfate) in 10 mL of water. Keep at room temperature to avoid precipitation in cold.

5. Phosphate buffer, pH 6.0 (0.5 M, 500 mL) for gel preparation.
   Dissolve 39.0 g $NaH_2PO_4 \cdot 2H_2O$ and 3.03 g NaOH in water at the final volume of 500 mL. Check the pH to assure the desired pH, and keep the buffer at room temperature to avoid possible precipitation in the cold (*see* **Note 10**).

6. Bromophenol blue (3',3",5',5"-tetrabromophenolsulfonphthalein) (1 % (w/v), 10 mL).
   Dissolve 0.10 g bromophenol blue in 10 mL of water.

7. SDS sample buffer (5 mL).
   Mix 2.5 mL of 0.5 M phosphate buffer pH 6.0, 0.5 mL of glycerol, 0.05 mL of 1 % (w/v) bromophenol blue, 0.25 mL of β-mercaptoethanol, 1.5 mL of 10 % (w/v) SDS solution, and 0.2 mL of water. Use a freshly prepared mixture, or you

may keep for a period at −20 °C in small aliquots and use after dissolving SDS-precipitate at room temperature.

8. Running buffer (0.125 M NaP$_i$ buffer at pH 6.0, 3 L).
Dissolve 58.5 g NaH$_2$PO$_4$·2H$_2$O, 4.55 g NaOH, and 3 g SDS in water at the final volume of 3 L. Check the pH to assure the desired pH, and keep the buffer at room temperature to avoid possible phosphate precipitation in the cold (*see* **Notes 10** and **11**).

9. Gel-fixing solution (5 % (v/v) methanol with 7 % (v/v) acetic acid, 3 L).

   Dilute 150 mL of 100 % methanol and 210 mL 100 % acetic acid into water and fill up to 3 L (*see* **Note 12**).

3 Methods

The phosphorylation/dephosphorylation assays for P-type ATPases are made with [γ-^{32}P]ATP and ^{32}P$_i$. The conditions for producing *E1* and *E2* states and phosphorylation/dephosphorylation depend on each member of P-type ATPases, and those described here are examples with a representative member SERCA1a.

3.1 Phosphorylation of the E1 State with [γ-^{32}P]ATP

1. Pipette 50 μL of reaction mixture 1 containing 20–25 μg/mL of microsomal protein (or SR vesicles) and 1 μM A23187 into 1.5 mL Eppendorf tubes and cool down to 0 °C (*see* **Note 7**).

2. Start phosphorylation at 0 °C (or other desired temperature) by adding a small volume of [γ-^{32}P]ATP to give 10 μM (if it is a large volume, mix with [γ-^{32}P]ATP dissolved in the above solution (reaction mixture 1) without the protein and A23187).

3. Terminate the phosphorylation reaction after incubation for a desired time period (e.g., 5–20 s at steady state) by the addition of the ice-cold 20 % trichloroacetic acid solution containing 0.1 M Pi to give 5–7.5 % acid followed by immediate vortex-mixing. The time course of the phosphorylation can be analyzed by performing the acid-quenching at a series of time points after the [γ-^{32}P]ATP addition.

4. Centrifuge at ~30,000 × g and 4 °C for 4 min, and discard the supernatant by aspiration or by careful pipetting. Be careful to not remove the acid-precipitated proteins attached on the tube wall, as the precipitates are not seen well.

5. Add 25 μL of the SDS-PAGE sample buffer into the tube and dissolve the precipitated proteins with vortex-mixing in about 2 min or longer (do not generate bubbles) at room temperature. Do not increase the temperature and do not boil the sample to avoid a possible formation of insoluble aggregates and to prevent a possible aspartyl phosphate hydrolysis.

6. Apply all 25 μL dissolved sample into one well in a polymerized 5 % SDS-acrylamide gel fixed on the electrophoresis apparatus (*see* Subheading 3.6). Run SDS-PAGE at pH 6.0 according to Weber and Osborn [11], then remove the gel and fix it by the gel-fixing solution, and determine the radioactivity associated with the separated ATPase chain by digital autoradiography, as described all in section "5 % SDS-PAGE at pH 6.0 According to Weber and Osborn [11] for $^{32}P_i$-Labeled ATPase".

3.2 Phosphorylation of the E2 State with $^{32}P_i$ to Form E2P

1. Pipette 50 μL of reaction mixture 2 containing 20–25 μg/mL of microsomal protein (or SR vesicles) and 1 μM A23187 into 1.5 mL Eppendorf tubes and incubate at 25 °C for few minutes (*see* **Note 13**).

2. Start the phosphorylation at 25 °C by adding $^{32}P_i$ in a small volume to give 0.1 mM and incubate for 15 s (*see* **Note 14**).

3. Terminate the phosphorylation reaction after 15 s incubation by the addition of the ice-cold 20 % trichloroacetic acid solution containing 0.1 M P_i to give 5–7.5 % acid followed by immediate vortex-mixing.

4. Perform the SDS-PAGE and autoradiography as described in **steps 4–6** of section "Phosphorylation of the $E1$ State with [γ-^{32}P]ATP".

3.3 Quantification of E1P and E2P

1. Phosphorylation with [γ-^{32}P]ATP.
   Prepare a pair of the samples; one for the determination of the total amount of EP ($E1P$ plus $E2P$) by the acid-quenching, and the other for the amounts of $E1P$ and $E2P$ by an ADP-addition followed by the acid-quenching. Phosphorylate the ATPase by adding [γ-^{32}P]ATP as described in **steps 1** and **2** of section "Phosphorylation of the $E1$ State with [γ-^{32}P]ATP".

2. Reverse dephosphorylation of $E1P$ by ADP.
   After a certain period of incubation for the phosphorylation, e.g., 15 s at steady state, add ADP to give a large excess (1 mM or higher) to dephosphorylate $E1P$ in the reverse reaction. The ADP solution should have the same composition as the phosphorylation medium (except for the absence of ATP, protein, and A23187, and in the case of Ca^{2+}-ATPase, the presence of excess EGTA (e.g., 5 mM at the final concentration) to remove Ca^{2+} to prevent further phosphorylation of $E1Ca_2$).

3. Acid-quenching.
   At 1 s after the ADP addition, add the trichloroacetic acid solution as described in **step 3** of section "Phosphorylation of the $E1$ State with [γ-^{32}P]ATP" to quench the reactions. The ADP-sensitive EP ($E1P$) disappears entirely within this period by the added ADP, whereas ADP-insensitive EP ($E2P$) is not decomposed in the period and can thus be determined. With the other phosphorylation sample in the pair, determine the total

amount of *EP* (*E*1P plus *E*2P) by the acid-quenching without the ADP addition at the time of ADP addition (*see* **Note 15**).

4. SDS-PAGE and autoradiography.

   Perform the SDS-PAGE and autoradiography as described in **steps 4–6** of section "Phosphorylation of the *E*1 State with [γ-^{32}P]ATP", and then determine the total amount of *EP* (*E*1P plus *E*2P) and the amounts of *E*2P and *E*1P.

3.4 Time Course of Forward Isomeric Transition E1P → E2P and EP Decay

Time course of the *E*1P → *E*2P isomeric transition can be followed by determining the total amount of *EP* and the amounts of *E*1P and *E*2P, i.e., the fractions of *E*1P and *E*2P in the following two ways: (1) If the *E*1P formation from the *E*1 state upon the addition of ATP is rapid enough to be kinetically isolated from the subsequent slow *E*1P → *E*2P isomeric transition, determine the increase in the *E*2P fraction at a series of time points after the [γ-^{32}P]ATP addition; you can follow the *E*2P formation from *E*1P (*E*1P → *E*2P kinetics). (2) In the second way of analysis, after a certain period of the phosphorylation reaction with [γ-^{32}P]ATP (e.g., after reaching the steady state), add nonradioactive ATP to initiate the "cold-chase" of ^{32}P-labeled *EP* and then determine the *E*1P and *E*2P fractions at a series of time points during the decay time course of the ^{32}P-labeled *EP*. In the case of the Ca^{2+}-ATPase, you may initiate the chase of ^{32}P-labeled *EP* by "EGTA-chase," i.e., by adding an excess EGTA to remove Ca^{2+} to prevent the phosphorylation in the subsequent catalytic cycle (*see* typical examples in Fig. 3 in ref. 12). Without the ADP addition, the overall time course of ^{32}P-labeled *EP* decay can be followed after the start of the "cold-ATP" chase with nonradioactive ATP, or after the "EGTA-chase" in the case of Ca^{2+}-ATPase.

To analyze the kinetics of the reverse isomeric transition from *E*2P to *E*1P, design experimental conditions to cause this conversion: For example, in the case of Ca^{2+}-ATPase, first produce *E*2P with ^{32}P$_i$ as described in **steps 1** and **2** of section "Phosphorylation of the *E*2 State with ^{32}P$_i$ to Form *E*2P", then dilute largely the phosphorylated sample to reduce Me$_2$SO concentration (to a negligible level, e.g., by 20-fold) and simultaneously add a very high concentration of Ca^{2+} in the presence of ionophore A23187. Thereby Ca^{2+} ions bind at the lumenally oriented low affinity transport sites in *E*2P and cause the reverse isomeric transition *E*2P + 2Ca^{2+} → *E*2PCa$_2$ → *E*1PCa$_2$. The concomitant addition of ADP together with Ca^{2+} further causes the dephosphorylation of *E*1PCa$_2$ to *E*1Ca$_2$, producing ATP.

3.5 Dephosphorylation (Hydrolysis) of E2P

1. Phosphorylation with ^{32}P$_i$.

   Produce *E*2P by phosphorylating the *E*2 state of ATPase with ^{32}P$_i$ in 50 μL of the reaction mixture as described in **steps 1** and **2** of section "Phosphorylation of the *E*2 State with ^{32}P$_i$ to

Form $E2P''$ for the Ca^{2+}-ATPase. Then cool down the mixture to 0 °C.

2. Hydrolysis of ^{32}P-labeled $E2P$ by "cold-P_i chase."
   To initiate the dephosphorylation (hydrolysis of aspartyl-^{32}phosphate), dilute the phosphorylation mixture 20-fold by adding 950 μL ice-cold chasing solution containing nonradioactive P_i (*see* **Note 16**).

3. Acid-quenching.
   At different time points after the start of "cold-P_i chase," quench the reaction with an ice-cold trichloroacetic acid solution containing P_i as described in **step 3** of section "Phosphorylation of the $E1$ State with [γ-^{32}P]ATP".

4. SDS-PAGE and autoradiography.
   Perform the SDS-PAGE and autoradiography as described in **steps 4–6** of section "Phosphorylation of the $E1$ State with [γ-^{32}P]ATP", and then determine the amount of $E2P$ remaining and obtain the hydrolysis time course.

3.6 5 % SDS-PAGE at pH 6.0 According to Weber and Osborn [11] for $^{32}P_i$-Labeled ATPase (See Note 17)

3.6.1 Casting Gels (130 mL) for Four Slab Gels with 18 cm × 8 cm × 2 mm Gel Cassette with a 26-Well Gel Comb (See **Note 18**)

1. Take 21.7 mL of 30 % acrylamide stock solution, 32.5 mL of 0.5 M phosphate buffer pH 6.0, 1.3 mL of 10 % SDS solution, and 72.4 mL water into a beaker and mix well (but gently not to produce bubbles) with magnetic stirrer.

2. Add 1.95 mL of 10 % ammonium persulfate solution and 0.195 mL TEMED, and mix well.

3. Pour the solution immediately and carefully into the gel cast without introducing air bubbles, insert the 26-well gel comb, and let the gel-polymerization proceed at room temperature for about 30 min. The amount (130 mL) of the gel solution is enough to prepare the 4 slab gels with 18 cm × 8 cm × 2 mm gel cassettes.

3.6.2 Running Gels

1. Set the cassette containing the polymerized gel on the electrophoresis apparatus, pour the running buffer into the lower and upper buffer chamber, and remove the comb carefully.

2. Load all 25 μL of the SDS-sample buffer containing the dissolved ATPase protein in an Eppendorf tube into one well.

3. Run the gels at constant current with 80 mA/plate for 2.5 h or until the bromophenol blue-front dye band reaches to 2–3 cm above the bottom of the gel.

4. Remove the gels from the apparatus and cassette, cut the gel above the broad front band of bromophenol blue dye, and discard the lower part.

5. Place the upper part of the gel into the gel-fixing solution and shake gently for 15 min by changing the fixing solution twice.

3.6.3 Gel Drying

Because water quenches the beta particle emission (from ^{32}P), the gels should be dried completely.

1. Place the gel strip on filter paper on a slab gel-dryer, cover with a nonporous plastic film (plastic food wrap) to avoid a contamination, and dry the gel completely for about 30 min at 80 °C.

3.6.4 Standards for Radioactivity Determination

1. Prepare several different concentrations of [γ-^{32}P]ATP or $^{32}P_i$ solution (e.g., 20, 10, 5, 2.5, 1.25, 0.625 µM) by a series of dilution of the solution with a known radioactivity.
2. Spot 2 µL of each on a filter paper and dry completely at room temperature, and wrap the paper by a plastic film.

3.6.5 Autoradiography and Data Analysis

1. Place the dried gel(s) on the filter paper covered by the plastic film together with the ^{32}P standards filter paper wrapped with the plastic film on a X-ray cassette with an imaging plate (together with enhancement screens for ^{32}P if desired). Be sure that the gels and standards do not move.
2. Expose the imaging plate overnight.
3. Quantitate the radioactivity associated with the separated ATPase band by digital autoradiography of the dried gels using, for example, Bio-Imaging Analyzer BAS2000 (Fuji Photo Film, Tokyo) or Typhoon FLA7000 (GE Healthcare Life Sciences) (*see* Fig. 2 in ref. 13 for an autoradiograph).
4. Draw the standard curve, and calculate the ^{32}P incorporated into the ATPase chain, i.e., the amount of aspartyl-^{32}phosphate using the standard curve. The radioactivity on the ATPase chain should be in the linear range of the standard (if not, make another series of the standards and repeat the exposure together with the dried gels). The amount of *EP* formed with the ATPase is obtained by subtracting the background digitized radioactivity, which is determined in an appropriate way depending on the ATPase type, e.g., in the absence of essential ligands (in the absence of Ca^{2+} with a large excess of EGTA in the case of ATP-induced *EP* formation in the Ca^{2+}-ATPase), or in the presence of a specific inhibitor that completely inhibits the phosphorylation (e.g., thapsigargin for the Ca^{2+}-ATPase or vanadate for all the P-type ATPases). In the case of experiments with the microsomes expressing ATPase protein, the background level can be determined with control microsomes that were prepared from the cells transfected with the expression vector containing no ATPase cDNA.

3.7 $^{32}P_i$ Purification (See Note 4)

The purification of carrier-free commercial $^{32}P_i$ (in about 10 µL) is carried out at room temperature using the solutions described in section "Solutions for $^{32}P_i$ Purification" and 15 mL plastic centrifuge tubes with a cap. The protocol described is based on refs. 4, 14, 15 with slight modifications. The protocol gives 50–80 % recovery of purified $^{32}P_i$.

1. Add 1 mL of 1 mM (nonradioactive) KP_i (KH_2PO_4) solution to the commercial carrier-free $^{32}P_i$.

2. Add 12 N HCl to give 1.1 N in order to acidify the solution. (For 10 μL $^{32}P_i$, add 102 μL 12 N HCl.)

3. Add 2 mL of isobutyl alcohol/benzene mixture (IBB) to the $^{32}P_i$ solution. To extract and thereby remove impurities from the $^{32}P_i$ solution into the IBB phase, mix the IBB phase and the aqueous phase vigorously by 200 times using a vortex mixer. Then let the tube stand for a while until the two phases separate again completely.

4. Discard the IBB (upper) phase (e.g., by pipetting or aspirating), and add 50 mM aqueous ammonium molybdate to give 12 mM (~350 μL) to the $^{32}P_i$ solution. This results in the formation of ammonium phosphomolybdate.

5. To extract the ammonium phosphomolybdate from the aqueous phase, add 2.5 mL of IBB and as described above in section "$^{32}P_i$ Purification", **step 3**, vortex vigorously, and then wait until the two phases separate completely. Take the IBB phase that contains the extracted ammonium phosphomolybdate, and keep it aside.

6. Repeat the ammonium phosphomolybdate extraction from the aqueous phase: Add the following to the above remaining aqueous solution; 12 N HCl (150 μL) to give 1.1 N, 50 mM aqueous ammonium molybdate (350 μL) to give 12 mM, and 2.5 mL IBB. Vortex vigorously and wait until the two phases separate completely, and take the IBB phase and combine with the previously extracted IBB phase obtained above.

7. If the color of the aqueous phase is still yellowish, repeat the IBB extraction as described above in section "$^{32}P_i$ Purification", **step 6**, until the color disappears.

8. Add 1 N HCl to the collected and combined IBB extract (in a volume ratio, 2 mL of 1 N HCl to 5 mL of the IBB extract). Vortex vigorously, and then wait until the two phases separate completely.

9. Discard the aqueous (lower) phase (e.g., by pipetting). Then add again 2 mL of 1 N HCl to the IBB phase, vortex vigorously, and let the two phases separate completely.

10. Discard the aqueous phase, and then add 1.5 mL of 7.5 N NH_4OH to the IBB phase (5 mL). Vortex vigorously, and wait until the two phases separate completely.

11. Collect the aqueous phase, and add 10 µL of 100 mM KP_i, and then 1.5 mL of the ice-cold Mg^{2+}/NH_4^+ mixture in order to precipitate the phosphate as $MgNH_4PO_4$ (which is "white" precipitate). Chill it on wet ice for 30 min and then spin down at $2000 \times g$ and 4 °C for 10 min.

12. Discard the supernatant and wash the precipitate with 1 mL of ice-cold 5 N NH_4OH. Vortex well and spin down the precipitate at $2000 \times g$ and 4 °C for 10 min.

13. Discard the supernatant, and then dissolve the precipitate completely in 20 µL of 6 N HCl.

14. Add 200 µL of 7.5 N NH_4OH to form and precipitate again the phosphate as $MgNH_4PO_4$. Chill it on wet ice for 30 min and spin down at $2000 \times g$ and 4 °C for 10 min.

15. Discard the supernatant and add 1 mL of 5 N NH_4OH to the precipitate. Vortex well and spin down at $2000 \times g$ and 4 °C for 10 min.

16. Discard the supernatant. Then remove ammonia from the precipitate in vacuum (usually it takes ~30 min with a vacuum pump).

17. Add 0.5 mL water and suspend the precipitate. Then by adding HCl, bring the pH of the solution between 2 and 4 to dissolve the P_i completely. Thus $^{32}P_i$ is purified and recovered with carrier nonradioactive P_i.

18. Determine the P_i concentration of the purified $^{32}P_i$ solution by using a photometric phosphate determination protocol. (The P_i concentration may be about 2 mM.)

19. Determine the radioactivity of the purified $^{32}P_i$ solution by diluting 1 µL of the solution into 10 mL of water in a scintillation vial, and measure the Cerenkov emission with a liquid scintillation counter. Calculate the radioactivity assuming 50 % counting efficiency.

20. Calculate the specific radioactivity of purified $^{32}P_i$ in the solution (e.g., as cpm/µmol P_i).

4 Notes

1. Working with ^{32}P isotopes should be carried out in laboratories designed for isotope handling and strictly following preventive measures. Consult the guides published by the suppliers for safe handling of radioactive materials before setting up your experiments!

2. Half-life of ^{32}P is 14.29 days; therefore perform your experiments within a period with high-enough radioactivity before an extensive decay.

3. KH$_2$PO$_4$ solution can be substituted by NaH$_2$PO$_4$ when you want to avoid K$^+$ in your assay medium.

4. The solutions required for ^{32}P$_i$ purification are hazardous for health! For using HCl and NH$_4$OH, carefully check safety sheets, follow instructions, and handle with a great care in a fume hood. HCl and NH$_4$OH are volatile and corrosive, and cause severe skin burns. Avoid inhalation, dermal and eye exposure. For eye and face protection, use tightly fitting safety goggles and a face shield. For skin protection, use proper gloves. Gloves must be inspected prior to use and removed carefully (without touching glove's outer surface) to avoid skin contact.

5. Isobutyl alcohol and benzene are flammable and hazardous; follow instructions and be careful with handling! Avoid breathing the vapor, wear proper protective clothes (gloves, eye protection, face protection)!

6. EGTA (ethylene glycol-bis(2-aminoethylether)-N,N,N',N'-tetraacetic acid) chelates Ca^{2+} with a high affinity at a ratio of 1:1. Its solubility increases with increasing pH. The free Ca^{2+} concentration can be adjusted precisely by a Ca^{2+}/EGTA buffer (appropriate concentrations of Ca^{2+} and EGTA). For calculation of free Ca^{2+} concentration, use programs such as the Calcon program.

7. Calcium ionophore A23187 is a divalent cation ionophore but highly selective for Ca^{2+}. Because this strongly hydrophobic reagent is dissolved in ethanol that is a protein-denaturant, do not exceed the ethanol concentration 1 % (v/v), preferably 0.5 % or less. The ionophore should be added to a reaction mixture that contains the membrane protein, so as to incorporate it into the membrane phase; otherwise it will be precipitated in an aqueous solution.

8. Trichloroacetic acid is extremely corrosive! Take great care when preparing the solution. Keep it in a tightly sealed dark bottle at 4 °C.

9. Important! Unpolymerized acrylamide is a neurotoxin; be extremely careful when handling it! Avoid inhalation of powder and skin contact!

10. Do not use potassium phosphate, because K$^+$ and dodecyl sulfate form an insoluble salt and precipitate.

11. Be sure that the pH of the running buffer is 6.0! Increase in pH causes a hydrolysis of the aspartyl phosphate. Also do not use potassium phosphate, because K$^+$ and dodecyl sulfate form an insoluble salt and precipitate.

12. Gel fixing solution: Be extremely careful when measuring methanol and acetic acid and follow instructions! Use protective clothes and handle in a fume hood. Dilution of

methanol results in heat; cool down the fixing solution to room temperature before use.

13. The addition of Me_2SO to an aqueous solution produces heat; therefore make first the medium containing Me_2SO and cool down to the desired temperature, and then add protein to the reaction medium.

14. The inclusion of Me_2SO extremely favors $E2P$ in the equilibrium $E2 + P_i \leftrightarrow E2P + H_2O$ [16, 17]; therefore under these conditions, practically all the active Ca^{2+}-ATPase phosphorylation sites in the preparation can be phosphorylated, and the content of the catalytic site can be determined. Note that in the absence of Me_2SO, you can analyze the equilibrium $E2 + P_i \leftrightarrow E2P + H_2O$ under physiological conditions. Also note that the Ca^{2+} ionophore A23187 is not required if you do not examine a possible effect of Ca^{2+} binding at the lumenally oriented transport sites in $E2P$ formed with $^{32}P_i$.

15. After the ADP addition, $E1P$ rapidly decomposes to the $E1$ state in the reverse reaction by the added ADP (within 1 s), and in the second phase the remaining EP (i.e., $E2P$) decomposes slowly to the $E2$ state by its hydrolysis. Thus the detailed time course of EP decay (the reverse $E1P$ decomposition and forward $E2P$ hydrolysis) can be followed by performing the acid-quenching at a series of time periods after the ADP addition.

16. The Ca^{2+}-ATPase and Na^+/K^+-ATPase (and probably other members) possess a specific K^+ binding site, and the K^+ binding markedly accelerates the $E2P$ hydrolysis. Replacing K^+ with Li^+ (thereby adjusting ionic strength), you may slow down the hydrolysis; in some cases the slowed hydrolysis kinetics may be more conveniently analyzed.

17. In order to avoid possible aspartyl phosphate decomposition, prepare SDS-PAGE gels before and have them ready to start immediately after the phosphorylation/dephosphorylation reaction.

18. The gel cassette size and the comb (the well number/size) can be modified according to the sample number you apply to the gel, but carefully design the cassette size and the comb with desired well number/size to completely separate (not to overlap) the signals between neighboring wells in autoradiograph as well as in protein staining.

Acknowledgment

This work was supported by JSPS KAKENHI Grant Number 23370058.

References

1. Toyoshima C (2008) Structural aspects of ion pumping by Ca^{2+}-ATPase of sarcoplasmic reticulum. Arch Biochem Biophys 476:3–11
2. Toyoshima C (2009) How Ca^{2+}-ATPase pumps ions across the sarcoplasmic reticulum membrane. Biochim Biophys Acta 1793:941–946
3. Møller JV, Olesen C, Winther A-ML, Nissen P (2010) The sarcoplasmic Ca^{2+}-ATPase: design of a perfect chemi-osmotic pump. Q Rev Biophys 43:501–566
4. Kanazawa T, Boyer PD (1973) Occurrence and characteristics of a rapid exchange of phosphate oxygens catalyzed by sarcoplasmic reticulum vesicles. J Biol Chem 248:3163–3172
5. Masuda H, de Meis L (1973) Phosphorylation of the sarcoplasmic reticulum membrane by orthophosphate. Inhibition by calcium ions. Biochemistry 12:4581–4585
6. Yamamoto T, Tonomura Y (1968) Reaction mechanism of the Ca^{2+}-dependent ATPase of sarcoplasmic reticulum from skeletal muscle. II. Intermediate formation of phosphoryl protein. J Biochem 64:137–145
7. Martonosi A (1969) Sarcoplasmic reticulum. VII. Properties of a phosphoprotein intermediate implicated in calcium transport. J Biol Chem 244:613–620
8. Makinose M (1969) The phosphorylation of the membranal protein of the sarcoplasmic vesicles during active calcium transport. Eur J Biochem 10:74–82
9. Nakamura S, Suzuki H, Kanazawa T (1994) The ATP-induced change of tryptophan fluorescence reflects a conformational change upon formation of ADP-sensitive phosphoenzyme in the sarcoplasmic reticulum Ca^{2+}-ATPase. J Biol Chem 269:16015–16019
10. Maruyama K, MacLennan DH (1988) Mutation of aspartic acid-351, lysine-352, and lysine-515 alters the Ca^{2+} transport activity of the Ca^{2+}-ATPase expressed in COS-1 cells. Proc Natl Acad Sci U S A 85:3314–3318
11. Weber K, Osborn M (1969) The reliability of molecular weight determinations by dodecyl sulfate-polyacrylamide gel electrophoresis. J Biol Chem 244:4406–4412
12. Yamasaki K, Daiho T, Danko S, Suzuki H (2013) Roles of long-range electrostatic domain interactions and K^+ in phosphoenzyme transition of Ca^{2+}-ATPase. J Biol Chem 288:20646–20657
13. Daiho T, Suzuki H, Yamasaki K, Saino T, Kanazawa T (1999) Mutations of Arg198 in sarcoplasmic reticulum Ca^{2+}-ATPase cause inhibition of hydrolysis of the phosphoenzyme intermediate formed from inorganic phosphate. FEBS Lett 444:54–58
14. De Meis L (1984) Pyrophosphate of high and low energy. Contributions of pH, Ca^{2+}, Mg^{2+}, and water to free energy of hydrolysis. J Biol Chem 259:6090–6097
15. Inesi G, Kurzmack M, Lewis D (1988) Kinetic and equilibrium characterization of an energy-transducing enzyme and its partial reactions. Methods Enzymol 157:154–190
16. de Meis L, Martins OB, Alves EW (1980) Role of water, hydrogen ion, and temperature on the synthesis of adenosine triphosphate by the sarcoplasmic reticulum adenosine triphosphatase in the absence of a calcium ion gradient. Biochemistry 19:4252–4261
17. Barrabin H, Scofano HM, Inesi G (1984) Adenosinetriphosphatase site stoichiometry in sarcoplasmic reticulum vesicles and purified enzyme. Biochemistry 23:1542–1548

Chapter 21

Tryptophan Fluorescence Changes Related to Ca^{2+}-ATPase Function

Pankaj Sehgal, Claus Olesen, and Jesper V. Møller

Abstract

Fluorescence measurements as monitored with either extrinsic or intrinsic probes constitute important ways with which to study the molecular properties of macromolecules. With high-quality spectrofluorimeters, it is, e.g., possible kinetically to follow local conformational changes, induced by ligands and inhibitors, with a sensitivity that is unsurpassed by any other physicochemical technique. We demonstrate here with Ca^{2+} and two specific inhibitors of SERCA how this can be done by measurements of the intrinsic fluorescence of the tryptophan residues of SERCA.

Key words Tryptophan fluorescence, Intrinsic fluorescence, Conformational change, Spectrofluorimetry, Calcium ATPase

1 Introduction

Fluorescence is the emission of light from the electrons in chromophoric groups of molecules that have attained an excited state by absorption of light. Compared to other spectroscopic techniques, e.g., light absorption and circular dichroism, fluorescence is characterized by a high degree of sensitivity of the fluorescence emissions due to subtle changes in the immediate chemical and physical environments of the excited molecules. Fluorescent molecules can therefore be used when present near the active sites of an enzyme to monitor the often minor conformational changes associated with functional properties. To detect fluorescence one can either employ proteins labeled with fluorescent probes or, as an alternative, make use of their own intrinsic fluorescent properties and thereby avoid probe-induced structural modifications of the proteins. Especially in proteins, tryptophan residues are widely used as an intrinsic probes to make inferences regarding local structure and dynamics. The fluorescence of a protein whether folded or unfolded is a mixture of the fluorescence from individual aromatic residues such as tryptophan, tyrosine, and phenylalanine. However, most of the intrinsic fluorescence emissions of a folded protein are caused by excitation of tryptophan residues (if present in reasonable quantities) due to a high quantum yield, while tyrosine and phenylalanine residues ordinarily make small contributions. The intrinsic fluorescence of proteins may be affected both by conformational changes of a global nature (such as the E1/E2 interconversions in P-type ATPases) as well as those that result from more localized changes in the surroundings of bound substrates and ligands. In the latter case decreased fluorescence (quenching) will be observed, if there is direct interaction between the bound compound and the fluorescent group (this is referred to as static quenching); whereas fluorescence changes resulting from conformational changes, rather than binding (i.e., complex formation), are referred to as dynamical quenching.

There are numerous reviews which describe how changes in fluorescence (tryptophan) properties such as fluorescence intensity, wavelength maximum (λ_{max}), band shape, anisotropy, fluorescence lifetimes, and energy transfer can be used to obtain information on the physicochemical properties of membrane proteins [1, 2]. Typically, tryptophan residues have a wavelength of maximum absorption at 280–295 nm, with emission peaks ranging from 325 to 350 nm, depending in particular on the polarity of the local environment. Intrinsic tryptophan fluorescence can also be used to probe the lipid environment of membranes from the quenching effect of bromosubstituted lipids when added to the membrane [3, 4]. If a protein containing a tryptophan in its "hydrophobic" core is denatured or exposed due to binding of denaturant, ligand or substrates or inhibitors, a red-shifted emission spectrum is likely to be observed due to the exposure of the tryptophan residues to an aqueous environment as opposed to the hydrophobic protein interior. Conversely, the emission from a tryptophan residue exposed to the aqueous solvent will become blue-shifted if the tryptophan is embedded or shielded from the medium by the substrate or inhibitors.

With respect to SERCA there are 14 tryptophan residues present per ATPase molecule, all of which except one are present in the membraneous domain [5]. The localization of almost all the tryptophan residues to the membraneous domain makes the Ca^{2+}-ATPase particularly suitable to studies with hydrophobic inhibitors as demonstrated in the protocol described below with two specific inhibitors of Ca^{2+}-ATPase that are bound with high affinity at two different membraneous sites in the E2 conformation.

2 Materials

2.1 Equipment

1. An advanced Spectrofluorometer, with recorder.
2. Water bath, with accurate temperature controller (+/− 0.1 °C).
3. 3.5 mL quartz disposable cuvettes.

2.2 Buffers and Reagents

All reagents should be of analytical grade.

1. Purified Ca^{2+}-ATPase (20–25 mg protein/mL) from rabbit skeletal muscle (*see* Chap. 3).
2. Basic assay medium: 50 mM MOPS (3-(4-Morpholino) propane sulfonic acid), adjusted to pH 7.2 with KOH, 50 mM KCl, 5 mM $MgCl_2$, and 0.05 mM Ca^{2+}.
3. Ethylene glycol-bis(2-aminoethylether)-N,N, N′,N′-tetra acetic acid (EGTA): 0.1 M solubilized in demineralized water and adjusted to pH 7.2.
4. Ca^{2+}-ATPase suspended at a protein concentration of 0.1 mg/mL in the basic assay medium (*see* **Note 1**).
5. Inhibitors: Thapsigargin (Alomone Laboratories or Sigma-Aldrich), Cyclopiazonic acid (CPA, Alomone Laboratories or Sigma-Aldrich) with a known working stock concentration dissolved in DMSO.

3 Methods

1. Turn on the spectrofluorometer and stabilize the lamp for about 1 h and set the excitation wavelength at 295 nm with slit size 5 nm, and the emission wavelength at 335 nm, with slit size 10 nm.
2. Incubate and thermoequilibrate appropriate aliquots of the assay components in the water bath at 23 ± 0.1 °C.
3. Transfer 3 mL of the 0.1 mg/mL Ca^{2+}-ATPase suspension to a quartz cuvette in the cuvette compartment. Close the lid, turn the recorder on and wait for the recorder to stabilize (*see* **Note 2**).
4. Start the experiment by adding EGTA to the cuvette to a final concentration of 1 mM. This will give rise to an instantaneous decrease in the fluorescence as a result of the conversion of the ATPase from the $E1Ca_2$ to the E2 conformation.
5. Wait for the baseline to stabilize apart from the slow downwards drift due to photolysis after the EGTA addition. When ready, test the effect of the inhibitor (thapsigargin or cyclopiazonic acid) by adding them to the cuvette compartments over a concentration range of 2–10 µM where the ATPase is now in the E2 conformation. Adding the inhibitors at a concentration of about 10 µM under conditions where the ATPase is in the E2 state will give rise to a decreased fluorescence arising from saturation of the specific high-affinity site for both thapsigargin and cyclopiazonic acid by a time-dependent (1–2 min) mechanism (*see* **Note 3**). To be able to obtain good recordings for a kinetic analysis of the specific site, the addition of the inhibitors, including mixing

and closing of the cuvette compartment, should be done swift and consistently.

6. From the change in the amplitude of fluorescence intensity vs. time plot, a rate constant (k) or half time ($t_{1/2}$) can be estimated by using the following equation:

$$k(\min^{-1}) = 0.693 / t_{1/2}$$

7. To test the effect of the inhibitors in the E1Ca$_2$ state, add the inhibitors to samples without prior addition of EGTA. Follow the changes for 10–15 min (*see* **Note 4**).

4 Notes

1. It is not recommended that measurements are done at higher protein concentrations than 0.1 mg/mL to minimize artifactual inner filter effects arising from the absorption and scattering of light by the membranes at the wavelength used for the excitation of tryptophan fluorescence.

2. This is expected to take approx. 5 min, but note that even when thermoequilibration has occurred the baseline will exhibit a slow downward drift due to an inevitable, but slow photolysis induced by the tryptophan residues by the UV radiation from the lamp.

3. But note that for cyclopiazonic acid the fluorescence response is biphasic, because this compound also reacts instantaneously with an additional number of nonspecific binding sites.

4. Nonspecific binding will not be affected by the presence of Ca^{2+} which only will prevent and retard specific binding of inhibitor which does not occur to the ATPase in the E1Ca$_2$ conformation. The changes in intrinsic fluorescence on E1Ca$_2$ are most easily observed by the use of high inhibitor concentrations, e.g., 10 μM.

References

1. Brand L, Witholt B (1967) Fluorescence measurements. Meth Enzymol 11:776–856
2. Lakowicz JR (2006) Principles of fluorescence spectroscopy, vol 3. Springer, New York
3. East JM, Lee AG (1982) Lipid selectivity of the calcium and magnesium ion dependent adenosinetriphosphatase, studied with fluorescence quenching by a brominated phospholipid. Biochemistry 21:4144–4151
4. De Foresta B, le Maire M, Orlowski S, Champeil P, Lund S, Møller JV, Michelangeli F, Lee AG (1989) Membrane solubilization by detergent: use of brominated phospholipids to evaluate the detergent-induced changes in Ca2+-ATPase/lipid interaction. Biochemistry 28:2558–2567
5. Møller JV, Juul B, le Maire M (1996) Structural organization, ion transport, and energy transduction of P-type ATPases. Biochim Biophys Acta 1286:1–51

Part IV

Ligand Binding Studies

Chapter 22

Determination of the ATP Affinity of the Sarcoplasmic Reticulum Ca^{2+}-ATPase by Competitive Inhibition of $[\gamma-^{32}P]$ TNP-8N$_3$-ATP Photolabeling

Johannes D. Clausen, David B. McIntosh, David G. Woolley, and Jens Peter Andersen

Abstract

The photoactivation of aryl azides is commonly employed as a means to covalently attach cross-linking and labeling reagents to proteins, facilitated by the high reactivity of the resultant aryl nitrenes with amino groups present in the protein side chains. We have developed a simple and reliable assay for the determination of the ATP binding affinity of native or recombinant sarcoplasmic reticulum Ca^{2+}-ATPase, taking advantage of the specific photolabeling of Lys^{492} in the Ca^{2+}-ATPase by $[\gamma-^{32}P]2',3'$-O-(2,4,6-trinitrophenyl)-8-azido-adenosine 5'-triphosphate ($[\gamma-^{32}P]$TNP-8N$_3$-ATP) and the competitive inhibition by ATP of the photolabeling reaction. The method allows determination of the ATP affinity of Ca^{2+}-ATPase mutants expressed in mammalian cell culture in amounts too minute for conventional equilibrium binding studies. Here, we describe the synthesis and purification of the $[\gamma-^{32}P]$TNP-8N$_3$-ATP photolabel, as well as its application in ATP affinity measurements.

Key words ATP binding affinity, Photoaffinity labeling, $[\gamma-^{32}P]$TNP-8N$_3$-ATP synthesis, Sarcoplasmic reticulum Ca^{2+}-ATPase, Ca^{2+}-ATPase mutants, P-type ATPase family

1 Introduction

P-type ATPases bind ATP with high affinity, typically displaying dissociation constants in the 0.1–10 μM range for the interaction of the Mg^{2+}·ATP complex with the high-affinity nucleotide-binding site of the $E1$ state of the enzyme. When a relatively pure source of the ATPase protein is available, ATP binding studies can be carried out directly in equilibrium binding assays [1–5]. However, in circumstances where the ATPase only represents a minor fraction of the total protein in the sample, as is typically the case when the ATPase is obtained by heterologous expression in mammalian cell culture, it is generally not feasible to measure ATP binding directly. An indirect measure of the ATP affinity of recombinant ATPase

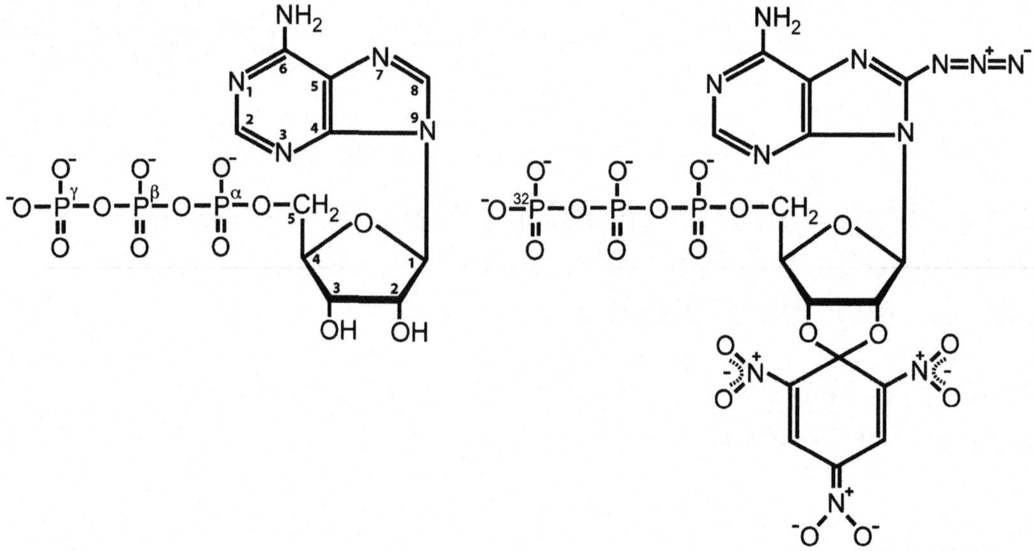

Fig. 1 Schematic structures of ATP (*left*) and [γ-^{32}P]TNP-8N$_3$-ATP (*right*). The [γ-^{32}P]TNP-8N$_3$-ATP photolabel is a derivative of ATP, formed by the addition of an azido group to the 8-position of the adenine and a TNP group to the 2′,3′-positions of the ribose, and by the exchange of the γ-phosphate with a ^{32}P-labeled phosphate

expressed in relatively low yield can often be obtained by studying the ATP dependence of a given functional characteristic, such as the phosphorylation by [γ-^{32}P]ATP (either the steady-state level or the transient kinetics) or the rate of overall ATP turnover [6–9]. Such measurements are, however, easily influenced by shifts in the protein conformational transitions towards or away from the ATP-reactive *E*1 state (often relevant for the study of mutant ATPases) and generally need to be interpreted in the light of a detailed investigation of other partial reaction steps in the ATPase cycle in order to permit trustworthy conclusions regarding the "true" ATP affinity.

We have developed an assay [10] that enables the direct determination of ATP affinity constants for the sarcoplasmic reticulum Ca^{2+}-ATPase, overcoming the technical difficulties associated with ATP binding measurement on the minute amounts of wild type or mutant Ca^{2+}-ATPase present in microsomes purified from transfected mammalian cell cultures. The assay makes use of the ATP-analogue [γ-^{32}P]TNP-8N$_3$-ATP, taking advantage of three unique features of this compound relative to ATP (Fig. 1): (a) a photoactivatable azido (N$_3$) group at the 8-position of the adenine, enabling the covalent coupling of the nucleotide to the ATPase [11–14], (b) a TNP moiety at the 2′,3′-positions of ribose, providing the nucleotide with an increased affinity and specificity for the ATPase relative to that of ATP [15, 16], and (c) a ^{32}P-labeled γ-phosphate, enabling radiometric detection of the labeled protein. Upon UV irradiation of the Ca^{2+}-ATPase in a reaction mix containing [γ-^{32}P]TNP-8N$_3$-ATP, this nucleotide becomes covalently attached via

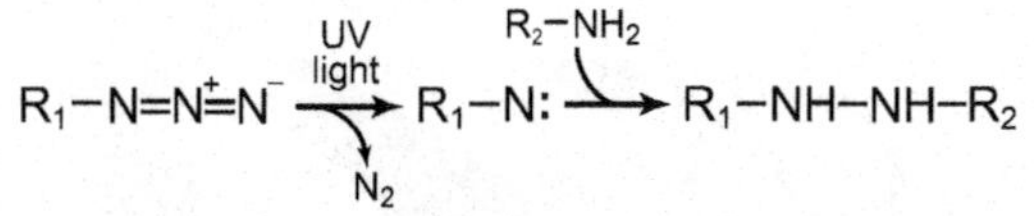

$$R_1-N=\overset{+}{N}=\overset{-}{N} \xrightarrow[N_2]{UV\ light} R_1-N: \xrightarrow{R_2-NH_2} R_1-NH-NH-R_2$$

Fig. 2 Photolabeling of the Ca^{2+}-ATPase by TNP-$8N_3$-ATP. The azido group of TNP-$8N_3$-ATP is activated by ultraviolet light, leading to the formation of a short-lived but highly reactive nitrene group. The nitrene can initiate reactions with neighboring reactive groups, such as the amino group of the Lys^{492} side chain in the Ca^{2+}-ATPase, thereby forming a stable covalent bond between the nucleotide and the protein [11–13]. R_1 represents the TNP-ATP part of TNP-$8N_3$-ATP (*see* Fig. 1), and R_2 is the Lys^{492} side chain

the azido group to a specific lysine, Lys^{492}, in the nucleotide-binding domain of the ATPase [11, 17, 18] (Fig. 2). Addition of non-radioactive ATP competitively inhibits the photolabeling reaction, and titration of the ATP dependence of the inhibition of photolabeling provides a means to determine binding of ATP to the enzyme [10]. Our studies have shown that the dissociation constants obtained by this method are remarkably similar to those obtained by conventional equilibrium binding methods when direct comparisons are made under identical conditions. This approach furthermore has the advantage over traditional direct and indirect nucleotide binding assays that it is not limited to the analysis of ATP binding to one specific intermediate state in the Ca^{2+}-ATPase reaction cycle, as is generally the case for assays measuring the ATP dependence of a given functional characteristic, but can be carried out on any reaction state that can be stabilized in vitro, owing to the high affinity of the Ca^{2+}-ATPase for TNP-$8N_3$-ATP throughout the reaction cycle. Thus, the assay enables the determination of ATP dissociation constants, not only for the catalytic high-affinity binding mode in the *E*1 state of the reaction cycle [10], but also for the regulatory binding modes in the *E*2 and *E*2P states [16, 19]. Our [γ-^{32}P]TNP-$8N_3$-ATP photolabeling studies of Ca^{2+}-ATPase mutants [10, 16, 19–28] have identified residues crucial to ATP binding, later confirmed by X-ray crystal structures of the Ca^{2+}-ATPase with bound ATP-analogs [29–32]. Hence, the importance of the conserved phenylalanine of the N-domain, Phe^{487}, in ATP binding by the Ca^{2+}-ATPase (Fig. 3 and Table 1) was first demonstrated in [γ-^{32}P]TNP-$8N_3$-ATP photolabeling studies [10].

The ability to undergo photolabeling from TNP-$8N_3$-ATP has, thus far, only been demonstrated for one member of the P-type ATPase family, namely the rabbit SERCA1a isoform of the sarcoplasmic reticulum Ca^{2+}-ATPase. It is well known, however, that trinitrophenylated nucleotides bind with high affinity to other P-type ATPases as well, including the Na^+,K^+-ATPase [33, 34] and the H^+,K^+-ATPase [35, 36]. The lysine that is labeled by TNP-$8N_3$-ATP in the Ca^{2+}-ATPase, Lys^{492} (Fig. 3), is highly conserved

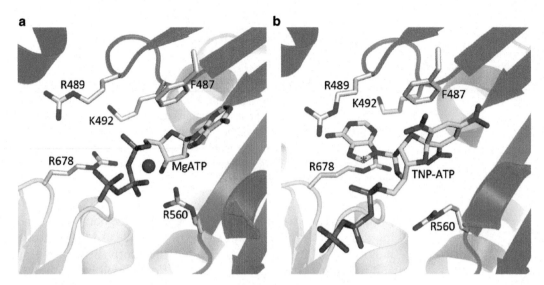

Fig. 3 Structural organization of the nucleotide binding site in the Ca^{2+}-ATPase. The structures shown are (**a**) E2·thapsigargin·Mg^{2+}·ATP (PDB 3AR4) and (**b**) E2·thapsigargin·TNP-ATP (PDB 3AR7) [47]. The N domain is shown in *red cartoon*, the A domain in *blue cartoon*, the P domain in *yellow cartoon*, the nucleotides and relevant side chains in *stick model* (*yellow* and *grey carbons*, respectively), and a Mg^{2+}-ion as a *purple sphere*. The TNP-moiety of TNP-ATP occupies the binding pocket that normally coordinates the adenine moiety of ATP. TNP-ATP is in the anti-conformation around the glycosidic bond, whereas TNP-$8N_3$-ATP would be expected to be mainly in the syn-conformation, owing to electrostatic repulsion between the azido-group and the phosphate-moiety [48]. The Lys^{492} side chain is positioned such that it would be in close proximity to the 8-position (marked by a *red asterisk* in panel **b**) of the adenine if the nucleotide were in the syn-conformation, in accordance with the fact that Lys^{492} is the residue that is photolabeled by TNP-$8N_3$-ATP [17]

throughout the P-type ATPase family (Table 1), being absent only in ATPases of the P_{1A} type (bacterial K^+-pumps; the lysine being replaced by an arginine), the P_{1B} type (heavy-metal pumps), and the P_5 type (ATPases with unassigned specificity) [37, 38]. With respect to the remaining seven P-type ATPase subfamilies, the lysine is conserved in 207 of 210 sequences analyzed [37]. It is therefore likely, that the ATP binding assay described here can be applied also with other P-type ATPases. Indeed, the assay is not only limited to transporters of the P-type ATPase family. Hence, both p-glycoprotein and Yor1p, two members of the family of ABC (ATP-binding cassette) transporters, are likewise able to become specifically photolabeled at their nucleotide-binding sites by TNP-$8N_3$-ATP, facilitating the determination of ATP-binding affinities in the same type of competition assays described here [39, 40].

The photolabeling assay itself is relatively easy and straightforward to carry out. The [γ-^{32}P]TNP-$8N_3$-ATP photolabel, however, is not commercially available, and needs to be prepared in the laboratory. In the following, we present a detailed protocol for the synthesis and purification of high specific activity [γ-^{32}P]TNP-$8N_3$-ATP, using commercially available $8N_3$-ATP as starting material. We subsequently describe the procedure for the application of

Table 1
Conservation in the family of P-type ATPases of the [487]Phe-Ser-Arg-Asp-Arg-Lys[492] nucleotide-binding motif, containing the lysine residue that is specifically labeled by TNP-8N$_3$-ATP in the Ca^{2+}-ATPase. The analysis is based on the protein sequences listed in the P-type ATPase database available on the World Wide Web [37, 38]

Subfamily	Substrate specificity	# Sequences[a]	Sequence conservation[b]
P$_{1A}$	Unassigned	6	FtAxtR
P$_{1B}$	Cu$^+$, Ag$^+$, Cu^{2+}, Cd^{2+}, Zn^{2+}, Pb^{2+}, Co^{2+}	71	-[c]
P$_{2A}$	Ca^{2+}, Mn^{2+} (incl. SERCA pumps)	57	FsrxrK
P$_{2B}$	Ca^{2+} (incl. PMCA pumps)	26	FnSxrK
P$_{2C}$	Na$^+$/K$^+$, H$^+$/K$^+$	43	FNStnK
P$_{2D}$	Unassigned	8	FDSxxK
P$_{3A}$	H$^+$	39	FxPvdK
P$_{3B}$	Mg^{2+}	3	FLDPPK
P$_4$	Phospholipids	34	FxsxrK
P$_5$	Unassigned	16	Fxsxlx

[a]The total number of protein sequences included in the analysis is listed for each subfamily.
[b]Amino acid sequence conservation within each subfamily of P-type ATPases. *Uppercase letter*, amino acid (single-letter code) conserved in > 95 % of the sequences in the database; *lower case letter*, amino acid conserved in 60–95 % of the sequences in the database; *x*, < 60 % amino acid conservation.
[c]The FxxxxK nucleotide-binding motif generally found in the P-type ATPase family is not present in ATPases of the P$_{1B}$ subfamily, which instead rely on a variant motif, where a histidine residue provides the π-stacking interaction with the adenine moiety of ATP [46] rather than a phenylalanine, as found at the corresponding position in the nine other P-type ATPase subfamilies (e.g., Phe[487] in the Ca^{2+}-ATPase; *see* Fig. 3).

[γ-^{32}P]TNP-8N$_3$-ATP in competition experiments, exemplified by the determination of the ATP affinity of COS-1-expressed wild type and mutant Ca^{2+}-ATPase stabilized in the E2·P phosphate transition state.

2 Materials

All solutions and dilutions are aqueous, prepared using deionized water filter-purified to a sensitivity of 18 MΩ. Unless stated otherwise, buffers and solutions are stored at room temperature. When indicated, solutions are filter sterilized using 0.22 μm pore size syringe filters.

2.1 Preparation of [γ-^{32}P]TNP-8N$_3$-ATP

1. 10 mM 8N$_3$-ATP pH 7.6 (BIOLOG Life Science Institute, Bremen, Germany; *see* **Note 1**). Store at –20 °C.
2. ^{32}P-exchange buffer (10×): 1 M Tris/HCl, pH 8.0, 60 mM MgCl$_2$. Filter sterilize and store in small aliquots at –20 °C.

3. 100 mM NaOH. Filter sterilize and store in small aliquots at −20 °C.
4. 500 mM L-cysteine. Store in small aliquots at −20 °C.
5. 100 mM d-(-)-3-phosphoglyceric acid. Store in small aliquots at −20 °C.
6. Glyceraldehyde 3-phosphate dehydrogenase. Supplied as a lyophilized powder. Dissolve in water to give a protein concentration of 27 mg/mL. Store at −20 °C.
7. 3-Phosphoglyceric phosphokinase. Supplied as an ammonium sulfate suspension, typically with a protein concentration of 3–4 mg/mL. Store at 4 °C. After first use, carefully seal the lid to prevent water evaporation.
8. Phosphorus-32 orthophosphoric acid in water, 10 mCi/mL. *See* **Note 2**.
9. Dilutions of HCl at concentrations of 1 mM, 10 mM, 30 mM, and 1 M.
10. 0.8 M Na_2CO_3/$NaHCO_3$, pH 9.5: Dissolve 0.91 g Na_2CO_3 and 0.63 g $NaHCO_3$ in 10 mL water by stirring and heating to ~37 °C. Store in aliquots of 0.5 mL at −20 °C.
11. 1 M 2,4,6-trinitrobenzenesulfonic acid (TNBS; sold by Sigma-Aldrich under the name *Picrylsulfonic acid solution, 1 M in H_2O*). Store at −20 °C.
12. 5,5′-Dithiobis(2-nitrobenzoic acid) (DTNB).
13. 0.2, 0.5, and 1 M ammonium formate, pH 8.2: Prepare the 1 M solution by dissolving 3.15 g ammonium formate in 45 mL water and adjusting the pH to 8.2 with ammonia. Add water to a final volume of 50 mL. Prepare the 0.2 and 0.5 mM solutions by diluting the 1 M solution in water (10 mL water + 10 mL 1 M ammonium formate, and 10 mL water + 2.5 mL 1 M ammonium formate, respectively). Prepare fresh prior to use.
14. 500 mM KH_2PO_4/K_2HPO_4, pH 6.7: Prepare 500 mM solutions of both KH_2PO_4 and K_2HPO_4, and then adjust the pH of the 500 mM KH_2PO_4 solution to 6.7 with the 500 mM K_2HPO_4 solution. Filter sterilize and store at 4 °C.
15. 10 mM KH_2PO_4/K_2HPO_4, pH 7.0: 0.4 mL 500 mM KH_2PO_4/K_2HPO_4, pH 6.7 + 19.6 mL water. Filter sterilize and store at 4 °C.
16. 60 % (v/v) acetonitrile: 30 mL 100 % acetonitrile + 20 mL water. Filter sterilize.
17. Whatman® DE52 pre-swollen microgranular anion exchange cellulose (GE Healthcare Life Sciences, Pittsburgh, PA, USA).
18. Two 10 cm × 0.5 cm (inner diameter) Glass Econo-Column® columns for gravity flow chromatography (Bio-Rad, Hercules, CA, USA) with outlet tubing of 1 and 2 mL dead-volume, respectively.

19. Sep-Pak® Plus C18 Cartridges for solid phase extraction (Waters, Milford, MA, USA).
20. Methanol.
21. 1, 2, 5, 10, and 20 mL plastic syringes.
22. pH indicator strips (not essential).
23. TLC PEI Cellulose F thin layer chromatography (TLC) sheets, 20×20 cm (Merck-Millipore, Billerica, MA, USA).
24. Mobile solvent for TLC: 1 M LiCl.
25. Hair dryer (for drying the TLC sheets).
26. Liquid nitrogen.
27. Nitrogen gas cylinder.
28. Freeze-drying equipment: We use a homemade freeze-drying setup consisting of a vacuum pump, capable of creating a vacuum of < 0.1 mbar (critical!), connected via a cold trap to a temperature- and vacuum-resistant Pyrex Quickfit FR50/3S 50 mL round-bottomed flask (Scilabware, Stoke-on-Trent, UK).
29. Incubator oven at 25 °C; not necessary if the room temperature is already around 25 °C (± 3–4°).
30. Temperature-regulated water bath.
31. A Geiger-Müller counter.
32. Thick (~1 cm) acrylic glass protection as well as several thick-walled lead containers for the protection against concentrated radioactivity (*see* **Note 3**).
33. Radiometric detection equipment: We use a PerkinElmer Cyclone Plus Phosphor Imager with MultiSensitive Phosphor Screens. The Optiquant software package (PerkinElmer) is used for quantification of the radioactive gel bands and TLC spots.
34. Spectrophotometric equipment, ideally with the option of creating a full absorbance spectrum between 200 and 600 nm.

2.2 [γ-^{32}P]TNP-8N$_3$-ATP Photolabeling of the Ca^{2+}-ATPase

1. 25 % tetramethyl ammonium hydroxide (TMAH) solution (*see* **Note 4**).
2. Ca^{2+}-ATPase storage buffer: 5 mM 4-(2-hydroxyethyl)-piperazine-1-ethanesulfonic acid (HEPES)/TMAH, pH 7.4, 0.3 M sucrose.
3. Sarco(endo)plasmic reticulum Ca^{2+}-ATPase: The protein source used in the present example is endoplasmic reticulum membranes (microsomes) purified from COS-1 cells transfected with an expression vector containing the gene encoding wild-type or mutant SERCA1a Ca^{2+}-ATPase isoform, according to established procedures [41]. The typical total protein

content of the stock samples is ~10 mg/mL, of which ~1 % is Ca^{2+}-ATPase, corresponding to a typical Ca^{2+}-ATPase concentration of ~1 μM. Dilute in Ca^{2+}-ATPase storage buffer to the desired concentration (*see* **Note 5**).

4. Ca^{2+}-ATPase preincubation buffer: In the present example, the photolabeling is carried out on Ca^{2+}-ATPase stabilized in an $E2 \cdot P$ phosphate transition state-like conformation by incubation of the Ca^{2+}-deprived enzyme with the phosphate-analog orthovanadate in the following buffer (2×): 50 mM 3-(N-morpholino)propanesulfonic acid (MOPS)/TMAH pH 7.0, 160 mM KCl, 10 mM $MgCl_2$, 4 mM EGTA.

5. 1 mM orthovanadate: 492 μL water/NaOH (pH ~ 10) + 8.17 μL 61.2 mM orthovanadate. The 61.2 mM orthovanadate stock solution is prepared according to established procedures [42].

6. Photolabeling buffer (4×): 100 mM 4-(2-hydroxyethyl)-piperazine-1-propanesulfonic acid (EPPS)/TMAH, pH 8.5, 8 mM EDTA.

7. 0.01–100 μM [γ-^{32}P]TNP-8N_3-ATP dilutions (10× relative to the final reaction mix during photolabeling). The specific activity of the [γ-^{32}P]TNP-8N_3-ATP immediately following its synthesis is much higher than necessary for the photolabeling experiments. Initially, we therefore prepare a 100 μM solution (this typically being the most concentrated solution needed for the photolabeling experiments), in which 1 per 20 mol comes from the newly synthesized radioactive [γ-^{32}P]TNP-8N_3-ATP batch and 19 per 20 mol come from a nonradioactive TNP-8N_3-ATP batch (*see* **Notes 6 and 7**). Less concentrated [γ-^{32}P]TNP-8N_3-ATP solutions (but, importantly, each with the same specific activity) are then prepared by dilution of the 100 μM solution in water. Store the [γ-^{32}P]TNP-8N_3-ATP dilutions on ice during the experiments and at −20 °C otherwise. Over time, as the radioactivity drops (the half-life of the ^{32}P isotope is 14.3 days), we retain a relatively high specific activity in the experiments by preparing new [γ-^{32}P]TNP-8N_3-ATP solutions, in which a larger proportion of the TNP-8N_3-ATP comes from the radioactive batch, until after ~2 months we finally use the radioactive [γ-^{32}P]TNP-8N_3-ATP batch without diluting the radioactivity with nonradioactive TNP-8N_3-ATP. In this manner, we find that each [γ-^{32}P]TNP-8N_3-ATP synthesis can be used reliably for photolabeling experiments for at least 3 months. The azido group is light sensitive, so take care to shield the [γ-^{32}P]TNP-8N_3-ATP solutions from direct light exposure at all times, e.g., by wrapping the tubes containing the [γ-^{32}P]TNP-8N_3-ATP stock dilutions in tinfoil.

8. 87 % glycerol.

9. 10 mM ATP/TMAH, pH 7.5. Filter sterilize and store in small aliquots at −20 °C.

10. Xenon arc light source. We use an LOT-QuantumDesign LSH102 Arc Light Source assembled with a 150 W ozone-free xenon arc lamp, a rear light reflector, an F/1.3 35-mm aperture quartz condenser, and a glass filter with a 295-nm wavelength cut-off (LOT-QuantumDesign, Darmstadt, Germany; see **Note 8**).

11. UV safety goggles. Should be worn at all times when operating the xenon arc light source.

12. 300–400 μL quartz cuvette with 1 mm path width and 10 mm path length. We use Hellma Suprasil 300 cuvettes. To permit the insertion of a pipette tip into the cuvette it is important that the cuvette is of the lidded type rather than of the type with a screw cap.

13. 200 μL pipette tips with long and thin (< 1 mm outer diameter) tip ends, to enable the photolabeling mix to be loaded into and extracted from the labeling cuvette.

14. Height-adjustable lab jack. At the center of one of the edges of the lab jack, place a 1 × 1 cm square piece of colored tape, defining the position of the cuvette during irradiation.

15. Table mini centrifuge (microfuge), for quick spin downs of Eppendorf tubes.

16. Temperature-regulated water bath.

2.3 Sodium Dodecyl Sulfate (SDS)-Polyacrylamide Gel Electrophoresis, Gel Drying, Radiometric Detection, and Data Analysis

1. 10 % (w/v) SDS: Dissolve 50 g SDS in water to a final volume of 500 mL.

2. Resolving gel buffer (4×): 1.5 M Tris–HCl, pH 8.8, 0.4 % SDS. Dissolve 90.8 g Tris in 400 mL water and adjust the pH to 8.8 with HCl. Add 20 mL 10 % SDS and water to a final volume of 500 mL.

3. Stacking gel buffer (4×): 0.5 M Tris–HCl, pH 6.8, 0.4 % SDS. Dissolve 15.2 g Tris in 200 mL water and adjust the pH to 6.8 with HCl. Add 10 mL 10 % SDS and water to a final volume of 250 mL.

4. Gel running buffer (5×): 0.125 M Tris, 0.96 M glycine, 0.5 % SDS. Dissolve 30.3 g Tris and 144.1 g glycine in 1.5 L water. Add 100 mL 10 % SDS and water to a final volume of 2 L.

5. 20 % SDS/bromophenol blue: Dissolve 10 g SDS and a "pinch" of bromophenol blue (see **Note 9**) in water to a final volume of 50 mL.

6. β-Mercaptoethanol.

7. Protein denaturation buffer (for 30 samples): Add 90 μL β-mercaptoethanol to 150 μL 20 % SDS/bromophenol blue and vortex. Prepare fresh for each experiment.

8. 30 % acrylamide/bis solution, 37.5:1.
9. N,N,N',N'-Tetramethyl-ethylenediamine (TEMED).
10. 10 % (w/v) ammonium persulfate. Store at 4 °C (*see* **Note 10**).
11. Polyacrylamide gel electrophoresis equipment and power supply: We use the Hoefer SE250 Mighty Small II Mini Vertical Electrophoresis Units with 1.5 mm thickness T-spacers and 1.5 mm thickness 10-well combs (Hoefer Inc., Holliston, MA, USA). Larger gel electrophoresis units may be used instead, such as the Hoefer SE600 Standard Dual Cooled Vertical Unit. The advantage of the larger unit is that more samples and higher sample volumes can be applied compared with the mini gels, at the disadvantage, however, of the electrophoresis being more time consuming (~2 h running time for the mini gels compared with ~4 h for the large gels).
12. Gel drying equipment: We use a Bio-Rad Model 583 Gel Dryer (Bio-Rad) connected via a cold trap to a vacuum pump. It is critical that the vacuum pressure in the gel dryer is < 5 mbar, otherwise the gel bands will get fuzzy.
13. Radiometric detection equipment: *See* **item 33** in Subheading 2.1.
14. Data analysis: We use the SigmaPlot program (Systat Software Inc.) for regression analysis, graphing, and statistical analysis of the quantified gel bands.

3 Methods

Carry out all procedures at room temperature unless otherwise specified. Be aware that the azido group is light sensitive and, thus, care should be taken at all times to shield the $8N_3$-ATP and TNP-$8N_3$-ATP nucleotides from light exposure, in particular from direct sun light (hence, if possible, draw the curtains in the laboratory, at least during the $[\gamma\text{-}^{32}P]$TNP-$8N_3$-ATP synthesis steps). The polypropylene material of regular laboratory plasticware is generally quite efficient in shielding against the photoactivation of the azido group by regular daylight.

3.1 Preparation of $[\gamma\text{-}^{32}P]8N_3$-ATP: The ^{32}P-Exchange Reaction

The first step in the synthesis of the $[\gamma\text{-}^{32}P]$TNP-$8N_3$-ATP photolabel is the exchange of the γ-phosphate of $8N_3$-ATP with ^{32}P-labeled phosphate (Fig. 4).

1. Mix the following, in this order, in a 1.5 mL Eppendorf tube (final volume: 0.5 mL): 66 μL water, 50 μL ^{32}P-exchange buffer (10×), 51 μL 100 mM NaOH, 2 μL 500 mM cysteine, 22.5 μL 10 mM $8N_3$-ATP, 5 μL 100 mM d-(−)-3-phosphoglyceric acid, 1.9 μL 27 mg/mL glyceraldehyde 3-phosphate dehydrogenase, 300 μL 10 mCi/mL phosphorus-32 orthophosphoric

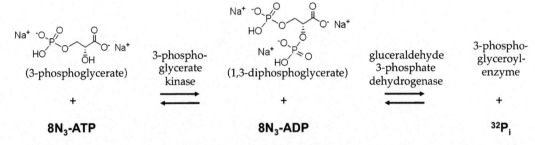

Fig. 4 Principles of the ^{32}P-exchange reaction. The exchange of the γ-phosphate of 8N$_3$-ATP with ^{32}P-labeled phosphate is made feasible by the actions of two enzymes, 3-phosphoglycerate kinase and glyceraldehyde 3-phosphate dehydrogenase, in the presence of the suitable substrates [49]. 3-Phosphoglycerate kinase catalyzes the phosphorylation of 3-phosphoglycerate by 8N$_3$-ATP, forming 1,3-diphosphoglycerate and 8N$_3$-ADP. A complex is then formed between 1,3-diphosphoglycerate and glyceraldehyde 3-phosphate dehydrogenase, leading to the derivatization of this enzyme with phosphoglycerate and concomitant release of phosphate. Because the equilibrium of the former reaction is far towards the left side, and since 8N$_3$-ATP is added in excess over phosphate (^{32}P$_i$), over time the vast majority of the ^{32}P-labeled phosphate included in the reaction mix will be incorporated into 8N$_3$-ATP, as verified by the results shown in Fig. 5a

acid (~3 mCi). Vortex and collect at the bottom of the tube, and then add 1.25 µL 4 mg/mL 3-phosphoglyceric phosphokinase to initiate the exchange reaction (if the 3-phosphoglyceric phosphokinase supplied by the manufacturer varies in concentration from that used in the present example, adjust the volume accordingly). Vortex, collect at the bottom of the tube, and incubate the tube in a lead container for 2½ h at room temperature with occasional mixing.

2. After 1½ h reaction check for ^{32}P-incorporation by TLC (continue the incubation of the reaction mix at room temperature while the TLC is running): Prepare a TLC sheet strip of 20 cm in length and a couple of centimeters in width. Draw a line lightly with a soft pencil ~2–3 cm from one end of the strip. Spot a minimum amount of the reaction mix (~0.2 µL) on the line. Lower the TLC strip (spotted end first) into a container with ~1 cm of 1 M LiCl (i.e., so that the liquid level does not reach the spot). Place a lid on the container and wait for the solvent to reach ~2 cm from the top of the strip (~30–40 min). Dry the strip with a hair dryer and check for radioactivity by phosphor imaging (Fig. 5a). At this stage, the reaction mix is highly radioactive, so the phosphor screen need only be placed on the TLC strip for ~10–20 s.

3. Prepare a 2 cm (resin height) × 0.5 cm (inner diameter) DE52 anion-exchange column with a 1 mL dead-volume outlet tubing (the column can be prepared while the TLC is running). Rinse the resin by several washings with water prior to its application to the column. Chromatography is carried out by gravity flow. Apply the ^{32}P-exchange reaction mix to the column.

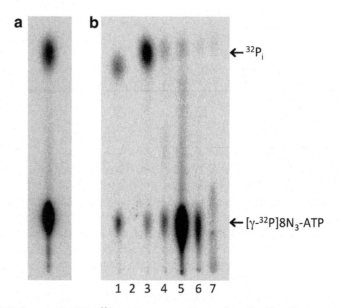

Fig. 5 TLC analysis of the ^{32}P-exchange reaction and purification eluates. (**a**) TLC analysis of the ^{32}P-exchange reaction mix after 1½ h incubation. At this stage, ~90 % of the ^{32}P-labeled phosphate has typically been incorporated into the 8N$_3$-ATP. (**b**) TLC analysis of the eluates from the DE52 anion exchange purification. The eluates correspond to the following additions to the column: *lane 1*, 0.5 mL reaction mix + 0.5 mL water + 5 mL water; *lane 2*, 5 mL 1 mM HCl; *lane 3*, 7 mL 10 mM HCl; *lane 4*, 1 mL 30 mM HCl; *lane 5*, 7 mL 30 mM HCl (main [γ-^{32}P]8N$_3$-ATP fraction); *lane 6*, 2 mL 30 mM HCl; *lane 7*, 5 mL 1 M HCl

Wash the reaction tube with 0.5 mL water and apply to the column. Wash the column with 5 mL water. Acidify the column with 5 mL 1 mM HCl. Elute the phosphate and 8N$_3$-ADP with 7 mL 10 mM HCl. Elute the [γ-^{32}P]8N$_3$-ATP with 30 mM HCl as follows: (a) add 1 mL to the column (dead-volume in tubing), (b) add 7 mL to the column (main [γ-^{32}P]8N$_3$-ATP fraction; collect in a 30 mL freeze-drying flask, wrapped in tinfoil to shield from light), (c) add 2 mL to the column (for control), and (d) add 5 mL 1 M HCl (for control). *See* Fig. 5b for a TLC analysis of the various eluates from the DE52 column.

4. Neutralize the pH of the 7 mL [γ-^{32}P]8N$_3$-ATP fraction with 105 μL 0.8 M Na$_2$CO$_3$/NaHCO$_3$, pH 9.5. The pH can be checked by spotting a small volume on a pH indicator strip.

5. Freeze-dry the [γ-^{32}P]8N$_3$-ATP fraction: Freeze the [γ-^{32}P]8N$_3$-ATP solution by lowering the freeze-drying flask into liquid nitrogen for 2–3 min. Place the freeze drying flask on ice in a lidded polystyrene foam thermobox with a small hole in the lid, allowing the vacuum tubing to reach the freeze drying flask. Connect the vacuum to the freeze drying flask and leave overnight.

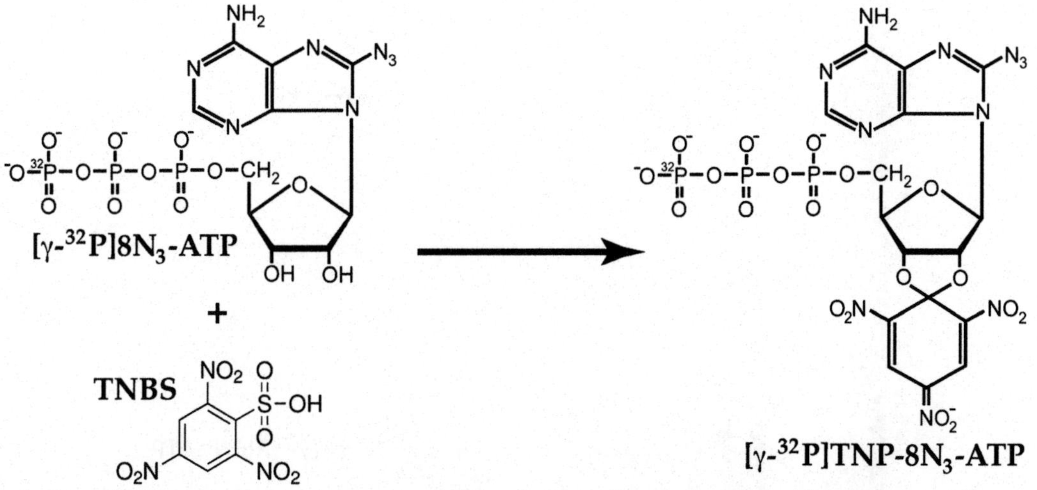

Fig. 6 Principles of the trinitrophenylation reaction. The reaction of the 2′,3′-hydroxyls of the ribose of [γ-^{32}P]8N$_3$-ATP with TNBS, leading to the formation of [γ-^{32}P]TNP-8N$_3$-ATP

3.2 Preparation of [γ-^{32}P]TNP-8N$_3$-ATP: Trinitrophenylation of [γ-^{32}P]8N$_3$-ATP

The second step in the synthesis of the [γ-^{32}P]TNP-8N$_3$-ATP photolabel is the addition of the TNP-group to the 2′,3′-hydroxyls of the ribose of [γ-^{32}P]8N$_3$-ATP (Fig. 6).

1. Check that all the water has evaporated from the freeze drying flask. Add 150 μL water to the flask and dissolve the precipitated [γ-^{32}P]8N$_3$-ATP by gently moving the water around. Adjust the freeze drying flask to room temperature (>20 °C).

2. Weigh ~13.5 mg DTNB into a 1.5 mL Eppendorf tube (*see* **Note 11**).

3. Thaw a tube of 500 μL 0.8 M Na$_2$CO$_3$/NaHCO$_3$, pH 9.5 in a 37 °C water bath. Vortex and repeat the incubation at 37 °C until all the salt crystals have dissolved. Adjust to room temperature.

4. Do the following in rapid succession (NB! at room temperature, not on ice): Transfer the 0.5 mL 0.8 M Na$_2$CO$_3$/NaHCO$_3$, pH 9.5 to the tube with DTNB and vortex. Add 40 μL 1 M TNBS to the dissolved DTNB and vortex. Transfer 83 μL of the DTNB/TNBS mix to the freeze-drying flask containing 150 μL [γ-^{32}P]8N$_3$-ATP and mix by moving the flask around. Incubate at 25 °C (or at room temperature if this is already around 25 ± 3–4 °C) in the dark with occasional mixing.

5. After 2½–3 h reaction check for trinitrophenylation by TLC (following the same procedure described in **step 2** in Subheading 3.1). An example of the typical outcome of this analysis is shown in Fig. 7a.

6. After 4 h reaction add 5 mL ice-cold water to the freeze drying flask. At this point, the reaction mix can be left at −20 °C overnight.

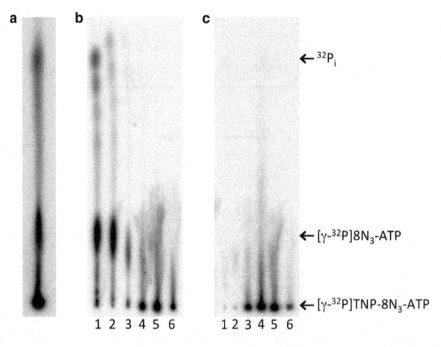

Fig. 7 TLC analysis of the trinitrophenylation reaction and purification eluates. (**a**) TLC analysis of the trinitrophenylation reaction mix after 3 h incubation. (**b**) TLC analysis of the eluates from the DE52 anion-exchange column purification. The eluates correspond to the following additions to the column: *lane 1*, 5.25 mL reaction mix + 6 mL water; *lane 2*, 10 mL 0.2 M ammonium formate pH 8.2; *lane 3*, 15 mL 0.5 M ammonium formate pH 8.2; *lane 4*, 4 mL 1 M ammonium formate pH 8.2; *lane 5*, 21 mL 1 M ammonium formate pH 8.2 (main [γ-^{32}P]TNP-8N$_3$-ATP fraction); *lane 6*, 5 mL 1 M ammonium formate pH 8.2. (**c**) TLC analysis of the eluates from the Sep-Pak purification. The eluates correspond to the following additions to the Sep-Pak cartridge: *lane 1*, 21 mL main [γ-^{32}P]TNP-8N$_3$-ATP fraction from the DE52 column; *lane 2*, 2 mL 10 mM KH$_2$PO$_4$/K$_2$HPO$_4$, pH 7.0; *lane 3*, 0.9 mL water; *lane 4*, 2 mL 60 % acetonitrile (main [γ-^{32}P]8N$_3$-ATP fraction; a 50-fold dilution of the final product after acetonitrile/water vaporization was spotted); *lane 5*, 2 mL 60 % acetonitrile; *lane 6*, 2 mL 60 % acetonitrile

7. Prepare a 4 cm (resin height) × 0.5 cm (inner diameter) DE52 anion-exchange column with a 2 mL dead-volume outlet tubing (this can be done during the 4 h reaction period). Rinse the resin by several washings with water prior to its application to the column. At this time also prepare the 0.2, 0.5, and 1 M ammonium formate, pH 8.2 solutions.

8. Chromatography is carried out by gravity flow. Apply the reaction mix (~5.25 mL) to the column. Wash the freeze-drying flask with 6 mL water and apply to the column. Elute with ammonium formate as follows: (a) 2× 5 mL 0.2 M ammonium formate, pH 8.2, (b) 3× 5 mL 0.5 M ammonium formate, pH 8.2, (c) 4 mL 1 M ammonium formate, pH 8.2 (dead volume in tubing), (d) 3× 5 mL + 1× 6 mL 1 M ammonium formate, pH 8.2 (main [γ-^{32}P]TNP-8N$_3$-ATP fraction; collect in a 50 mL plastic tube on ice), and (e) 5 mL 1 M ammonium

formate, pH 8.2 (for control). *See* Fig. 7b for a TLC analysis of the various eluates from the DE52 column. At this point, the 21 mL main [γ-^{32}P]TNP-8N$_3$-ATP fraction can be left at −20 °C overnight.

9. Prime a Sep-Pak® Plus C18 solid-phase extraction cartridge by applying 6 mL methanol followed by 10 mL water to the cartridge using plastic syringes.

10. Apply the 21 mL main [γ-^{32}P]TNP-8N$_3$-ATP fraction to the cartridge using a plastic syringe. This is best done as follows: (a) remove the piston from the syringe, (b) connect the syringe to the cartridge, (c) pour the 21 mL main [γ-^{32}P]TNP-8N$_3$-ATP fraction into the syringe, (d) carefully reinsert the piston into the syringe, and (e) push the piston down and collect the eluate in a suitable container (be aware that the eluate is radioactive).

11. Apply 2 mL 10 mM KH$_2$PO$_4$/K$_2$HPO$_4$, pH 7.0 to the cartridge using a plastic syringe.

12. Load a 1 mL plastic syringe with 1 mL water and slowly apply water to the cartridge until the [γ-^{32}P]TNP-8N$_3$-ATP starts eluting. This is easily observable/measurable, as the drops start to turn strongly yellow/orange and start to become highly radioactive. Stop pushing the piston of the syringe immediately after the first highly radioactive yellow/orange drop has been released from the cartridge. This happens after ~0.9 mL of water has been applied.

13. Apply 2 mL 60 % acetonitrile to the cartridge, and collect the eluate (main [γ-^{32}P]TNP-8N$_3$-ATP fraction) in a 5 mL radioisotope bottle in a lead container.

14. Apply further 2× 2 mL 60 % acetonitrile to the cartridge (for control). *See* Fig. 7c for a TLC analysis of the various eluates from the Sep-Pak purification.

15. Vaporize the acetonitrile, as well as some of the water, from the 2 mL main [γ-^{32}P]TNP-8N$_3$-ATP fraction by directing a stream of nitrogen gas into the radioisotope bottle through a Pasteur pipette connected via tubing to a nitrogen gas cylinder. Proceed until the final volume is ~300–400 µL, at which point the [γ-^{32}P]TNP-8N$_3$-ATP concentration will typically be ~200–300 µM, corresponding to a typical molar yield of ~40 % [γ-^{32}P]TNP-8N$_3$-ATP relative to the amount of 8N$_3$-ATP (225 nmol) applied at the start of the procedure. The time it takes to reach a volume of 300–400 µL from a starting volume of 2 mL is typically ~1 h.

16. Determine the concentration of [γ-^{32}P]TNP-8N$_3$-ATP by measuring the light absorbance of the TNP-moiety at 408 and 468 nm in a 100-fold dilution of the [γ-^{32}P]TNP-8N$_3$-ATP

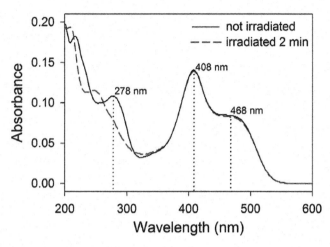

Fig. 8 Absorbance spectrum of [γ-^{32}P]TNP-8N$_3$-ATP. The absorbance spectrum of the final product of the [γ-^{32}P]TNP-8N$_3$-ATP synthesis/purification procedure was measured on a 100-fold dilution in 10 mM KH$_2$PO$_4$/K$_2$HPO$_4$, pH 7.0, before (*solid curve*) and after (*broken curve*) irradiation for 2 min. The peak at 278 nm is attributed to the azido group (which breaks down upon irradiation) and those at 408 and 468 nm to the TNP moiety

solution in 10 mM KH$_2$PO$_4$/K$_2$HPO$_4$, pH 7.0, using extinction coefficients of 28,000 and 19,300 M^{-1} cm^{-1} for the two wavelengths, respectively (Fig. 8). We define the [γ-^{32}P]TNP-8N$_3$-ATP concentration as the average of the values measured at 408 and 468 nm.

17. Store the [γ-^{32}P]TNP-8N$_3$-ATP solution at −20 °C.

3.3 Time Dependence of TNP-8N$_3$-ATP Photolysis

The rate-limiting step in the photolabeling reaction is the photoactivation of the azido group of TNP-8N$_3$-ATP, resulting in the formation of the reactive nitrene (Fig. 2; left half of the reaction scheme). The subsequent chemical reaction between the nitrene of the photoactivated nucleotide and Lys^{492} of the Ca^{2+}-ATPase (Fig. 2; right half of the reaction scheme) is likely a much faster reaction, given the typical short life time and high reactivity of nitrene intermediates [12–14]. The irradiation setup we have applied in recent years is shown in Fig. 9. A quartz cuvette containing the labeling reaction mix is placed at a fixed position (e.g., marked by a 1 × 1 cm square piece of colored tape on a lab jack) in front of the light source, with the collimated light beam centered at the part of the cuvette containing the reaction mix. The diameter of the light beam should cover the entire reaction volume contained in the cuvette. Irradiation is then carried out by manually opening the shutter and then closing the shutter again after an appropriate time interval, determined as described in the following. The photoactivation should be performed under pre-steady-state conditions, with an irradiation time interval compatible with

Fig. 9 The irradiation setup. The xenon arc light source irradiation setup with closed (*left panel*) and open (*right panel*) shutter: *a*, lamp housing; *b*, light condenser; *c*, 295-nm wavelength cut-off glass filter; *d*, manually operated light shutter (a square piece of cardboard wrapped in tinfoil); *e*, filter holder; *f*, labeling cuvette; *g*, height-adjustable lab jack

70–80 % photoactivation of the TNP-8N$_3$-ATP photolabel. This time interval is determined by measuring the rate of photoactivation, which can be optimized to individual preferences by adjusting either the intensity of the light source or the distance between the light source and the labeling cuvette. We find that an irradiation time interval of 30–40 s is optimal for the practical handling of the samples during the experiments. With our standard irradiation setup, placing the labeling cuvette at a fixed position 5 cm from the tip of the filter holder mounted in front of the condenser (Fig. 9), setting the power supply of the light source to 38 W, and irradiating the samples for 35 s is compatible with ~75 % photoactivation. This, however, should be optimized (by trial and error) for each individual irradiation setup. The rate of photoactivation of TNP-8N$_3$-ATP is determined as follows:

1. Prepare a 10 µM solution of nonradioactive TNP-8N$_3$-ATP by dilution of the stock solution in 10 mM KH$_2$PO$_4$/K$_2$HPO$_4$, pH 7.0.
2. Irradiate aliquots of 100 µL of the 10 µM TNP-8N$_3$-ATP solution for varying time intervals.
3. Dilute in 10 mM KH$_2$PO$_4$/K$_2$HPO$_4$, pH 7.0 to an appropriate volume for spectrophotometrical analysis (e.g., to 400 µL, for absorbance measurements in a 400 µL quartz cuvette) and measure the light absorbance at 278 nm (the absorbance peak of the 8N$_3$ group; Fig. 10a).
4. Plot the light absorbance as a function of the irradiation time and extract the rate constant (v) for photoactivation of TNP-8N$_3$-ATP by fitting a monoexponential decay function, $A = A_0 + A\infty \cdot e^{-v \cdot t}$, to the data (Fig. 10b). The time interval giving 75 % labeling is then calculated by the equation $t_{75\%} = -(\ln 0.25)/v$.

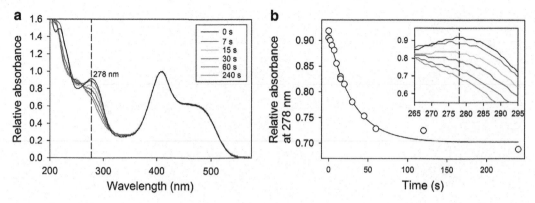

Fig. 10 Time dependence of the photolysis of TNP-8N$_3$-ATP. (**a**) The absorbance spectrum of TNP-8N$_3$-ATP was recorded following irradiation for varying time intervals. The absorbance maximum at 278 nm, contributed by the azido group, decreases upon irradiation owing to its degradation (*see* Fig. 2). For comparison, the maximum absorbance at 408 nm (contributed by the TNP-moiety) was set to 1 for all time intervals, and spectra are shown for selected times. (**b**) Graph showing the relative absorbance at 278 nm as a function of the time of irradiation. The *inset* shows a close-up view of the absorbance around 278 nm from panel **a**. These experiments were originally published in The Journal of Biological Chemistry. Clausen JD, McIntosh DB, Woolley DG, and Andersen JP. (2011) Modulatory ATP binding affinity in intermediate states of *E*2P dephosphorylation of sarcoplasmic reticulum Ca^{2+}-ATPase. J Biol Chem 286: 11792–11802. © 2011 by The American Society for Biochemistry and Molecular Biology

3.4 [γ-^{32}P]TNP-8N$_3$-ATP Dependence of Photolabeling of the Ca^{2+}-ATPase

The ultimate goal of the photolabeling assay is to measure the affinity of the Ca^{2+}-ATPase for ATP, and the first step in the process involves the determination of the $K_{0.5}$ for TNP-8N$_3$-ATP binding. In the present example, photolabeling is carried out on wild type or mutant F487S Ca^{2+}-ATPase stabilized in an *E*2·P phosphate transition state-like conformation by incubation of Ca^{2+}-deprived enzyme with orthovanadate [16, 43, 44], and the photolabeling buffer contains excess EDTA, to chelate Ca^{2+} and Mg^{2+}, in a pH 8.5 buffer. The assay may, however, be carried out with enzyme stabilized in a multitude of different reaction states and at various buffer conditions (e.g., *see* [10, 16, 19]). Buffers in the pH 8–9 range are generally preferable for photolabeling experiments, owing to reduced labeling levels outside this range and increased levels of unspecific labeling, in particular for the COS-1 expressed enzyme, at pH below 8 [10]. The photolabeling should be carried out one sample at a time. Keep all constituents on ice during the experiment. The following is the protocol for an experiment comprising ten samples with varying concentrations of [γ-^{32}P]TNP-8N$_3$-ATP (Fig. 11).

1. Stabilization of the enzyme in the *E*2·orthovanadate state: 20 μL Ca^{2+}-ATPase preincubation buffer (2×), 12 μL water, 4 μL 1 mM orthovanadate, 4 μL 250 nM Ca^{2+}-ATPase (to give a final concentration of Ca^{2+}-ATPase in the photolabeling mix of 1 nM). Vortex and incubate for 30 min at 25 °C, followed

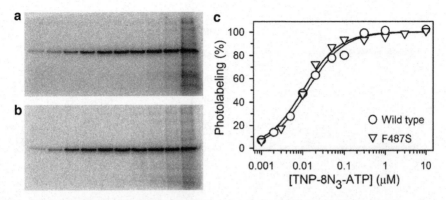

Fig. 11 The [γ-^{32}P]TNP-8N$_3$-ATP dependence of photolabeling. (**a, b**) Autoradiographs of the gels from experiments studying the [γ-^{32}P]TNP-8N$_3$-ATP dependence of photolabeling on COS-1-expressed wild type (**a**) and mutant F487S (**b**) Ca^{2+}-ATPase stabilized in the E2·orthovanadate state. The concentrations of [γ-^{32}P]TNP-8N$_3$-ATP applied are as indicated on the graph in panel **c**. Note the extraordinary specificity of the labeling reaction. Hence, up to ~1 μM [γ-^{32}P]TNP-8N$_3$-ATP, i.e., far beyond the $K_{0.5}$ for the photolabel (~10 nM), only one band can be distinguished in the gels, namely that corresponding to the Ca^{2+}-ATPase, regardless of the fact that the Ca^{2+}-ATPase only constitutes ~1 % of the total protein in the sample. At [γ-^{32}P]TNP-8N$_3$-ATP concentrations above 1 μM other weaker bands start to appear, corresponding to the low-specificity labeling of various other proteins in the sample. (**c**) Analysis of the experiments corresponding to panels **a** and **b** by nonlinear regression. The $K_{0.5}$ (TNP-8N$_3$-ATP) values extracted from the analysis are wild type, $K_{0.5}$ = 12.7 nM; mutant F487S, $K_{0.5}$ = 10.1 nM

by at least 10 min on ice (once the E2·orthovanadate state is formed, it is stable for hours).

2. Sample premix: 18.75 μL photolabeling buffer (4×), 30.75 μL water. Prepare a batch mix for all samples and aliquot 49.5 μL into 1.5 mL Eppendorf tubes. Incubate on ice.

3. Photolabeling reaction mix: In rapid succession, add 15 μL 87 % glycerol, 7.5 μL [γ-^{32}P]TNP-8N$_3$-ATP dilution (10×), and 3 μL of the orthovanadate-inhibited 25 nM Ca^{2+}-ATPase to the sample premix (final volume: 75 μL).

4. Vortex, collect at the bottom of the tube, and transfer to the quartz cuvette (pre-cooled on ice). Wipe the labeling side of the cuvette with tissue and place the cuvette on the 1 × 1 cm piece of tape on the lab jack (Fig. 9). Open the shutter of the light source and irradiate the reaction mix for the time interval determined to be compatible with ~75 % photoactivation of photolabel (as determined in the experiments described in Subheading 3.3). Close the shutter of the light source and transfer the reaction mix from the cuvette back to the same Eppendorf tube from which it came. Leave on ice until all samples have been irradiated.

5. Proceed to **step 3** in Subheading 3.6 for gel electrophoretic separation of the samples.

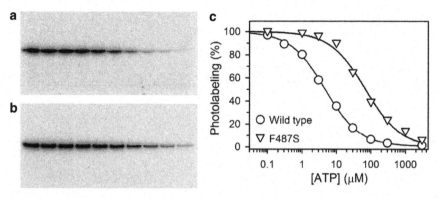

Fig. 12 The ATP dependence of the inhibition of [γ-^{32}P]TNP-8N$_3$-ATP photolabeling. (**a, b**) Autoradiographs of the gels from experiments studying the ATP dependence of [γ-^{32}P]TNP-8N$_3$-ATP photolabeling on COS-1-expressed wild type (**a**) and mutant F487S (**b**) Ca^{2+}-ATPase stabilized in the $E2$·orthovanadate state. The concentrations of ATP applied are as indicated on the graph in panel **c** (note that, for both wild type and mutant the first data point on the gel corresponds to 0 μM ATP and is, thus, not visible on the graph owing to the logarithmic scale of the abscissa). (**c**) Analysis of the experiments corresponding to panels **a** and **b** by nonlinear regression. The $K_{0.5}$ (ATP) values and Hill coefficients extracted from the analysis are: wild type, $K_{0.5}$ = 4.63 μM, n = 0.86; mutant F487S, $K_{0.5}$ = 69.88 μM, n = 0.84. The experiments were carried out at the standard [γ-^{32}P]TNP-8N$_3$-ATP concentration of 3× $K_{0.5}$ (TNP-8N$_3$-ATP) and, hence, the true dissociation constants for ATP binding are as follows (calculated as described in **step 2** of Subheading 3.7): wild type, K_D = 1.16 μM; mutant F487S, K_D = 17.5 μM

3.5 Competitive Inhibition by ATP of [γ-^{32}P]TNP-8N$_3$-ATP Photolabeling of the Ca^{2+}-ATPase

In the present example we study the binding of ATP to wild-type and mutant F487S Ca^{2+}-ATPase stabilized in the $E2$·orthovanadate state. The assay is, however, not limited to ATP affinity measurements, but may be applied to measure the affinity of any nucleotide, nucleotide analog, or other substance that competes with TNP-8N$_3$-ATP for binding. The photolabeling is generally carried out at a concentration of [γ-^{32}P]TNP-8N$_3$-ATP in the final reaction mix of 3 × $K_{0.5}$ for TNP-8N$_3$-ATP binding, with the $K_{0.5}$ for TNP-8N$_3$-ATP being as determined in the experiments described in Subheading 3.4. The photolabeling should be carried out one sample at a time. Keep all ingredients on ice during the experiment. The following is the protocol for an experiment comprising ten samples with varying concentrations of ATP (Fig. 12).

1. Stabilize the enzyme in the $E2$·orthovanadate state as described in **step 1** of Subheading 3.4.

2. Sample premix: 18.75 μL photolabeling buffer (4×), 23.25 μL water. Prepare a batch mix for all samples and aliquot 42 μL into 1.5 mL Eppendorf tubes. Incubate on ice.

3. Photolabeling reaction mix: In rapid succession, add 15 μL 87 % glycerol, 7.5 μL ATP dilution (10×), 7.5 μL [γ-^{32}P]TNP-8N$_3$-ATP at a concentration of 30× the $K_{0.5}$ for TNP-8N$_3$-ATP (as determined in the experiments described in Subheading 3.4),

and 3 μL of the orthovanadate-inhibited 25 nM Ca^{2+}-ATPase to the sample premix (final volume: 75 μL).

4. Irradiate the photolabeling reaction mix as described in **step 4** of Subheading 3.4.

5. Proceed to **step 3** in Subheading 3.6 for gel electrophoretic separation of the samples.

3.6 7 % SDS-Polyacrylamide Gel Electrophoresis, Gel Drying, and Phosphor Imaging

In the present example, the gels are prepared using the Hoefer SE250 Mighty Small II Mini Vertical Electrophoresis Units. We typically run four gels per experiment. For convenience, the gels can be prepared 1–2 days in advance (in which case, store the gels in 1× gel running buffer at 4 °C). If other gel running systems are applied, or less/more than four gels are needed, adjust the gel volumes, gel running times, and power supply settings accordingly.

1. Resolving gel: 10 mL resolving gel buffer (4×), 20.4 mL water, 9.33 mL 30 % acrylamide/bis solution, 29 μL TEMED, 234 μL 10 % ammonium persulfate (final volume: 40 mL). Mix gently (do not shake) and pour the gel into the gel caster. Allow space for the stacking gel (the stacking gel need not be very large, 4–5 mm between the tip of the comb and the resolving gel is fine). Overlay with water by pipetting 150 μL of water slowly down along the spacer at both sides of the gel. Wait for the resolving gel to polymerize (15–20 min). Pour the water off the surface of the resolving gel.

2. Stacking gel: 3.75 mL stacking gel buffer (4×), 9.6 mL water, 1.5 mL 30 % acrylamide/bis solution, 9.6 mL water, 12 μL TEMED, 95 μL 10 % ammonium persulfate (final volume: 15 mL). Mix gently (do not shake) and pour the stacking gel on top of the polymerized resolving gel. Insert the 10-well combs. Wait for the stacking gel to polymerize (15–20 min). Top with stacking gel mix (this can be kept on ice to delay polymerization) if the gel level drops during the polymerization.

3. Sample denaturation: Add 8 μL protein denaturation buffer to each 75 μL sample and vortex.

4. Load 35 μL in each lane (*see* **Note 12**) and run the gels in 1× gel running buffer at 15 mA per gel at 4 °C until the bromophenol blue band is ~0.5–1 cm from the bottom of the gel (~2 h).

5. Cut off and discard the bottom ~1.5 cm of the gel. Be aware that this part of the gel is highly radioactive, as it contains the non-reacted [γ-^{32}P]TNP-8N$_3$-ATP. Hence, be careful not to smear it across the main gel. The Ca^{2+}-ATPase bands will be roughly at the center of the remaining part of the resolving gel.

6. Dry the gel and measure the radioactivity associated with the Ca^{2+}-ATPase gel bands by phosphor imaging.

7. Typical results obtained in experiments measuring the [γ-^{32}P] TNP-8N$_3$-ATP dependence of photolabeling and the ATP-dependence of the inhibition of photolabeling are shown in Figs. 11 and 12, respectively, exemplified by experiments with COS-1-expressed wild-type and mutant F487S Ca^{2+}-ATPase.

3.7 Data Analysis

The photolabeling data is analyzed by nonlinear regression, applying the previously validated equations [10].

1. The analysis of the TNP-8N$_3$-ATP dependence of photolabeling (Fig. 11) is based on the hyperbolic function $Y = Y_{max} \cdot [\text{TNP-8N}_3\text{-ATP}]/(K_{0.5} + [\text{TNP-8N}_3\text{-ATP}])$, in which Y is the amount of photolabeled Ca^{2+}-ATPase, Y_{max} is the maximum amount of photolabeled Ca^{2+}-ATPase, and $K_{0.5}$ is the concentration of TNP-8N$_3$-ATP giving half-maximum labeling. In some cases, typically when working with poorly expressed protein or in situations where the affinity for TNP-8N$_3$-ATP is particularly low, it may be necessary to include a linear component to account for secondary/unspecific labeling occasionally observed at high TNP-8N$_3$-ATP concentrations: $Y = Y_{max} \cdot [\text{TNP-8N}_3\text{-ATP}]/(K_{0.5} + [\text{TNP-8N}_3\text{-ATP}]) + m \cdot [\text{TNP-8N}_3\text{-ATP}]$, where m is the slope of the linear component.

2. The analysis of the data obtained from the ATP inhibition of TNP-8N$_3$-ATP photolabeling (Fig. 12) is based on the Hill equation modified to describe inhibition, $Y = Y_{max} \cdot (1 - [\text{ATP}]^n)/(K_{0.5}^n + [\text{ATP}]^n)$, in which Y and Y_{max} are defined as above, $K_{0.5}$ is the concentration of ATP giving half-maximum inhibition of labeling, and n is the Hill coefficient (typically displaying a value of around 0.7–1.0, in accordance with the presence of one binding site for ATP). The "true" dissociation constant, K_D, for ATP binding is calculated from the equation K_D (ATP) = $K_{0.5}$ (ATP)/(1 + [TNP-8N$_3$-ATP]/$K_{0.5}$ (TNP-8N$_3$-ATP)), where [TNP-8N$_3$-ATP] is the concentration of TNP-8N$_3$-ATP applied in the ATP titration experiment. Given that the ATP dependence of photolabeling is generally performed at a concentration of TNP-8N$_3$-ATP of 3 × $K_{0.5}$ (TNP-8N$_3$-ATP), the equation simply becomes K_D (ATP) = $K_{0.5}$ (ATP)/(1 + 3/1) = $K_{0.5}$ (ATP)/4 (*see* **Note 13**).

4 Notes

1. As an alternative to using the commercially available 8N$_3$-ATP, this nucleotide can be synthesized from ATP by bromination of the adenine moiety at the 8-position, followed by replacement of the bromine with an azido group, according to established procedures [11, 45].

2. It is critical that the ^{32}P-labeled orthophosphate used is carrier-free, i.e., not supplemented with nonradioactive phosphate. We use product number NEX053H (10 mCi/mL) from PerkinElmer.

3. The procedure requires the handling of a rather large amount (~3 mCi = 111 Mbq) of ^{32}P, a beta particle-emitting radioactive isotope. Hence, it is essential to take the necessary precautions, such as frequently monitoring the workspace with a Geiger-Müller counter and shielding the radioactive material behind thick acrylic glass protection and/or in thick-walled lead containers at all times. Diligently follow all waste disposal regulations when disposing the radioactive materials.

4. It is critical that TMAH is used instead of Tris for the pH adjustment of all reaction buffers applied in the photolabeling experiments, because the presence of Tris will lead to increased background labeling levels, owing to the reaction of the amino group of Tris with the azido group of [γ-^{32}P]TNP-8N$_3$-ATP upon UV irradiation. Be aware that TMAH is toxic.

5. It is advisable to have a good and reliable estimate of the Ca^{2+}-ATPase concentration of the protein sample applied, to ensure that the [γ-^{32}P]TNP-8N$_3$-ATP is in excess of the Ca^{2+}-ATPase during photolabeling. It is critical that the Ca^{2+}-ATPase concentration is kept well below the $K_{0.5}$ for [γ-^{32}P]TNP-8N$_3$-ATP, which under some reaction conditions may get as low as 5–10 nM. Hence, it is sometimes necessary to keep the Ca^{2+}-ATPase concentration in the final photolabeling reaction mix as low as 0.5–1 nM.

6. The nonradioactive TNP-8N$_3$-ATP is prepared in the same way as the radioactive [γ-^{32}P]TNP-8N$_3$-ATP, except that the ^{32}P-exchange reaction step is skipped, and that the reaction is scaled up by applying 6.7-fold more 8N$_3$-ATP as starting material, i.e., starting from **step 2** in Subheading 3.2 with 150 μL 10 mM 8N$_3$-ATP (in this case, the reaction need not be done in the freeze drying flask, but can simply be carried out in a 1.5 mL Eppendorf tube).

7. Example of the preparation of a 100 μM [γ-^{32}P]TNP-8N$_3$-ATP working solution for photolabeling experiments: Assuming we have a 200 μM radioactive [γ-^{32}P]TNP-8N$_3$-ATP batch and a 1 mM non-radioactive TNP-8N$_3$-ATP batch, and that we want to prepare 200 μL of a 100 μM [γ-^{32}P]TNP-8N$_3$-ATP solution with the specific activity being diluted 20-fold relative to the newly synthesized [γ-^{32}P]TNP-8N$_3$-ATP batch. 200 μL × 100 μM = 20 nmol total TNP-8N$_3$-ATP, of which 1 nmol must come from the radioactive batch and 19 nmol must come from the non-radioactive batch (corresponding to a 20-fold dilution of the radioactivity). Hence, to 176 μL of

water add 5 μL 200 μM [γ-^{32}P]TNP-8N$_3$-ATP (=1 nmol) and 19 μL 1 mM TNP-8N$_3$-ATP (=19 nmol).

8. It is important that the UV light is filtered, because short wavelength UV light is harmful to the protein (*see* Supplementary Fig. S3 in [16]). As an alternative to the 295-nm wavelength cut-off glass filter one can also filter the UV light by placing a quartz cuvette, with a 10 mm path length and a > 1 mm path width, containing toluene in front of the labeling cuvette (toluene has a rather steep light absorbance cut-off around 270–275 nm).

9. The exact amount of bromophenol blue is not very important. A few mg, e.g., corresponding to the amount that will stick to a (sterile) plastic pipette tip when dipped into the jar of bromophenol blue, is typically fine. The purpose of the bromophenol blue is just to give color to the electrophoretic front during gel running, thus indicating visibly how far the gel has run. Since the bromophenol blue gets stacked during the run, even small amounts of bromophenol blue are easily visible in the gels.

10. It is commonly stated that the 10 % ammonium persulfate used for the preparation of polyacrylamide gels must be prepared fresh. We, however, find that the solution, when stored at 4 °C, can easily be used for a month without any noticeable decrease in the quality of the gels.

11. The amount of DTNB does not need to be very precise, as anywhere between 12 and 15 mg DTNB will give the same result. The oxidizing agent DTNB prevents the reduction of the azido group by sulfonic acid and also increases the yield of the trinitriphenylation reaction. Other oxidizing agents, such as H_2O_2, are also applicable; however DTNB was found to give the best yield [11].

12. The final sample volume after adding the protein denaturation buffer is 83 μL. With the Hoefer SE250 Mighty Small II Mini Vertical Electrophoresis Units with 1.5 mm spacers/combs it is possible to load a maximum of ~45 μL per lane. We generally load 35 μL per lane, but run two gels per 10-sample experiment (thus loading each sample on two separate gels). In our experience, the step in the procedure, where problems are most likely to occur, is the gel running (individual samples that for one reason or another run atypically on the gel). By running two gels for each experiment we minimize the risk of an effect on the final result by any samples "lost" on the gels.

13. In cases where the affinity for TNP-8N$_3$-ATP is particularly low, it can be an advantage to carry out the experiments at a lower concentration of [γ-^{32}P]TNP-8N$_3$-ATP, e.g., at 1 × $K_{0.5}$ (TNP-8N$_3$-ATP) instead of at 3 × $K_{0.5}$ (TNP-8N$_3$-ATP), to prevent background labeling levels from affecting the result.

Be aware, though, that if a concentration of [γ-^{32}P]TNP-8N$_3$-ATP other than $3 \times K_{0.5}$ (TNP-8N$_3$-ATP) is applied, this should be taken into account when calculating the K_D (ATP). For example, if a concentration of $1 \times K_{0.5}$ (TNP-8N$_3$-ATP) is applied, the K_D (ATP) is given by $K_{0.5}$ (ATP)$/(1+1/1) = K_{0.5}$ (ATP)$/2$.

Acknowledgements

This work was supported by the Lundbeck Foundation and the Centre for Membrane Pumps in Cells and Disease—PUMPKIN, Danish National Research Foundation (to JDC), the National Research Foundation, South Africa, and the University of Cape Town, South Africa (to DBM and DGW), and the Danish Medical Research Council and the Novo Nordisk Foundation (to JPA).

References

1. Lacapere JJ, Bennett N, Dupont Y, Guillain F (1990) pH and magnesium dependence of ATP binding to sarcoplasmic reticulum ATPase. Evidence that the catalytic ATP-binding site consists of two domains. J Biol Chem 265(1):348–353
2. Lacapere JJ, Guillain F (1993) The reaction mechanism of Ca(2+)-ATPase of sarcoplasmic reticulum. Direct measurement of the Mg.ATP dissociation constant gives similar values in the presence or absence of calcium. Eur J Biochem 211(1-2):117–126
3. Norby JG, Jensen J (1971) Binding of ATP to brain microsomal ATPase. Determination of the ATP-binding capacity and the dissociation constant of the enzyme-ATP complex as a function of K$^+$ concentration. Biochim Biophys Acta 233(1):104–116
4. Hegyvary C, Post RL (1971) Binding of adenosine triphosphate to sodium and potassium ion-stimulated adenosine triphosphatase. J Biol Chem 246(17):5234–5240
5. Fedosova NU, Champeil P, Esmann M (2003) Rapid filtration analysis of nucleotide binding to Na,K-ATPase. Biochemistry 42(12):3536–3543
6. Vilsen B, Andersen JP, MacLennan DH (1991) Functional consequences of alterations to amino acids located in the hinge domain of the Ca(2+)-ATPase of sarcoplasmic reticulum. J Biol Chem 266(24):16157–16164
7. Sorensen T, Vilsen B, Andersen JP (1997) Mutation Lys758 → Ile of the sarcoplasmic reticulum Ca^{2+}-ATPase enhances dephosphorylation of E2P and inhibits the E2 to E1Ca2 transition. J Biol Chem 272(48):30244–30253
8. Sorensen TL, Dupont Y, Vilsen B, Andersen JP (2000) Fast kinetic analysis of conformational changes in mutants of the Ca(2+)-ATPase of sarcoplasmic reticulum. J Biol Chem 275(8):5400–5408
9. Vilsen B (1993) Glutamate 329 located in the fourth transmembrane segment of the alpha-subunit of the rat kidney Na$^+$,K$^+$-ATPase is not an essential residue for active transport of sodium and potassium ions. Biochemistry 32(48):13340–13349
10. McIntosh DB, Woolley DG, Vilsen B, Andersen JP (1996) Mutagenesis of segment 487Phe-Ser-Arg-Asp-Arg-Lys492 of sarcoplasmic reticulum Ca^{2+}-ATPase produces pumps defective in ATP binding. J Biol Chem 271(42):25778–25789
11. Seebregts CJ, McIntosh DB (1989) 2′,3′-O-(2,4,6-Trinitrophenyl)-8-azido-adenosine mono-, di-, and triphosphates as photoaffinity probes of the Ca^{2+}-ATPase of sarcoplasmic reticulum. Regulatory/superfluorescent nucleotides label the catalytic site with high efficiency. J Biol Chem 264(4):2043–2052
12. Chowdhry V, Westheimer FH (1979) Photoaffinity labeling of biological systems. Annu Rev Biochem 48:293–325
13. Kotzyba-Hibert F, Kapfer I, Goeldner M (1995) Recent trends in photoaffinity labeling. Angew Chem Int Ed Engl 34:1296–1312

14. Potter RL, Haley BE (1983) Photoaffinity labeling of nucleotide binding sites with 8-azidopurine analogs: techniques and applications. Methods Enzymol 91:613–633
15. Suzuki H, Kubota T, Kubo K, Kanazawa T (1990) Existence of a low-affinity ATP-binding site in the unphosphorylated Ca2(+)-ATPase of sarcoplasmic reticulum vesicles: evidence from binding of 2′,3′-O-(2,4,6-trinitrocyclohexadienylidene)-[3H]AMP and -[3H]ATP. Biochemistry 29(30): 7040–7045
16. Clausen JD, McIntosh DB, Woolley DG, Andersen JP (2011) Modulatory ATP binding affinity in intermediate states of E2P dephosphorylation of sarcoplasmic reticulum Ca^{2+}-ATPase. J Biol Chem 286(13):11792–11802
17. McIntosh DB, Woolley DG, Berman MC (1992) 2′,3′-O-(2,4,6-Trinitrophenyl)-8-azido-AMP and -ATP photolabel Lys-492 at the active site of sarcoplasmic reticulum Ca(2+)-ATPase. J Biol Chem 267(8): 5301–5309
18. McIntosh DB, Woolley DG (1994) Catalysis of an ATP analogue untethered and tethered to lysine 492 of sarcoplasmic reticulum Ca(2+)-ATPase. J Biol Chem 269(34):21587–21595
19. McIntosh DB, Woolley DG, MacLennan DH, Vilsen B, Andersen JP (1999) Interaction of nucleotides with Asp(351) and the conserved phosphorylation loop of sarcoplasmic reticulum Ca(2+)-ATPase. J Biol Chem 274(36): 25227–25236
20. Clausen JD, McIntosh DB, Woolley DG, Andersen JP (2001) Importance of Thr-353 of the conserved phosphorylation loop of the sarcoplasmic reticulum Ca^{2+}-ATPase in MgATP binding and catalytic activity. J Biol Chem 276(38):35741–35750
21. Clausen JD, McIntosh DB, Vilsen B, Woolley DG, Andersen JP (2003) Importance of conserved N-domain residues Thr441, Glu442, Lys515, Arg560, and Leu562 of sarcoplasmic reticulum Ca^{2+}-ATPase for MgATP binding and subsequent catalytic steps. Plasticity of the nucleotide-binding site. J Biol Chem 278(22):20245–20258
22. McIntosh DB, Clausen JD, Woolley DG, MacLennan DH, Vilsen B, Andersen JP (2003) ATP binding residues of sarcoplasmic reticulum Ca(2+)-ATPase. Ann N Y Acad Sci 986:101–105
23. McIntosh DB, Clausen JD, Woolley DG, MacLennan DH, Vilsen B, Andersen JP (2004) Roles of conserved P domain residues and Mg^{2+} in ATP binding in the ground and Ca^{2+}-activated states of sarcoplasmic reticulum Ca^{2+}-ATPase. J Biol Chem 279(31):32515–32523
24. Clausen JD, McIntosh DB, Woolley DG, Anthonisen AN, Vilsen B, Andersen JP (2006) Asparagine 706 and glutamate 183 at the catalytic site of sarcoplasmic reticulum Ca^{2+}-ATPase play critical but distinct roles in E2 states. J Biol Chem 281(14):9471–9481
25. Clausen JD, McIntosh DB, Anthonisen AN, Woolley DG, Vilsen B, Andersen JP (2007) ATP-binding modes and functionally important interdomain bonds of sarcoplasmic reticulum Ca^{2+}-ATPase revealed by mutation of glycine 438, glutamate 439, and arginine 678. J Biol Chem 282(28):20686–20697
26. Clausen JD, McIntosh DB, Woolley DG, Andersen JP (2008) Critical interaction of actuator domain residues arginine 174, isoleucine 188, and lysine 205 with modulatory nucleotide in sarcoplasmic reticulum Ca^{2+}-ATPase. J Biol Chem 283(51):35703–35714
27. Clausen JD, Bublitz M, Arnou B, Montigny C, Jaxel C, Moller JV, Nissen P, Andersen JP, le Maire M (2013) SERCA mutant E309Q binds two Ca(2+) ions but adopts a catalytically incompetent conformation. EMBO J 32(24): 3231–3243
28. Clausen JD, Anthonisen AN, Andersen JP (2014) Critical role of interdomain interactions for modulatory ATP binding to sarcoplasmic reticulum Ca^{2+}-ATPase. J Biol Chem 289(42): 29123–29134
29. Sorensen TL, Moller JV, Nissen P (2004) Phosphoryl transfer and calcium ion occlusion in the calcium pump. Science 304(5677): 1672–1675
30. Toyoshima C, Mizutani T (2004) Crystal structure of the calcium pump with a bound ATP analogue. Nature 430(6999):529–535
31. Jensen AM, Sorensen TL, Olesen C, Moller JV, Nissen P (2006) Modulatory and catalytic modes of ATP binding by the calcium pump. EMBO J 25(11):2305–2314
32. Olesen C, Picard M, Winther AM, Gyrup C, Morth JP, Oxvig C, Moller JV, Nissen P (2007) The structural basis of calcium transport by the calcium pump. Nature 450(7172):1036–1042
33. Moczydlowski EG, Fortes PA (1981) Characterization of 2′,3′-O-(2,4,6-trinitrocyclohexadienylidine)adenosine 5′-triphosphate as a fluorescent probe of the ATP site of sodium and potassium transport adenosine triphosphatase. Determination of nucleotide binding stoichiometry and ion-induced changes in affinity for ATP. J Biol Chem 256(5):2346–2356
34. Moczydlowski EG, Fortes PA (1981) Inhibition of sodium and potassium adenosine triphosphatase by 2′,3′-O-(2,4,6-trinitrocyclo-

hexadienylidene) adenine nucleotides. Implications for the structure and mechanism of the Na:K pump. J Biol Chem 256(5):2357–2366
35. Faller LD (1989) Competitive binding of ATP and the fluorescent substrate analogue 2′,3′-O-(2,4,6-trinitrophenylcyclohexadienylidine) adenosine 5′-triphosphate to the gastric H+, K+-ATPase: evidence for two classes of nucleotide sites. Biochemistry 28(16): 6771–6778
36. Faller LD (1990) Binding of the fluorescent substrate analogue 2′,3′-O-(2,4,6-trinitrophenylcyclohexadienylidene)adenosine 5′-triphosphate to the gastric H+,K+-ATPase: evidence for cofactor-induced conformational changes in the enzyme. Biochemistry 29(13):3179–3186
37. Axelsen KB (2014) The P-type ATPase Database. http://traplabs.dk/patbase/. Accessed 24 Aug 2014
38. Axelsen KB, Palmgren MG (1998) Evolution of substrate specificities in the P-type ATPase superfamily. J Mol Evol 46(1):84–101
39. Decottignies A, Grant AM, Nichols JW, de Wet H, McIntosh DB, Goffeau A (1998) ATPase and multidrug transport activities of the overexpressed yeast ABC protein Yor1p. J Biol Chem 273(20):12612–12622
40. de Wet H, McIntosh DB, Conseil G, Baubichon-Cortay H, Krell T, Jault JM, Daskiewicz JB, Barron D, Di Pietro A (2001) Sequence requirements of the ATP-binding site within the C-terminal nucleotide-binding domain of mouse P-glycoprotein: structure-activity relationships for flavonoid binding. Biochemistry 40(34):10382–10391
41. Maruyama K, MacLennan DH (1988) Mutation of aspartic acid-351, lysine-352, and lysine-515 alters the Ca^{2+} transport activity of the Ca^{2+}-ATPase expressed in COS-1 cells. Proc Natl Acad Sci U S A 85(10):3314–3318
42. Ko YH, Bianchet M, Amzel LM, Pedersen PL (1997) Novel insights into the chemical mechanism of ATP synthase. Evidence that in the transition state the gamma-phosphate of ATP is near the conserved alanine within the P-loop of the beta-subunit. J Biol Chem 272(30): 18875–18881
43. Pick U (1982) The interaction of vanadate ions with the Ca-ATPase from sarcoplasmic reticulum. J Biol Chem 257(11):6111–6119
44. Dupont Y, Bennett N (1982) Vanadate inhibition of the Ca^{2+}-dependent conformational change of the sarcoplasmic reticulum Ca^{2+}-ATPase. FEBS Lett 139(2):237–240
45. Owens JR, Haley BE (1984) Synthesis and utilization of 8-azidoguanosine 3′-phosphate 5′-[5′-32P]phosphate. Photoaffinity studies on cytosolic proteins of Escherichia coli. J Biol Chem 259(23):14843–14848
46. Gourdon P, Liu XY, Skjorringe T, Morth JP, Moller LB, Pedersen BP, Nissen P (2011) Crystal structure of a copper-transporting PIB-type ATPase. Nature 475(7354):59–64
47. Toyoshima C, Yonekura S, Tsueda J, Iwasawa S (2011) Trinitrophenyl derivatives bind differently from parent adenine nucleotides to Ca^{2+}-ATPase in the absence of Ca^{2+}. Proc Natl Acad Sci U S A 108(5):1833–1838
48. Sarma RH, Lee CH, Evans FE, Yathindra N, Sundaralingam M (1974) Probing the interrelation between the glycosyl torsion, sugar pucker, and the backbone conformation in C(8) substituted adenine nucleotides by 1H and 1H-(31P) fast Fourier transform nuclear magnetic resonance methods and conformational energy calculations. J Am Chem Soc 96(23):7337–7348
49. Glynn IM, Chappell JB (1964) A simple method for the preparation of 32-P-labelled adenosine triphosphate of high specific activity. Biochem J 90(1):147–149

Chapter 23

Ca^{2+} Binding and Transport Studied with Ca^{2+}/EGTA Buffers and $^{45}Ca^{2+}$

Pankaj Sehgal, Claus Olesen, and Jesper V. Møller

Abstract

The chapter describes procedures useful for determination of Ca^{2+} binding by membranous Ca^{2+}-ATPase based on the correction for the removal of Ca^{2+} present in a non-bound state in the suspension medium. This is done by a filtration procedure that retains the membranous material on the Millipore filters. With suitable sucking devices it is possible to gently remove without dehydration nearly all medium from the Ca^{2+} containing membranes, except that required for wetting of the filters on which they are deposited. Correction for this effect can be done with a double-filter where the radioactive content of the lower (protein-free) filter is subtracted from that present in the upper filter for calculation of Ca^{2+} binding. This methodology can be used to study the effect of inhibitors on Ca^{2+} binding and –transport, and with Ca^{2+}/EGTA buffers to explore the Ca^{2+} binding affinities and cooperative aspects of the two transport sites.

Key words $^{45}Ca^{2+}$, Calcium binding, Calcium transport, SERCA, Filtration assay

1 Introduction

Studies on the binding and transport of Ca^{2+}-ATPase present in skeletal muscle (SERCA1a) benefit from the large amounts of the protein that can be prepared from skeletal muscle (see SR Ca^{2+}-ATPase purification Chap. 3), and from the availability of radioactively labelled calcium ($^{45}Ca^{2+}$) with suitable properties such as a relatively high half-life, making it possible to study details of the transport mechanism in a way that is unparalleled by other P-type ATPases. The analyses are also simple, since binding and transport of Ca^{2+} by membranous ATPase can be measured from the amount of radioactivity retained on a Millipore filter after filtration of unbound Ca^{2+} present in the suspension medium. The present protocol considers measurements of Ca^{2+} binding performed at the low concentrations relevant for the specific and reversible binding to the two calcium transport-binding sites of SERCA1a.

In order to do this it is necessary to use specific Ca^{2+} complexing agents such as EGTA. This is also required to buffer the free

$$H_4EGTA \underset{pK\ 2.0}{\overset{H^+}{\longleftrightarrow}} H_3EGTA^- \underset{pK\ 2.68}{\overset{H^+}{\longleftrightarrow}} H_2EGTA^{2-} \underset{pK\ 8.85}{\overset{H^+}{\longleftrightarrow}} H_2EGTA^{3-} \underset{pK\ 9.46}{\overset{H^+}{\longleftrightarrow}} EGTA^{4-}$$

Scheme 1 pK values for dissociation constants of protons from EGTA

concentration of Ca^{2+} and eliminate the effect of the ubiquitous presence of contaminating Ca^{2+} that otherwise would make it difficult or impossible to study details of the Ca^{2+} binding at the micromolar and submicromolar level. However, it should be understood that the almost universal use of EGTA for this purpose introduces problems of its own. The reason is that the formation of the Ca^{2+}-EGTA complex is affected by a number of environmental variables such as pH, the presence of Mg^{2+}, other Ca^{2+} complexing agents, and ionic strength. In particular small variations in pH lead to appreciable changes in the apparent dissociation constant of the Ca^{2+}-EGTA complex. The reason for this is that EGTA with four carboxyl groups is present in four different protonated states, with the following pK values ranging from 2.0 to 9.5 as shown in Scheme 1 below.

At physiologically relevant pH values around 7 the predominant form is H_2EGTA^{2-} which after deprotonization reacts with Ca^{2+} according to the following scheme:

$$H_2EGTA^{2-} <----------> EGTA^{4-} + 2H^+$$

$$Ca^{2+} + EGTA^{4-} <------------> Ca\text{-}EGTA^{2-}$$

There are two important consequences of these reactions. Firstly, it emphasizes the need to bear in mind that significant pH changes may occur when mixing Ca^{2+} and EGTA solutions that are not sufficiently well pH buffered. Secondly, that the release of two protons associated with the formation of the Ca-EGTA^{2-} complex strongly influences the apparent, pH-dependent dissociation constant of CaEGTA, (pK'_{Ca}) for the physiologically predominant species (H_2EGTA^{2-}), it in fact doubles the effect such that a change of, e.g., + 0.1 pH unit leads to a change of pK'_{CaEGTA} of + 0.2 units. Hence when working with EGTA frequent checks of pH are important. There is also a third consideration to bear in mind which is the necessity to have a critical look at the equivalence of the Ca^{2+} and EGTA solutions being used: SERCA is most sensitive to changes in the levels of free Ca^{2+} taking place in buffers with nearly equimolar concentrations of EGTA and total Ca^{2+}, and it cannot be assumed that the nominal concentrations of EGTA and Ca^{2+} that are calculated even by accurate weighing from analytical grade chemicals are absolutely equivalent without testing, if this is actually the case (*see* **Note 1** about how to do this by acid-base titration).

2 Materials

All reagents should be of analytical grade.

1. A radioactivity scintillation counter.
2. A fine-tipped tweezer.
3. Nitrocellulose filters (Millipore HAWP02500).
4. A vacuum filtration device.
5. Ca^{2+} stock solutions: 1 M $CaCl_2$ (as bought from a commercial firm) and as a 0.100 M solution prepared from this by dilution with demineralized water.
6. Purified Ca^{2+}-ATPase or sarcoplasmic reticulum (SR) vesicles (20–25 mg protein/mL), from rabbit skeletal muscle.
7. A $^{45}Ca^{2+}$ stock solution: 1 mCi $^{45}CaCl_2$ (> 370 GBq/g) solubilized in 1 mL of 1 M $CaCl_2$.
8. EGTA stock solution: 0.100 M EGTA, solubilized in deionized water and neutralized with KOH to pH 7.2. Before use standardized to equivalence with 0.100 M Ca^{2+} (*see* **Note 1**).
9. $^{45}Ca^{2+}$ reaction medium with ATPase or SR; 0.1 mg/mL (for binding assay) or 0.2 mg/mL (for inhibition assay) ATPase or SR, 50 μM $^{45}Ca^{2+}$, 50 mM MOPS, 100 mM KCl, 1 mM $MgCl_2$, adjusted to pH 7.20 with KOH (*see* **Note 2**).
10. $^{45}Ca^{2+}$-EGTA buffer series for binding assay: Prepare from 50 mL aliquots of reaction medium $^{45}Ca^{2+}$/EGTA buffers by addition of the following EGTA concentrations from the stock solution of 100 mM EGTA (pH 7.2): (a) 400 μM EGTA; (b) 200 μM EGTA; (c) 100 μM EGTA; (d) 80 μM EGTA; (e) 70 μM EGTA; (f) 60 μM EGTA; (g) 50 μM EGTA; (h) 40 μM EGTA; (i) 20 μM EGTA.
11. Liquid monoscintillator (ethoxylated alkylphenol) (Perkin Elmer), suitable for solubilisation of aqueous samples.
12. 100 mM ATP stock solution of pH 7.2.

3 Methods

3.1 Calcium Binding Assay as a Function of pCa

1. Wet the required number of filters by transferring them to a Petri dish with deionized water. With the tweezer align two filters accurately on top of each other in a central position on the porous plate of the filtration device and clamp the filters with the upper "chimney" part to make a watertight seal with the base plate.
2. Start suction with the vacuum pump and add 2 mL samples of the ATPase-EGTA dilution series samples to the filters. The

addition should be done gently aiming at the deposition of the membranes in an even layer on the exposed part of the upper filter. The vacuum applied should be moderate, yet sufficient for the filtration to proceed effortlessly, without evidence of filter clogging and result in the virtually complete removal of fluid between the lower filter and the base plate.

3. When finished with the deposition of the sample on the filter assembly remove the clamp, and with the tweezer transfer each of the two filters to separate scintillation vials and add 5 mL scintillator (*see* **Note 3**).

4. After rinsing the base plate with nonradioactive reaction medium continue with the next sample.

5. As standards, prepare 6 × 25 μL samples of the reaction medium without protein that together with dry filters are put in scintillation vials along with 5 mL scintillation fluid.

6. After thorough shaking of the scintillation vials count them on the scintillation counter. Bound $^{45}Ca^{2+}$ is calculated by subtraction of the radioactive counts in the lower filter from those present in the upper filter, to correct for the contribution of unbound $^{45}Ca^{2+}$ that is caused by wetting of the filter (*see* **Note 4**). From the counts of the standards (CPM_{sample}) calculate the specific radioactivity (SRA) in terms of CPM_{stand}/nmol Ca^{2+} and use this value to calculate bound Ca^{2+} (Ca^{2+}_{bound}) as Ca^{2+}_{bound} (nmol/mg) = ($CPM_{upper\ filter}$ − $CPM_{lower\ filter}$)/SRA.

7. Use the Maxchelator (http://maxchelator.stanford.edu) to obtain the free concentrations of Ca^{2+} in the reaction media under the prevailing concentrations of Ca^{2+}, EGTA, pH, and ionic strength (ignore the vanishingly small correction for Ca^{2+} bound in the ATPase-containing samples, *see* **Note 2**). Convert the free concentrations of Ca^{2+} indicated in the Maxchelator in molar units to *p*Ca values according to:

$$pCa = -\log[Ca^{2+}]_f$$

Use *p*Ca units for presentation of Ca^{2+} binding as a function of *p*Ca for graphical representation as shown in Fig. 1.

3.2 Inhibition of Calcium Binding

1. This assay makes simultaneous use of at least 5 or 6 of the filtration devices, assembled together with a double layer of filters. As a control (time zero sample with full Ca^{2+} binding capacity) deposit 2 mL $^{45}Ca^{2+}$ assay buffer with ATPase on one double filter and suck the sample through the filter at a uniform speed sufficient to remove extraneous fluid.

2. Start the reaction by adding 2 mL solution containing ATPase and inhibitor (10 μM). After timed intervals rapidly filter 2 mL of ATPase with 10 μM inhibitor.

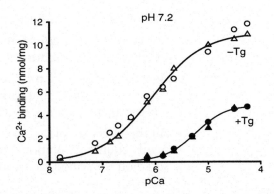

Fig. 1 Illustrative example exploring Ca^{2+} binding as a function of pCa of purified Ca^{2+}-ATPase in the presence of thapsigargin, a specific inhibitor of SERCA, and AMPPCP, an ATP analog that is bound to, but does not phosphorylate Ca^{2+}-ATPase. Explanation of symbols: thapsigargin-reacted (*filled circle, filled triangle*) and -unreacted (*open circle, open triangle*) Ca^{2+}-ATPase. Note that Ca^{2+} binding is only partially inhibited despite thapsigargin binding and is not affected by the presence of AMPPCP (*open triangle, filled triangle*). Reproduced from Jensen et al. [2]

3. After the filtration, remove the filters from Millipore device. Upper and lower filters are placed in separate counting vials followed by the addition of 5 mL scintillator and counting solution.

4. As standards, prepare 6 × 25 µL samples of the reaction medium without protein that are put on the filters together with 5 mL scintillation fluid in the scintillation vial.

5. From the standards calculate SRA (the specific radioactivity) and from the samples the amount of Ca^{2+} bound to the ATPase as obtained by subtracting the counts on the lower, protein-free filters from the Ca^{2+}-ATPase-deposited filters. With the aid of the standards calculate SRA (the specific radioactivity) and the amount of bound calcium to the samples in terms of nmol/mg protein.

3.3 Calcium Transport and Inhibition

1. Suspend SR vesicles (0.1 mg protein/mL) with a known concentration of inhibitor in 10 mL of ^{45}Ca^{2+} reaction medium supplemented with 0.07 mM EGTA and incubate for 5 min at the desired temperature.

2. To initiate the transport reaction add 0.1 mM ^{45}Ca^{2+} (diluted from stock solution 7) and 0.8 mM MgATP (from stock solution 11) together added to the ATPase sample.

3. Take out 0.2 mL reactions mixtures at different time intervals (10 s to 20 min) and mix with 3 mL of ice cold transport assay buffer. Immediately, pass the sample through the filter.

4. Wash twice with 3 mL ice cold transport assay buffer.

5. For comparison, perform the same Ca^{2+} transport experiment on SR without inhibitor.

6. As standards, prepare 6×25 μL samples of the reaction medium with 0.1 mM $^{45}Ca^{2+}$, but without protein that are put on the filters together with 5 mL scintillation fluid in the scintillation vial.

7. All the filters are placed in counting vials with the addition of 5 mL counting solution and their radioactive content measured on the scintillation counter.

4 Notes

1. The equivalence with the Ca^{2+} stock solution is tested by adding a surplus of the Ca^{2+}-stock solution, e.g., 2 mL 1 M $CaCl_2$ to 10 mL of the EGTA 0.1 M stock solution, which prior to this has been adjusted to pH 6.50 with approx. 0.5 mL HCl 0.1 N. The protons released by the addition of Ca^{2+} are titrated back to pH 6.50 with base for determination of the EGTA concentration.

2. Note that to take contaminating Ca^{2+} into account we often estimate the total concentration of this solution to be 55 μM.

3. Note that the filters are not to be rinsed after the filtration — otherwise bound Ca^{2+} will be washed away from the upper filter.

4. As an alternative wetting of the filter has been estimated on single filters with ^3H-labelled sugar added together with $^{45}Ca^{2+}$ and where both isotopes have been measured simultaneously by double-channel counting, *see*, e.g., Orlowski and Champeil [1].

References

1. Orlowski S, Champeil P (1991) Kinetics of calcium dissociation from its high-affinity transport sites on sarcoplasmic reticulum ATPase. Biochemistry 30:352–361

2. Jensen A-ML, Sørensen TL-M, Olesen C, Møller JV, Nissen P (2006) Modulatory and catalytic modes of ATP binding by the calcium pump. EMBO J 25:2305–2314

Chapter 24

Assay of Copper Transfer and Binding to P$_{1B}$-ATPases

Teresita Padilla-Benavides and José M. Argüello

Abstract

P$_{1B}$-type ATPases transport transition metals across biological membranes. The chemical characteristics of these substrates, as well as, physiological requirements have contributed to the evolution of high metal binding affinities (fM) in these enzymes. Metal binding determinations are consequently facilitated by the stable metal–protein interaction, while affinity measurements require careful analysis of metal levels. In the cell, transition metals are associated with chaperone proteins. Metals reach the ATPase transport sites following specific protein–protein interactions and ligand exchange enabling the metal transfer from the chaperone to the transporter. Here, we describe methods for analyzing the binding of Cu$^+$ to Cu$^+$-ATPases, as well as the approach to monitor Cu$^+$ transfer from soluble Cu$^+$-chaperones donors to and from membrane Cu$^+$-ATPases.

Key words P$_{1B}$-type ATPases, Cu$^+$-ATPases, Cu$^+$-chaperones, Transition metals, Copper, Binding, Affinity, Stoichiometry

1 Introduction

P$_{1B}$-type ATPases are polytopic membrane proteins that selectively transport specific transition metals (Cu$^+$, Cu^{2+}, Zn^{2+}, Mn^{2+}, Co^{2+}, or Ni^{2+}). They drive efflux of their substrates to maintain cytoplasmic metal levels, redox homeostasis, and the assembly of periplasmic and secreted metalloproteins [1–5]. These ATPases follow the Post–Albers catalytic cycle. They alternate between E1 and E2 conformations and in the presence of ATP, upon binding of the outwardly transported metal, are phosphorylated at conserved DKTGT sequences [2]. However, transition metals are seldom free in cellular compartments but rather bound to specific metallochaperones [2, 5–7]. Therefore, the metal binding and release involves substrate transfer between soluble metal chaperones and membrane ATPases [8, 9].

Structurally, P$_{1B}$-ATPases have a transmembrane core consisting of six or eight helices [3]. As all P-type ATPases, they have the characteristic nucleotide, phosphorylation, and actuator domains.

Many of them have also cytosolic amino- and carboxy-terminal metal binding domains [10], which play a regulatory role controlling the transport rate [2]. Invariant residues in the last three transmembrane segments form the metal binding sites responsible for transmembrane transport [11–14]. These signature sequences enabled predicting metal specificity and the classification in various protein subfamilies with close structural and functional features [3]. Characterization of metal binding to P_{1B}-type ATPases has been instrumental to establish their transport stoichiometry, high affinity binding, the structure of the metal binding sites, and the mechanism of metal transfer from and to chaperones [8, 9, 11, 13–15].

In this chapter, we describe protocols for metal binding and protein–protein metal transfer using the Cu^+-ATPase as a model. Cu^+-ATPases bind two Cu^+ ions in a planar trigonal coordination, each with $K_a = 1-2$ fM^{-1} [11]. These sites are constituted by the invariant amino acid motifs CPC in helix 4 (H4), NY in H5 and MXXS in H6. The ATPase's Cu^+ binding stoichiometry was determined using methods similar to those developed earlier by Glynn, Richards, Karlish, and others to characterize the binding of alkali metal substrates to the Na^+, K^+, and Ca^{2+} ATPases [16]. These required the cation "occlusion" by the protein; that is, placing the enzyme in a conformation where cation release was prevented or at least proceeded at a slow rate. Subsequently, fast passage of the cation-bound enzyme through a Sephadex or cation exchange column allowed separation of the unbound metal. Radioisotopes ($^{45}Ca^{2+}$, $^{23}Na^+$, or $^{86}Rb^+$) were used to measure the amount of bound metal. In the case of transition metal ATPases, the high binding affinities allowed the use of Sephadex columns at room temperature with no major concern for the rate of protein elution [11, 15]. In fact, X-ray spectroscopy studies required the concentration of metal bound ATPases up to millimolar level, which was achieved by lengthy centrifugations through Centricon filters [11, 12, 14]. While the high binding affinity enables easy binding and the removal of the free metal, determination of K_a requires assays of metal competition between the protein and a colorimetric high affinity metal indicator [15, 17]. Bicinchoninic acid (BCA) is used the case of Cu^+ [9, 11].

A unique feature of Cu^+-ATPase transport is the mechanism of metal access to the transmembrane metal binding sites. Cellular Cu^+ is bound to soluble transcription factors, chaperones, cuproenzymes and other small molecules [18]. Therefore, Cu^+-ATPases do not receive free metal ions. Instead, the substrate is delivered to transmembrane metal sites by cytosolic Cu^+-chaperones [9, 19, 20]. We have shown that this is mediated by the interaction of the Cu^+-chaperone with an electropositive platform in the structure of the ATPase [19, 21]. This electrostatic interaction places the Cu^+ close to invariant carboxyl and thiol groups providing ligands for the metal exchange between the proteins and subsequent entrance into the permeation path. This interaction is of course transient, as after presenting the first Cu⁺, the apo-chaperone must leave to allow a second holo-protein to deliver Cu⁺. Considering the exit of the metal from the ATPase in vivo, again a partnering chaperone takes Cu⁺ during a transient interaction [8]. Characterization of these transient interactions is possible by using different tags ((His)$_6$ or Strep) on the ATPase and the chaperones, and reconstituting the membrane ATPase in micellar form. This allows co-incubation of apo- (acceptor) and holo- (donor) forms of these proteins, followed by separation using affinity columns, and monitoring Cu⁺ in the various eluates containing either the ATPase or the chaperone.

2 Materials

Prepare all solutions using ultrapure water (18 mΩ cm) and analytical grade reagents. Extensively rinse glassware with ultrapure water to prevent metal contamination. Use new plastic ware when possible.

2.1 Protein Purification

1. Washing buffer: 100 mM KCl, 25 mM Tris–HCl, pH 7.4.
2. Buffer A: 25 mM Tris–HCl, pH 7.0, 100 mM sucrose, 1 mM phenylmethylsulfonyl fluoride.
3. Buffer B: 25 mM Tris–HCl, pH 8.0, 100 mM sucrose, 500 mM NaCl, 1 mM phenylmethylsulfonyl fluoride.
4. ATPase washing buffer 1: 5 mM imidazole, 0.01 % asolectin, 0.01 % dodecyl-β-d-maltoside (DDM) in Buffer B.
5. ATPase washing buffer 2: 20 mM imidazole, 0.01 % asolectin, 0.01 % DDM in Buffer B.
6. Buffer C: 25 mM Tris–HCl, pH 8.0, 100 mM sucrose, 50 mM NaCl, 0.01 % DDM, 0.01 % asolectin, 1 mM phenylmethylsulfonyl fluoride.
7. ATPase elution buffer: 150 mM imidazole in Buffer C.
8. Buffer W: 100 mM Tris–HCl, pH 8.0 and 150 mM NaCl.
9. Buffer E: 100 mM Tris–HCl, pH 8.0, 150 mM NaCl and 2.5 mM desthiobiotin, keep refrigerated.
10. Regeneration Buffer: 100 mM Tris–HCl, pH 8.0, 150 mM NaCl, and 1 mM 2-[4′-hydroxy-benzeneazo] benzoic acid.
11. 10 % DDM stock solution.
12. Ni^{2+}-nitriloacetic acid resin (Qiagen) or Strep-Tactin Sepharose resin (IBA Life Sciences).
13. Plastic Gravity and Spin Chromatography Column (Bio-Rad).

2.2 Cu⁺ Binding and Cu⁺ Transfer

1. Buffer H: 50 mM HEPES, pH 7.5, 200 mM NaCl, 0.01 % DDM, 0.01 % asolectin, 100 mM ascorbic acid.
2. Buffer T: 25 mM HEPES, pH 8.0, 100 mM sucrose, 500 mM NaCl, 0.01 % DDM, 0.01 % asolectin, 10 mM ascorbic acid.
3. Buffer T20: 20 mM imidazole in Buffer T.
4. Buffer T300: 300 mM imidazole in Buffer T.
5. Cu⁺ loading buffer: 25 mM HEPES, pH 8.0, 150 mM NaCl, and 10 mM ascorbic acid.
6. $CuSO_4$ 100 µM stock solution.
7. BCA disodium salt (4,4′-dicarboxy-2,2′-biquinoline) 100 µM stock solution.
8. Freshly prepared 40 µM ascorbic acid.
9. BCA solution: 3.5 µM BCA disodium salt (4,4′-dicarboxy-2,2′-biquinoline), 18 mM NaOH, 13 mM HEPES acid.
10. Sephadex G-25 or G-10 resin (GE Healthcare Life Sciences).
11. Plastic Gravity and Spin Chromatography Column (Bio-Rad).

3 Methods

3.1 Preparation of the Ni^{2+}-Nitrilotriacetic Acid Resin

1. Wash the Ni^{2+}-nitrilotriacetic acid resin in an appropriate size column with 5 volumes of deionized H_2O to eliminate ethanol present in the stock supplied by the manufacturer. Equilibrate the resin by passing 5 volumes of 5 mM imidazole in Buffer B.
2. After using the resin, this can be washed with 15 volumes of 0.5 M NaOH (0.5 mL/min flow rate) followed by 10 volumes of 50 mM NaH_2PO_4, 300 mM NaCl, 10 mM imidazole, pH 8.0.
3. If poor performance is observed, i.e., co-elution of contaminant bands, the resin can be stripped and recharged. For this, strip the resin with 10 volumes of 50 mM NaH_2PO_4, 300 mM NaCl, 100 mM EDTA, pH 8.0 and wash with 20 volumes of deionized H_2O. Recharge the resin with 2 volumes of 100 mM $NiSO_4$. Wash the resin with 20 volumes of deionized H_2O. Do not let the resin dry. Store in 20 % ethanol at 4 °C.

3.2 Preparation of Strep-Tactin Resin for Purification of Cu⁺-Chaperones

1. Load the Strep-Tactin Sepharose resin in an appropriate size column and wash with 5 volumes of deionized H_2O.
2. Equilibrate the resin by passing 5 volumes of Buffer W.
3. The resin can be regenerated by washing three times with 5 volumes of Regeneration Buffer and extensively washing the resin with Buffer W (15–20 volumes). Strep-Tactin resin is reusable up to 15 times. Do not let the resin dry. Store in 20 % ethanol at 4 °C.

3.3 Expression and Purification of P_{1B}-ATPases

The following protocol was optimized for preparation of bacterial and archaebacterial P_{1B}-type ATPases heterologously expressed in *Escherichia coli*. The methodology involves cell disruption in a French Press (alternatively bead beater or high pressure homogenizer), isolation of cell membranes by centrifugation, solubilization of membrane proteins with detergents, affinity purification, and reconstitution in micellar form. All purification procedures should be carried out at 0–4 °C.

1. Induce protein expression according to protocols for the selected plasmid (*see* **Note 1**).

2. Harvest cells by centrifugation at $3200 \times g$ for 10 min. Resuspend cell pellet in 30–35 mL Washing buffer. Centrifuge the cells at $3200 \times g$ for 10 min and resuspend the cell pellet in 30–35 mL of Buffer A.

3. Disrupt the cells by passing them through a French press at 20,000 p.s.i.

4. Centrifuge the lysed cells at $8000 \times g$ for 30 min and discard pellets. Centrifuge the supernatant at $229,000 \times g$ for 1 h. Discard supernatant and resuspend the membrane pellets by adding several small volumes (approx. 1 mL each) of Buffer A and using a glass rod.

5. Transfer the membranes to a 15 mL teflon-glass hand homogenizer and homogenize the membranes with 5 piston strokes. Transfer to a Falcon tube.

6. Estimate the protein concentration by Bradford's method [22].

7. Dilute the membrane to 3 mg/mL protein with Buffer B (25 mL maximum volume) in a small erlenmeyer flask. While gently stirring the membrane suspension with a magnetic stirrer, add dropwise a 10 % stock solution of DDM until the suspension is 0.75 % DDM. Continue the slow stirring of the membranes for 1 h.

8. Centrifuge the solubilized membrane at $229,000 \times g$ for 1 h. Discard the pellet. Measure the protein concentration in the supernatant.

9. Transfer the supernatant to a fresh Falcon tube and add the pre-equilibrated Ni^{2+}-nitrilotriacetic acid resin (1 mL of resin per 100 nmol of His-tagged protein) (*see* **Note 2**). Incubate 1 h with gentle rocking. Load the resin in a column and wash with 10 volumes of ATPase washing buffer 1, followed by 10 volumes of ATPase washing buffer 2. Elute the ATPase with 10 volumes of ATPase elution buffer while collecting 1 mL fractions.

10. Identify those fractions containing the ATPase by SDS-PAGE and pool them. Exchange the buffer by diluting the sample one in five with Buffer C and concentrate the protein to its

original volume using a 30 kDa molecular weight cut-off Centricon (Millipore); repeat this step 3–5 times. Finally, the protein should be concentrated to 1 mg/mL or higher. Measure the protein concentration and aliquot. Most P_{1B}-Type ATPases can be stored in 20 % glycerol in Buffer C at –20 °C.

3.4 Expression and Purification of Cu^+-Chaperones

The following protocol was optimized for preparation of bacterial and archaebacterial soluble Cu^+-chaperones heterologously expressed in *E. coli*. The methodology involves cell disruption by sonication, isolation of cytoplasmic fraction and purification by affinity chromatography. All purification procedures should be carried out at 0–4 °C.

1. Induce protein expression according to protocols for the selected plasmid (*see* **Note 3**).

2. Harvest the cells by centrifugation at $3200 \times g$ for 10 min. Wash the cells by resuspending the pellet in 5 volumes of Buffer W and centrifugation at $3200 \times g$ for 10 min. Resuspend the cell pellet in 5 volumes of Buffer W.

3. Disrupt the cells by sonication 6 times for 30 s, placing the sample on ice for 30 s between repeats.

4. Centrifuge the cell lysate at $8000 \times g$ for 30 min. Transfer the supernatant to a fresh tube and discard the pellet.

5. Load the supernatant into a pre-equilibrated Strep-Tactin column (1 mL per 50–100 nmol of tagged protein). Wash the column with 6 volumes of Buffer W. Elute the protein with 3–4 column volumes of Buffer E. Collect 6–8 eluate fractions of 1-mL each.

6. Identify the fractions containing the chaperone by SDS-PAGE and pool them. Exchange the buffer by diluting the sample one in five with Buffer C and concentrating in a 3 kDa molecular weight cut-off Centricon; repeat this step 3–5 times. Finally, concentrate the protein to 1 mg/mL or higher, measure protein concentration, aliquot, and store in 20 % glycerol in Buffer C at –20 °C.

3.5 Preparation of Sephadex G-25 Columns for Cu^+ Binding

1. Hydrate the Sephadex G-25 resin with 10–20 volumes of deionized H_2O for at least 3 h at 20 °C.

2. Place 1 mL of settled resin per reaction in a plastic Gravity and Spin Chromatography Column. Equilibrate the resin by passing 5 volumes of Buffer H. No air bubbles should be present since these may lead to sample loss.

3. After each use, the resin can be washed with 2 volumes of 0.2 M NaOH, rinsed with water, and stored in 20 % ethanol. Re-equilibrate with 2–3 volumes of Buffer H.

3.6 Binding of Cu$^+$ to Cu$^+$-ATPases

The assay consists of incubating the target protein in the presence of the cognate metal plus desired effectors followed by separation in a gel permeation column.

1. Prepare the reaction in an Eppendorf tube. Include 10 μM protein (70–80 μg in the case of Cu$^+$-ATPases) and 100 μM CuSO$_4$. Prepare the various effectors to be tested in Buffer H and add to the reaction mix. Adjust all tubes to 100 μL final volume with Buffer H. Include a control with no protein and a separate control with no CuSO$_4$.

2. Allow the system to reach equilibrium for 5 min at room temperature.

3. Remove unbound Cu$^+$ by loading each reaction on a 1 mL Sephadex G-25 column prepared as described above (Subheading 3.4).

4. Elute the column with 1.5 volumes of Buffer H. Collect fifteen 100 μL fractions in Eppendorf tubes (*see* **Note 4**).

5. Quantify protein by Bradford [22] in each fraction to determine the fractions containing the ATPase and measure Cu$^+$ content by BCA (*see* below). Subtract values obtained from control reactions.

6. Verify that the unbound free Cu$^+$ is eluted from the resin (fractions 7–9) by BCA (*see* below).

3.7 Cu$^+$ determination by the Bicinchoninic Acid Method

The bicinchoninic acid method to measure Cu$^+$ described by Brenner and Harris [23] was modified to work with reduced amounts of proteins in small reaction volumes. The assay is initiated by denaturation of the target protein and separation by centrifugation, followed by the colorimetric determination of Cu$^+$ in the supernatant.

1. Take 60 μL of the protein sample, i.e., a fraction from metal binding experiments. Prepare a Cu$^+$ standard curve with tubes containing 60 μL of 0, 5, 10, 20, 40, 60 μM CuSO$_4$ in Buffer H.

2. Add 20 μL of 30 % trichloroacetic acid to the samples and standard tubes. Vortex. Spin samples at 10,000 × g for 5 min at room temperature in a microcentrifuge.

3. Take 65 μL of supernatant without disrupting the protein pellet and place in a fresh Eppendorf tube (*see* **Note 5**).

4. Add 5 μL of freshly prepared 40 μM ascorbic acid. Vortex.

5. Add 30 μL of BCA solution (*see* **Note 6**). Vortex and incubate for 5 min at room temperature.

6. Measure the absorbance at 354.5 nm of the standards and samples. Using the standard curve, determine the BCA-Cu$^+$ concentration in the samples (*see* **Notes 7** and **8**).

3.8 Determination of Protein Cu^+ Binding Affinities by Competition with BCA

The assay takes advantage of the relatively similar affinities of Cu^+ binding proteins and BCA and the spectral properties of $BCA_2 \cdot Cu^+$ to perform a titration of the protein in the presence of the competitor.

1. Prepare the assay tube containing 20 µM protein (140–160 µg for P_{1B}-ATPases), 25 µM BCA in a final volume of 200 µL Buffer H and transfer to a spectroscopic cuvette. The volume can be modified according to available cuvette; however, if smaller volumes are used the dilution when adding Cu^+ might be significant (*see* below).

2. Obtain a baseline absorbance at 359 nm.

3. Titrate the protein with Cu^+ by successively adding 1 µL of 200 µM $CuSO_4$ in Buffer H; after each addition mix by gently pipetting up and down with a 200 µL pipette and measure the absorbance at 359 nm (*see* **Note 9**).

4. Using $\varepsilon_{359} = 43{,}000$ M^{-1} cm^{-1} for the Cu^+-bound BCA complex calculate the concentration of the complex after each addition. Take into account the dilution associated with the addition of $CuSO_4$.

5. Free metal concentrations are calculated using $K_a = [BCA_2Cu^+]/[BCA_{free}]^2[Cu^+]_{free}$ for the BCA_2Cu^+ complex, where $[BCA_{free}] = [BCA_{total}] - [BCA_2 \cdot Cu^+]$ and $K_a = 4.6 \times 10^{14}$ M^{-2} [9, 24].

6. Calculate the Cu^+-protein K_a and the number of metal-binding sites (n) by fitting a curve of v vs. $[Cu^+]_{free}$ to the equation $v = n[Cu^+]_{free}K_a/(1 + K_a[Cu^+]_{free})$, where K_a is the association constant for the protein·Cu^+ complex, v is $[Cu^+ \cdot protein]/[protein]$ in moles and $[Cu^+ \cdot protein] = [Cu^+]_{total} - ([BCA_2 \cdot Cu^+] + [Cu^+]_{free})$ [9].

3.9 Cu^+ Transfer from Soluble Chaperones to Cu^+-ATPases

Cu^+ chaperones and Cu^+-ATPases should be cloned with different tags for purification and separation (*see* **Note 10**).

1. Load the necessary amount of purified apo-form of the chaperone with Cu^+ by incubating 10 µM protein, 20 µM $CuSO_4$ in Cu^+ loading buffer (100 µL per reaction) for 10 min at room temperature with gentle agitation. Remove the unbound Cu^+ by passing through a Sephadex G-10 column. Follow protocols similar to those described for Sephadex G-25 preparation (Subheading 3.5) and Cu^+ binding to the ATPase (Subheading 3.6). Maintain the Cu^+-chaperone on ice and plan the experiments to use immediately following loading.

2. Verify the amount of Cu^+ bound to the chaperone by the BCA method (Subheading 3.7).

3. Dilute 10 µM of the His_6-tagged Cu^+-ATPase (Cu^+ acceptor) in 100 µL of Buffer T (per reaction). Bind the enzyme to 100 µL of Ni^{2+}-nitrilotriacetic acid resin pre-equilibrated in Buffer T. For this, incubate for 30–45 min with gentle agitation at 4 °C.

4. Place the enzyme-loaded resin in a 1 mL plastic Gravity and Spin Chromatography Column and wash the unbound protein with 5 volumes of 5 mM imidazole Buffer T.

5. Cu^+-transfer from the Strep-tagged chaperone to His_6-tagged Cu^+-ATPase is initiated by adding a 3 molar excess of Cu^+-loaded Strep-tagged chaperone (Cu^+ donor) to the column and allow to interact for 10 min at room temperature.

6. Wash the chaperone protein from the column with 5 volumes of Buffer T20 and collect 5 fractions of 100 μL. Elute the Cu^+-loaded ATPase with 10 volumes of Buffer T300 and collect 10 fractions of 100 μL (see **Note 11**). The absence of the ATPase in the wash fractions and the chaperone in the elution fractions must be verified by SDS-PAGE and Western Blot using specific antibodies against either the protein of interest or the strep/His_6 tags.

7. Determine protein concentration by Bradford method [22] and Cu^+ content of each fraction by the BCA method described above [23].

4 Notes

1. The pBAD-TOPO vector (Invitrogen) has been routinely used for expression of an array of P_{1B}-type ATPases in our laboratory. This vector adds a His_6-tag to the resulting protein.

2. Molar protein concentrations are estimates based on protein determinations by Bradford and the molecular weight of the corresponding protein.

3. pPRIBA vector (IBA Life Sciences) has been routinely used for expression of soluble Cu^+-chaperones in our laboratory. This vector adds a Strep-tag to the resulting protein.

4. The protein typically elutes in fractions 3 and 4.

5. This represents 81.25 % of the volume in the tube and should be considered in the calculations.

6. Dissolve BCA first and then add other reagents in the solution.

7. TRIS, DTT and TCEP may interfere with BCA reaction, preventing the formation of the BCA-Cu^+ complex. We recommend using 1 mM ascorbic acid (as in Buffer H) to reduce Cu.

8. The BCA solution reacts exclusively with Cu^+ to form a pink BCA-Cu^+ complex, with maximum absorbance at 354.5 and 562 nm. However, the composition of the solution may affect the position of the peaks, so a UV-Visible spectrum should be

measured from 300 to 700 nm to determine the wavelength yielding the maximal absorbance.

9. Note that the concentration of CuSO$_4$ might have to be adapted depending on the particular metal affinity of the tested protein. The goal is to produce changes in absorbance after each addition.

10. We suggest using Strep-tag (pPRIBA vector, from IBA Life Sciences) for the chaperone and His$_6$ (pBAD-TOPO vector, from Invitrogen) for the ATPase.

11. It is necessary to determine the number of washes needed for each different protein. For this, the Cu$^+$ content on each washing fraction should be determined. The optimal number of washes will be when the Cu$^+$ content in the wash fraction approaches zero.

References

1. Solioz M, Abicht HK, Mermod M, Mancini S (2010) Response of Gram-positive bacteria to copper stress. J Biol Inorg Chem 15(1):3–14
2. Argüello JM, González-Guerrero M, Raimunda D (2011) Bacterial transition metal P$_{1B}$-ATPases: transport mechanism and roles in virulence. Biochemistry 50(46):9940–9949
3. Argüello JM (2003) Identification of ion-selectivity determinants in heavy-metal transport P$_{1B}$-type ATPases. J Membr Biol 195(2):93–108
4. Rensing C, Fan B, Sharma R, Mitra B, Rosen BP (2000) CopA: an Escherichia coli Cu(I)-translocating P-type ATPase. Proc Natl Acad Sci U S A 97(2):652–656
5. Osman D, Cavet JS (2008) Copper homeostasis in bacteria. Adv Appl Microbiol 65:217–247
6. Argüello JM, Raimunda D, González-Guerrero M (2012) Metal transport across biomembranes: emerging models for a distinct chemistry. J Biol Chem 287(17):13510–13517
7. Robinson NJ, Winge DR (2010) Copper metallochaperones. Annu Rev Biochem 79:537–562
8. Padilla-Benavides T, George Thompson AM, McEvoy MM, Argüello JM (2014) Mechanism of ATPase-mediated Cu$^+$ export and delivery to periplasmic chaperones: the interaction of Escherichia coli CopA and CusF. J Biol Chem 289(30):20492–20501
9. González-Guerrero M, Argüello JM (2008) Mechanism of Cu$^+$-transporting ATPases: soluble Cu$^+$ chaperones directly transfer Cu$^+$ to transmembrane transport sites. Proc Natl Acad Sci U S A 105(16):5992–5997
10. Argüello JM, Eren E, González-Guerrero M (2007) The structure and function of heavy metal transport P$_{1B}$-ATPases. Biometals 20(3–4):233–248
11. González-Guerrero M, Eren E, Rawat S, Stemmler TL, Argüello JM (2008) Structure of the two transmembrane Cu$^+$ transport sites of the Cu$^+$-ATPases. J Biol Chem 283(44):29753–29759
12. Raimunda D, Subramanian P, Stemmler T, Argüello JM (2012) A tetrahedral coordination of Zinc during transmembrane transport by P-type Zn^{2+}-ATPases. Biochim Biophys Acta 1818(5):1374–1377
13. Padilla-Benavides T, Long JE, Raimunda D, Sassetti CM, Argüello JM (2013) A novel P$_{1B}$-type Mn^{2+}-transporting ATPase is required for secreted protein metallation in mycobacteria. J Biol Chem 288(16):11334–11347
14. Zielazinski EL, Cutsail GE III, Hoffman BM, Stemmler TL, Rosenzweig AC (2012) Characterization of a cobalt-specific P(1B)-ATPase. Biochemistry 51(40):7891–7900
15. Liu J, Dutta SJ, Stemmler AJ, Mitra B (2006) Metal-binding affinity of the transmembrane site in ZntA: implications for metal selectivity. Biochemistry 45(3):763–772
16. Glynn IM, Karlish SJ (1990) Occluded cations in active transport. Annu Rev Biochem 59:171–205
17. Eren E, Kennedy DC, Maroney MJ, Argüello JM (2006) A novel regulatory metal binding

domain is present in the C terminus of Arabidopsis Zn^{2+}-ATPase HMA2. J Biol Chem 281(45):33881–33891
18. Rae TD, Schmidt PJ, Pufahl RA, Culotta VC, O'Halloran TV (1999) Undetectable intracellular free copper: the requirement of a copper chaperone for superoxide dismutase. Science 284(5415):805–808
19. Padilla-Benavides T, McCann CJ, Argüello JM (2013) The mechanism of Cu^+ transport ATPases: interaction with Cu^+ chaperones and the role of transient metal-binding sites. J Biol Chem 288(1):69–78
20. González-Guerrero M, Hong D, Argüello JM (2009) Chaperone-mediated Cu^+ delivery to Cu^+ transport ATPases: requirement of nucleotide binding. J Biol Chem 284(31): 20804–20811
21. Gourdon P et al (2011) Crystal structure of a copper-transporting P_{IB}-type ATPase. Nature 475(7354):59–64
22. Bradford MM (1976) A rapid and sensitive method for the quantitation of microgram quantities of protein utilizing the principle of protein-dye binding. Anal Biochem 72:248–254
23. Brenner AJ, Harris ED (1995) A quantitative test for copper using bicinchoninic acid. Anal Biochem 226(1):80–84
24. Yatsunyk LA, Rosenzweig AC (2007) Cu(I) binding and transfer by the N terminus of the Wilson disease protein. J Biol Chem 282(12): 8622–8631

Part V

Electrophysiology

Part V

Chapter 25

Voltage Clamp Fluorometry of P-Type ATPases

Robert E. Dempski

Abstract

Voltage clamp fluorometry has become a powerful tool to compare partial reactions of P-type ATPases such as the Na+,K+-ATPase and H+,K+-ATPase with conformational dynamics of these ion pumps. Here, we describe the methodology to heterologously express membrane proteins in *X. laevis* oocytes and site-specifically label these proteins with one or more fluorophores. Fluorescence changes are measured simultaneously with current measurements under two-electrode voltage clamp conditions.

Key words Fluorescence, Distance constraints, Donor photobleaching

1 Introduction

Voltage clamp fluorometry has become a facile method to detect site-specific conformational changes of membrane proteins in real time under physiological conditions. This is achieved by combining fluorescence and two electrode voltage clamp (voltage clamp fluorometry). In contrast to other approaches, voltage clamp fluorometry can be used to obtain time-resolved and spatially well-defined information on conformational rearrangements of membrane proteins. Voltage clamp fluorometry has been used to investigate the conformational dynamics of K+ channels, Na+ channels, and glutamate transporters [1-4]. Equally, this approach has also been used to elucidate distance constraints for multiple membrane proteins [5, 6]. More relevant to this book, voltage clamp fluorometry was initially used in the P-type ATPase field to study the conformational dynamics of the extracellular loop between transmembrane five and six of the alpha subunit of the Na+,K+-ATPase [7]. Analysis of these experiments eloquently demonstrated that fluorescence intensity changes measured after site-specific fluorescence labeling within this loop could be used to investigate the conformational state and kinetics of the Na+,K+-ATPase. Analysis of subsequent experiments examined the conformational dynamics of the alpha, beta and gamma subunits as a function of ion transport as well as elucidated distance constraints between the three subunits of this ion pump [8–11]. More recently, voltage clamp fluorometry was used to examine the electroneutral H^+,K^+-ATPase [12, 13]. Combined, analyses of these experiments have provided new information about site-specific molecular movements of P-type ATPases during ion transport.

The methods underpinning this approach are multifaceted. First, the gene of interest is inserted into a plasmid which can be translated by *X. laevis* oocytes. All free extracellular cysteine residues are removed from the target protein and a single cysteine residue is inserted at the interface of the transmembrane-extracellular domains as predicted by hydropathy analysis or through inspection of crystal structures. mRNA is synthesized *in vitro* from this plasmid and injected into oocytes isolated from *X. laevis*. Oocytes are incubated for 2–7 days at 18 °C to enable protein synthesis and surface expression. Surface expressed protein is then site-specifically labeled with one or more fluorophores. Surface expression of membrane proteins can be validated with biotinylation of the expressed proteins with subsequent Western blotting with an antibody specific for the protein of interest [14]. Functional expression of the target protein is validated using a two electrode voltage clamp set-up. Changes in fluorescence intensity are measured concurrently with ion transport using a fluorescence microscope with a tungsten lamp, filters, and photomultiplier tube upon activation of the protein of interest. Finally, while not utilized for the investigation of P-type ATPases, it is important to note that parallel approaches to study the conformational dynamics of membrane proteins on the surface of cells have more recently included patch clamp fluorometry and the utilization of lathanides to obtain distance constraints [15, 16]. Equally, it is possible to obtain more detailed information in regard to changes in fluorescence intensity through the use of charge-coupled device cameras [2].

Combined, voltage clamp fluorometry enables examination of changes in the conformational dynamics of P-type ATPases during ion transport. Equally, it has enabled investigation of the electroneutral ion pump H^+,K^+-ATPase.

2 Materials

All solutions should be prepared using ultrapure water. All solutions required for DNA and/or mRNA synthesis should be autoclaved with diethylpyrocarbonate (DEPC). All reagents are analytical grade.

2.1 mRNA Synthesis

1. DNA can be dissolved in DEPC-treated H_2O. The pTLN vector has been successfully used to synthesize mRNA for functional expression in *X. laevis* oocytes [17]. With this vector, the restric-

tion enzyme *MluI* can be used to linearize the plasmid DNA prior to mRNA synthesis as long as the restriction site is not present in the gene of interest. Following linearization, High-Pure PCR Product Purification Kit (Roche Applied Science: 11732 676 001) can be used to purify the linearized DNA.

2. mMessage mMachine kit (Life Technologies) can be used to synthesize mRNA according to the manufacturer's instructions.
3. TAE buffer: 40 mM Tris-HCl (pH 8.0–8.5), 20 mM acetic acid, and 1 mM EDTA.
4. Agarose.
5. Cell-Porator (Life Technologies).

2.2 X. laevis Oocytes and mRNA Components

1. Mature female *X. laevis* frogs (Nasco, Cat. No. LM00535MX) can be used with good results. Maintenance of these frogs should be in accordance with institutional requirements. Alternatively, oocytes can be purchased (Ecocyte Bioscience).
2. Anesthetic solution: 0.5–3.0 g MS-222 in water, pH 7.0–8.0 w/sodium bicarbonate.
3. Diaper pads.
4. Microdissection scissors, forceps, and hooks (World Precision Instruments).
5. 3-0 to 4-0 Ethicon Vicryl sutures (Johnson and Johnson).
6. Collagenase Type II (Worthington Biochemical). Sample batches of different collagenase type II lots can be used to probe which lot works best for oocyte digestion protocol.
7. Ringer's solution (−) Ca^{2+}: 90 mM NaCl, 2 mM KCl, 2 mM $MgCl_2$, 5 mM MOPS pH 7.4.
8. Ringer's solution (+) Ca^{2+}: 90 mM NaCl, 2 mM KCl, 2 mM $CaCl_2$, 5 mM MOPS pH 7.4.

2.3 Voltage Clamp Fluorometry

1. Filter sets (Semrock, Inc. or Chroma Technology Corporation).
2. Loading buffer: 110 mM NaCl, 2.5 mM sodium citrate, 10 mM MOPS/Tris, pH 7.4.
3. Post-loading buffer: 100 mM NaCl, 1 mM $CaCl_2$, 5 mM $BaCl_2$, 5 mM $NiCl_2$, 10 mM MOPS/Tris, pH 7.4.
4. Tetramethyl-6-maleimide or fluorescein maleimide (Life Technologies).
5. RC-10 chamber (Warner Instruments).
6. Polarized filters (Linos Photonics).
7. Axio Examiner Fluorescence Microscope (Zeiss).
8. Photomultiplier tube detection system for fluorescence measurements (Cairn Research).

9. Turbo-Tec-05X amplifier (NPI Instruments).
10. Axioscope 1440 Digitizer (Axon Instruments).
11. pCLAMP Software (Axon Instruments).

3 Methods

3.1 mRNA Synthesis

1. Digest 3 μg of plasmid DNA in 50 μL DEPC-treated H_2O with an appropriate restriction enzyme (1 h, 10 U of enzyme) and ensure RNase-free conditions (wear gloves, change them frequently).

2. To purify the linearized template, use the High-Pure PCR Product Purification Kit. Here, add 350 μL binding buffer to the digest vial and vortex.

3. Place spin columns in collection tubes, load the columns with sample and centrifuge for 20 s at full-speed (eppendorf-centrifuge, room temperature). Discard the flow through.

4. Add 500 μL of wash buffer to the column, centrifuge for 20 s at full speed, and discard the flow through.

5. Add 200 μL of wash buffer to the column, centrifuge for 30 s at full speed (the column should be dry) and discard the flow through.

6. Place the spin column into a new autoclaved eppendorf tube, add 50 μL of DEPC-treated water on the column material and wait 1 min. Elute by 30 s full-speed centrifugation.

7. To concentrate the DNA solution, place the sample in a SpeedVac until the concentration of the DNA is approximately 166 ng/μL.

8. To synthesize mRNA *in vitro* use the mMessage mMachine Kit. Here, choose the appropriate kit according to the promoter available on the plasmid DNA (ex. SP6 or T7). Normally, a 10 μL reaction is sufficient. Combine the following reagents: 5 μL NTP-cap mix (usually pre-aliquoted), 3 μL linearized template DNA solution (from **step 7**), 1 μL 10× transcription buffer and 1 μL (appropriate) RNA polymerase. The sample should be incubated for 1-5–2 h at 37 °C. Do not extend the reaction beyond 2 h as the yield does not increase. Furthermore, the integrity of the mRNA might be impaired.

9. To purify the mRNA, combine 12.5 μL LiCl solution from the kit (stored at –20 °C) and 15 μL DEPC treated water (also from the kit). Mix briefly, do not vortex, centrifuge briefly, and place in –20 °C 30 min or overnight for precipitation. Precool a centrifuge to + 4 °C. Centrifuge for 15 min at full speed (4 °C). Here, a white pellet should be visible at bottom of tube. Cautiously remove the supernatant and add 150 μL

70 % RNA grade ethanol (stored at −20 °C). Centrifuge for 5 min at full speed at 4 °C. Remove the supernatant completely and dry the pellet in SpeedVac briefly until all the ethanol has evaporated (2–7 min at RT). Redissolve the pellet in 12 μL DEPC treated water.

10. To quantify the yield, load 0.3 μL RNA solution on a 1 % agarose gel (freshly cast, fresh TAE buffer in electrophoresis chamber). mRNA should show a clear band structure (several bands are possible due to secondary structure). Degradation will show up as a broad smear. Use a NanoDrop to determine the mRNA concentration.

3.2 Isolation of X. laevis *oocytes* and mRNA Injection

1. Prior to surgery, the frog should be fasted for 12 h to prevent vomiting during the operation. To begin surgery, the frog is immersed in anesthetic solution (0.5–3.0 g MS-222 in water, pH to 7.0–8.0 with sodium bicarbonate) until the frog becomes unresponsive to a toe pinch (*see* **Note 1**).

2. The frog is removed from the anesthetic solution, and placed dorsal side down on the absorbent side of a diaper pad. A wet paper towel is placed on the frog to keep it moist.

3. A small coelomic incision on one side of the frog's midline is made. The ovary is identified and removed. The incision is sutured using an interrupted suture pattern. The frog is returned to its housing to recover from anesthesia.

4. The isolated ovarian lobe is cut with scissors into smaller parts in a petri dish containing Ringer's solution − Ca^{2+}. The oocytes are transferred into Ringer's Solution − Ca^{2+} with ~3 mg/mL collagenase. The sample is gently shaken in solution for 2 h at 18 °C. When most of the oocytes are separated, the oocytes are washed with Ringer's Solution − Ca^{2+}. The oocytes are washed exhaustively in Ringer's + Ca^{2+} solution and transferred to Ringer's Solution + Ca^{2+} supplemented with 1 mg/mL gentamycin.

5. The oocytes are injected with a total volume of 50 nL. For the Na^+,K^+-ATPase, best results have been observed with 25 ng of the sheep α_1 subunit and 1 ng of the β_1 subunit. When required, 1 ng of the sheep γ subunit is injected along with the α_1 and β_1 subunits (*see* **Note 2**).

6. The oocytes are incubated in Ringer's Solution + (Ca^{2+}) and antibiotic (ex. 1 mg/mL gentamycin) at 18 °C in the dark for 3–7 days. This allows the protein of interest to be expressed within the oocyte plasma membrane.

3.3 Voltage Clamp Fluorometry

1. On the day of the experiment, the oocytes are removed from the incubator and placed in loading buffer for 45 min and then in post-loading buffer for 45 min to increase the intracellular Na^+ concentration to facilitate measuring the Na^+,K^+-ATPase [11].

2. The oocytes are incubated in post-loading buffer that contains 5 μM of the desired fluorophore, such as tetramethylrhodamine-6-maleimide (TMRM) or fluorescein-5-maleimide (FM), for at least 5–10 min in the dark at room temperature. After fluorophore labeling, the oocytes are washed exhaustively with dye-free post-loading buffer and are kept in the dark until the experiment.

3. To prepare for the two-electrode voltage clamp measurements, microelectrodes are filled with 3 M KCl where the resistance of each micro-electrode is between 0.5 and 1.5 MΩ. Our laboratory uses a Zeiss Axio Examiner Fluorescence Microscope with a Cairn photomultiplier tube detection system for fluorescence measurements (*see* **Note 3**). In addition, the fluorescence microscope contains a 535DF50 excitation filter, a 565 EFLP emission filter and a 570DRLP dichroic mirror for TMRM.

4. An oocyte is placed in a RC-10 chamber on the fluorescence microscope stage. The oocyte is then impaled by the microelectrodes. Using the amplifier to hold the membrane potential at a constant value, ion flux across the membrane is measured following solution exchange or by changing the membrane potential. For concurrent fluorescence measurements, a 100 W tungsten light source equipped with an emission filter excites the fluorophore and a photomultiplier tube detects changes in fluorescence intensity (*see* Fig. 1). Measurements under Na^+/Na^+-exchange conditions have been performed in Na^+ test

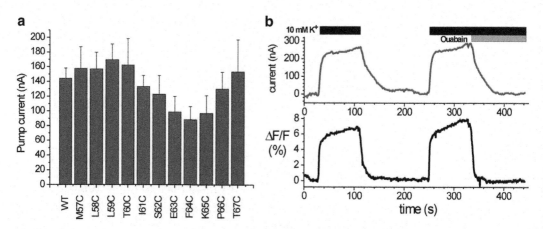

Fig. 1 Stationary current and fluorescence measurements of Na^+,K^+-ATPase α/β complexes containing different β subunit constructs upon expression in *Xenopus laevis* oocytes. (**a**) Stationary pump currents of the Na^+,K^+-ATPase expressed with wild-type and single cysteine mutants of the β subunit at 0 mV holding potential in response to 10 mM K^+. Data originated from 5 to 11 oocytes; values are means ± SEM. (**b**) Parallel recording of pump current (*top*) and fluorescence change (*bottom*) from an oocyte co-injected with the sheep Na^+,K^+-ATPase $α_1$ subunit (where all extracellular cysteine residues have been replaced with conservative mutations which do not affect the function of the ion pump) and the sheep $β_1$ subunit with F64C mutation in response to 10 mM K^+ and 10 mM ouabain at 0 mV holding potential. Reprinted with permission from ref. [10]. ©2005 Dempski et al. Journal of General Physiology. 125:505–520. doi: 10.1085/jgp.200409186

solution (100 mM NaCl, 5 mM $BaCl_2$, 5 mM $NiCl_2$, 10 μM ouabain, 5 mM Mops, Tris, pH 7.4) (*see* **Note 4**). For Na^+/K^+ titration experiments and stationary pump current recordings, 0.1–10 mM K^+ have been replaced for equimolar amounts of Na^+. For K^+/ *N*-methyl-d-glucamine (NMDG) titration experiments, 0.1–1 mM K^+ have been combined with NMDG for a final ionic concentration of 100 mM. Heterologously expressed Na^+,K^+-ATPase was largely inhibited by 10 mM ouabain due to two mutations (Q111R and N122D) on the α subunit, which confers a lower binding affinity for the heterologously expressed Na^+,K^+-ATPase when compared to the endogenous *X. laevis* oocyte Na^+,K^+-ATPase [18].

5. To measure distance constraints, two accessible extracellular cysteine residues are labeled. Here, the oocytes are labeled in post loading buffer containing 1 μM FM (the donor fluorophore) and 4 μM TMRM (the acceptor fluorophore) for 30 min on ice in the dark, or only 1 μM FM. This allows for measurements to be made of holoenzymes labeled with and without an acceptor fluorophore. For these experiments, the following filter set is used: 525AF45 excitation filter, 565ALP emission filter, and 560DRLP dichroic mirror. At these concentrations and with the assumption that both fluorophores react with the free cysteine residues at the same rate, it is predicted that cysteine residues will be labeled in the following ratios: 4 % FM–FM, 32 % with TMRM–FM, and 64 % with TMRM–TMRM (*see* **Note 5**). Maintaining a continuous solution flow, the time dependence of donor bleaching is measured in the presence and absence of the acceptor fluorophore (*see* **Note 6** and Fig. 2). These results can be used to calculate the distance between two residues on the Na^+,K^+-ATPase utilizing Förster's equation (*see* **Note 7**).

6. Anisotropy is calculated alongside distance measurements to determine the range of κ^2 values (*see* **Note 8**). Anisotropy measures the relative mobility of the fluorophore due to rotation. This can be measured using polarized filters during fluorophore excitation and emission.

7. To measure the relative movement of protein subunits, double cysteine constructs are employed that are labeled with acceptor and donor fluorophores. For these experiments, the microscope is equipped with the following filter set: 475AF40 excitation filter, 595AF60 emission filter, and 505DRLP dichroic mirror.

8. Here, the donor is excited with a 100 W tungsten light source and the fluorescence intensity of the acceptor fluorophore is measured. Using an appropriate method such as voltage or solution exchange, changes in the distance between the two fluorophores (donor and acceptor) can be correlated to movement of the labeled residues towards or away from each other (*see* Fig. 2).

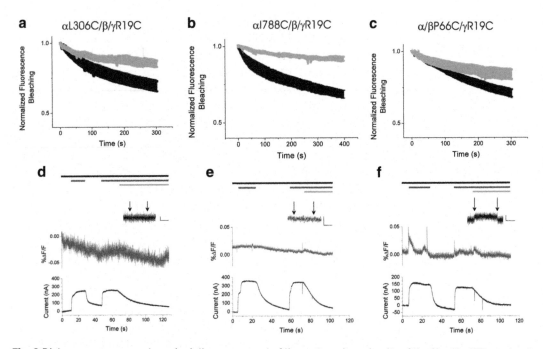

Fig. 2 Distance measurements and relative movement of the α, β, and γ subunits of the Na$^+$,K$^+$-ATPase. (**a–c**) Normalized ΔF_{donor} photobleaching for the Na$^+$,K$^+$-ATPase αL306C/β/γR19C, αI788C/β/γR19C, and α/βP66C/γR19C constructs, respectively, labeled with donor fluorophore alone (*black line*) or donor plus acceptor fluorophore (*grey line*) in a 1:4 ratio. The results are the means of 4–7 oocytes ± standard error. The photodestruction rates for αL306C/β/γR19C are 99.7 ± 0.4 s^{-1} for the donor only and 144.3 ± 0.2 s^{-1} for the donor in the presence of acceptor. The photodestruction rates for αI788C/β/γR19C are 156.3 ± 0.9 s^{-1} for the donor only and 434 ± 2 s^{-1} for the donor in the presence of acceptor. The photodestruction rates for α/βP66C/γR19C are 161.1 ± 0.6 s^{-1} for the donor only and 291.0 ± 0.6 s^{-1} for the donor in the presence of acceptor. (**d–f**) Relative movement of the αL306C residue (M3–M4 loop) of the α subunit, the αI788C residue (M5–M6 loop) of the α subunit, and the βP66C residue (extracellular–transmembrane interface of the β subunit) in relation to the γR19C residue (extracellular–transmembrane interface of the γ subunit), respectively. Parallel recording of the pump current (*lower, black*) and fluorescence resonance energy transfer change (*upper, grey*) at −40 mV under Na$^+$/Na$^+$ exchange conditions (*black bars*) in response to 10 mM K$^+$ (*dark grey bars*) and 10 mM ouabain (*light grey bars*). Results are from a single oocyte, but the experiments were repeated, with essentially identical results, on at least four occasions. The *inset* of each figure (in *black*) shows the fluorescence response to a change in membrane potential from 0 to −200 mV (*arrows* indicate the onset and offset of the membrane potential pulse) where the scale bars on the *x* and *y* axes are 0.01 % $\Delta F/F$ and 200 ms, respectively. Reprinted with permission from ref. [8]. Copyright 2008 American Chemical Society

4 Notes

1. This protocol has been approved by WPI's Institutional Animal Care and Use Committee.

2. Our laboratory has successfully injected oocytes manually using a Nanoject Injector (Drummond Scientific) or through an automated system Roboinject (MultiChannel Systems).

3. We have found the most success using an upright fluorescence microscope with water immersion lens. However, inverted fluorescence microscopes have also been effectively used [19].

4. *X. laevis* oocytes have an endogenous Na^+,K^+-ATPase which can be inhibited upon addition of 10 µM ouabain. The sheep Na^+,K^+-ATPase used in our studies contains two mutations within the α_1 subunit (Q111R and N122D) which confers a reduced ouabain sensitivity (mM range) [18]. Therefore, the endogenous ion pump can be selectively inhibited with 10 µM ouabain and the heterologously expressed ion pump can be inhibited upon addition of 10 mM ouabain.

5. Under these conditions, for the donor photobleaching experiments, the major fluorophore pair will be FM–TMRM. The TMRM–TMRM pair is not measured due to the selection of excitation/emission filters. There will be a small population of FM:FM pairs. This will result in a faster observed decay rate and this could result in a smaller distance calculated than is actually present.

6. A continuous solution flow is required as the light source generates heat which may result in heat-induced fluorescence changes. The constant solution flow maintains constant temperature during the experiment. Alternatively, one can use a chamber with temperature control to maintain constant temperature during the donor photodestruction experiments.

7. To determine the distance constraints between two residues, the following protocol was used. First, the efficiency of energy transfer (E) was calculated using the following equation: $E = 1 - \tau_{DA}/\tau_D$ (where τ_{DA} and τ_D are the time constants of donor photobleaching in the presence and the absence of acceptor, respectively) [20]. Fluorescence resonance energy transfer (FRET) efficiency was converted to distance using the Förster equation: $E = 1/(1 + R^6/R_o^6)$, where R is the distance between the donor and acceptor and R_o is the distance corresponding to 50 % efficiency for a specific donor–acceptor pair (for the fluorescein–tetramethylrhodamine pair: 55 Å) [20]. R_o is defined by the equation: $R_o = (9.7 \times 10^3 J \Phi_D n^{-4} \kappa^2)^{1/6}$ (in Å), where J is the normalized spectral overlap of the donor emission and acceptor absorption, Φ_D is the donor emission quantum yield in absence of the acceptor, n is the index of refraction, and κ^2 is the orientation factor for a dipole-dipole interaction [21].

8. The anisotropy (r) of the donor and acceptor fluorophores can be measured on oocytes which represent a quasiplanar system and evaluated through the use of the following equation: $r = (I^{\parallel} - I^{\perp})/(I^{\parallel} + 2I^{\perp})$, where $I^{\parallel}$ is the parallel and $I^{\perp}$ is the perpendicular-emitted light with respect to the polarized excitation light at each amino acid [22, 23]. These components

can be measured sequentially with polarized filters placed adjacent to the emission and excitation filters. To calculate the error in the distance measurements due to anisotropy, the following equations can be used: $\kappa^2_{max} = 2/3 \ (1 + F_{rd} + F_{ra} + 3 F_{rd} \times F_{ra})$ and $\kappa^2_{min} = 2/3(1 - (F_{rd} + F_{ra})/2)$ where $F_{rd} = (r_d/r_o)^{0.5}$ and $F_{ra} = (r_a/r_o)^{0.5}$; r_d is the anisotropy of FM, r_a is the anisotropy of TMRM, and r_o is the fundamental anisotropy of each fluorophore (FM: 0.4 and TMRM: 0.38) [24, 25].

Acknowledgements

The Dempski laboratory gratefully acknowledges the WPI Research Foundation and NIH (grant GM105964) for support. The author thanks Ryan Richards for critical reading of this manuscript.

References

1. Mannuzu LM, Moronne MM, Isacoff EY (1996) Direct physical measure of conformational rearrangement underlying potassium channel gating. Science 271:213–216
2. Cha A, Bezanilla F (1997) Characterizing voltage-dependent conformational changes in the Shaker K+ channel with fluorescence. Neuron 19:1127–1140
3. Oelstrom K, Goldschen-Ohm MP, Holmgren M, Chanda B (2014) Evolutionarily conserved intracellular gate of voltage-dependent sodium channels. Nat Commun 5:3420
4. Larsson HP, Tzingounis AV, Koch HP, Kavanaugh MP (2004) Fluorometric measurements of conformational changes in glutamate transporters. Proc Natl Acad Sci U S A 101:3951–3956
5. Koch HP, Larsson HP (2005) Small-scale molecular motions accomplish glutamate uptake in human glutamate transporters. J Neurosci 25:1730–1736
6. Glauner KS, Mannuzu LM, Gandhi CS, Isacoff EY (1999) Spectroscopic mapping of voltage sensor movement in the Shaker potassium channel. Nature 402:813–817
7. Geibel S, Kaplan JH, Bamberg E, Friedrich T (2003) Conformational dynamics of the Na+/K+-ATPase probed by voltage clamp fluorometry. Proc Natl Acad Sci U S A 100:964–969
8. Dempski RE, Lustig J, Friedrich T, Bamberg E (2008) Structural arrangement and conformational dynamics of the gamma subunit of the Na+/K+-ATPase. Biochemistry 47:257–266
9. Dempski RE, Hartung K, Friedrich T, Bamberg E (2006) Fluorometric measurements of intermolecular distances between the alpha and beta subunits of the Na+/K+-ATPase. J Biol Chem 281:36338–36346
10. Dempski RE, Friedrich T, Bamberg E (2005) The beta subunit of the Na+/K+-ATPase follows the conformational state of the holoenzyme. J Gen Physiol 125:505–520
11. Geys S, Dempski RE, Bamberg E (2009) Ligand-dependent effects on the conformational equilibrium of the Na+, K+-ATPase as monitored by voltage clamp fluorometry. Biophys J 96:4561–4570
12. Dürr KL, Tavraz NN, Zimmermann D, Bamberg E, Friedrich T (2008) Characterization of Na, K-ATPase and H, K-ATPase enzymes with glycosylation-deficient b-subunit variants by voltage-clamp fluorometry in Xenopus oocytes. Biochemistry 47:4288–4297
13. Durr KL, Tavraz NN, Friedrich T (2012) Control of gastric H, K-ATPase activity by cations, voltage and intracellular pH analyzed by voltage clamp fluorometry in Xenopus oocytes. PLoS One 7:e33645
14. Wagner CA, Friedrich B, Setiawan I, Lang F, Broer S (2000) The use of Xenopus laevis oocytes for the functional characterization of heterologously expressed membrane proteins. Cell Physiol Biochem 10:1–12
15. Cha A, Snyder GE, Selvin PR, Bezanilla F (1999) Atomic scale movement of the voltage-sensing region in a potassium channel measured via spectroscopy. Nature 402:809–813

16. Biskup C, Kusch J, Schulz E, Nache V, Schwede F, Lehmann F, Hagen V, Benndorf K (2007) Relating ligand binding to activation gating in CNGA2 channels. Nature 446:440–443
17. Lorenz C, Pusch M, Jentsch TJ (1996) Heteromultimeric CLC chloride channels with novel properties. Proc Natl Acad Sci U S A 92:13362–13366
18. Price EM, Lingrel JB (1988) Structure-function-relationships in the Na, K-ATPase alpha subunit-Site-directed mutagenesis of glutamine-111 and asparagine-122 to aspartic acid generates a ouabain-resistant enzyme. Biochemistry 27:8400–8408
19. Mourot A, Bamberg E, Rettinger J (2008) Agonist- and competitive antagonist-induced movement of loop 5 on the alpha subunit of the neuronal alpha4beta4 nicotinic acetylcholine receptor. J Neurochem 105:413–424
20. Forster T (1948) Intermolecular energy migration and fluorescence. Ann Phys 2:55–75
21. Van Der Meer BW, Coker G, Chen SYD (1994) Resonance energy transfer: theory and data. Wiley, New York
22. Dale RE, Eisinger J, Blumberg WE (1979) The orientational freedom of molecular probes. Biophys J 26:161–194
23. Lakowicz JR (1999) Principles of fluorescence spectroscopy. Kluwer Academic/Plenum Publishers, New York
24. Cha A, Bezanilla F (1998) Structural implications of fluorescence quenching in the Shaker K+ channel. J Gen Physiol 112:391–408
25. Chen RF, Bowman RL (1965) Fluorescence polarization: measurement with ultraviolet-polarizing filters in spectrophotofluormeter. Science 147:729–732

Chapter 26

Electrophysiological Measurements on Solid Supported Membranes

Francesco Tadini-Buoninsegni and Gianluca Bartolommei

Abstract

The solid supported membrane (SSM) represents a convenient model system for a biological membrane with the advantage of being mechanically so stable that solutions can be rapidly exchanged at the surface. The SSM consists of a hybrid alkanethiol–phospholipid bilayer supported by a gold electrode. Proteoliposomes, membrane vesicles, or membrane fragments containing the transport protein of interest are adsorbed on the SSM surface and are subjected to a rapid substrate concentration jump. The substrate concentration jump activates the protein and the charge displacement concomitant with its transport activity is recorded as a current transient. Since this technique is well suited for the functional characterization of electrogenic membrane transporters, it is expected to become a promising platform technology for drug screening and development.

Key words Solid-supported membrane, Concentration jump, Current transient, Gold sensor, Solution exchange, Membrane transport protein, Charge displacement

1 Introduction

A convenient model system for a biological membrane is a solid-supported membrane (SSM). The SSM, in combination with a rapid solution exchange technique, has been employed for the investigation of charge translocation processes in a number of primary and secondary active membrane transporters [1–3]. In particular, P-type ATPases have been conveniently investigated on an SSM using a wide range of preparations: membranes from native tissues (sarcoplasmic reticulum Ca^{2+}-ATPase from rabbit skeletal muscle [4, 5] and H^+,K^+-ATPase from pig gastric mucosa [6]), purified membrane fragments (Na^+,K^+-ATPase from pig [7, 8] or rabbit kidney [9, 10]), reconstituted proteoliposomes (Na^+,K^+-ATPase from shark rectal gland [11]), and heterologous expression in mammalian cells (human Cu^+-ATPases ATP7A and ATP7B in microsomes from COS-1 cells[12, 13]).

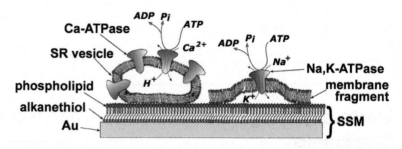

Fig. 1 Cartoon of a sarcoplasmic reticulum vesicle containing Ca-ATPase and of a membrane fragment containing Na,K-ATPase adsorbed on a SSM and subjected to ATP activation (not drawn to scale). Reprinted with permission from ref. [10]. Copyright 2009 American Chemical Society

The SSM consists of a hybrid alkanethiol–phospholipid bilayer supported by a gold electrode. In particular, the SSM is formed by an octadecanethiol monolayer covalently bound to the gold electrode surface via the sulfhydryl group and a second phosphatidylcholine monolayer on top of the thiol layer [7, 14] (Fig. 1). Membrane preparations, i.e., proteoliposomes, membrane vesicles, or membrane fragments, containing the transport protein of interest are adsorbed on the SSM surface (Fig. 1) and are activated using a fast solution exchange technique [7]. By rapidly changing from a solution containing no substrate for the protein to one that contains a substrate, the protein can be activated and a current transient can be detected, which is related to the movement of charged substrates within the protein. The transient nature of the observed electrical currents is a consequence of the capacitively coupled system formed by the SSM and the membrane entities adsorbed to it [1, 15]. Adsorption of proteoliposomes or membrane fragments on the SSM surface occurs in a simple spontaneous process. This experimental approach is much easier and more effective than direct incorporation of the protein in a planar lipid membrane, which requires complicated reconstitution procedures, leading to a superior signal-to-noise ratio and time resolution of the electrical measurement.

Semi-automated and fully automated analysis devices for SSM-based electrophysiological measurements are now commercially available. These instruments are based on the SURFE²R (Surface Electrogenic Event Reader) technology, which is expected to become an attractive platform technology for drug screening and development [6, 16].

2 Materials

Prepare all solutions using ultrapure water (obtained by purifying deionized water, 18.2 MΩ cm at 25 °C) and analytical grade reagents. Store solutions at 4 °C, unless otherwise specified.

2.1 Solutions for Formation of the SSM

1. Alkanethiol solution: 1 mM 1-octadecanethiol in 2-propanol. Solubilize 14.3 mg 1-octadecanethiol in 50 mL 2-propanol (min. 99.7 %) (see **Note 1**). Briefly (~1 or 2 min) sonicate the solution in an ultrasonic bath sonicator to dissolve octadecanethiol in propanol.

2. Phospholipid solution: 7.5 mg/mL diphytanoylphosphatidylcholine in n-decane. Add 75 µL diphytanoylphosphatidylcholine (10 mg/mL in chloroform) or 37.5 µL diphytanoylphosphatidylcholine (20 mg/mL in chloroform) in a carefully cleaned and dry glass vial (see **Note 2**). Evaporate the chloroform under a continuous nitrogen gas flow for about 4–5 min. Redissolve the phospholipid by adding 100 µL of n-decane in the vial and mixing with a pipette and shaking. Store the phospholipid solution at −20 °C.

2.2 Preparation of the Measuring Solutions

In the basic solution exchange protocol two solutions are required: the non-activating solution and the activating solution containing the substrate of the transport protein immobilized on the SSM. Exchanging from non-activating to activating solution produces a substrate concentration jump at the SSM surface, initiating transport and concomitant charge displacement by the protein [15].

Because of the strong interaction of solutes with the lipid surface of the SSM [17], electrical artifacts may be generated by exchange of one buffer solution by another buffer of different composition at the SSM surface [7]. These electrical artifacts are difficult to circumvent and can only be minimized by the appropriate choice of solutes (see **Note 3**). A few simple guidelines are here provided in order to minimize solution exchange artifacts [15, 18].

1. The basic buffer (solution with background salts but without activating substrate) for the activating solution and non-activating solution must be the same. Prepare the basic buffer in one batch, adjust the pH and then divide the basic buffer into two aliquots. The activating substrate is added to one aliquot (activating solution).

2. A compensating inert compound, which is known not to affect the protein, is added to the non-activating solution at the same concentration as the activating substrate (see **Note 4**).

3. A high salt background, e.g., 100–150 mM NaCl or KCl, is recommended to reduce electrical artifacts.

4. It is preferable to use activating compounds which are well water soluble (see **Note 5**).

5. If the solutions are stored at 4 °C, make sure that all solutions reach room temperature before starting a measurement.

2.3 Protein Preparations

Typically, 5–20 µg of membrane transporter are needed for an SSM-based experiment. The target protein can be analyzed in a wide range of membrane systems:

1. Membrane vesicles or enriched membrane fragments from native tissues.

2. Membrane vesicles/fragments with overexpressed protein from recombinant expression in bacteria and mammalian cell lines.

3. Recombinantly produced protein reconstituted into proteoliposomes.

The protein preparation can be diluted with the non-activating solution. The dilution factor has to be optimized empirically for each protein preparation (see **Note 6**).

For long term storage, the protein preparations can be flash frozen in a resting buffer and stored in liquid nitrogen or in a freezer at −80 °C for months.

2.4 Cleaning Solutions

1. 30 % (v/v) ethanol in ultrapure water. Use ethanol puriss. p.a. absolute (≥ 99.8 %).

2. 1 % (w/v) Edisonite (universal detergent) in ultrapure water. Weight 1 g Edisonite and dissolve in 100 mL ultrapure water. Warm up the Edisonite solution to 43 °C before use.

3 Methods

We now describe the typical procedures for an SSM-based electrophysiological experiment. Commercial instruments for SSM-based electrophysiology are available (Nanion Technologies, Munich, Germany). As mentioned above, these instruments are based on the SURFE^{2}R technology, which was originally developed by IonGate Biosciences (Frankfurt am Main, Germany). A description of the SURFE^{2}R technology can be found in refs. [6, 16]. Here we report the procedures followed in our laboratory using the SURFE^{2}R^{one} instrument.

3.1 SSM Preparation

The gold sensor (Nanion Technologies, Munich, Germany) consists of a circular gold electrode of 3 mm diameter on a glass base (Fig. 2).

1. Fill the sensor well with 50 μL of 1 mM octadecanethiol solution in propanol.

2. Incubate for 30 min at room temperature in a closed petri dish.

3. Remove the octadecanethiol solution by tapping the sensor on a tissue, holding it upside down.

4. Rinse the sensor with ultrapure water. Repeat this step three times.

5. Dry the sensor thoroughly in a stream of nitrogen gas.

6. Apply 3.5 μL of 7.5 mg/mL diphytanoylphosphatidylcholine solution in decane to the surface of the thiolated sensor with-

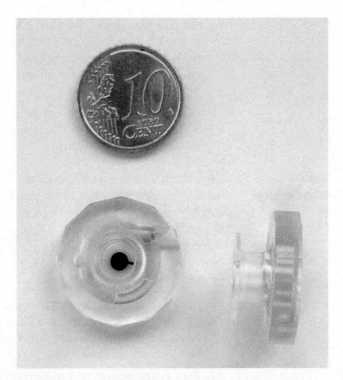

Fig. 2 Gold sensor for the SURFE²R^{one} instrument. The figure shows a circular gold electrode of 3 mm diameter on a glass base in the sensor well. The sensor chip is designed to fit the SURFE²R^{one} flow cell

out touching the gold surface with the pipette tip. Make sure that the gold surface is fully covered by the phospholipid (*see* **Note 7**).

7. Immediately after adding the phospholipid solution, fill the sensor well with 50 μL of non-activating solution. The composition of the applied solution depends on the target protein.

8. Incubate for ~2 h at room temperature (*see* **Note 8**).

3.2 Measuring the SSM Electrical Parameters

To check the correct formation of the SSM, its capacitance and conductance are measured using the SURFE²R^{one} system. Each step described below and in the following sections is performed by selecting the respective command from the sequence pull-down menu in the SURFE²R^{one} main window (*see* **Note 9**).

1. Fill the system with the non-activating and the activating solutions. All measuring solutions have to be at room temperature before starting the experiment to avoid electrical artifacts.

2. Mount the sensor in the flow cell.

3. Determine capacitance and conductance (*see* **Note 10**).

4. Repeat capacitance and conductance measurements three times (every 2–3 min) to check reproducibility. The capacitance and

the conductance of the SSM have characteristic values ranging from 0.2 to 0.5 μF/cm² and from 20 to 50 nS/cm², respectively. A low capacitance below 0.1 μF/cm² and a high conductance above 100 nS/cm² indicates that the SSM is not properly formed.

5. If the electrical parameters are not in the optimal range, the SSM sensor should be discarded and another SSM prepared, using a new gold electrode.

3.3 Control for Solution Exchange Artifacts

After the electrical parameters of the SSM have been measured, the measuring solutions should be tested on the protein-free SSM to check the presence of solution exchange artifacts. The solutions are driven through the flow cell containing the SSM sensor by applying pressure to the solution containers connected to the cell (*see* **Note 11**). The solution flow is controlled by electromechanically operated valves.

1. Applying the solution exchange protocol which you want to use in your experiment, perform the substrate concentration jump on the protein-free SSM and acquire the baseline and background electrical noise.

2. Repeat the substrate concentration jump at least four times. Check the presence of electrical artifacts due to solution exchange.

3. Apart from the artifacts due to mechanical valve switching, ideally no electrical artifacts should be observed or they should be much smaller than the expected protein signals (*see* **Note 12**).

3.4 Addition of Protein Sample

Usually the protein samples, e.g., membrane fragments/vesicles or proteoliposomes, are stored at −80 °C. Before use, the protein samples, typically aliquots of 10–20 μL, are thawed on ice.

1. Unmount the SSM sensor from the flow cell.

2. Dilute the protein containing suspension with the non-activating solution (used for the assembly of the SSM) to a final concentration of 0.2–1 mg/mL. As mentioned above, the dilution factor has to be optimized empirically for each membrane preparation (*see* **Note 13**).

3. Sonicate the protein suspension with a tip sonicator or ultrasonic bath sonicator (*see* **Note 14**).

4. Carefully remove the non-activating solution from the sensor well with a pipette. Leave approximately 40–50 μL of solution in the sensor well, so that the gold surface is covered by a solution layer.

5. Add 40 μL of the diluted protein suspension to the SSM sensor using a pipette. Do not touch the sensor surface with the pipette tip. Avoid bubble formation at the tip and near the sensor surface.

6. The membrane fragments/vesicles or proteoliposomes are allowed to adsorb on the SSM for about 1 h at room temperature (*see* **Note 15**). Make sure that the SSM sensor does not dry out during incubation, e.g., by placing the sensor in an air-tight box or a box with a damp tissue.

3.5 Concentration Jump Experiment and Signal Recording

After adsorption of the membrane entities (containing the protein of interest) on the SSM surface, the protein can be activated by a concentration jump of a suitable substrate [1, 7, 15]. Charge displacement by the protein following the substrate concentration jump is recorded as a current transient (Fig. 3) and analyzed with the SURFE^{2}R^{one} system.

1. Mount the sensor (SSM with adsorbed membrane entities) in the flow cell.
2. Flush the flow cell with the non-activating solution.
3. Measure capacitance and conductance of the SSM with adsorbed membrane entities, as described in Subheading 3.2.
4. Apply the solution exchange protocol. Generally, a solution flow rate of 290 μL/s is used by applying a pressure of 0.2 bar. In most cases, the basic solution exchange protocol is employed, which consists of three steps (Fig. 3): (1) washing the flow cell with the non-activating solution (1 s); (2)

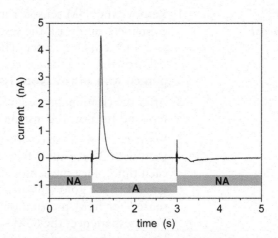

Fig. 3 Current transient following an ATP concentration jump on a SSM using the basic solution-exchange protocol. The displayed current is from native vesicles incorporating Ca-ATPase of sarcoplasmic reticulum. Exchange from non-activating (NA) to activating solution (A), which contains ATP, produces an ATP concentration jump at the SSM that induces charge displacement, resulting in a current transient (current amplitude of 4.5 nA at ~1.2 s) which can be detected in the external circuit. When ATP is removed, a small negative current transient is observed (at ~3.3 s), that is related to capacitance discharge [19]. Only the current upon addition of the substrate (ATP) is used for the analysis. The electrical artifacts at 1 and 3 s are due to mechanical valve switching

injecting the activating solution into the cell (1 or 2 s, depending on the target protein); (3) removing the activating solution from the cell with the non-activating solution (2 s) (*see* **Note 16**).

5. At the beginning of each set of measurements, ATP concentration jumps (initial control) can be carried out to test the activity of the ATPase protein adsorbed on the SSM. The first two measurements are usually rejected until the peak current remains constant.

6. Perform the concentration jump experiment with the activating substrate and verify the reproducibility of the current transients. Each measurement should be repeated at least four times and then averaged to improve the signal-to-noise ratio.

7. The ATP concentration jump can also be performed at the end of the set of measurements, and the initial and final ATP-induced current transients are then compared to rule out any loss of activity during the time of the experiment. If differences between the two transients are greater than ± 10 % the set should be discarded.

8. If possible try to validate your results by inhibiting the protein of interest at the end of the experiment. If a residual small signal is observed after protein inhibition, it is probably a solution exchange artifact (*see* **Note 17**).

3.6 After the Experiment

1. Remove the SSM sensor from the flow cell, empty all solution containers, and wash the system with plenty of ultrapure water and 30 % ethanol/water. The ethanolic solution avoids bacterial growth when the system is not in use. The system is finally purged with air (*see* **Note 18**).

2. After the cleaning procedure, release the pressure from the system and turn off the instrument.

3. Wash the sensor first with ultrapure water (four times) and then with 2-propanol (four times, using 50 µL of 2-propanol each time). Dry the sensor thoroughly in a stream of nitrogen gas. Store the gold sensors away from light. The sensors can be reused for SSM preparation as described in Subheading 3.1. From experience, the SSM sensors can usually be reused for at least two-three times.

4 Notes

1. Use only carefully cleaned glass containers, e.g., a volumetric flask or vial. Plastic containers should be avoided. Also use 2-propanol provided in glass bottles.

2. Glass containers for alkanethiol and phospholipid solutions should be cleaned with Piranha solution (the typical composi-

tion is 3:1 concentrated sulfuric acid to 30 % hydrogen peroxide), washed with plenty of ultrapure water and dried in a oven at 120 °C for approximately 3 h.

3. As a general rule, special care should be taken to keep non-activating and activating solutions as similar as possible. Small differences in ionic strength, pH and temperature may generate current artifacts of comparable magnitude and shape as the protein induced signals.

4. For instance, if Na^+ binding to the Na^+,K^+-ATPase is triggered by 100 mM NaCl in the activating solution, 100 mM choline chloride should be used in the non-activating solution as an inert compensation. However, if ATP is used as activating compound (µM concentration), a compensating inert compound is not required in the non-activating solution.

5. Lipophilic and amphiphilic compounds may generate large artifacts if used at high concentrations.

6. A proteoliposome suspension suitable for the SSM experiment has a typical lipid concentration of ~10 mg/mL and a lipid-to-protein ratio of ~5–10 (w/w). Membranes from native tissues are usually prepared at a total protein concentration of ~2–10 mg/mL.

7. Make a droplet of lipid solution at the pipette tip and carefully deposit the lipid droplet on the gold surface. As soon as the droplet touches the gold electrode, it spreads over the surface.

8. You may also prepare the SSM sensor the day before as described in **steps 1–7** and incubate the sensor in the non-activating solution overnight at room temperature.

9. The sequence menu in the main window of the SURFE^{2}R^{one} control software lists several commands, each of them performing a specific function.

10. To measure the SSM capacitance the system applies a triangular AC voltage (50 mV peak-to-peak amplitude, 0.5 Hz frequency) to the sensor circuit while recording the current. Ideally, the resulting current trace should resemble a square waveform with the amplitude being proportional to the capacitance of the SSM. To determine the SSM conductance, the system applies a 100 mV voltage jump to the sensor circuit while recording the current. The current decay after 1 s is used as a measure for the conductance of the SSM.

11. The solution flow rate can be chosen by the user via varying the pressure (0.1–0.5 bar) within a range of about 160–530 µL/s. Typically, a flow rate of 290 µL/s is used by applying a pressure of 0.2 bar.

12. If solution exchange artifacts are observed, try to optimize the solution composition before continuing the experiment, *see* Subheading 2.2.

13. From experience, it is favorable to use between 5 and 20 μg of protein for the adsorption on one SSM sensor.

14. In our laboratory we prefer to sonicate the protein containing suspension in an ultrasonic bath sonicator for 1 min. After the sonication step, the protein sample is briefly put back on ice.

15. The required time for a sufficient adsorption depends on the type of target protein and applied membrane entities. In general, 1 h incubation at room temperature is sufficient. However, in some cases an overnight incubation at 4 °C may be more convenient.

16. More complicated solution exchange protocols can also be established. For example, a double solution exchange protocol employs an additional resting solution, which is conducted through the flow cell after the non-activating/activating/non-activating solution exchange. Thus, in between experiments the SSM is incubated in resting solution, allowing the establishment and/or maintenance of ion gradients.

17. The signal generated by the protein should be corrected for the observed solution exchange artifact, i.e., the small current artifact can be subtracted from the protein-induced signal.

18. We advise to clean monthly the system using the Edisonite solution, whose preparation is described in Subheading 2.4. The solution should be used at a temperature of 43 °C.

Acknowledgments

The authors acknowledge with gratitude financial support from the Italian Ministry of Education, University and Research (MIUR) and Ente Cassa di Risparmio di Firenze.

References

1. Tadini-Buoninsegni F, Bartolommei G, Moncelli MR, Fendler K (2008) Charge transfer in P-type ATPases investigated on planar membranes. Arch Biochem Biophys 476:75–86

2. Ganea C, Fendler K (2009) Bacterial transporters: charge translocation and mechanism. Biochim Biophys Acta—Bioenerg 1787:706–713

3. Grewer C, Gameiro A, Mager T, Fendler K (2013) Electrophysiological characterization of membrane transport proteins. Annu Rev Biophys 42:95–120

4. Tadini-Buoninsegni F, Bartolommei G, Moncelli MR, Guidelli R, Inesi G (2006) Pre-steady state electrogenic events of Ca^{2+}/H^+ exchange and transport by the Ca^{2+}-ATPase. J Biol Chem 281:37720–37727

5. Bartolommei G, Tadini-Buoninsegni F, Moncelli MR, Gemma S, Camodeca C, Butini S et al (2011) The Ca^{2+}-ATPase (SERCA1) is inhibited by 4-aminoquinoline derivatives through interference with catalytic activation by Ca^{2+}, whereas the ATPase E_2 state remains functional. J Biol Chem 286:38383–38389

6. Kelety B, Diekert K, Tobien J, Watzke N, Dörner W, Obrdlik P et al (2006) Transporter assays using solid supported membranes: a novel screening platform for drug discovery. Assay Drug Dev Technol 4:575–582

7. Pintschovius J, Fendler K (1999) Charge translocation by the Na$^+$/K$^+$-ATPase investigated on solid supported membranes: rapid solution exchange with a new technique. Biophys J 76:814–826

8. Pintschovius J, Fendler K, Bamberg E (1999) Charge translocation by the Na$^+$/K$^+$-ATPase investigated on solid supported membranes: Cytoplasmic cation binding and release. Biophys J 76:827–836

9. Gramigni E, Tadini-Buoninsegni F, Bartolommei G, Santini G, Chelazzi G, Moncelli MR (2009) Inhibitory effect of Pb^{2+} on the transport cycle of the Na$^+$, K$^+$-ATPase. Chem Res Toxicol 22:1699–1704

10. Bartolommei G, Moncelli MR, Rispoli G, Kelety B, Tadini-Buoninsegni F (2009) Electrogenic ion pumps investigated on a solid supported membrane: comparison of current and voltage measurements. Langmuir 25:10925–10931

11. Tadini-Buoninsegni F, Pintschovius J, Cornelius F, Bamberg E, Fendler K (2000) K$^+$ induced charge translocation in the phosphoenzyme formed from inorganic phosphate. In: Kaya S, Taniguchi K (eds) Na/K-ATPase and related ATPases. Elsevier, Amsterdam, pp 341–348

12. Tadini-Buoninsegni F, Bartolommei G, Moncelli MR, Pilankatta R, Lewis D, Inesi G (2010) ATP dependent charge movement in ATP7B Cu$^+$-ATPase is demonstrated by pre-steady state electrical measurements. FEBS Lett 584:4619–4622

13. Tadini-Buoninsegni F, Bartolommei G, Moncelli MR, Inesi G, Galliani A, Sinisi M et al (2014) Translocation of platinum anticancer drugs by human copper ATPases ATP7A and ATP7B. Angew Chem Int Ed Engl 53:1297–1301

14. Seifert K, Fendler K, Bamberg E (1993) Charge transport by ion translocating membrane proteins on solid supported membranes. Biophys J 64:384–391

15. Schulz P, Garcia-Celma JJ, Fendler K (2008) SSM-based electrophysiology. Methods 46:97–103

16. Geibel S, Flores-Herr N, Licher T, Vollert H (2006) Establishment of cell-free electrophysiology for ion transporters: application for pharmacological profiling. J Biomol Screen 11:262–268

17. Garcia-Celma JJ, Hatahet L, Kunz W, Fendler K (2007) Specific anion and cation binding to lipid membranes investigated on a solid supported membrane. Langmuir 23:10074–10080

18. Bazzone A, Costa WS, Braner M, Călinescu O, Hatahet L, Fendler K (2013) Introduction to solid supported membrane based electrophysiology. J Vis Exp 75:e50230

19. Tadini Buoninsegni F, Bartolommei G, Moncelli MR, Inesi G, Guidelli R (2004) Time-resolved charge translocation by sarcoplasmic reticulum Ca-ATPase measured on a solid supported membrane. Biophys J 86:3671–3686

Chapter 27

Electrophysiological Characterization of Na,K-ATPases Expressed in *Xenopus laevis* Oocytes Using Two-Electrode Voltage Clamping

Florian Hilbers and Hanne Poulsen

Abstract

The transport of three Na^+ per two K^+ means that the Na,K-ATPase is electrogenic, and though the currents generated by the ion pump are small compared to ion channel currents, they can be measured using electrophysiology, both steady-state pumping and individual steps in the transport cycle. Various electrophysiological techniques have been used to study the endogenous pumps of the squid giant axon and of cardiac myocytes from for example rabbits. Here, we describe the characterization of heterologously expressed Na,K-ATPases using two-electrode voltage clamping (TEVC) and oocytes from the *Xenopus laevis* frog as the model cell. With this system, the effects of particular mutations can be studied, including the numerous mutations that in later years have been found to cause human diseases.

Key words Two-electrode voltage clamping, TEVC, Electrophysiology, Steady-state currents, Pre-steady-state currents, *Xenopus laevis* oocytes

1 Introduction

The asymmetric distribution of sodium and potassium across the plasma membrane of animal cells depends on the Na,K-ATPase [1]. The transport cycle of the pump is described by the Post-Albers scheme [2], according to which the pump alternates between two states, E1 and E2, with high affinity for Na^+ and K^+, respectively, that enable three Na^+ to be exported from the cell and two K^+ to be imported into it at the expense of one ATP molecule (Fig. 1). The 3:2 ratio has made it obvious to study the activity and ion transport mechanisms of the Na,K-ATPase with electrophysiological methods, and for the past 30 years, patch clamping of single cardiac myocytes [3], insertion of voltage clamp electrodes into the giant squid axon [4], and heterologous expression of cloned cDNAs in oocytes from the African clawed frog *Xenopus laevis* [5] have been used. The heterologous expression of Na,K-ATPases

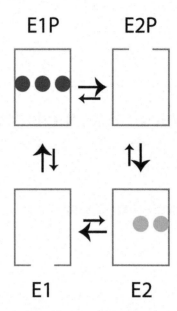

Fig. 1 Simple Post-Albers scheme for the Na,K-ATPase. In the E1 state, the Na,K-ATPase is open to the cytoplasm for binding of three Na$^+$ (*bottom left*). Upon binding and phosphorylation from ATP, the E1P state with three occluded Na$^+$ (*red spheres*) is reached (*top left*). Opening to the extracellular side and release of the three Na$^+$ leaves a pump ready for binding of two K$^+$ (*top right*). Upon binding and dephosphorylation, two K$^+$ are occluded in the E2 state (*bottom right*), and with K$^+$ release, the pump is ready for another cycle. When pre-steady-state currents are determined, the pump is restricted to the E1P-E2P part of the cycle (Fig. 3). Steady-state currents reflect full cycling (Fig. 2)

allows the effects of specific mutations to be examined, which has shed light on many aspects of the basic pump mechanism, but has also been of particular interest in recent years, since it provides an opportunity to study the molecular mechanisms of various diseases caused by mutations in the Na,K-ATPase genes [6–8].

The currents specifically attributable to the pump can be determined by the use of Na,K-ATPase inhibitors, which have been developed by both plants and animals as defense mechanisms. Some pumps have been found to have much lower sensitivity to the inhibitors than others, and the amino acid differences in the inhibitor binding pocket that cause the difference can be used for heterologously expressed pumps: with the naturally occuring resistance mutations inserted into the pump, millimolar concentrations of the inhibitor ouabain are required for inhibition, while the endogenous Na,K-ATPase expressed in the oocytes can be silenced by ouabain in nanomolar concentrations [9].

The steps in the catalytic cycle are in principle reversible, but the low cytoplasmic concentration of ADP blocks an E1P to E1 transition (Fig. 1) in oocytes. If K^+ is omitted from the extracellular side, the pump is therefore restricted to binding and release of Na^+. The main electrogenic event in the catalytic cycle is the release of the third Na^+, while the two other Na^+ travel through a smaller part of the membrane potential when released. Hyperpolarization promotes Na^+ binding (the positive charges are dragged into the pump by the negative membrane potential), and depolarization promotes Na^+ release. Therefore, the membrane potential will determine the equilibrium between pumps in the E1P and E2P states. Between the two conformations an exit pathway opens, which may be viewed as a restricting channel that the ions have to pass before they can reach the extracellular side.

The main drawback of the TEVC technique is that the time resolution is only down to 1–2 ms, while other electrophysiology techniques offer sub-ms resolution, but the release and binding of the third sodium ion is slow enough for reliable measurement by TEVC. With a K^+-free, Na^+-containing buffer on the extracellular side, pre-steady-state currents can therefore be determined and the E1P-E2P equilibrium analyzed.

With K^+ on the extracellular side, the pump will be fully active. The Na,K-ATPase turnover rate depends on the isoform used, the experimental setup, the ionic conditions and the temperature. For optimal pumping, the intracellular Na^+ and extracellular K^+ concentrations should be high, the membrane potential depolarized and the temperature 37 °C. Under these conditions, the turnover may be more than 100/s, but the high expression levels in oocytes allow measurements of smaller currents as well. Because of the voltage dependence of the E1P-E2P equilibrium, extracellular Na^+ can compete with K^+ for binding at hyperpolarized potentials and inhibit pumping. Na^+ can be omitted from the perfusion buffer by replacing it with a non-transported cation, e.g., NMDG or TMA. The measured K^+ affinity is higher in the absence of Na^+, and the inhibition of the pump current by hyperpolarization smaller. However, an inwardly rectifying proton current is pronounced in the absence of extracellular Na^+ and K^+. To determine the forward pumping, it is therefore important to determine currents in the presence of K^+ relative to currents in the presence of ouabain and not relative to currents in the absence of K^+.

Thus, from heterologous expression of Na,K-ATPases in *Xenopus* oocytes, the effects on pump activity and kinetics of different isoforms and mutations can be determined.

In the following sections, we describe how to prepare oocytes from *Xenopus laevis*, inject mRNA encoding Na,K-ATPase α- and β-subunits, measure their currents using TEVC, and analyze the data.

2 Materials

Prepare all solutions using ultrapure water (resistivity of 18 MΩ cm at 25 °C) and analytical grade reagents. Prepare and store all reagents at room temperature unless otherwise noted. Follow waste disposal regulations when disposing waste materials.

2.1 Oocyte Preparation

1. Immersion anesthetic for *Xenopus laevis*: 0.2 % (w/v) ethyl-m-aminobenzoate-methanesulfonate salt. Add 1 g of the anesthetic to 500 mL water.
2. ND96: 96 mM NaCl, 2 mM KCl, 1 mM $MgCl_2$, 1.8 mM $CaCl_2$, 5 mM Hepes, pH 7.4. Add about 600 mL water to a 1 L glass beaker and stir with a magnetic stirrer. Weigh NaCl, KCl and Hepes and transfer to the beaker. Adjust the pH with NaOH and add water to about 950 mL. Add 1 mL of a 1 M $MgCl_2$ solution and 1.8 mL of a 1 M $CaCl_2$ solution. Check pH and adjust the volume to 1 L. In the ND96, we further include 0.5 mL 40 mg/mL gentamicin to counter bacterial infection, and 2.5 mM pyruvate as energy supply for the ATPase expressed (*see* **Note 1**).
3. Liberase solution: add 2 mL of water directly to 5 mg Liberase (Roche Diagnostics). Keep all components on ice and make 70 μL aliquots for storage at −20 °C. (*see* **Note 2**).

2.2 Generation of mRNA for Injection

2.2.1 Plasmid Digest

1. Two plasmids encoding the ATPase α and β subunits with a T7 (or SP6) promoter in front.
2. An appropriate restriction enzyme and its reaction buffer.

2.2.2 Plasmid Purification by Phenol Extraction

1. Tris saturated phenol pH 7.9 (*see* **Notes 3** and **4**).
2. Chloroform.
3. Ethanol (EtOH), p.a.
4. 70 % EtOH.
5. 3 M NaAc pH 5.2.
6. RNAse-free water.
7. The mMessage mMachine® Ultra transcription kit (Ambion), to generate mRNA from the linearized plasmid.

2.3 Solutions for TEVC Measurements

Before use, filter solutions through a 0.22 μm filter.

1. Na buffer: 115 mM NaOH, 110 mM sulfamate, 1 mM $MgCl_2$, 0.5 mM $CaCl_2$, 5 mM $BaCl_2$, 10 mM Hepes, 1 μM ouabain, pH 7.4. To about 800 mL water in a 1 L glass beaker, NaOH, sulfamate, $BaCl_2$ and Hepes are added and stirred. Adjust the pH to 7.4 with sulfamate, add 1 mL 1 M $MgCl_2$, 0.5 mL $CaCl_2$ and 100 μL 10 mM ouabain. Adjust the volume to 1 L and check the pH (*see* **Notes 5** and **6**).

2. NMDG buffer: the same as the sodium buffer but with 115 mM NMDG replacing the NaOH.
3. Potassium buffer: the same as the sodium buffer but with 115 mM KOH replacing the NaOH.
4. 15 mM potassium solutions: mix 13 mL of the potassium buffer with 87 mL of either the sodium (giving the Na + K buffer) or the NMDG buffer (giving the NMDG + K buffer).
5. 10 mM ouabain solutions: Add 0.3 g ouabain to 40 mL of the Na buffer, the Na + K buffer, the NMDG buffer and the NMDG + K buffer. Leave to stir until all of the ouabain has been dissolved, this may take 30 min.
6. Sodium loading buffer: 95 mM NaOH, 90 mM sulfamate, 5 mM Hepes, 10 mM TEACl, 0.1 mM EGTA, pH 7.6.

2.4 Hardware for RNA Injection and TEVC Measurements

1. Glass capillaries for Nanoliter 2000 (#4878 WPI).
2. Glass capillaries that are 1.5 OD, 1.17 mm × 100 mm (Harvard Apparatus GC150TF-10).
3. A PC-10 puller (Narishige).
4. Mineral oil.
5. A syringe with a 0.8 mm needle.
6. 20 µL microloader pipette tips.
7. Fine-graded sand paper.
8. An OC-725C voltage-clamp apparatus (Warner Instruments Corp.), Clampex 10.4, and a Digidata 1440A (Molecular Devices).

2.5 Data Analysis Software

1. Clampfit 10.4 and GraphPad Prism 6.

3 Methods

Carry out all procedures at room temperature unless otherwise specified.

3.1 Preparation of Oocytes (See Note 8)

1. Place an adult female *Xenopus laevis* in the anesthetic solution for 15 min. If the frog is not fully sedated, it will react to being placed on its back and must be returned for an additional 5 min to the anesthetic solution.
2. Clean the surfaces, use gloves, use sterile instruments, have a petri dish with ND96 ready and take a vial of Liberase from the freezer.
3. Place the frog on its back on an absorbent pad, keeping the back wet. Using scalpel and forceps, lift the skin on one side of the lower third of the body and make an incision of about two

centimeters. Make an additional incision through the muscle layer underneath the skin, extract the ovaries using two forceps and place them in the petri dish with ND96 buffer (*see* **Note 9**). To sacrifice the animal, make an incision just beneath the chest and remove the heart (*see* **Note 10**).

4. Using two sharp forceps, separate the ovaries into fragments with approximately ten oocytes in each. Transfer the oocytes to a 50 mL tube and discard the buffer above, which will be dirty and cloudy.

5. Wash the oocytes by adding 20 mL ND96, gently invert the tube and discard the buffer. Repeat until the buffer is clear after inversion (typically two to four times). Discard most of the buffer, leaving 2–5 mL above the oocytes.

6. Add 70 µL of 2.5 mg/mL Liberase.

7. Shake the oocytes mildly for 1–3 h.

8. Wash the oocytes with ND96 until the buffer is clear, transfer the oocytes to a petri dish and inspect them using a stereo microscope. If the majority of the oocytes are still wrapped in the follicular layer (a transparent membrane, possibly with visible blood veins), repeat the Liberase incubation and check after 30 min that this layer has been removed on the majority of the oocytes (*see* **Note 11**).

9. Select stage V and VI oocytes using a stereo microscope. With a plastic pasteur pipette, large (1 mm or more in diameter), round oocytes can be picked and transferred to a novel petri dish with ND96 buffer for storage. One half of an oocyte, the animal pole, is darker than the other half, the vegetal pole. Select oocytes with uniform coloring of the two hemispheres.

10. The selected oocytes are ready for injection and can be stored for a couple of days at 16–18 °C before injection (*see* **Note 12**).

3.2 Plasmid Digest

1. Two plasmids encoding the ATPase α and β subunits with a T7 (or SP6) promoter in front are linearized after the stop codon. In a typical reaction, 3 µg of plasmid is digested in a 30 µL reaction with 3 µL 10× buffer and 1 µL of fast digest enzyme at 37 °C for 30 min.

3.3 Plasmid Purification by Phenol Extraction

1. Add 170 µL water to the restriction reaction and 200 µL Tris saturated phenol pH 7.9 (*see* **Notes 3** and **4**).

2. Shake or vortex vigorously, centrifuge at $10,000 \times g$ for 5 min, remove the top (aqueous) phase to a new vial and discard the phenol. Be careful not to mix the phases and to take only the aqueous phase.

3. Add 200 µL chloroform, shake or vortex vigorously and centrifuge at 13,000 rpm for 3 min. Carefully remove the top phase to a new vial and discard the chloroform.

4. Add 20 μL 3 M NaAc pH 5.2, 500 μL EtOH and incubate at −20 °C for 15 min or longer.

5. Centrifuge at 4 °C, 13,000 rpm for 20 min. A small pellet of the precipitated DNA may be visible.

6. Carefully remove the solution, add 300 μL 70 % EtOH, centrifuge at 4 °C, 13,000 rpm for 5 min and carefully remove the ethanol.

7. Let the pellet air dry and then redissolve it in 8 μL RNAse-free water. Be sure that no residual ethanol is present.

8. Determine the DNA concentration using for example a NanoDrop spectrophotometer. The digested DNA can be stored at −20 °C.

9. Using the mMessage mMachine® Ultra transcription kit (Ambion), mRNA is generated from the linearized plasmid according to the manufacturers guidelines. Quarter reactions with 0.5–1 μg DNA typically yield 5–15 μg injection-ready mRNA. For injections, a mix of the two subunits with final concentrations of 0.2 g/L of α and 0.04 g/L of β are typically used. Aliquots of the mRNA are stored at −80 °C.

3.4 Preparing Glass Capillaries

Settings on the puller will need to be adjusted for the individual instrument, and the parameters may change over time. The capillaries rest on a line of plasticine.

1. For mRNA injection, we pull glass capillaries for Nanoliter 2000 on a PC-10 puller. After pulling, the capillary has a long (1 cm) narrowing of the tube and is sealed at the tip. Under a stereomicroscope, two capillaries are used to break the tip and open the capillary. With the slightly thicker part of one capillary, the outermost part of the other capillary is hit, and the shortened capillary is then used to shorten the other. A small hole makes it harder to draw the mRNA into the capillary, but also makes the subsequent injection less invasive for the oocyte, so the size of the hole is a trade-off that needs optimization.

2. A capillary is filled with mineral oil using a syringe with a 0.8 mm needle. The capillary is mounted on a Nanoliter 2000, and 2 μL of the mRNA mix is drawn.

3. For TEVC measurements, glass capillaries that are 1.5 OD, 1.17 mm × 100 mm are pulled on the PC-10 puller. Two capillaries are filled half way with a 3 M NaCl solution using 20 μL microloader tips mounted on a plastic pasteur pipette with the electrolyte and sealed with Parafilm. Air bubbles at the tip should carefully be avoided and removed, since they will hinder current flow.

4. The electrode silver wire is cleaned with fine-grained sand paper and chlorided by letting a current (0.15 mAmp) pass

through it in a 1 M KCl bath (*see* **Note 7**). After less than a minute, the chloridation will have given the wire a light white or grey color indicating that an AgCl layer has been formed on the wire.

5. The capillary is mounted on the micromanipulator, ensuring that the silver wire is in contact with the electrolyte. The tip resistance of the measuring electrodes should be around 0.7 MΩ.

3.5 Injection of mRNA

1. Oocytes are placed in a petri dish with ND96 and either a mesh at the bottom or a plate with a groove to stabilize their position. Using a stereo microscope, inject 50 nL of mRNA into each oocyte. Aim for injecting in the vegetal pole (the oocytes will naturally align with the vegetal pole at the top), but not at the very top, where there is a higher risk of hitting the nucleus rather than the cytoplasm.

2. The oocytes are incubated for 2–7 days at 12–16 °C to allow for expression of the Na,K-ATPase.

3.6 Measurement of Pump-Generated Currents

1. Raising the intracellular concentration of Na^+ in the oocytes before the measurements will typically give a larger current. 0.5–2 h before measurements, the injected oocytes can therefore be placed in sodium loading buffer.

2. Set the rate of all perfusion buffers to 1 drop/s and ensure that there are no bubbles in the tubings of the perfusion system. The eight perfusion buffers are Na, Na+K, Na+ouabain, Na+K+ouabain, NMDG, NMDG+K, NMDG+ouabain, NMDG+K+ouabain.

3. Offset of the current and potential electrodes are adjusted to zero and appropriate resistances (approximately 0.7 MΩ) ensured.

4. Place an oocyte in the recording chamber in Na buffer, approach with the electrodes until it gives a slight indent in the membrane. Gently tip the micromanipulator to insert the electrodes and turn on the clamp with a holding potential of −30 mV. The current should be close to 0 nAmp (−40–0 nAmp). Within the first couple of minutes, the seal between the oocyte membrane and the electrode may improve.

5. Test if changing to superfusion with the Na+K buffer gives an increase in current (*see* **Note 13**). If it does, return to the Na buffer. Record the current response to a series of voltage jumps starting with 20 mV steps between −140 and +20 mV. A broader range of voltage jumps may be allowed after the first series of recordings. Record the response to the same voltage jump series when superfusing the oocyte with buffers in this order: Na, Na+K, Na, Na+ouabain, Na+K+ouabain, Na.

Allow full response to each buffer before recording, this may take a couple of minutes. Check that the I/V curves for the three recordings in Na are identical. The series may be repeated with a wider range of voltages.

6. Record the response to the voltage jump series when superfusing the oocyte with buffers in this order: NMDG, NMDG+K, NMDG, NMDG+ouabain, NMDG+K+ouabain, NMDG (*see* **Note 14**). Confirm that the I/V curves for the three recordings in NMDG are identical.

3.7 Data Analysis

3.7.1 Analysis of Steady-State Currents

1. Generate I/V curves in Clampfit for the recordings at the different conditions, making sure that all Na or NMDG curves overlay.

2. The differences between the measurements in the buffers Na+K and Na+K+ouabain give the Na,K-ATPase specific steady-state current. For normalization, we use a value close to the maximal forward pumping, typically the value at 20 mV. For an example of steady state currents *see* Fig. 2.

3. The Na,K-ATPase specific steady-state current in the absence of extracellular sodium is similarly determined as the difference between the I/V curves in NMDG+K and NMDG+K+ouabain. Furthermore, the inward current is determined from the I/V curves in NMDG and NMDG+ouabain, which are also normalized to the maximal forward pumping [10] (*see* **Note 15**).

4. The voltage-dependent turnover may be determined as the pumping current divided by the total number of pumps expressed in that given oocyte. The latter can be estimated from the pre-steady-state currents, since the total charged moved, Q_{tot}, is equal to the number of available pumps, N, times the charge that each of them moves, $Z_q \times e$, where the elementary charge is 1 for one transported sodium ion, and the steepness factor Z_q is determined from the charge translocation curve (cf. **step 7** in Subheading 3.7.2).

3.7.2 Analysis of Pre-steady-state Currents

1. Pre-steady state currents are determined from the difference curves obtained by subtracting Na+ouabain from Na measurements (Fig. 3). For reliable data, it is preferable to use a Na recording and an immediately following Na+ouabain recording.

2. There are pre-steady-state currents both at the beginning of a voltage jump, and when the voltage returns to the holding potential. These are called the 'on' and 'off' transient currents.

3. Both the 'on' and 'off' transient currents can be fitted with single exponentials: $f(t) = \sum_{n_i}^{i=1} A_i t e^{-t/\tau_i} + C$ with A being the amplitude, t the time, τ the time constant, and C a constant.

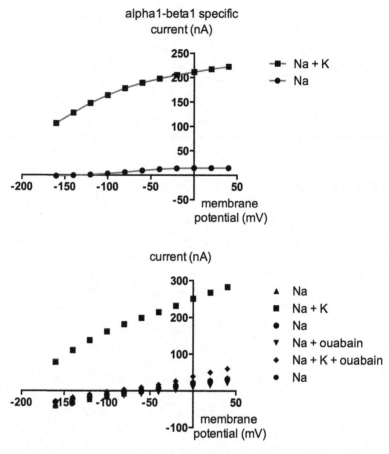

Fig. 2 Steady-state currents of human α1-β1 Na,K-ATPase. *Top*: the I/V curves measured as described in Subheading 3.4 in the indicated buffers. The three measurements in Na perfusion buffer are sufficiently similar. Note that at depolarized potentials, the Na + K + ouabain curve is slightly above the Na and Na + ouabain curves, indicating the presence of voltage-gated K+ channels in the oocyte. *Bottom*: the steady-state current of the human α1-β1. The graphs were calculated from the data shown in the top graph, Na + K: (Na + K) – (Na + K + ouabain) and Na: Na – (Na + ouabain)

A fit is made between two cursors: one is placed as close to the voltage jump as possible, but after the immediate effects of the relatively slow clamp, the other is placed towards the end of the slope. Carefully note that the fitted curves follow the recorded curves.

4. The 'on' rate constants (located at $x = 108$ ms in Fig. 3) are determined from the τ values, usually as $1000/\tau$ (τ given in ms) (*see* **Note 16**).

5. The 'off' rate constants (located at $x = 308$ ms in Fig. 3) should give a similar value, since all jumps are to the same potential, i.e., to the same equilibrium. This value can be used in the

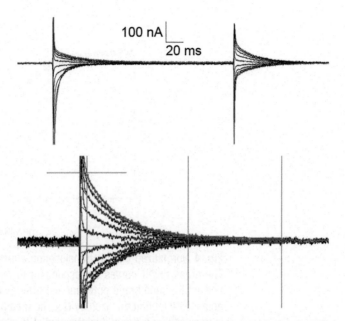

Fig. 3 Pre-steady-state currents of human α1-β1 Na,K-ATPase. *Top*: trace from a Na recording minus a (Na+ouabain) recording. The 'on' pulse started at 108 ms, the 'off' pulse at 308 ms. A 20 mV voltage step protocol from −140 mV to 40 mV (*red*) was applied. *Bottom*: zoom in of the 'off' currents shown above with exponential fitting in *blue*. The fittings were done between the two *middle* cursors and extended to the two *outer* cursors (the *left* is at the start of the 'off' pulse)

curve of rate constants at the x-value corresponding to the holding potential.

6. The charge translocation curve is determined by plotting the area under the pre-steady-state curves against the membrane potential. The area may be calculated as the A value at the time of the voltage jump (possibly corrected for C if the y-offset is significant) times τ. Charge translocation curves for 'on' and 'off' currents should be similar.

7. Fitting of the charge translocation curve with a Boltzmann equation yields the total amount of charge being transported, Q_{tot}, the steepness factor, Z_q, which reflects the penetration depth of the third sodium ion through the dielectric of the membrane (typically 0.8), and the midpoint potential V_{50}. With F, R and T for the Faraday constant, the molar gas constant and absolute temperature: $Q/Q_{tot} = 1/(1 + \exp((Z_q \times F \times (V - V_{50}))/(R \times T)))$.

8. A reasonable Boltzmann fit depends on both the top and bottom plateaus being approached by the raw data. In the pre-steady-state curves, this is seen by the curves coming closer together at both sides (*see* **Note 17**).

9. To compare different oocytes, the charge translocation values are normalized by Q_{tot} (Fig. 4).

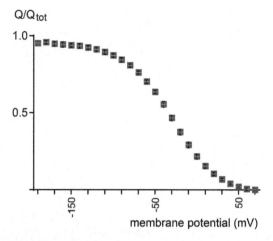

Fig. 4 Normalized charge translocation curve for human α1-β1 Na,K-ATPase. The areas under curves corresponding to the blue fits in Fig. 3 were plotted against the membrane potential and fitted with a Boltzmann equation, and the areas were then normalized by Q_{tot}. The membrane potential at which half of Q_{tot} is translocated is the midpoint potential V_{50}

4 Notes

1. A 10× stock of ND96 can be prepared for ease.
2. Removal of the follicular layer with approximately 1 % (v/v) of a 2.5 mg/mL Liberase solution does not require calcium-free buffers. ND96 can therefore be used throughout the oocyte preparation and storage.
3. Do not use phenol if it is turning pink. Oxidation of the phenol gives a reddish compound that can cause DNA nicking. When preparing the phenol, it is a good idea to aliquot it into 15 or 50 mL tubes and store them in the freezer. Frozen phenol will take a long time to thaw; it is therefore preferable to leave it at room temperature over night.
4. It is important to use phenol saturated with a buffer of slightly alkaline pH (usually Tris–HCl pH 7.9) when extracting DNA. Acidic phenol can cause DNA denaturation and partition of the DNA into the organic phase, causing the DNA to be lost in the procedure.
5. When preparing the solutions for measurements, adding less than the calculated amount of sulfamate and using a sulfamate solution to lower the pH to the desired value ensures that the concentrations of sodium, potassium and NMDG are precise.
6. Using a 1 M solution of $CaCl_2$ and adding it in the end reduces precipitation.

7. Alternatively, silver wires may be chlorided by immersion in bleach for 15–30 min.
8. Oocytes ready for injection can be bought from EcoCyte Bioscience, that delivers day-to-day in USA and Europe.
9. Ovaries can be removed from both sides if needed, but one side typically contains thousands of oocytes.
10. If the frog is not to be sacrificed, close the incisions with simple interrupting suturing (1–3 stitches) using Vicryl sutures, first the muscle layer, then the skin layer. Place the frog in a small water bath with its head elevated to avoid drowning. After about 2 h, the frog will have regained consciousness and can be returned to the animal housing. After a month, the frog can be operated again on the other side.
11. Prolonged incubation with Liberase may destabilize the oocytes in later applications and should be avoided.
12. We routinely keep the oocytes slightly colder at 12 °C for increased durability.
13. If no expression is seen, the RNA injected may be damaged and should be checked on a gel. However, expression levels can vary significantly between oocyte batches, so it is a good idea to include a positive control by injecting an mRNA known to give robust currents.
14. Washing away potassium and ouabain typically takes longer with NMDG buffers than with Na buffers.
15. With Na^+ in the perfusion buffer, the inward current is usually negligible, but some mutations cause a significant inward current even in the presence of 115 mM Na^+. It is therefore always advisable to include the Na−Na+ouabain I/V curves, although they will generally lie close to the x-axis.
16. It may be difficult to fit the 'on' rates for a Na,K-ATPase with a very low V_{50}, since only little charge will be moved by a jump from a holding potential of −50 mV to 30 mV, if the pump has already released most of its Na^+ at −50 mV. A holding potential of −110 mV may destabilize the oocyte, but with an initial jump to −110 mV 100 ms before the voltage jump protocol, more charge will be released, and the fit will be more accurate.
17. If the V_{50} of a Na,K-ATPase is around 0 mV or higher, it may be difficult to obtain trustworthy charge translocation curves, since oocytes are often unstable at membrane potentials above 40 or 60 mV. This may be circumvented by using half the sodium concentration (i.e., 75.5 mM Na and 75.5 mM NMDG), which left-shifts the Q/V curve 15–20 mV, depending on the construct [4].

References

1. Skou JC (1957) The influence of some cations on an adenosine triphosphatase from peripheral nerves. J Am Soc Nephrol 23:394–401
2. Albers RW, Fahn S, Koval GJ (1963) The role of sodium ions in the activation of electrophorus electric organ adenosine triphosphatase. Proc Natl Acad Sci U S A 50(3):474–481
3. Glitsch HG (2001) Electrophysiology of the sodium-potassium-ATPase in cardiac cells. Physiol Rev 81(4):1791–1826
4. Holmgren M et al (2000) Three distinct and sequential steps in the release of sodium ions by the Na+/K+-ATPase. Nature 403(6772): 898–901
5. Noguchi S, Mishina M, Kawamura M, Numa S (1987) Expression of functional (Na+ + K+)-ATPase from cloned cDNAs. FEBS Lett 225(1-2):27–32
6. Poulsen H et al (2010) Neurological disease mutations compromise a C-terminal ion pathway in the Na(+)/K(+)-ATPase. Nature 467(7311):99–102
7. Azizan EA et al (2013) Somatic mutations in ATP1A1 and CACNA1D underlie a common subtype of adrenal hypertension. Nat Genet 45(9):1055–1060
8. Li M et al (2015) A functional correlate of severity in alternating hemiplegia of childhood. Neurobiol Dis 77:88–93
9. Price EM, Lingrel JB (1988) Structure-function relationships in the Na, K-ATPase alpha subunit: site-directed mutagenesis of glutamine-111 to arginine and asparagine-122 to aspartic acid generates a ouabain-resistant enzyme. Biochemistry 27(22): 8400–8408
10. Li C, Geering K, Horisberger JD (2006) The third sodium binding site of Na, K-ATPase is functionally linked to acidic pH-activated inward current. J Membr Biol 213(1):1–9

Part VI

Functional Studies by Cell Culture and Transgenic Animals

Chapter 28

Functional Studies of Na⁺,K⁺-ATPase Using Transfected Cell Cultures

Elena Arystarkhova and Kathleen J. Sweadner

Abstract

The properties of different combinations of Na,K-ATPase subunits or their mutations can be studied in stably transfected mammalian cells. As a specific example, the methods here are for transfection of a modulatory subunit into cells with endogenous α and β subunits. Renal Na,K-ATPase is tightly bound to a small single-span membrane protein, the γ subunit, or FXYD2. The protein co-localizes and co-immunoprecipitates with the α/β complex, however it is not required for basic enzyme properties. Functional consequences of association with FXYD2 were investigated in stably transfected cells. The outcome was that FXYD2 reduced activity of Na,K-ATPase at the level of apparent affinity for Na⁺ and to a smaller extent for K⁺. Moreover, expression of FXYD2 reduced cell growth. Here we describe the methodologies as well as potential pitfalls.

Key words Gamma subunit, FXYD, Stable transfectants, Na,K-ATPase activity, Affinity, Cell growth

1 Introduction

The crystal structure of Na,K-ATPase in both the sodium-bound and the potassium-bound states clearly revealed a three-subunit complex [1, 2]. A catalytic α subunit and a glycoprotein β subunit are obligatory for enzyme function and proper assembly (reviewed in [3]). Both subunits are encoded by multigene families, and their expression is developmentally regulated and species- and tissue-specific [4]. In several experimental systems exchange of either α or β isoforms affected enzymatic properties of the complex, most notably affinities for K⁺ and Na⁺ [5]. The significance of the third subunit, originally called the γ subunit, was a mystery for a long time. The key properties of the Na,K-ATPase (ouabain binding, enzymatic activity, and cation transport) can be obtained without it. It is also not required for α/β assembly and trafficking of functional units to the plasma membrane [6].

Based on structural homology, a family of different genes for γ subunit-like proteins was discovered. They are small (7–20 kDa)

single-span membrane proteins, named FXYD after a signature motif PFxYD [7]. There are seven FXYD proteins in mammals, distributed in a tissue- or cell-specific manner. Association of each of them with Na,K-ATPase changes functional properties of the enzyme in a different way (reviewed in [8]).

FXYD2, or the γ subunit, is expressed throughout the nephron except upper cortical thick ascending limb and collecting duct [9]. Notably, while the protein is abundant in kidney, it is absent from any established renal (or any other) cell line. Thus assessment of FXYD2's functional role for Na,K-ATPase was done in transfected cells. Most experiments described here were performed in the normal rat kidney epithelial cell line, NRK-52E cells [10, 11] or in C6 glioma cells, but the methods should also work in HEK 293, HeLa cell, and COS-1 cell lines.

2 Materials

2.1 Plasmid Construction

1. Total RNA isolation kit (Qiagen).
2. DEPC-treated (RNAse free) water.
3. cDNA synthesis kit (Life Technologies).
4. Agarose.
5. TAE buffer containing 40 mM Tris-acetate, pH 8.0, and 0.1 mM ethylenediaminetetraacetic acid (EDTA).
6. Ethidium bromide.
7. DNA gel-purification kit.
8. Plasmid suitable for mammalian cell transfection.
9. Restriction enzymes and buffers.
10. PCR- specific primers.
11. Ligation system.
12. Competent *E. coli* cells.
13. Bacterial agar.
14. Antibiotic for selection of transformants.
15. Bacterial LB broth.
16. Miniprep plasmid purification kit.
17. 37 °C *E. coli* incubator with shaker.

2.2 Transfection Protocol for Generation of Stable Cell Lines

1. DMEM medium (high glucose, without L-glutamine).
2. Fetal bovine serum.
3. Penicillin/streptomycin mixture: 100× stock contains 10,000 IU of penicillin and 10 mg of streptomycin.
4. L-glutamine.

5. Trypsin–EDTA.
6. Dulbecco's phosphate buffered saline, Ca^{2+} and Mg^{2+} free (here called simply PBS).
7. Opti-MEM medium.
8. Lipofectamine.
9. Geneticin (G418) antibiotic (0.5 mg/mL final concentration).
10. Tissue culture plates and flasks (see **Note 1**).
11. Hemacytometer.
12. Trypan blue.
13. Dimethylsulfoxide.

2.3 Characterization of Stable Clones

1. Nu-PAGE gel electrophoresis system (Life Technologies).
2. Reducing agent : 0.5 M dithiothreitol.
3. Nu-Page^R 4x-LDS sample buffer containing lithium dodecyl sulfate, pH 8.4.
4. Nu-Page MES SDS running buffer: 1×: 50 mM MES 2-(N-morpholino)ethanesulfonic acid, 50 mM Tris Base, 0.1 % SDS, 1 mM EDTA, pH 7.3.
5. Precast 4–12 % polyacrylamide gel plates.
6. Blot transfer buffer: 20 mM Tris-Cl, 150 mM glycine, 20 % methanol.
7. Nitrocellulose.
8. TBS buffer: 50 mM Tris–HCl, pH 7.4, 150 mM NaCl.
9. Blocking buffer: 5 % nonfat milk in TBS buffer.
10. 0.1 % Tween 20 in TBS.
11. Primary and HRP-conjugated secondary antibodies of interest.
12. Chemiluminescence detection system.

2.4 Purification of Na,K-ATPase

1. ISE buffer: 25 mM imidazole, 250 mM sucrose, 1 mM EDTA, pH 7.3.
2. Protease inhibitor cocktail (Roche Diagnostics).
3. Motor-driven Teflon-glass homogenizer.
4. SET buffer: 250 mM sucrose, 1 mM EDTA, 10 mM Tris, pH 7.4.
5. Sodium dodecyl sulfate, 20 % stock in water.
6. ATP-Tris salt.
7. "ATP stabilization solution" (10× stock contains 30 mM ATP-Tris salt, 500 mM imidazole, and 20 mM EDTA).
8. 7–30 % sucrose gradients made with 10 mM histidine, 1 mM EDTA, pH 7.3.

9. 2× stock of HEA solution: 60 mM histidine, 2 mM EDTA, 6 mM ATP-Tris salt, pH 7.3.

2.5 Functional Characterization of Na,K-ATPase

1. Complete ATPase reaction mixture: 120 mM NaCl, 20 mM KCl, 4 mM $MgCl_2$, 3 mM ATP-Tris salt (*see* **Note 2**), 30 mM histidine, pH 7.3.

2. Partial reaction medium without NaCl contains 20 mM KCl, 3 mM Tris-ATP, 4 mM $MgCl_2$, and 30 mM histidine, pH 7.3. Concentration of NaCl will vary from 0 up to 100 mM per tube.

3. Partial reaction medium to measure affinity to K^+: 140 mM NaCl, 3 mM Tris-ATP and 4 mM $MgCl_2$, 30 mM histidine, pH 7.3 with various concentrations of KCl (0–4 mM).

4. Partial reaction medium to measure affinity for ATP: 120 mM NaCl, 20 mM KCl, 4 mM $MgCl_2$, 30 mM histidine, pH 7.3. Concentration of ATP will vary from 0 up to 1.5 mM per tube.

5. Ouabain stock solution: 240 mM in dimethylformamide (protect from light). Store the reaction mixture and ouabain stock at −20 °C.

6. "Quenching solution": one part stock 5 % molybdate in water, one part stock 10 N H_2SO_4, eight parts of water, make fresh.

7. Reducing agent Fiske–Subbarow powder: stock solution 50 mg/mL in water.

8. WST-1 tetrazolium salt (Roche Diagnostics).

3 Methods

3.1 Plasmid Construction for Stable Transfection

1. Synthesize cDNA by reverse transcriptase from total rat kidney RNA.

2. Design specific primers for PCR (*see* **Note 3**). The primers should include restriction sites for unidirectional cloning. In our case, *Eco*RI and *Bam*HI restriction sites were added to the 5′ and 3′ of forward and reverse primers, respectively (*see* **Note 4**). In addition, the forward primer should include the Kozak consensus sequence for initiation of translation if no bases 5′ to the initiation methionine are included.

3. Run a PCR reaction and gel-purify the obtained DNA fragment.

4. Perform restriction and gel purification of the plasmid vector (for our studies, a pIRES vector with dual internal ribosome binding sites (IRES) and the neomycin resistance gene was used).

5. Ligate the PCR fragment into the vector.

6. Transform competent *E. coli* cells; select colonies under antibiotic pressure (we used ampicillin).

7. Propagate selected colonies in liquid cultures containing antibiotic.

8. Prepare plasmid DNA ('Minipreps').

9. Sequence clones to confirm the integrity of the DNA insert (*see* **Note 5**).

10. Estimate the concentration and purity of DNA by the ratio of OD at 260 and 280 nm. This requires a UV/visible spectrophotometer and quartz cuvette, or a dedicated instrument such as NanoDrop.

3.2 Transfection Protocol for Generation of Stable Cell Lines

Production of stable clones allows unlimited experimentation on uniform preparations. The disadvantage of stable transfectants is a random DNA integration site that can affect transcription efficiency. This complicates determination of the effect of a gene of interest on total specific activity of the enzyme. Utilization of an engineered recombination site such as the Flp-In system may be used to overcome this problem. Another limitation is that expression levels in stably transfected cells may be lower compared to transient transfection because there may be one copy instead of many per cell. However, many expression plasmids employ strong promoters, leading to more than enough protein production even when integrated into genomic DNA.

All manipulations should be performed in a tissue culture biological cabinet under sterile conditions. All incubations of cells at 37 °C are in a sterile 5 % CO_2 incubator.

1. Plate cells at density of 4×10^5 per well in a 6-well tissue culture dish in DMEM medium supplemented with fetal bovine serum. Expect 70–80 % confluency the next morning. Replace the medium with serum- and antibiotic-free medium and return to the incubator while setting up (*see* **Note 6**).

2. Prepare master stocks for transfection:

    (a) DNA dilutions. A1, A2, A3, ... stocks containing different concentrations of expression vector (from 2.5 to 15 µg) in 125 µL of serum-free medium (for instance, Opti-MEM medium) with no antibiotics added. Negative control must contain vector with no expression cassette at the same DNA concentration.

    (b) Transfection reagent dilutions. B1, B2, B3, ... stocks containing different concentration of transfection reagent in OptiMEM medium (6–15 µL/125 µL). Good results were with Lipofectamine 2000, which has been efficiently used with a wide variety of eukaryotic cells.

3. Add diluted DNA to diluted transfection reagent. Try different ratios of DNA to lipid from 1:1 to 1:4 (*see* **Note 7**).

4. Incubate the mixture at room temperature for 5–30 min.

5. Add the DNA-lipid complex mixtures to the cells, replacing the medium, and incubate cells for 3–18 h at 37 °C.

6. Replace DNA and lipid-containing medium in the dish with complete medium. Grow cells until they reach 90 % confluency.

7. Trypsinize cells and split 1:6 into new 6-well tissue culture dishes and allow them to adhere to the plastic.

8. Next day, add G418 for cells selection.

9. Cells with no DNA plasmid with the antibiotic resistance gene will start to die as soon as day 3. Depending on the efficiency of transfection, the first colonies containing stable transfectants will start to appear in 10–12 days (*see* **Note 8**).

10. Single isolated colonies should be trypsinized using sterile cloning cylinders, transferred to 96-well tissue culture plates, passaged into 24- and 6-well dishes, and finally into 25 cm^2 flasks.

11. Final collection and preservation of individual clones should be done by trypsinization from the 25 cm^2 flask. Remove medium from the flask and rinse 1–2 times with phosphate buffered saline with no Ca^{2+} and Mg^{2+}.

12. Add trypsin–EDTA solution (take care that all cells are covered) and incubate for 2–3 min at 37 °C. Trypsinization time varies depending on the cell type; use a microscope to monitor the process.

13. Collect detached cells in a 15 mL conical tube and dilute the contents with 3–4× volume of medium containing 10 % fetal bovine serum. Fetal bovine serum contains alpha-1-antitrypsin which inhibits trypsin.

14. Centrifuge at $200 \times g$ for 5 min at room temperature.

15. Resuspend the pellet in 1 mL of fetal bovine serum containing 10 % dimethylsulfoxide and transfer to a sterile cryotube.

16. Place the tube immediately in a −20 °C freezer for 1–2 h, then transfer to a deep freezer (−80 °C) overnight, and finally to liquid nitrogen for storage.

3.3 Authentication of Stable Clones

1. Take out cells from liquid nitrogen and defrost quickly at 37 °C.

2. Put the contents of a tube into a conical sterile 15 mL tube, and add 9× volume of serum-free DMEM medium.

3. Centrifuge cells 5 min at $500 \times g$, discard the supernatant, resuspend the pellet in 5 mL of serum-containing DMEM supplemented with fetal bovine serum.

4. Plate cells in a 25 cm^2 tissue culture flask.

5. At 90 % confluency, split the cells 1:5, and expand culture into another 25 cm² flask and a 35 × 16 mm tissue culture plastic dish.
6. Grow cells until confluency.
7. Take a dish, aspirate culture medium and rinse with PBS.
8. Add 0.2 mL/dish of 1× NuPage LDS sample buffer supplemented with a reducing agent and incubate the dish for 30 min at room temperature with gentle rotation.
9. Scrape the cells, collect them in a tube, and centrifuge for 10 min at 6000 ×g.
10. Take 50 µL of supernatant, run the sample on a NuPage SDS gel, transfer to nitrocellulose, and analyze the blot with specific antibodies.
11. Since DNA integration into the genome was not controlled, there should be variability between clones in expression yield. Choose several clones with similar levels of expression and use them for kinetic analysis.

3.4 Purification of Na,K-ATPase from Stable Clones

A peculiarity of membrane-bound Na,K-ATPase is its resistance to SDS. Treatment with relatively high concentration of SDS (up to 0.6 mg/mL at 1.4 mg/mL of total protein) leads to solubilization of most of membrane proteins while Na,K-ATPase remains active in a membrane-bound state [12]. We applied this method to transfected cells to get partially purified enzyme for further kinetic analysis [11].

1. Grow cells in 75 cm² flasks until confluency.
2. Aspirate the medium, rinse cells twice with PBS and add 3 mL/flask of ISE buffer supplemented with protease inhibitor cocktail.
3. Scrape the cells, collect them in a 15 mL conical tube, and centrifuge for 5 min at 500 ×g.
4. Discard the supernatant, and resuspend the pellet in 1 mL of ISE buffer with protease inhibitors.
5. Homogenize the pellet with a motor-driven teflon-glass homogenizer (10–15 strokes on ice).
6. Spin down the contents at 3000 ×g, 15 min, 4 °C (to remove unbroken cells and nuclear fraction) followed by ultracentrifugation at 100,000 ×g, 30 min, 4 °C.
7. Resuspend the final pellet (crude membranes) in SET buffer.
8. Measure the protein concentration either by BCA or Lowry method. Typical recovery is 1 mg of total protein/75 cm² flask.
9. In a pilot experiment, determine the SDS–protein ratio to obtain the most active Na,K-ATPase. In final volumes of 100 µL, combine crude membranes (14 µg), 10 µL of "ATP stabilization solution" and various concentrations of SDS from

0.03 to 0.08 mg/mL diluted in SET buffer. SDS is added last and mixed immediately.

10. Incubate for 30 min at room temperature with occasional tapping. To stop, put the reaction tubes on ice and add 400 µL of the complete ATPase assay medium.

11. Transfer 25–50 µL of this mixture to another to 350–375 µL of reaction mixture and measure ATPase activity with and without addition of ouabain (*see* below). Expect to observe a bell-shaped curve dependent on SDS concentration. Choose the concentration corresponding to the peak of the curve, and use it for full-size preparation.

12. Combine 1.1 mg of protein, 400 µL of 2× stock of HEA solution and adjust the volume up to 800 µL with SET buffer. Add the amount of SDS predetermined in the pilot experiment and mix immediately. Let it stir gently with a magnetic stir bar for 30 min at room temperature, and then keep on ice until use.

13. Create 7–30 % sucrose continuous gradients in 5 mL centrifuge tubes (*see* **Note 9**). Total volume of the gradient should be 4.2 mL. Chill at 4 °C.

14. Apply 0.8 mL of SDS-treated crude membranes to the top of the sucrose gradient, and sediment at $100,000 \times g$ for 3 h at 4 °C with slow acceleration and no brake.

15. Collect the turbid zones with a glass Pasteur pipette. The band is usually located in the top half of the gradient.

16. Dilute with 4 volumes of sucrose-free buffer, and sediment at $100,000 \times g$ for 1 h at 4 °C.

17. Discard the supernatant and resuspend the pellet in SET buffer. This pellet represent a partially purified Na,K-ATPase. Measure protein concentration either by BCA or Lowry method.

3.5 Functional Characterization of Na,K-ATPase from Stable Transfectants

Activity of Na,K-ATPase is the measure of its function. The final readout is ATP hydrolysis, i.e., measurement of liberated inorganic phosphate. One can look at the total specific activity of the pump or apparent affinities for the ligands and substrate (Na^+, K^+, and ATP). Another variable is sensitivity of Na,K-ATPase to ouabain, a specific inhibitor of the enzyme.

3.5.1 Total Specific Activity of Na,K-ATPase from Rat Cell Lines

1. Distribute 0.4 mL of complete ATPase reaction medium to glass tubes, with duplicates or triplicates for each point. Pre-incubate tubes at 37 °C.

2. Add 5 µL of ouabain stock solution to half of the tubes. The final concentration of ouabain should be 3 mM to get complete inhibition of rodent α1 subunit.

3. Add 5–10 µg of crude membrane fraction to the first tube, vortex, and put it back in the water bath at 37 °C.

4. Keep a fixed interval between tubes, 20 or 30 s.
5. Incubate for 20 or 30 min at 37 °C.
6. Set up a standard curve. The tubes (in duplicates) will have various aliquots of 10 mM KH_2PO_4 equal to 0, 5, 10, 15, 20, and 25 nmol of P_i. The standards will be incubated at 37 °C for the same time as the experimental tubes.
7. Stop the reaction by taking out a tube from the water bath and adding 0.6 mL "quenching solution." Keep the same time interval between tubes.
8. After quenching all tubes, add 10 μL of reducing Fiske–Subbarow agent and vortex. Incubate for 30 min at room temperature. The resulting phosphomolybdate complexes will turn blue. Read OD at 700 nm (*see* **Note 10**).
9. Specific activity of Na,K-ATPase is calculated as the difference between total activity (without ouabain added) and ouabain-resistant activity, and will be expressed in μmol P_i/mg/h or nmol P_i/mg/min.

3.5.2 Measurement of the Affinity for Na^+

1. To measure affinity for Na^+, use partial reaction medium. Add NaCl separately so that its final concentration in tubes varies from 0 to 40 mM. One can also add enough choline chloride to keep ionic strength constant.
2. Perform reaction as above for 30 min at 37 °C and stop with quenching solution.
3. Perform color development with reducing agent for 30 min.
4. Analyze the data by nonlinear regression using software such as GraphPad Prism 6 program. Fit Na^+-activation curves according to the Hill model for ligand binding. Expect the $K_{0.5}$ for Na^+ to be within the 5–15 mM range (Fig. 1a).

3.5.3 Measurement of the Affinity for K^+

1. Activity is measured in partial ATPase reaction medium containing various concentrations of KCl (0–4 mM) with [Na^+] fixed at 140 mM.
2. Perform reactions for 30 min at 37 °C with and without 3 mM ouabain, and measure ouabain-sensitive P_i release colorimetrically at OD 700 nm.
3. Fit K^+-activation curves according to the Hill model for ligand binding. Expect $K_{0.5}$ K^+ to be 0.5–0.9 mM.

3.5.4 Measurement of the Affinity for ATP

1. To measure affinity to ATP, use partial reaction medium. Add ATP (Tris salt) in the 50 μM to 1.5 mM range, and assay as above.
2. K_mATP can be derived from the Michaelis–Menten equation, and $K_{0.5}$ by Hill equation or nonlinear regression. The expected range is 0.3–0.8 mM.

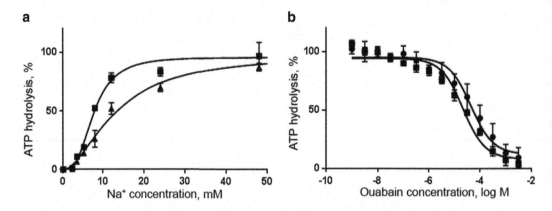

Fig. 1 Effects of FXYD2 on Na$^+$ and ouabain affinity of Na,K-ATPase in stably transfected C6 glioma cells. (**a**) activity of partially purified Na,K-ATPase from mock-transfected (*closed squares*) or FXYD2-transfected (*closed triangles*) cells was tested as a function of Na$^+$ concentration at a fixed K$^+$ concentration of 20 mM. Each set of data points is the mean ± SD from at least three independent experiments (with duplicate determinations) expressed as percentage of maximal Na,K-ATPase activity. Na$^+$ affinity of Na,K-ATPase was altered from 7.7 ± 0.35 to 13.2 ± 0.55 in mock-versus FXYD2 transfected C6 cells, respectively. (**b**) preparations of Na,K-ATPase from mock-transfected (*closed circles*) and FXYD2 transfected cells (*closed squares*) were assayed for ATPase activity as a function of ouabain concentration. Each set of data points shown is the mean ± SD from at least four independent experiments (with duplicate determinations) expressed as percentage of the ATPase activity at 0 mM ouabain. Affinity for ouabain was not significantly different in mock and FXYD2 transfected cells: 4.5 ± 10^{-5} and 2.2 ± 10^{-5}, respectively

3.5.5 Measurement of Affinity for Ouabain

1. To measure affinity to ouabain, use complete ATPase reaction medium supplemented with various concentrations of ouabain: from 0 to 3 mM per tube. Start the reaction by adding protein to tubes. Incubate for 20 or 30 min at 37 °C.
2. Stop and quench the reaction as described above.
3. Analyze the curve by nonlinear regression (Fig. 1b).

3.5.6 Cell Growth

1. Trypsinize cells and determine the density of viable cells with a hemacytometer.
2. Seed cells, either wild type or transfectants, into wells of 96-well flat-bottomed plates at a density of 5×10^3 cells/well in quadruplicate. Make one plate for each intended time point. Use medium containing 10 % fetal bovine serum.
3. Assay cell proliferation over the following 2–6 days with WST-1 reagent, a tetrazolium salt that is cleaved by dehydrogenases in viable cells, producing a red color.
4. Quantify with a microtiter plate reader at OD 450 nm (*see* **Note 11**).

4 Notes

1. Tissue culture plastic is surface-modified for adhesion of mammalian cells. Other sterile plasticware is suitable for growing bacteria. Plastic composition differs in ability to adsorb proteins and drugs and should be chosen mindfully.
2. Tris salt of ATP is less stable than sodium salt of ATP and should be stored frozen as a solution < pH 6.8. Magnesium salt of ATP is another alternative to sodium ATP, but only when presence of Mg^{2+} is intended, and with adjustment of buffers for it. ATP should be vanadium-free because vanadate is a Na,K-ATPase inhibitor.
3. Be sure that the chosen restriction sites are absent in the sequence of the gene of interest.
4. Choose restriction sites in the multiple cloning site of the plasmid distanced as far as possible. This will enhance ligation efficiency.
5. Sequencing of plasmid is an obligatory step before starting transfection to eliminate cloning errors.
6. Avoid pH and cold shock by using room temperature or warm, CO_2-equilibrated medium.
7. DNA–lipid ratio may be different for every cell line. Optimization is a necessary step to get good efficiency of transfection.
8. After selection with G418, one has to decide what to do next: combine the whole population of stable transfectants, or pick single colonies to get "monoclonal" cell lines. The main advantage of using the whole population is that results will not depend so much on the integration site of your construct. The main advantage of individual cell lines is that one can choose clones that differ (or are similar) in their level of expression of the protein of interest. The disadvantage of individual clones is that it takes additional weeks to grow and analyze the lines.
9. Important: dissolve the sucrose in 10 mM histidine, 1 mM EDTA, pH 7.3 at higher concentration and bring it up to the final volume (i.e., w/v not w/w).
10. Alternatively, color development can be done using the so-called "yellow method," which tolerates more lipid. After quenching, the unreduced phosphomolybdate complex is extracted into isobutanol (1.4 mL) by intensive vortexing (15–30 s) and centrifugation at $300 \times g$ for 5 min. The organic phase is collected and absorbance is read at 380 nm. The "yellow method' for ATPase activity assay may be difficult to perform if humidity is high because the organic phase becomes turbid. In the event that ATPase activity is low,

radioactive analysis based on release of ^{32}P from [γ-^{32}P]ATP may be utilized.

11. Alternatively, seed and grow cells in 24-well plates for 2–6 days. Harvest by trypsinization and count cells with a hemacytometer. Use trypan blue exclusion as a criterion of cell viability.

Acknowledgements

This work was supported by NIH grants HL036271, NS045283, EY014390, NS050696, and NS081558.

References

1. Shinoda T, Ogawa H, Cornelius F, Toyoshima C (2009) Crystal structure of the sodium-potassium pump at 2.4 A resolution. Nature 459:446–450
2. Morth JP, Pedersen BP, Toustrup-Jensen MS, Sorensen TL-M, Petersen J, Andersen JP, Vilsen B, Nissen P (2007) Crystal structure of the sodium-potassium pump. Nature 450:1043–1049
3. Kaplan JH (2002) Biochemistry of Na, K-ATPase. Annu Rev Biochem 71:51–535
4. Blanco G, Mercer RW (1998) Isozymes of the Na-K-ATPase: heterogeneity in structure, diversity in function. Am J Physiol 275:F633–F650
5. Geering K (2008) Functional roles of Na, K-ATPase subunits. Curr Opin Nephrol Hyperten 17:56–532
6. Beguin P, Wang X, Firsov D, Puoti A, Claeys D, Horisberger JD, Geering K (1997) The γ subunit is a specific component of the Na, K-ATPase and modulates its transport function. EMBO J 16:4250–4260
7. Sweadner KJ, Rael E (2000) The FXYD gene family of small ion transport regulators or channels: cDNA sequence, protein signature sequence, and expression. Genomics 68:41–56
8. Geering K (2006) FXYD proteins: new regulators of Na-K-ATPase. Am J Physiol 290:F241–F250
9. Wetzel RK, Sweadner KJ (2001) Immunocytochemical localization of the Na, K-ATPase α and γ subunits in the rat kidney. Am J Physiol 281:F531–F545
10. Arystarkhova E, Wetzel RK, Asinovski NK, Sweadner KJ (1999) The γ subunit modulates Na+ and K+ affinity of the renal Na, K-ATPase. J Biol Chem 274:33183–33185
11. Arystarkhova E, Donnet C, Asinovski NK, Sweadner KJ (2002) Differential regulation of renal Na, K-ATPase by splice variants of the γ subunit. J Biol Chem 277:10162–10172
12. Jorgensen PL (1974) Purification and characterization of (Na$^+$ + K$^+$)-ATPase. III. Purification from the outer medulla of mammalian kidney after selective removal of membrane components by sodium dodecylsulphate. Biochim Biophys Acta 356:36–52

Chapter 29

HPLC Neurotransmitter Analysis

Thomas Hellesøe Holm, Toke Jost Isaksen, and Karin Lykke-Hartmann

Abstract

High performance liquid chromatography (HPLC) is a powerful tool to measure neurotransmitter levels in specific tissue samples and dialysates from patients and animals. In this chapter, we list the current protocols used to measure neurotransmitters in the form of biogenic amines from murine brain samples.

Key words Mouse model, HPLC, Neurotransmitter, Brain structures

1 Introduction

Neurotransmitters are endogenous molecules that transmit signals across a synapse from the presynaptic neuron to the postsynaptic neuron. The release and subsequent removal of neurotransmitters must be precisely timed for postsynaptic neurons to be able to reengage in new signaling. This complex process is affected in many central nervous system (CNS) pathologies. Examples of this are Parkinson's disease, where dopamine (DA) producing neurons in the substantia nigra undergo apoptosis [1], and attention deficit hyperactivity disorder, which for some patients have been shown to be caused by mutations in the dopamine transporter [2].

For many neurotransmitters, reuptake is dependent on ion gradients generated and maintained by ion pumps and channels [3]. Part of the plasma membrane's sodium/potassium gradient is generated by the Na^+/K^+-ATPase. Mammals express four Na^+/K^+-ATPase α isoforms ($α_{1-4}$), which differ both in functionality and expression pattern. Whereas $α_1$ is ubiquitously expressed, $α_2$ is predominantly expressed by astrocytes and $α_3$ by neurons [4–6].

The $α_2$ Na^+/K^+-ATPase is associated with the astroglial-specific glutamate-dependent transport of lactate [7]. Thus, $α_2$ Na^+/K^+-ATPase is essential for glutamate and lactate uptake in astrocytes, and is strongly implicated in glutamate clearance from the synaptic cleft.

Rapid-onset dystonia-parkinsonism (RDP) is a neurological disorder caused by mutations in the *ATP1A3* gene encoding the

α_3 Na$^+$/K$^+$-ATPase. Levels of the DA metabolite, homovanillic acid was decreased in severely affected RDP patients, suggesting a relation between α_3 Na$^+$/K$^+$-ATPase and neurotransmitter homeostasis [8].

Despite several evidence from genetic, in vitro and in vivo research associated with mutations in the *ATP1A3* gene encoding the α_3 subunit of Na$^+$/K$^+$-ATPase as a cause of rapid-onset dystonia parkinsonism (RDP) [9], alternating hemiplegia of childhood (AHC) [10, 11] and CAPOS [12], as well as the established role of the *ATP1A2* gene in Familial Hemiplegic Migraine type 2 (FHM) [13] the underlying mechanism remain largely unknown, and HPLC-based neurotransmitter analysis would help to elucidate the underlying signaling pathways affected in these diseases. We are currently investigating knock-in mouse models harboring specific disease-mutations related to FHM2 and AHC (Lykke-Hartmann, unpublished).

2 Materials

2.1 Equipment

1. A automated and inert HPLC system fitted with electrochemical detection.
    (a) A biocompatible micro pump fitted with an online degasser (*see* **Note 1**).
    (b) A biocompatible analytical autosampler.
    (c) An electrochemical detector fitted with a coulometric cell (*see* **Note 2**).
2. Column C18 150 mm, particle size 2.1 μm.
3. Instrument analysis software.
4. Scientific scale with 0.1 mg precision.
5. Refrigerated table top centrifuge.
6. Pestle and motor mixer.
7. Surgical microdissection tools.
8. Vacuum suction.
9. 0.2 μm sterile filter 47 mm.
10. 0.2 μm PFTE syringe filter 4 mm.

2.2 Solutions

1. Prepare 0.3 N perchloric acid (PCA) by diluting 2.5 mL 70 % PCA to 100 mL with 18 MΩ water.
2. Prepare standard stock solutions as 1 mg/mL in 50 % 0.3 N PCA, 50 % Methanol. These can be stored at stored at −20 °C for several months (*see* **Note 3**).
3. Prepare serial dilutions with all standards combined in 0.03 N PCA to cover the concentration range of the samples (*see* **Note 4**).

4. Prepare mobile phase MD-TM as 75 mM sodium dihydrogen phosphate, 1.7 mM 1-octanesulfonic acid, 100 µL/L triethylamine, 25 µM EDTA, 10 % acetonitrile, adjust to pH = 3 with phosphoric acid reagents (should be HPLC-grade or higher) vacuum filter solution using 0.2 µm filter.

3 Methods

3.1 Sample Preparation

1. Mice are usually kept at a daily 12 + 12 light–dark cycle with access to water and food ad libitum.
2. The mice are euthanized humanly by cervical dislocation.
3. Using surgical tools to remove brains and to dissect brain areas of interest (AOI), e.g., cortex, striatum, or cerebellum.
4. Weigh the AOIs and keep on ice throughout the remaining procedure (optionally the samples can be frozen in liquid N_2 and stored at −80 °C until use).
5. Add 100 µL of cold 0.3 N PCA acid per 10 mg wet weight brain tissue.
6. Then homogenize tissue at 4 °C using pestle mixer.
7. After homogenization, centrifuge at 15.7 × g for 10 min 4 °C.
8. Remove the supernatant after the centrifugation step and filter supernatant using 4 mm 0.2 µm teflon filters.
9. Transfer 50 µL of the filtered sample(s) to HPLC autosampler vial(s) and place in the autosampler, set to 8 °C.

3.2 Analytical Conditions

1. The components of the Ultimate 3000 HPLC system from Thermo Scientific are shown in Fig. 1.
2. The mobile phase and programming of the instrument for injection, separation, and analysis of biogenic amines is standardized and typically will only require minor adjustments depending on the sample conditions.
3. Below follows a description of the analytical conditions used to set up quantitative analysis of biogenic amines in a brain homogenate using the HPLC setup from Thermo Scientific.
4. Turn on the individual HPLC components required to perform the analysis (Degasser, pump, autosampler and electrochemical detector).
5. Turn on the PC and open the Chromeleon software.
6. Place a flask of MD-TM mobile phase on solvent rack and connect it to the autosampler's inlet using the customized blue-cap lids.
7. Ensure that the flask containing wash buffer is filled. Otherwise replenish with 10 % MeOH.

Fig. 1 UltiMate® 3000 HPLC ELC system. Please visit http://www.dionex-uhplc.com/ for a comprehensive description

8. Open the pump front cover and untighten the *purge valve knob*.
9. In the Chromeleon software: Navigate to the Pump tab and activate "purge" for 3–4 min to ensure that the tubing leading into the HPLC has been flushed with the correct mobile phase.
10. Turn off "purge" and close the purge valve by tightening the *purge valve knob* (should be finger tight).
11. Close pump front cover.
12. Turn on pump and set flow rate to 0.4 mL/min.
13. Navigate to the autosampler tab: Turn on temperature control and set to 8 °C.
14. Navigate to the ECDRS detector tab: Set column oven temperature to 32 °C, turn on cells and set oxidation potential of the first electrode to 400 mV.
15. Allow a minimum of 30 min for the system to reach equilibrium—progress can be monitored by monitoring pressure and current from the first electrode (*see* **Note 5**).
16. Ensure that tube leaving the detector with mobile phase goes to waste.

17. Once the system has been equilibrated, navigate to the Chromeleon Data screen.
18. In the Data screen, create a data-stamped folder to contain the experiment.
19. Within the current experiment folder create a date-stamped instrument sequence.
20. Create a "normal" injection method. Using the *create instrument method* wizard specify the same conditions as used for equilibration of the system, i.e., flowrate = 0.4 mL/min, min/max pressure = 5/250 bar, column oven temperature = 32 °C, oxidation potential of first electrode = 400 mV and a run time of 22 min. Save the instrument method in current experiment folder.
21. Assign the individual sample positions and set injection volume to 20 μL.
22. Assign the newly created instrument method to all injections.
23. If no leaks have been observed press start to begin experiment by pressing the *Start* button located above Chromeleon's instrument sequence.

3.3 Software Analysis

3.3.1 Relative Quantitation

1. Double-clicking the thumbnail of a completed chromatogram in the instrument sequence will open Chromeleon's analysis module in a separate window.
2. Chromatogram peaks of interest should be identified based on the retention times obtained from previous injections of single amine standards.
3. Peak integration can be performed automatically by using the peak detection wizard (for further reading please see http://www.dionex.com/en-us/webdocs/114606-TN-Chromeleon-Intelligent-Integration-TN70698_E.pdf).
4. Integrated peak areas for the individual analytes correspond to amount of analyte in the sample (reported as nA × min) will be shown in the interactive results window (*see* **Note 6**).
5. Relative amine concentrations should be normalized to the total amount of dissected tissue.

3.3.2 Absolute Quantitation

1. Absolute quantitation is performed as described previously for relative quantitation but with additional injection of standards containing defined concentration of specific amines.
2. To perform absolute quantitation, the standards must be identified as calibration standards in Chromeleon's Data screen.
3. Each standard dilution is assigned a unique *level*, which in Chromeleon's analysis module, can be defined as part of the *Processing method*.

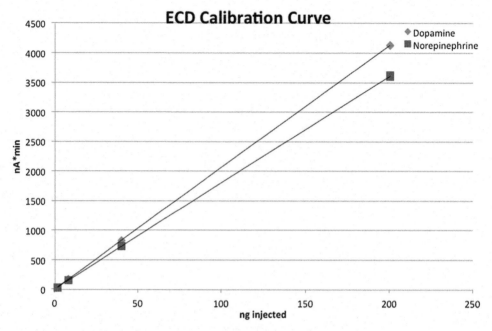

Fig. 2 Standard calibration curve for dopamine and norepinephrine separated by isocratic HPLC and detected by ECD. Linear detection was observed when injecting defined calibration standards containing 200–1.6 ng amines ($R^2 = 1$)

4. Calibration curves are automatically generated by the software and can be visualized at any time by pressing the *Calibration plot* in the analysis module.

5. Having performed **steps 1–4**, all samples will automatically be integrated and quantitated. All results are reported in the interactive results window as peak areas (nA × min) and absolute values, e.g., ng/mL depending on the information entered in the *Processing method*.

6. Absolute amine concentrations should be normalized to the total amount of dissected tissue (*see* Fig. 2).

4 Notes

1. An online degasser is preferred to rid the mobile phase and wash buffer of air bubbles (alternatively, the wash fluid should be degassed daily using compressed helium). The mobile phase should not be more than 2 weeks old.

2. Coulometry is based on the exhaustive oxidation of analyte passing through the detector. Using controlled-potential coulometry, a fixed potential of 400 mV is used to completely oxidize the amines. The total charge passing through the detector is proportional to the absolute amount of analyte by Faraday's law.

3. Degradation of amine stock solutions can be tested by HPLC analysis. A reduction in expected peak area and/or the appearance unknown peaks with higher retention times is clear indicators of degradation. We recommend injecting a known amine standard in triplicate before performing absolute quantitation.

4. Typical standards for biogenic amines include norepinephrine (NE), dopamine (DA), serotonin (5HT), dihydroxyphenyl acetic acid (DOPAC), 5-hydroxyindole acetic acid (5HIAA), homovanillic acid (HVA). Preparation of standard solutions is required for determining the retention time of the individual amines and for absolute quantification. Here it is important to note, that most neurotransmitters are purchased as salts. As an example, for every gram of Norepinephrine bitartrate salt, only 53 % is norepinephrine (the molecular weight of NE *bitartrate salt* is approximately 319 g/mol of which the NE is 169 g/mol and the bitartrate salt 150 g/mol). Hence, 1 mg/mL stock solutions should be prepared by resuspending approximately 1.89 mg NE bitartrate salt per mL. Dilutions should be prepared to contain amine concentrations covering the 10 μg/mL to 10 ng/mL range. Serial dilutions of for example 10, 2, 0.4, 0.08, 0.016 μg/mL containing all six standards listed above should cover the most of the dynamic detection range of the system.

5. The coulometric detector is sensitive to changes in pressure and flow rate. In order to establish a stable base line, the system must be allowed to equilibrate until base line noise fluctuations reach 0.10 nA and the base line drift is less than 3.0 nA/h.

6. The electric charge generated during the time an analyte passes the detector is reported as nA×min. If oxidized completely, nA×min is directly proportional to the total amount of analyte.

References

1. Pujol J, Junque C, Vendrell P, Grau JM, Capdevila A (1992) Reduction of the substantia nigra width and motor decline in aging and Parkinson's disease. Arch Neurol 49(11): 1119–1122
2. Hansen FH, Skjorringe T, Yasmeen S, Arends NV, Sahai MA, Erreger K et al (2014) Missense dopamine transporter mutations associate with adult parkinsonism and ADHD. J Clin Invest 124(7):3107–3120
3. Kristensen AS, Andersen J, Jorgensen TN, Sorensen L, Eriksen J, Loland CJ et al (2011) SLC6 neurotransmitter transporters: structure, function, and regulation. Pharmacol Rev 63(3):585–640
4. Brines ML, Robbins RJ (1993) Cell-type specific expression of Na+, K(+)-ATPase catalytic subunits in cultured neurons and glia: evidence for polarized distribution in neurons. Brain Res 631(1):1–11
5. Cameron R, Klein L, Shyjan AW, Rakic P, Levenson R (1994) Neurons and astroglia express distinct subsets of Na, K-ATPase alpha and beta subunits. Brain Res Mol Brain Res 21(3-4):333–343
6. McGrail KM, Phillips JM, Sweadner KJ (1991) Immunofluorescent localization of three Na, K-ATPase isozymes in the rat central nervous system: both neurons and glia can express more than one Na, K-ATPase. J Neurosci 11(2):381–391

7. Kleene R, Loers G, Langer J, Frobert Y, Buck F, Schachner M (2007) Prion protein regulates glutamate-dependent lactate transport of astrocytes. J Neurosci 27(45):12331–12340
8. Brashear A, Butler IJ, Hyland K, Farlow MR, Dobyns WB (1998) Cerebrospinal fluid homovanillic acid levels in rapid-onset dystonia-parkinsonism. Ann Neurol 43(4):521–526
9. de Carvalho AP, Sweadner KJ, Penniston JT, Zaremba J, Liu L, Caton M et al (2004) Mutations in the Na+/K+-ATPase alpha3 gene ATP1A3 are associated with rapid-onset dystonia parkinsonism. Neuron 43:169–175
10. Rosewich H, Thiele H, Ohlenbusch A, Maschke U, Altmuller J, Frommolt P et al (2012) Heterozygous de-novo mutations in ATP1A3 in patients with alternating hemiplegia of childhood: a whole-exome sequencing gene-identification study. Lancet Neurol 11(9):764–773
11. Heinzen EL, Swoboda KJ, Hitomi Y, Gurrieri F, Nicole S, de Vries B et al (2012) De novo mutations in ATP1A3 cause alternating hemiplegia of childhood. Nat Genet 44(9):1030–1034
12. Demos MK, van Karnebeek CD, Ross CJ, Adam S, Shen Y, Zhan SH et al (2014) A novel recurrent mutation in ATP1A3 causes CAPOS syndrome. Orphanet J Rare Dis 9:15
13. De Fusco M, Marconi R, Silvestri L, Atorino L, Rampoldi L, Morgante L et al (2003) Haploinsufficiency of ATP1A2 encoding the Na+/K+ pump alpha2 subunit associated with familial hemiplegic migraine type 2. Nat Genet 33(2):192–196

Chapter 30

Behavior Test Relevant to α_2/α_3 Na$^+$/K$^+$-ATPase Gene Modified Mouse Models

Toke Jost Isaksen, Thomas Hellesøe Holm, and Karin Lykke-Hartmann

Abstract

The behavioral phenotypes of mice are the result of a complex interplay between overall health, sensory abilities, learning and memory, motor function as well as developmental milestones, feeding, sexual, parental, and social behaviors. This chapter lists a selected number of key behavioral tests, specifically designed to assay fundamental behavioral features such as memory, activity, and motor skills in mice models.

Key words Mouse model, Animal behavior, Open field, Passive avoidance, Barnes Maze, Balance beam, Grip strength

1 Introduction

The field of behavioral neuroscience is often used to study genotype–phenotype relations in transgenic and gene modified mice. Mutations in the *ATP1A2* and *ATP1A3* genes, encoding the α_2 and α_3 isoforms of the Na$^+$/K$^+$-ATPase are associated with neurological pathologies. Mutations in the *ATP1A2* gene cause Familial hemiplegic migraine type 2 (FHM2) [1] whereas mutations in *ATP1A3* gene cause a spectrum of symptoms, which presently are categorized into three diagnoses, Rapid onset of Dystonia Parkinsonism (RDP) [2], Alternating Hemiplegia of Childhood (AHC) [3, 4], and the cerebellar ataxia with areflexia, pes cavus, optic atrophy, and sensorineural deafness (CAPOS) syndrome [5]. Interestingly, while sharing some symptoms, patients exhibit a range of manifestations that are specific to their diagnosis (reviewed in [6]). This phenomenon has also been reported for different Na$^+$/K$^+$-ATPase mouse models (reviewed in[7]). Genotype–phenotype correlation thus requires the use of extensive behavioral tests to distinguish subtle differences and to reveal specific manifestations. The neurological deficits reported for patients suffering from *ATP1A3* mutations mainly affect motor control and memory. Standard neurobehavioral phenotyping of mouse models has been extensively described and is explained in detail elsewhere [8]. We are currently investigating knock-in mouse models harboring specific disease-mutations related to FHM2 and AHC (Lykke-Hartmann, unpublished).

1.1 Open Field: Exploratory Behavior and General Activity

The Open field test is the benchmark test to assess exploratory behavior and general activity in rodents [9, 10].

The open field apparatus is an enclosure, generally square in shape, with high walls to prevent escaping. The animal is placed in the open field and tracked with the help of specialized tracking software for a fixed time period. Some of the basic activity parameters obtained from the open field are total distance moved, time spent moving and change in activity over time. Emotional features such as anxiety also can be evaluated by parameters such as freezing, defecation and time spent in the periphery of the arena, i.e., thigmotaxis [11].

1.2 Passive Avoidance: Memory

Passive avoidance is a fear-motivated test, used to assess short-term or long-term memory on rodents [12]. The test exploits the tendency in rodents to escape from an illuminated area into a dark one. The instrument consists of two compartments, divided by a sliding door. The compartment where the animal is placed will be illuminated, whereas the adjacent compartment room will be dark. On the first meeting with this situation (the trial day), the animal will rush into the dark room. However upon this, the animal receives a small electrical shock. Thus the next time (the test day), usually the following day, where the animal is put in the same situation, it will remember the unpleasant experience and therefore fight its instinct to seek the dark place. The latency to go into the dark room is evaluated and can be correlated to the animal's learning capabilities [12].

1.3 Barnes Maze: Spatial Learning and Memory

The Barnes maze consists of a 1.2 m diameter round brightly lit platform, with a series of evenly spaced holes around the edge of the platform. Only one of the holes exits to a dark tunnel underneath the platform. As the test starts the animal is placed in the center of the platform. The animal will try to escape the unpleasant open brightly light platform but can only do so by entering the tunnel. During several trials the animal learns the position of the tunnel [13, 14]. The latency to enter the tunnel and the efficiency by which the animal does so is tracked and used a measurement of spatial learning and memory [15].

1.4 Balance Beam: Motor Coordination and Balance

The Balance beam test is used to assess deficits in motor coordination and balance of rodents [16]. The animals' ability to cross a narrow elevated beam is tested, motivated to do so by a safe dark cage placed at the opposite end of the beam. The test is normally performed over 3 consecutive days, with 2 training days and a final test day. Latency to cross the beam and the number of slips during the crossing is scored [16].

1.5 Grip Strength: Muscle Strength

Muscle strength is affected both by actual muscle function and by the neuronal input to the muscle. A severe lack of muscle strength is observed in many neuromuscular diseases, which is often caused by a degradation or miss-function of motor neurons. The Grip strength test is used to very explicitly determine the muscle strength in rodents. The animal is allowed to grip a metal bar and is then steadily pulled backwards until it lets go of the bar, the force applied to the bar just before the animal loses its grip is recorded via a grip strength meter [17].

2 Materials

2.1 Open Field

1. 10–15 age-matched animals per group (treatment and/or genotype) (*see* **Note 1**).
2. Open Field apparatus for mice (Stoelting Co.). Constructed of transparent or opaque walls mounted on a grey, non-reflective base plate. Base plate dimensions are 50 cm × 50 cm. The walls are 50 cm tall. The setup is illustrated in Fig. 1a.
3. Sound isolated test room, with consistent lighting of the Open Field apparatus (*see* **Note 2**).
4. Video camera connected to a computer.
5. ANY-maze video tracking software (Stoelting Co.).
6. 70 % ethanol for cleaning.

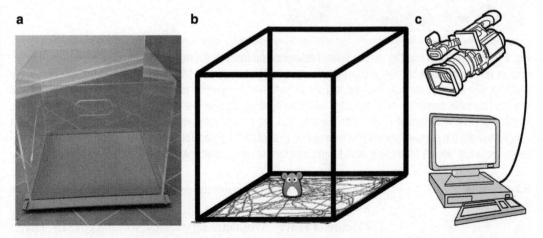

Fig. 1 The Open Field Test is an experiment used to assay general locomotor activity levels and anxiety in rodents. (**a**) Principally, the open field is an enclosure generally square in shape with surrounding walls that prevent escape. The open field can be of different sizes; small (38 × 38 cm), or large (72 × 72 cm). (**b**) Distance moved, time spent moving, rearing, and change in activity over time are among many measures that can be tabulated and reported, e.g., the track plot noted beneath the mouse. (**c**) The Open Field Test provides simultaneous measures of locomotion, exploration and anxiety, all monitored by a computer and analysis is performed using ANY-maze (www.anymaze.com/)

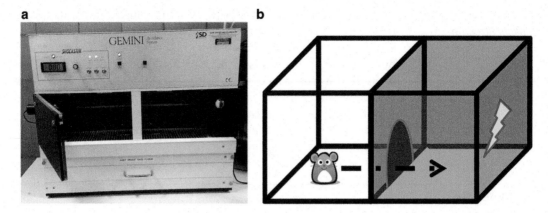

Fig. 2 The *Passive Avoidance* task is a fear-aggravated test used to evaluate learning and memory in rodent models. (**a**) The chamber is divided into a lit and a dark compartment, with a gate between the two. Animals explore both compartments on the first day. On the following day, they are given a mild foot shock in one of the compartments (often the dark compartment). (**b**) In this test, subjects learn to avoid an environment in which an aversive stimulus (such as a foot-shock) was previously delivered

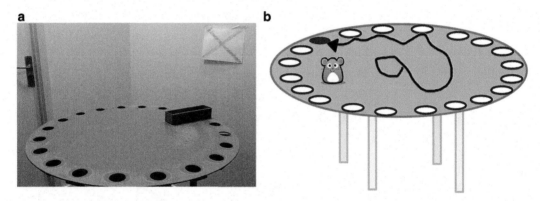

Fig. 3 Barnes maze is used to assess spatial reference memory in the mice. (**a**) A circular open platform surface to one small dark recessed chamber located under one of the 18 holes around the perimeter of the platform. Visual cues, such as colored shapes or patterns, are placed around the table in plain sight of the animal. The table surface is brightly lit by overhead lighting. (**b**) The test mouse is place in the middle of the Barnes maze, and a normal rodent will learn to find the escape box within four to five trials and will head directly toward the escape box without attempting to escape via incorrect holes. Various parameters are measured including latency to escape, path length, number of errors, and velocity

2.2 Passive Avoidance

1. 10–15 age-matched animals per group (treatment and/or genotype) (*see* **Note 1**).
2. Gemini Passive Avoidance apparatus for mice (San Diego Instruments). Illustrated in Fig. 2a.
3. Gemini analysis software (San Diego Instruments).
4. 70 % ethanol for cleaning.

2.3 Barnes Maze

1. 10–15 age-matched animals per group (treatment and/or genotype) (*see* **Note 1**).
2. Barnes Maze (Stoelting Co.). Illustrated in Fig. 3a.

3. Visual cues printed on A4 papers.
4. Video camera connected to a computer.
5. ANY-maze video tracking software (Stoelting Co.).
6. A fan.
7. 70 % ethanol for cleaning.

2.4 Balance Beam

1. 10–15 age-matched animals per group (treatment and/or genotype) (*see* **Note 1**).
2. A square wooden beam with a smooth surface. The length is 1 m and the width between 6 and 24 mm (*see* **Note 3**). The beam has two marks delimiting the center 80 cm of the beam.
3. Support stands to hold the beam horizontal approximately 40 cm above the ground surface.
4. Protective foam pads.
5. A dark cage mounted on a support stand so that it will fit to the end of the beam (*see* **Note 4**).
6. Video camera and tripod.
7. 70 % ethanol for cleaning.

2.5 Grip Strength

1. 10–15 age matched animals per group (treatment and/or genotype) (*see* **Note 1**).
2. Grip strength meter (Bioseb).
3. Bar grip or grid for mice (Bioseb).
4. 70 % ethanol for cleaning.
5. Scale with resolution of 0.1 g.

3 Methods

3.1 Open Field

3.1.1 Preparation

1. Keep the mice at a daily 12 + 12 light–dark cycle.
2. Perform the test during the light cycle.
3. Excess stress to the animals, such as cage change, should be avoided 1–2 days before testing.
4. Transfer the mice to the test room 1 h prior to testing; this is to allow the mice to acclimate.
5. Set up the video camera in a central position over the open field and connect it to the PC with tracking software.
6. Adjustments in the tracking software should include specifying fur color, zones in which to track the mice and tracking time (*see* **Note 5**).

3.1.2 Test

1. Place a mouse gently into the open field arena facing a corner and set the tracking software immediately to record via the video camera (*see* Fig. 1b, c).

2. Leave the room while the test is running.
3. Return the mouse to its cage upon completion.
4. Clean the cage thoroughly with 70 % ethanol between animals to avoid scents affecting the behavior (*see* **Note 2**).

3.2 Passive Avoidance

3.2.1 Preparation

1. Keep the mice at a daily 12 + 12 light–dark cycle.
2. Perform the test during the light cycle.
3. Perform the test over 2 consecutive days, during this period excess stress to the animals, such as cage change, should be avoided.
4. On both days, transfer the mice to the test room 1 h prior to training; this is to allow the mice to acclimate.

3.2.2 Trial Day

1. Place the mouse gently into the passive avoidance apparatus and allowed to acclimate there for 15 s (*see* Fig. 2b).
2. The start button turns on the compartment light and opens the sliding door to the dark compartment.
3. Give the mouse a 0.4 mA foot shock (*see* **Note 6**) when it enters the dark compartment and the latency to enter is recorded.
4. Leave the mouse for 30 s in the dark compartment before being returning it to its home cage.
5. Clean the apparatus thoroughly between animals with 70 % ethanol.

3.2.3 Test Day

1. Place the mouse gently into the passive avoidance apparatus and allow it to acclimate for 15 s.
2. The start button turns on the compartment light and opens the sliding door to the dark compartment.
3. Record the latency to enter the dark compartment.
4. No shock is applied during test day.
5. Return the mouse to its home cage just after it enters the dark compartment.
6. Thoroughly clean the apparatus between animals with 70 % ethanol.

3.3 Barnes Maze

3.3.1 Preparation

1. Keep the mice at a daily 12 + 12 light–dark cycle.
2. Perform the test during the light cycle.
3. Perform the test over 5 days with a sixth trial day separated by 1 week. During this period excess stress to the animals should be minimized. Cage changes should be done 1–2 days before experiments and during the first couple of days in week 2.
4. Transfer the mice to the test room 1 h prior to testing; this is to allow the mice to acclimate.
5. Place visual cues in the form of A4 printouts of a square, a cross and a circle on the walls in the test room.

3.3.2 Test Days 1–4

1. Gently place the mouse under an opaque cup in the middle of the table to ensure random orientation (*see* Fig. 3b).
2. Lift the cup and activate the tracking software.
3. Guide the mouse into the escape tunnel if it fails to enter within 3 min (day 1 only) (*see* **Note 7**).
4. After 30 s in the tunnel, return the mouse to its home cage.
5. Record the latency to enter the tunnel.
6. Between tests, clean the Barnes maze table and removable cups thoroughly in 70 % ethanol.
7. Identify two pairs of cups using a random number generator and swap them.
8. Test each mouse three times per day.
9. Mice of the same gender should be used for this experiment to minimize scents obscuring results.

3.3.3 Test Days 5 and 12

1. On the 2 last days of the test, the mice are only tested once.
2. Record the latency to find the tunnel and also the path to the tunnel via the tracking software.
3. Return the mouse to its home cage after 30 s in the tunnel.

3.4 Balance Beam

3.4.1 Preparation

1. Keep the mice at a daily 12 + 12 light–dark cycle.
2. Perform the test during the light cycle.
3. Perform the test over 3 consecutive days, during this period excess stress to the animals, such as cage change, should be avoided.
4. On all days, transfer the mice to the test room 1 h prior to training to allow the mice to acclimate.
5. Place the balance beam on the support stands and the dark cage at one end of the beam (which makes it the goal end). Place protective foam pads underneath the beam to soften any potential falls.
6. Set up a video recorder to record all trials; record both training and test trials. Position the video recorder ensuring that the full length of the beam is in the viewfinder. The hind limb slips are best visualized by placing the video recorder at an angle facing the dark cage.

3.4.2 Training Days

1. Place the mouse at the start end of the beam and allow it to cross the beam to the goal end and enter the safety of the dark cage. If a mouse stalls during the training, motivate it to move by gently pushing it forward (*see* **Note 8**).
2. After a successful crossing, allow the mouse to rest for 2 min before next trial.
3. Clean the beam with 70 % ethanol between each trial.

4. Perform three trials per training day for each mouse, after the last trial return the mouse to its home cage.

5. Between animals, clean the dark cage and beam with 70 % ethanol.

6. Two training days are normally enough to familiarize the mice with the task (*see* **Note 9**).

3.4.3 Test Day

1. Place the mouse at the start end of the beam and allow it to cross the beam to the goal end and enter the dark cage. No motivation should be necessary on the test day due to the prior training. If the mouse stalls during a trial on the test day, that trial is omitted and repeated.

2. After a successful crossing, allow the mouse to rest for 2 min before next trial.

3. Clean the beam with 70 % ethanol between each trial.

4. Perform three trials on the test day for each mouse, return the mouse to its home cage after the last trial.

5. Between animals, clean the dark cage and beam with 70 % ethanol.

6. Using the recorded videos, determine the latency to cross the center 80 cm of the beam and calculate the average of three successfully trials for each mouse.

7. The number of foot slips can also be evaluated from the recorded videos. A foot slip is defined a slip of the hind paw from the horizontal surface of the beam during the crossing of the center 80 cm of the beam.

3.5 Grip Strength

3.5.1 Preparation

1. Keep the mice at a daily 12 + 12 light–dark cycle.
2. Perform the test during the light cycle.
3. Excess stress to the animals, such as cage change, should be avoided on the day of testing.

3.5.2 Test

1. Holding the mouse by its tail, lower the mouse to the metal bar and allow it to grasp the bar with both front paws. Obtain grip strength of all four paws by using the grid.

2. Bring the animal to an almost horizontal position.

3. Pull the mouse away, following the angle of the grip strength meter.

4. Record the force at the moment of release via the grip strength meter as the peak tension (given in grams).

5. Test each mouse in five sequential trials, and use the highest grip strength recorded as the grip strength (*see* **Note 10**).

6. Before the mouse is returned to its home cage, determine its body weight.

7. Normalize the grip strength parameter to the body weight of each mouse.
8. Clean the bar and grid with 70 % ethanol between animals.

4 Notes

1. Mice may respond differently to behavioral setups depending on their genetic background, age and sex, etc. Data from male and female mice may be pooled if shown to be statistically identical. Also animals previously subjected to other behavioral tests may respond differently. Thus in order to reduce variance, the reuse of animals should be kept to a minimum.
2. The open field test is very susceptible to altered behavior due to scents, shadows, light intensity and external sounds. Carefully cleaning the open field between each animal is essential for reproducible results, it is furthermore recommended that male and female mice not be tested on the same day. Any external noise should be reduced to a minimum. Typical lighting is in the range of 50–150 lx and should not be flickering [11]. It is also important that the whole open field apparatus is evenly illuminated, otherwise the animals tend to seek dark areas.
3. The width of the balance beam determines the complexity of the test. For mice models with severe motor deficits a wider beam might be required. It is recommended to test different beam sizes, e.g., 6/12/24 mm, in a preliminary setup to determine the optimal size for the applied mice model.
4. In order to motivate the mice to cross the beam, the goal cage should feel like a safe spot for the mice. The darkness that the goal cage provides might in some instances be sufficient, however some nesting material from the home cage can further motivate the mice to enter the goal cage.
5. The duration of a single test is highly variable and depends on many parameters including strain, genotype, sex and size and the purpose of the test. For a normal overall test it is commonly set between 20 and 60 min. It is recommended to do preliminary long lasting tests, 1–2 h, and from the result of these adjusts the time duration for the main tests.
6. If control mice, e.g., unmanipulated wild type mice, do not show any latency to enter the dark chamber on the test day, first ensure that the apparatus applies the correct current. If so, the electrical shock on the trial day might have been too low; it can then be increased to 0.8 mA.
7. An important aspect of Barnes maze is the fear-induced motivation to seek away from the bright light platform and into the tunnel. However the mice can be so afraid and tense that they

simply cannot understand the concept and do not seek out the tunnel at all during the initial test day. Examples of this are that the mice do not explore the platform at all and that they continually freeze on the platform or in the normal cups for longer periods. If this is experienced, try to reduce fear in the mice before the test starts. This is easiest done by extensively handling the mice every day and by letting them explore on larger surface such as tables in the weeks leading up to the test.

8. If the mice continuously fail to understand the task, i.e., to traverse the beam and reach safety, it is recommended to perform initial trials where the mice initially are placed closer to the cage.

9. If the mice on day 2 of training still stall, additional training days might be required. However it should be noted that too much training can cause the mice to become too familiarized with the test and therefore lose motivation to seek safety in the dark end cage.

10. The mice will often let go of the bar before it has reached its maximum grip strength, thus to get a measure as close to the maximum grip strength as possible only the highest measurement is used [18]. To further confirm that this is the maximum grip strength, the test can be repeated after a week of rest.

References

1. De Fusco M, Marconi R, Silvestri L, Atorino L, Rampoldi L, Morgante L et al (2003) Haploinsufficiency of ATP1A2 encoding the Na+/K+ pump alpha2 subunit associated with familial hemiplegic migraine type 2. Nat Genet 33(2):192–196
2. de Carvalho AP, Sweadner KJ, Penniston JT, Zaremba J, Liu L, Caton M et al (2004) Mutations in the Na+/K+-ATPase alpha3 gene ATP1A3 are associated with rapid-onset dystonia parkinsonism. Neuron 43(2):169–175
3. Heinzen EL, Swoboda KJ, Hitomi Y, Gurrieri F, Nicole S, de Vries B et al (2012) De novo mutations in ATP1A3 cause alternating hemiplegia of childhood. Nat Genet 44(9):1030–1034
4. Rosewich H, Thiele H, Ohlenbusch A, Maschke U, Altmuller J, Frommolt P et al (2012) Heterozygous de-novo mutations in ATP1A3 in patients with alternating hemiplegia of childhood: a whole-exome sequencing gene-identification study. Lancet Neurol 11(9):764–773
5. Demos MK, van Karnebeek CD, Ross CJ, Adam S, Shen Y, Zhan SH et al (2014) A novel recurrent mutation in ATP1A3 causes CAPOS syndrome. Orphanet J Rare Dis 9:15
6. Heinzen EL, Arzimanoglou A, Brashear A, Clapcote SJ, Gurrieri F, Goldstein DB et al (2014) Distinct neurological disorders with ATP1A3 mutations. Lancet Neurol 13(5):503–514
7. Bottger P, Doganli C, Lykke-Hartmann K (2012) Migraine- and dystonia-related disease-mutations of Na+/K+-ATPases: relevance of behavioral studies in mice to disease symptoms and neurological manifestations in humans. Neurosci Biobehav Rev 36(2):855–871
8. Picciotto MR, Wickman K (1998) Using knockout and transgenic mice to study neurophysiology and behavior. Physiol Rev 78(4):1131–1163
9. Walsh RN, Cummins RA (1976) The Open-Field Test: a critical review. Psychol Bull 83(3):482–504
10. Gould TD, Dao DT, Kovacsics CE. The open field test. Springer Protocols. 2009;42(Mood and Anxiety Related Phenotypes in Mice. Neuromethods). 1–20

11. Gould T, Dao T, Kovacsics C (2009) The open field test. Mood and anxiety related phenotypes in mice. Humana, New York, pp 1–20
12. McGaugh JL (1966) Time-dependent processes in memory storage. Science 153(3742): 1351–1358
13. Barnes CA (1979) Memory deficits associated with senescence: a neurophysiological and behavioral study in the rat. J Comp Physiol Psychol 93(1):74–104
14. Sunyer B, Patil S, Höger H, Lubec G (2007) Barnes maze, a useful task to assess spatial reference memory in the mice. Nat Protoc (Protocol Exchange)
15. Mayford M, Bach ME, Huang YY, Wang L, Hawkins RD, Kandel ER (1996) Control of memory formation through regulated expression of a CaMKII transgene. Science 274(5293):1678–1683
16. Carter RJ, Morton J, Dunnett SB (2001) Motor coordination and balance in rodents. In: Jacqueline N. Crawley et al. Current protocols in neuroscience. Chapter 8:Unit 8 12
17. Cabe PA, Tilson HA, Mitchell CL, Dennis R (1978) A simple recording grip strength device. Pharmacol Biochem Behav 8(1):101–102
18. Lambertsen KL, Gramsbergen JB, Sivasaravanaparan M, Ditzel N, Sevelsted-Moller LM, Olivan-Viguera A et al (2012) Genetic KCa3.1-deficiency produces locomotor hyperactivity and alterations in cerebral monoamine levels. PLoS One 7(10), e47744

Chapter 31

Zebrafish Whole-Mount *In Situ* Hybridization Followed by Sectioning

Canan Doganli, Jens Randel Nyengaard, and Karin Lykke-Hartmann

Abstract

In situ hybridization is a powerful technique used for locating specific nucleic acid targets within morphologically preserved tissues and cell preparations. A labeled RNA or DNA probe hybridizes to its complementary mRNA or DNA sequence within a sample. Here, we describe RNA in situ hybridization protocol for whole-mount zebrafish embryos.

Key words In situ hybridization, Zebrafish, DIG-labeled RNA antisense probe, Pre-hybridization, NBT/BCIP staining, Sectioning

1 Introduction

The zebrafish is increasingly acknowledged for its ability to model human diseases, and as a useful organism in drug discovery and development [1]. Zebrafish models of the Na^+/K^+-ATPase α_2 and α_3 isoforms aim to further highlight the development of the FHM2 and RDP diseases and enable pathology and later drug screenings in larger scale prior to trials on mammalian models.

1.1 Animal Models

The development of gene modified Na^+/K^+-ATPase mouse models has revealed insight into part of the pathology of FHM2 and RDP through a subset of biobehavioral assays and biochemical approaches [2]. Although reverse genetics are feasible in the mouse, it is time and money consuming, and in this context, the zebrafish (*Danio rerio*) has emerged as a model system that allows rapid reverse genetic experiments and robust assays in which a range of hypomorphic phenotypes can be generated in a rapid, cost-effective, and gene-specific manner [3]. Therefore, to assess mechanisms of human diseases in model systems, several advantages are united through the combination of different model organisms, such as mouse and zebrafish, which offer different experimental options.

1.2 Zebrafish as Model System

Zebrafish have evolved as an excellent embryologically and genetically tractable vertebrate model system. High genetic and organ system homology to humans, external development with optical clarity, small size, short generation time (2–3 months), production of high number of embryos (over 200/female/week), and less expensive maintenance comprise the multifactorial advantages of zebrafish as a model organism over classical vertebrate models [1, 4].

The gross architecture of many zebrafish brain areas, e.g., the retina, the olfactory bulb, the hypothalamus, the cerebellum, and the spinal cord, is similar to human brain structures, although differences between teleosts and mammals do exist [5]. The main neurotransmitter systems, such as the cholinergic, dopaminergic, and noradrenergic pathways are present and have been mapped throughout the zebrafish brain [6]. Indeed, several neurotransmitter systems have been assessed using zebrafish, e.g., the cocaine- and amphetamine-regulated transcript peptidergic system [7], glutamatergic, serotonergic and dopaminergic neurotransmitter systems [8–13].

Indeed, it has served as a model to study several neurological disorders, e.g., Parkinson's disease (PD) [14–16], Alzheimer's disease [17, 18], Huntington's disease [19], and Schizophrenia [20]. Thus, there are several reports supporting that zebrafish models of neurological disorders can serve to reveal the pathogenic mechanisms underlying human diseases [21–24].

Zebrafish have been efficiently used to study ion transport, and zebrafish Na^+/K^+-ATPase models have been previously addressed in different contexts. A recent review suggested zebrafish embryos as a powerful in vivo model for understanding body fluid ion homeostasis and hormonal control in association with four types of ionocyte expressing distinct sets of transporters H^+-ATPase-rich, Na^+/K^+-ATPase-rich, Na^+-Cl^- cotransporter-expressing and K^+-secreting cells [25].

The zebrafish model system offers unique properties and serves to further help unravel the pathologies for the neurological disorders caused by Na^+/K^+-ATPase dysfunction [26]. For instance, zebrafish Na^+/K^+-ATPase has been associated with schizophrenia, where Na^+/K^+-ATPase activity is reduced in animals exposed to NMDA antagonist, MK-801, which mimic schizophrenia via NMDA hypofunction [27].

Recent studies assessing the functions of α_2 and $\alpha_3 Na^+/K^+$-ATPase by morpholino-induced knock-down of ATP1α_2 and ATP1α_3 have established functions for the $\alpha_2 Na^+/K^+$-ATPase in skeletal and heart muscles as well as astrocytes [28]. The roles in regulation of left-right (LR) patterning in conjugation with Ncx4a [29, 30] for $\alpha_2 Na^+/K^+$-ATPase, and regulation of brain ventricle volume and embryonic motility for $\alpha_3 Na^+/K^+$-ATPase, are likely linked to a depolarization of the resting membrane potential [28, 31].

Detecting specific RNA sequences within cells and tissues by in situ hybridization (ISH) using chemically labeled antisense RNA probes emerged in 1990s [32, 33]. ISH is a valuable method for assessing the distribution of the expression of a gene of interest and developing hypotheses about its functions. The optically clear nature of zebrafish embryos makes ISH more readily applicable to zebrafish embryos, and indeed ISH is one of the most commonly used techniques for zebrafish researchers. Fixed, pre-hybridized zebrafish embryos are hybridized with a digoxigenin uridine-5′-triphosphate (DIG) labeled RNA probe specific to the gene of interest. Fab fragments from polyclonal anti-DIG antibodies, conjugated to alkaline phosphatase are added to bind to DIG. The staining is performed by adding a combination of NBT (nitro-blue tetrazolium chloride) and BCIP (5-bromo-4-chloro-3′-indolylphosphate p-toluidine salt), which then yields an insoluble black-purple precipitate when reacted with alkaline phosphatase.

Here, we outline a step by step protocol, adapted from Thisse and Thisse, 2008 [34], for in situ hybridization on whole-mount zebrafish embryos, also used for detecting *Atp1a2*, *Atp1a3a*, and *Atp1a3b* transcripts [28, 31].

2 Materials

All solutions should be prepared using autoclaved ultrapure water (resistivity levels of 18.2 MΩ cm at 25 °C).

2.1 RNA Probe Synthesis

1. Template DNA harboring RNA probe coding sequence.
2. T3 or T7 RNA polymerase (20 U/μL).
3. 10× transcription buffer.
4. DTT (0.1 M).
5. DIG-RNA labeling mix.
6. RNAse inhibitor (40 U/μL).
7. DNAseI (RNase-free).
8. 0.5 M ethylenediaminetetraacetic acid (EDTA).
9. RNA cleanup column, e.g., SigmaSpin post-reaction cleanup column.
10. RNAlater®.
11. Nuclease free water.

2.2 Egg Collection/ Fixation

1. 1× E3 embryonic medium: 5 mM NaCl, 0.17 mM KCl, 0.33 mM $CaCl_2$, 0.33 mM $MgSO_4$ in water. Add 100 μL of 1 % methylene blue as a fungicide to 1 L of medium.
2. Pronase (protease from *Streptomyces griseus*): To prepare 1 % (w/v) pronase solution dissolve 1 g of pronase in 100 mL of

1× E3 medium, incubate for 2 h at 37 °C, aliquot and store at −20 °C.

3. 0.003 % 1-Phenyl-2-thiourea (PTU) in 1× E3 medium.

4. Anesthetic solution (0.008 % 3-amino benzoic acid ethylester (*tricaine*)): To prepare the tricaine stock solution dissolve 400 mg tricaine powder in 97.9 mL water. Add 2.1 mL of 1 M Tris–HCl, pH 9.0. Adjust pH to 7.0 and store the tricaine stock in aliquots at −20 °C. To prepare working solution add 2 mL of tricaine stock solution to 98 mL of 1× E3 medium.

5. 4 % paraformaldehyde (PFA) (*see* **Note 1**).

6. Methanol dilutions; prepare 25, 50, and 75 % methanol in 1× PBS.

7. A 100 mL beaker.

8. 100-mm petri dishes (some of them coated with 2 % agarose).

9. 35-mm petri dishes coated with 2 % agarose.

10. 1.5 mL microcentrifuge tubes.

11. Incubator (28.5 °C).

2.3 Hybridization

1. 1× PBS-T: 1× PBS (phosphate-buffered saline) with 0.1 % Tween 20 (v/v).

2. Proteinase K (20 mg/mL).

3. 20× SSC: 0.3 M sodium citrate, 3 M NaCl, pH 7.0. Make 2× and 0.2× SSC dilutions in water.

4. Hybridization mix minus (HM−): 50 % formamide, 5× SSC, 0.1 % Tween 20, adjust pH to 6.0 with 1 M citric acid.

5. Hybridization mix plus (HM+): HM⁻ with 500 μg/mL RNA from Torula yeast, 50 μg/mL heparin.

6. Blocking buffer: 2 % sheep serum (v/v), 2 mg/mL bovine serum (BSA) in 1× PBS-T.

7. HM− dilutions in 2× SSC (75, 50, 25 %).

8. 100 % 2× SSC.

9. 100 % 0.2× SSC.

10. Sheep anti-digoxigenin-AP Fab fragments.

11. Alkaline Tris buffer: 0.1 M Tris–HCl, pH 9.5, 0.1 M NaCl, and 0.1 % Tween 20 (v/v) (*see* **Note 2**).

12. Labeling solution: 1:50 dilution of NBT/BCIP stock solution (solution of 18.75 mg/mL nitro blue tetrazolium chloride and 9.4 mg/mL 5-bromo-4-chloro-3-indolyl-phosphate, toluidine-salt in 67 % (DMSO) (v/v)).

13. Stop solution: 1× PBS, pH 5.5, 1 mM EDTA, 0.1 % Tween 20 (v/v).

14. Hybridization oven (70 °C).

2.4 Sectioning	1. Ethanol solutions; 70, 96 and 99 %.
2. 5 % agarose.
3. 2-Hydroxyethyl methacrylate (Technovit 7100).
4. Dehydrating machine Citadel 1000.
5. Rotatory microtome Microm HM 355 with Ralph glass knives.

3 Methods

3.1 Probe Synthesis	1. Generate a template DNA harboring the coding sequence for antisense RNA probe. Sense RNA probe can also be generated as a control. This can be done by amplifying the sequence to be targeted from cDNA and subcloning into a vector with a T3 or T7 RNA polymerase promoter located at 3′ end (for antisense probes) or 5′ end (for sense probes). Another way to generate a template would be PCR amplifying this sequence and including T3 or T7 RNA polymerase promoter in the 5′ end of the reverse primer (for antisense probes) or the forward primer (for sense probes). Table 1 lists primers used to generate the ISH probes as well as the ISH probe sequences.
2. Prepare the mix indicated in Table 2 below and incubate for 2 h at 37 °C.
3. Add 2 μL RNAse free DNAse I and 18 μL nuclease-free water. Mix and incubate for 30 min (min) at 37 °C.
4. Stop the reaction by adding 1 μL sterile 0.5 M EDTA and 9 μL nuclease-free water.
5. Purify the probe using an RNA cleanup column, e.g., SigmaSpin post-reaction cleanup column.
6. Add 1 μL sterile 0.5 M EDTA and 9 μL of RNAlater to the sample; this protects the RNA from degradation. Store at −80 °C.
3.2 Preparation of Zebrafish Embryos	1. Collect zebrafish eggs and place them in a 100 mL beaker covered with a minimal amount of 1× E3 medium.
2. To remove the chorion, add pronase (1 % w/v) to a dilution of 1:10 depending on the medium volume in the beaker. Incubate for 5 min at 28.5 °C swirling occasionally (*see* **Note 3**).
3. Gently rinse the eggs three times with 1× E3 medium. Avoid the contact of the embryos with the medium/air interface.
4. Place the embryos into an agarose coated petri dish and grow them to the desired stage at 28.5 °C.
5. To prevent pigmentation, change the buffer to 0.003 % PTU solution (*see* **Note 4**) at the end of gastrulation (defined by the 50 %-epiboly stage, which begins at 5.25 h of development at 28.5 °C).

Table 1
List of primers used to generate the ISH probes, and well as the sequence of the ISH probes

1. List of primers and probes	
Primers used for generating probes	
AS *Atp1a2*	T7, 5′-TAATACGACTCACTATAGGG-3′
	DR-atp1a2-probeR, 5′-AAGCAGGAGGATGGCACCGAT-3′
AS *Atp1a3a*	DR-atp1a3-p-R-T7, 5′-TAATACGACTCACTATAGGGATTGCTCCGATCCACAGCA-3′
	DR-atp1a3-probeF, 5′-AACCACGCAGGACAATGATG-3′
AS *Atp1a3b*	DRatp1a3b-probeF, 5′-TCTCGCCCCACAATTTCCTA-3′
	atp1a3b-probeRT7, 5′-TAATACGACTCACTATAGGGAGCCACCAAAATCCACCAGA-3′
Probe	*Sequence*
Atp1a2 (316 bp)	5′-GGAAAGGGTACGGACATGAAAGCAGCCGGAGGGGCACCAACGGGGAAAGA GGAAGAAGAAGGATAAAGATTTAGATGAGCTGAAGAAAGAGGTGTCACTGGATGAT CATAAGCTGACTCTGGATGAGCTCAGTACTCGTTATGGAGTTGACCTTGCCAGAGGT TTGACCCATAAGAGAGCGATGGAAATCCTGGCGGGTGACGGTCCAAATGCGCTGACC CCTCCGCCCACACCCGGAGTGGGTGAAGTTCTGCAAGCAGCTGTTGGAGGATTC TCCATCCTGCTCTGGATCGGTGCCATCCTCTGCTT-3′
Atp1a3a (292 bp)	5′-AACCACGCAGGACAATGATGATAAGGAATCTCCAAAAGGGAAAGGGAGGGAA AGACCTTGATGACTTGAAGAAGAAGTGCCACTGACAGAGCATAAGATGTCGATTG AGGAAGTGTGCCGAAAGTACAACACTGACATCGTACAGGGTCTGACTAATGCCAGG GCTGCAGAGTACCTGCTCGGGATGGTCCCAATGCTCTCACCCCTCCACCCACCACC CCGGAGTGGGTCAAATTCTGCCGACAGTCTTCGGAGGTTTCTCCATCCTGCTGTGG ATCGGAGCAATC-3′
Atp1a3b (330 bp)	5′-TCTCGCCCCACAATTTCCTATATATACTCTTTGCATGTATTTTCACACGGCTGTGTACATAAAGTCTACTA TATATCTAAAACGGCAGTGCATAACCATCATGTAGGTTGGAAAGGAGCTTTGTGCTATAAAGGGAA TATGTACGCTTGCCTATTTAAACAAAGACTGTCCCTCCATAAGCCCATGCCTTTCCCTC CAGTCACACACACAGTTCACACAGAACACCTCACACATCTACACTACTCTAGACTGCGTTTGAC ATAATAGTTTCTCGTTTTCATGTTTGTTCTTAATTGAGCTCTTTAATCTCTGGTGGAT TTTGGTGGCT-3′

Table 2
List of components for RNA synthesis

Component	Amount per reaction	Final
PCR product or linearized plasmid DNA		100–200 ng
Transcription buffer (10×)	0.5 μL	1×
DTT (0.1 M)	0.5 μL	10 mM
DIG RNA labeling mix (10×)	0.5 μL	1×
RNAse inhibitor (40 U/μL)	0.25 μL	10 U
T3 or T7 RNA polymerase (20 U/μL)	0.25 μL	5 U
H_2O	Up to 5 μL	

6. If the embryos are older than 24 h post-fertilization (hpf), anesthetize them in 0.008 % tricaine.
7. Fix dechorionated embryos in 4 % PFA, overnight (ON) at 4 °C in 35 mm petri dishes.
8. Collect the embryos in 1.5 mL microcentrifuge tubes, 30–50 embryos per tube, (the rest of the protocol can be performed in these tubes) and dehydrate in 100 % methanol for 3× 5 min at room temperature (RT).
9. Place the embryos to -20 °C in fresh 100 % methanol for at least 2 h. Embryos can be stored in methanol at −20 °C for several months.

3.3 Pre-hybridization

1. Rehydrate the embryos by 5 min washes in serial dilutions of methanol in 1× PBS: 75, 50 and 25 % methanol (v/v).
2. Wash 4×5 min in 1× PBS-T.
3. Permeabilize embryos in proteinase K solution (10 μg/mL in 1× PBS-T) at RT for a period of time depending on the stage of the embryo:
   (a) 0–18 somite — 0 min.
   (b) 18 somite to 30 hpf — 5 min.
   (c) 36 hpf — 10–15 min.
   (d) 48 hpf — 35–40 min.
   (e) 60 hpf — 45 min.
   (f) 72 hpf to 6 dpf — 60 min.
4. Stop the proteinase K digestion by replacing proteinase K solution with 4 % PFA solution, for 20 min at RT.
5. Discard the 4 % PFA and wash the embryos 4×5 min in 1× PBS-T to remove the traces of PFA.

6. Wash the embryos 5 min in HM−.
7. Replace HM− with 700 µL of HM+ and incubate the embryos in an hybridization oven set to 70 °C for 2–5 h. Embryos in HM+ can be stored at −20 °C for several weeks.

3.4 Hybridization and Staining

1. Discard HM+ and replace it with fresh HM+ containing 30–50 ng antisense DIG-labeled RNA probe (*see* **Note 5**), incubate the embryos at 70 °C ON (*see* **Note 6**).
2. Pre-warm the following buffers at 70 °C prior to use: HM−, HM− dilutions in 2× SSC (75, 50, 25 %), 100 % 2× SSC, 100 % 0.2× SSC.
3. Remove the probe (probe can be saved at −80 °C for several usages), add HM− and incubate at 70 °C for 10 min with gentle agitation. Gradually change HM− to 2× SSC by 10 min washes at 70 °C with gentle agitation: 75, 50 and 25 % HM− in 2× SSC, and 100 % 2× SSC.
4. Wash 2 × 30 min in 0.2× SSC at 70 °C.
5. Perform the following washes at RT. Gradually change 0.2× SSC to 1× PBS-T by 10 min washes with gentle agitation: 75, 50 and 25 % 0.2× SSC in 1× PBS-T, and 100 % 1× PBS-T.
6. Incubate the embryos in blocking buffer at RT for 3–4 h.
7. Replace the blocking buffer with fresh blocking buffer including anti-DIG antibody (1:5000 to 1:10,000) and incubate ON at 4 °C with gentle agitation.
8. Remove the antibody solution and wash the embryos 6 × 15 min in 1× PBS-T at RT.
9. Wash 3 × 5 min in alkaline Tris buffer.
10. Replace the alkaline Tris buffer with 700 µL of labeling solution (1:50 dilution of NBT/BCIP stock solutions). Protect embryos from exposure to light and incubate at RT until desired staining intensity is reached.
11. Observe the embryos occasionally for staining and when the desired staining intensity is reached (*see* **Notes 5** and **6**), stop the reaction by replacing the labeling solution with stop solution.
12. Wash 3 × 15 min in stop solution.
13. For embryos at early stages (younger than 15-somite stage), in order to clear the yolk that darkens upon staining and affects visualization of the tissues on it, embryos can be incubated in an acidic buffer (1× PBS, pH 3.0) for 5 min.
14. Transfer the embryos to 35 mm petri dishes with 100 % glycerol carrying a minimal amount of stop solution. Store at 4 °C in dark.

3.5 Sectioning

1. Wash the embryos in 1× PBS-T to remove glycerol.
2. Embed and orient the embryos in 5 % agarose, so they could be cut transversely with respect to the length direction.
3. Dehydrate the agarose block including the embryos in ascending alcohol solutions 1½ h in 70 % EtOH, 1¼ h in 96 % EtOH, and 1¼ h in 99 % EtOH.
4. Infiltrate the agarose block for 2 h in 2-hydroxyethyl methacrylate, then followed by infiltration in 2-hydroxyethyl methacrylate. After addition of hardener, the time of workability with samples is approximately 5–8 min and the specimens will cure within a couple of hours at RT.
5. Take a 4- and 20-μm-thick section on a rotatory microtome equipped with Ralph glass knives for each 200 μm, collecting sections in a systematic, uniformLy and random manner.
6. Image and analyze the sections using light microscopy.

A representative image, whole-mount and sections of it, of a zebrafish embryo stained by this protocol for *Atp1a3a* transcripts [26] is illustrated in Fig. 1.

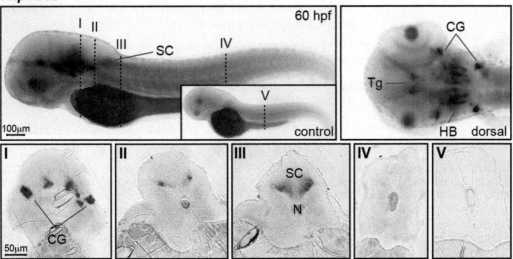

Fig. 1 Expression of *Atp1a3b* mRNA in zebrafish embryos. *Atp1a3b* mRNA expression analyzed by whole-mount in situ hybridization in 60 hpf zebrafish embryos; the *inset* shows sense probe hybridized control embryo. *Atp1a3b* is expressed in specific brain regions. The *numbered vertical dashed lines* show the positions of the transverse sections shown below in *sections I–V*. The abbreviations used are: *CG* cranial ganglia, *HB* hindbrain, *N* notochord, *Tg* tegmentum, *SC* spinal cord. Scale bars represent 100 μm in whole-mount images and 50 μm in sections

4 Notes

1. Prepare the PFA solution in a ventilated hood. Add 4 g PFA into 80 mL of 1× phosphate buffered saline (PBS) that is heated approximately to 60 °C (be careful not to boil) on a stir plate. Keep stirring while slowly raising the pH by adding 1 N NaOH drops until the solution clears (PFA will not dissolve unless the pH is basic). Once the PFA is dissolved, adjust the volume of the solution to 100 mL with 1× PBS and the pH to approximately 7.4 by diluted HCl. Cool the solution and filter it using a Millipore 0.22 μm filtration system. The solution can be aliquoted and frozen at −20 °C or stored at 4 °C for up to 1 month.

2. For DIG system applications of Roche NBT/BCIP stock solution (*see* below), $MgCl_2$ is not included in this buffer as this might lead to spotty background after the detection procedure.

3. This step is not needed if the protocol will be performed on embryos at a stage upon hatching (48–72 h post-fertilization (hpf)).

4. If the embryos are not raised in PTU, embryos can be bleached by incubation in 3 % H_2O_2/0.5 % KOH at RT until the pigmentation disappears. This procedure can be applied after fixing the embryos in 4 % PFA and before dehydrating in methanol. Upon bleaching, wash the embryos in 1× PBS and proceed with the methanol dehydration step.

5. The probe concentration can be optimized if embryos are over/under stained.

6. The reaction time varies depending on the expression level of the assessed gene in the range of 15 min up to 10 h.

References

1. Lieschke GJ, Currie PD (2007) Animal models of human disease: zebrafish swim into view. Nat Rev Genet 8(5):353–367
2. Bottger P, Doganli C, Lykke-Hartmann K (2012) Migraine- and dystonia-related disease-mutations of Na+/K+-ATPases: relevance of behavioral studies in mice to disease symptoms and neurological manifestations in humans. Neurosci Biobehav Rev 36(2):855–871
3. Cheng KC, Levenson R, Robishaw JD (2003) Functional genomic dissection of multimeric protein families in zebrafish. Dev Dyn 228(3):555–567
4. Barut BA, Zon LI (2000) Realizing the potential of zebrafish as a model for human disease. Physiol Genomics 2(2):49–51
5. Friedrich RW, Jacobson GA, Zhu P (2010) Circuit neuroscience in zebrafish. Curr Biol 20(8):R371–R381
6. Rink E, Wullimann MF (2004) Connections of the ventral telencephalon (subpallium) in the zebrafish (Danio rerio). Brain Res 1011(2):206–220
7. Mukherjee A, Subhedar NK, Ghose A (2012) Ontogeny of the cocaine- and amphetamine-regulated transcript (CART) neuropeptide system in the brain of zebrafish, Danio rerio. J Comp Neurol 520(4):770–797
8. Dahlbom SJ, Backstrom T, Lundstedt-Enkel K, Winberg S (2012) Aggression and monoamines: effects of sex and social rank in zebrafish (Danio rerio). Behav Brain Res 228(2):333–338

9. Rico EP, Rosemberg DB, Seibt KJ, Capiotti KM, Da Silva RS, Bonan CD (2011) Zebrafish neurotransmitter systems as potential pharmacological and toxicological targets. Neurotoxicol Teratol 33(6):608–617

10. Buske C, Gerlai R (2011) Early embryonic ethanol exposure impairs shoaling and the dopaminergic and serotoninergic systems in adult zebrafish. Neurotoxicol Teratol 33(6):698–707

11. Buske C, Gerlai R (2012) Maturation of shoaling behavior is accompanied by changes in the dopaminergic and serotoninergic systems in zebrafish. Dev Psychobiol 54(1):28–35

12. Cognato Gde P, Bortolotto JW, Blazina AR, Christoff RR, Lara DR, Vianna MR et al (2012) Y-Maze memory task in zebrafish (Danio rerio): the role of glutamatergic and cholinergic systems on the acquisition and consolidation periods. Neurobiol Learn Mem 98(4):321–328

13. Vuaden FC, Savio LE, Piato AL, Pereira TC, Vianna MR, Bogo MR et al (2012) Long-term methionine exposure induces memory impairment on inhibitory avoidance task and alters acetylcholinesterase activity and expression in zebrafish (Danio rerio). Neurochem Res 37(7):1545–1553

14. Son OL, Kim HT, Ji MH, Yoo KW, Rhee M, Kim CH (2003) Cloning and expression analysis of a Parkinson's disease gene, uch-L1, and its promoter in zebrafish. Biochem Biophys Res Commun 312(3):601–607

15. Anichtchik OV, Kaslin J, Peitsaro N, Scheinin M, Panula P (2004) Neurochemical and behavioural changes in zebrafish Danio rerio after systemic administration of 6-hydroxydopamine and 1-methyl-4-phenyl-1,2,3,6-tetrahydropyridine. J Neurochem 88(2):443–453

16. Flinn L, Mortiboys H, Volkmann K, Koster RW, Ingham PW, Bandmann O (2009) Complex I deficiency and dopaminergic neuronal cell loss in parkin-deficient zebrafish (Danio rerio). Brain 132(Pt 6):1613–1623

17. Musa A, Lehrach H, Russo VA (2001) Distinct expression patterns of two zebrafish homologues of the human APP gene during embryonic development. Dev Genes Evol 211(11):563–567

18. Tomasiewicz HG, Flaherty DB, Soria JP, Wood JG (2002) Transgenic zebrafish model of neurodegeneration. J Neurosci Res 70(6):734–745

19. Lumsden AL, Henshall TL, Dayan S, Lardelli MT, Richards RI (2007) Huntingtin-deficient zebrafish exhibit defects in iron utilization and development. Hum Mol Genet 16(16):1905–1920

20. Burgess HA, Granato M (2007) Sensorimotor gating in larval zebrafish. J Neurosci 27(18):4984–4994

21. Sager JJ, Bai Q, Burton EA (2010) Transgenic zebrafish models of neurodegenerative diseases. Brain Struct Funct 214(2-3):285–302

22. Bandmann O, Burton EA (2010) Genetic zebrafish models of neurodegenerative diseases. Neurobiol Dis 40(1):58–65

23. Flinn L, Bretaud S, Lo C, Ingham PW, Bandmann O (2008) Zebrafish as a new animal model for movement disorders. J Neurochem 106(5):1991–1997

24. Xi Y, Noble S, Ekker M (2011) Modeling neurodegeneration in zebrafish. Curr Neurol Neurosci Rep 11(3):274–282

25. Hwang PP, Chou MY (2013) Zebrafish as an animal model to study ion homeostasis. Pflugers Arch 465(9):1233–1247

26. Doganli C, Oxvig C, Lykke-Hartmann K (2013) Zebrafish as a novel model to assess Na+/K(+)-ATPase-related neurological disorders. Neurosci Biobehav Rev 37(10 Pt 2):2774–2787

27. Seibt KJ, da Luz OR, Rosemberg DB, Savio LE, Scherer EB, Schmitz F et al (2012) MK-801 alters Na+, K+-ATPase activity and oxidative status in zebrafish brain: reversal by antipsychotic drugs. J Neural Transm 119(6):661–667

28. Doganli C, Kjaer-Sorensen K, Knoeckel C, Beck HC, Nyengaard JR, Honore B et al (2012) The alpha2Na+/K+-ATPase is critical for skeletal and heart muscle function in zebrafish. J Cell Sci 125(Pt 24):6166–6175

29. Shu X, Huang J, Dong Y, Choi J, Langenbacher A, Chen JN (2007) Na, K-ATPase alpha2 and Ncx4a regulate zebrafish left-right patterning. Development 134(10):1921–1930

30. Shu X, Cheng K, Patel N, Chen F, Joseph E, Tsai HJ et al (2003) Na, K-ATPase is essential for embryonic heart development in the zebrafish. Development 130(25):6165–6173

31. Doganli C, Beck HC, Ribera AB, Oxvig C, Lykke-Hartmann K (2013) alpha3Na+/K+-ATPase deficiency causes brain ventricle dilation and abrupt embryonic motility in Zebrafish. J Biol Chem 288(13):8862–8874

32. Schulte-Merker S, Ho RK, Herrmann BG, Nusslein-Volhard C (1992) The protein product of the zebrafish homologue of the mouse T gene is expressed in nuclei of the germ ring and the notochord of the early embryo. Development 116(4):1021–1032

33. Thisse C, Thisse B, Schilling TF, Postlethwait JH (1993) Structure of the zebrafish snail1 gene and its expression in wild-type, spadetail and no tail mutant embryos. Development 119(4):1203–1215

34. Thisse C, Thisse B (2008) High-resolution in situ hybridization to whole-mount zebrafish embryos. Nat Protoc 3(1):59–69

Chapter 32

Whole-Mount Immunohistochemistry for Anti-F59 in Zebrafish Embryos (1–5 Days Post Fertilization (dpf))

Canan Doganli, Lucas Bukata, and Karin Lykke-Hartmann

Abstract

Immunohistochemistry (IHC) is a powerful method to determine localization of tissue components by the interaction of target antigens with labeled antibodies. Here we describe an IHC protocol for localizing the myosin heavy chain of zebrafish embryos at 1–2 and 3–5 days post fertilization (dpf).

Key words Zebrafish, F59, Myosin heavy chain, Immunohistochemistry

1 Introduction

The Na^+/K^+-ATPase generates ion gradients across the plasma membrane, essential for multiple cellular functions. In mammals, four different Na^+/K^+-ATPase α-subunit isoforms are associated with characteristic cell-type expression profiles and kinetics. Using the zebrafish as a model system to evaluate gene functions has several advantages (please *see* the introduction to Chap. 31 for a more detailed description of the zebrafish model).

More than 70 years ago, Coons and colleagues published the first IHC study describing the use of FITC-labeled antibodies to identify pneumococcal antigens in infected tissue [1]. Since then, immunohistochemistry became a routine and essential tool in diagnostic and research laboratories with the improvements made in tissue fixation methods, protein conjugation, label detection, as well as imaging technologies. IHC can provide spatial and temporal information for proteins of interest and it has been efficiently serving zebrafish researchers. In IHC, the samples are exposed to labeled antibodies that are directed against epitopes in the target protein and thus, the target can be visualized by a marker such as fluorescent dye and enzyme. Antibodies can be applied directly or indirectly, i.e., by binding of a marker-conjugated antibody to its target or by adding primary antibody to the target, and subsequent binding of a labeled secondary antibody to the primary antibody. While there are multiple approaches and modifications in IHC methodology depending on the antibody and the embryonic stage, the steps involved can be overall divided into two main groups: sample preparation and labeling. A previous study found the zebrafish α_2Na$^+$/K$^+$-ATPase associated with striated muscles and that knockdown causes a significant depolarization of the resting membrane potential in slow-twitch fibers of skeletal muscles [2]. Moreover, the α_2Na$^+$/K$^+$-ATPase deficiency reduced the heart rate and caused a loss of left–right asymmetry in the heart tube.

Here we describe IHC methodology step by step in detection of myosin heavy chain, using anti-F59 antibody, which detects primarily slow-twitch muscle, in whole mount zebrafish embryos at 1–2 dpf and 3–5 dpf. By analyzing the muscle development in zebrafish embryos bearing Na$^+$/K$^+$-ATPase mutations, the morphological consequences of the above mentioned knockdown effects can be analyzed. Moreover, it allows further to evaluate other muscle-specific factors, and how their distribution might be altered by the knockdown of Na$^+$/K$^+$-ATPase subunit isoforms.

2 Materials

All solutions should be prepared using autoclaved ultrapure water (resistivity levels of 18.2 MΩ cm at 25 °C).

2.1 Egg Collection/Fixation Components

1. 1× E3 embryonic medium: 5 mM NaCl, 0.17 mM KCl, 0.33 mM CaCl$_2$, 0.33 mM MgSO$_4$ in water. Add 100 µL of 1 % methylene blue as a fungicide to 1 L of medium.
2. Pronase (protease from *Streptomyces griseus*): To prepare 1 % (w/v) pronase solution dissolve 1 g of pronase in 100 mL of 1× E3, incubate for 2 h at 37 °C, aliquot and store at −20 °C.
3. Anesthetic solution (0.008 % 3-amino benzoic acid ethylester (tricaine)): To prepare the tricaine stock solution dissolve 400 mg tricaine powder in 97.9 mL of water. Add 2.1 mL of 1 M Tris–HCl, pH 9.0. Adjust pH to 7.0 and store the tricaine stock in aliquots at −20 °C. To prepare the working solution add 2 mL of tricaine stock solution to 98 mL of 1× E3 medium.
4. 0.003 % 1-phenyl-2-thiourea (PTU) in 1× E3 medium.
5. 4 % paraformaldehyde (PFA) (*see* **Note 1**).
6. A 100 mL beaker.
7. 100-mm petri dishes (some of them coated with 2 % agarose).
8. 35-mm petri dishes.
9. Incubator (28.5 °C).

2.2 Labeling Components

1. 0.1 % PBX: 1× PBS with 0.1 % Triton X-100 (v/v).
2. 1 % PBX: 1× PBS with 1 % Triton X-100 (v/v).
3. 2 % PBX: 1× PBS with 2 % Triton X-100 (v/v).
4. Blocking solution: 10 % goat serum (v/v) in 1 % PBX.
5. 2 % BSA solution: Dissolve 0.2 g BSA in 10 mL 0.1 % PBX.
6. Ab-F59 cell concentrate (Developmental Studies Hybridoma Bank).
7. Alexa Fluor® 488 goat anti-mouse IgG (H+L).

2.3 Imaging

1. For high-resolution imaging, confocal laser microscopy is advised. The Alexa Fluor® 488 has emission maxima properties of 493 and 518 nm, respectively (*see* **Note 3**).

3 Methods

3.1 Preparation of Zebrafish Embryos

1. Collect zebrafish eggs, and place them in a 100 mL beaker covered with a minimal amount of 1× E3 embryonic medium.
2. To remove the chorion, add pronase (1 % w/v) to a dilution of 1:10 depending on the medium volume in the beaker. Incubate for 5 min at 28.5 °C swirling occasionally (*see* **Note 2**).
3. Gently rinse the eggs three times with 1× E3 medium. Avoid the contact of the embryos with the medium/air interface.
4. Place embryos into an agarose-coated petri dish and grow to desired stage at 28.5 °C.
5. To prevent pigmentation, change the buffer to 0.003 % PTU solution at the end of gastrulation (defined by the 50 %-epiboly stage, which begins at 5.25 h of development at 28.5 °C).

3.2 Labeling of 1–2 dpf Embryos

1. Start by fixing dechorionated embryos in 4 % PFA for 1.5 h at room temperature (RT) in 35 mm petri dishes or in 1.5 mL microcentrifuge tubes (~15 embryos per tube) placed on their sides with gentle agitation.
2. Then wash embryos 3×5 min in 1 % PBX.
3. Block the embryos in blocking solution for 1 h at RT.
4. Then, add primary antibody F59 cell supernatant at 1:10 or concentrate at 1:100 dilution in blocking solution, and incubate the embryos overnight at 4 °C with gentle agitation.
5. After antibody incubation, wash embryos 6×15 min in 1 % PBX at RT with gentle agitation.
6. Then, add secondary antibody, e.g., Alexa Fluor 488 goat-anti mouse 1:400 in blocking solution, and incubate for 4 h at RT or overnight (ON) at 4 °C with gentle agitation in dark. Embryos should be protected from exposure to light during and post this step.

7. Wash embryos 6×15 min in 1 % PBX at RT with gentle agitation.
8. Carefully mount in 0.5 % low melting point agarose right before imaging.
9. Embryos can be transferred gradually to 100 % glycerol (*see* **Note 4**) and stored in dark to preserve the fluorescent signal.

3.3 Labeling of 3–5 dpf Embryos

1. Anesthetize embryos in 0.008 % tricaine until their heart beat is undetectable.
2. Fix dechorionated embryos in 4 % PFA for ~8 h at 4 °C in 35 mm petri dishes or in 1.5 mL microcentrifuge tubes (~15 embryos per tube) placed on their sides with gentle agitation.
3. Wash 3×5 min in 0.1 % PBX.
4. Incubate the embryos in 2 % PBX ON at RT with gentle agitation.
5. Wash 3×5 min in 0.1 % PBX.
6. Permeabilize the embryos with prechilled acetone for 7 min at −20 °C.
7. Wash 3×5 min in 0.1 % PBX and incubate the embryos in 0.1 % PBX, ON, at 4 °C.
8. Block the embryos in 2 % BSA in 0.1 % PBX for 1 h at RT with gentle agitation.
9. Add primary antibody F59 concentrate at 1:30 dilution in 2 % BSA solution at 4 °C for 2 days in a rotator.
10. Wash 5×10 min in 0.1 % PBX at RT with gentle agitation.
11. Add secondary antibody, e.g., Alexa Fluor 488 goat anti-mouse 1:200 in 2 % BSA solution, and incubate ON at 4 °C in a rotator in dark. Embryos should be protected from light exposure during this step and onwards.
12. Wash 5×10 min in 0.1 % PBX at RT with gentle agitation.
13. Mount in 0.5 % low melting point agarose right before imaging.
14. Embryos can be transferred gradually to 100 % glycerol and stored in dark.
15. Imaging analysis: Whole-mount embryos and sections are observed using an inverted microscope, Olympus IX71 or Zeiss LSM 710 T-PMT confocal microscope.

Examples of F59 staining in sections illustrated in Fig. 1 [2].

4 Notes

1. Prepare the PFA solution in a ventilated hood. Add 4 g PFA into 80 mL of 1× phosphate buffered saline (PBS) that is heated approximately to 60 °C (be careful not to boil) on a stir

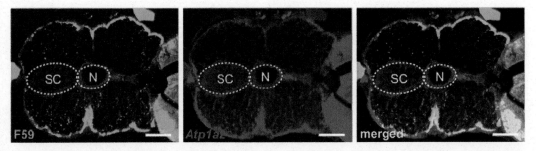

Fig. 1 Transverse sections of embryos double stained by in situ hybridization for *Atp1a2* (*red* fluorescence) in combination with immunostaining using F59 (primarily slow-twitch muscle; *green* fluorescence). *N* notochord, *SC* spinal cord. Scale bars: 100 mm

plate. Keep stirring while slowly raising the pH by adding 1 N NaOH drops until the solution clears (PFA will not dissolve unless the pH is basic). Once the PFA is dissolved, adjust the volume of the solution to 100 mL with 1× PBS and the pH to approximately 7.4 by diluted HCl. Cool the solution and filter it using a Millipore 0.22 μm filtration system. The solution can be aliquoted and frozen at −20 °C or stored at 4 °C for up to 1 month.

2. Imaging of green fluorescent protein (GFP) chromophores have a single excitation peak centered at about 488 nm, with an emission peak wavelength of 509 nm. Red Fluorescent Protein (RFP) chromophores can be excited by the 488 or 532 nm laser line and is optimally detected at 588 nm. Spectral excitation and emission maxima properties of dyes used should be investigated before confocal imaging to ensure optimal imaging.

3. This step is not needed if the protocol will be performed on embryos at a stage upon hatching (48–72 hpf).

4. This is achieved by adding the embryo to increasing concentration of glycerol (10, 20, 60, and 80 %) and eventually into 100 % glycerol, usually leaving the embryo approximately 1 h in each glycerol solution. After this, the embryos can be transferred to a 100 % glycerol solution.

References

1. Coons AH, Creech HJ, Norman Jones R, Berliner E (1942) The demonstration of pneumococcal antigen in tissues by the use of fluorescent antibody. J Immunol 45(3):159–170

2. Doganli C, Kjaer-Sorensen K, Knoeckel C, Beck HC, Nyengaard JR, Honore B et al (2012) The alpha2Na+/K+-ATPase is critical for skeletal and heart muscle function in zebrafish. J Cell Sci 125 (Pt 24):6166–6175

Chapter 33

Cell-Based Lipid Flippase Assay Employing Fluorescent Lipid Derivatives

Maria S. Jensen, Sara Costa, Thomas Günther-Pomorski, and Rosa L. López-Marqués

Abstract

P-type ATPases in the P4 subfamily (P4-ATPases) are transmembrane proteins unique for eukaryotes that act as lipid flippases, i.e., to translocate phospholipids from the exofacial to the cytofacial monolayer of cellular membranes. While initially characterized as aminophospholipid translocases, studies of individual P4-ATPase family members from fungi, plants, and animals show that P4-ATPases differ in their substrate specificities and mediate transport of a broader range of lipid substrates. Here, we describe an assay based on fluorescent lipid derivatives to monitor and characterize lipid flippase activities in the plasma membrane of cells, using yeast as an example.

Key words Back-exchange, Flow cytometry, Fluorescence, Heterologous expression, Phospholipid probes, *Saccharomyces cerevisiae*

1 Introduction

Cellular membranes, notably eukaryotic plasma membranes, are equipped with special proteins that actively translocate lipids from one leaflet to the other and thereby help generate membrane lipid asymmetry [1, 2]. Among these ATP-driven transporters, the P4 subfamily of P-type ATPases (P4-ATPases) comprises lipid flippases that operate as heterodimers in combination with protein subunits from the ligand-effect modulator (LEM)3/cell division cycle (CDC)50 family to flip phospholipids from the exofacial to the cytosolic side of cell membranes [3]. As these transporters are trapped in an environment (cellular membranes) formed by their own substrate (lipids), analyzing their activity is not a trivial task and most assays are based on the use of fluorescent lipid analogs, typically 7-nitrobenz-2-oxa-1,3-diazol-4-yl (NBD)-labeled lipids (Fig. 1). These analogs have a fluorescent reporter group attached to a short fatty acid chain and maintain most of the properties of endogenous phospholipids, except that they are more water-

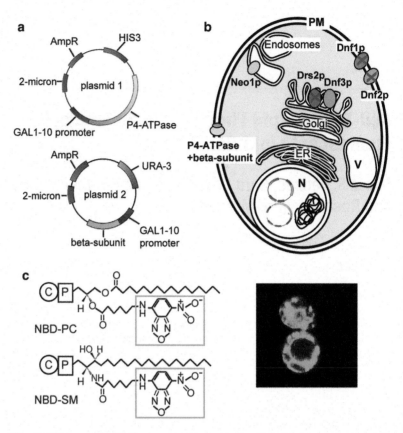

Fig. 1 Cell-based lipid flippase assay employing heterologous expression in yeast and fluorescent lipid analogs. (**a**) A pair of vectors based on plasmids of the pRS42X-GAL series [16] allows expression of P4-ATPases with their β-subunits under the control of a strong inducible galactose promoter (Gal1-10). (**b**) The plasmid pair is transformed into yeast Δdrs2Δdnf1Δdnf2 cells that lack three of their five P4-ATPases. *ER* endoplasmic reticulum, *N* nucleus, *PM* plasma membrane, *V* vacuole. (**c**) C6 acyl-NBD-labeled fluorescent lipid analogs are then used to analyze lipid uptake in the transformed cells, containing either a glycerol backbone (e.g., NBD-PC) or a ceramide backbone (e.g., NBD-SM). The NBD group is boxed in *green*. *P* phosphate group, *C* choline. As an example, co-expression of the plant P4-ATPase ALA2 with the β-subunit ALIS5 results in cells displaying C6-NBD-PS uptake, as shown here by an intense labeling of intracellular membranes visualized by confocal microscopy. Scale bar, 10 μm

soluble, which facilitates incorporation from the medium into the outer monolayer of the plasma membrane. Short-chain NBD-lipids have been extensively used as reporters of phospholipid flip-flop processes and the data obtained with them are qualitatively similar to that for other phospholipid analogs (e.g., radiolabeled dibutyryl phospholipids or acyl-spin-labeled phospholipids [4, 5] and natural phospholipids [6, 7], which justifies their choice for the lipid flippase assay described here).

Transmembrane transport of short-chain NBD-lipids is usually monitored by taking advantage of their ability to be extracted from the membrane by defatted bovine serum albumin (BSA) as well as to be chemically reduced by dithionite, a membrane-impermeant dianion. Here we only describe the use of BSA in such experiments. BSA cannot penetrate intact cells and will therefore only deplete the short-chain NBD-lipids present on the cell surface. Thus, when cells have been incubated with short-chain NBD-lipids and the excess removed by washing with BSA, the remaining cell-associated fluorescence corresponds to the lipid fraction internalized by flippases and this can be quantitatively measured by fluorimetric methods.

1.1 Choice of Lipid Analogs

Lipid analogs to be used in lipid flippase assays must comply with several requirements. Firstly, they cannot be modified at their polar head group, as this usually defines substrate specificity. Secondly, they should incorporate readily into the outer plasma membrane leaflet upon their addition to cells. Thirdly, they must be readily extractable from membranes by defatted BSA. These requirements are best met by analogs having the NBD group attached to a short C6 acyl chain in the *sn*-2 position of the glycerol backbone or linked by an amide bond to the ceramide backbone (Fig. 1c).

1.2 Cellular Systems

The assay described here has been optimized for yeast cells. When using other cells, modifications must be introduced with respect to medium and buffer conditions. Special attention must be put on the assay medium not to contain serum albumin or calf serum as these bind NBD-lipids. Lipid translocation assays have been successfully carried out in both yeast and mammalian cells [8–11].

Our standard lipid flippase assay makes use of a *S. cerevisiae* mutant strain deleted in three P4-ATPases (Drs2p, Dnf1p, and Dnf2p; Fig. 1b). This mutant displays a severe defect in ATP-dependent phospholipid transport across the plasma membrane [12]. Thus, even small amounts of plasmid-borne functional flippases at the plasma membrane might result in detectable internalization of phospholipid analogs added to the exterior of the cells. In order to avoid putative toxicity problems, the gene of interest is cloned into a yeast expression plasmid under the control of an inducible promoter, e.g., galactose promoter (GAL10). The resulting construct is used to transform the yeast *Δdrs2Δdnf1Δdnf2* strain. As controls, cells transformed with an empty vector and yeast expressing nonfunctional P4-ATPases are used.

1.3 Assay Conditions

Experiments on intact cells are complicated by endocytosis and metabolism of the lipid probes inserted into the plasma membrane. For example, a common modification of the C6-NBD analogs in eukaryotic cells is the hydrolysis of the glycerophospholipid analogs into lysophospholipids and free NBD-fatty acid which is released into the medium [13]. It is thus important either to suppress endocytosis and conversion of the NBD-lipids (e.g., by lowering the temperature) or to quantify both processes by independent measurements before drawing conclusions from the experimental data.

1.4 Flow Cytometry

Flow cytometry measures the fluorescence of individual cells and offers the possibility to detect and quantify various cellular properties for each cell by using different marker molecules simultaneously. This approach has the advantage over fluorimetry of allowing the exclusion of dead cells (which readily absorb NBD-lipids) from the analysis. The most widely used dye to label nonviable cells is propidium iodide (PI), which is generally excluded from viable cells. PI emits a red fluorescence when excited at 488 nm and can be used in combination with other fluorochromes excited at the same wavelength, such as NBD.

1.5 NucleoCounter® NC-3000™

The NucleoCounter® NC-3000™ (ChemoMetec A/S, Allerod, Denmark) is a microscope-based cell analyzer allowing multiple images of individual cells in dark field and up to four different fluorescent channels. The four LED channels are selected from a range of different excitation/emission filters and are fixed for a customized instrument. As flow cytometry, it allows the analysis of individual cells and their properties, in this case based on imaging of an entire cell population under different excitation and detection conditions, followed by mathematical interpolation of the individual pixels in each image set. The advantage of this system is that samples are mounted on especially designed microscopic slides, thus eliminating the need of a liquid flow that might derive in clotting problems and extensive cleaning.

2 Materials

Prepare all solutions using ultrapure water (prepared by purifying deionized water to attain a sensitivity of 18 MΩ cm at 25 °C).

2.1 Preparation of NBD-Lipids

1. NBD-lipids with a short C6 acyl chain in the *sn*-2 position are available from Avanti Polar Lipids (Alabaster, AL). In addition, synthesis of several NBD-lipid derivatives has been previously described [13–15]. Lipids stocks: 1 mM in chloroform, stored at −20 °C using 1 mL glass vials with Teflon-lined screw caps and periodically monitored for purity by thin-layer chromatography.
2. A glass Hamilton syringe and disposable 12-mL glass tubes.
3. A Nitrogen flow system.
4. Dimethyl sulfoxide (DMSO).

2.2 Preparation of Yeast Cells	1. Vectors from the pRS42X-GAL series [16] bearing the cDNAs coding for the gene(s) of interest under the control of a galactose inducible promoter (Fig. 1a) (*see* **Note 1**).
2. *S. cerevisiae* mutant strain ZHY709 (*MATα his3 leu2 ura3 met15 dnf1Δ dnf2Δ drs2::LEU2*) [17] freshly transformed with the plasmids of interest.
3. Rich synthetic glucose (SD) or galactose (SG) media: 0.7 % (w/v) Yeast Nitrogen Base (YNB), 2 % (w/v) glucose or galactose supplemented with yeast synthetic dropout medium lacking the desired amino acid(s). Prepare stocks solutions of 7 % (w/v) YNB, 20 % (w/v) sugar and 10× yeast drop out medium without the desired amino acid(s), and keep at 4 °C for long-term storage. For 1 L of SD/SG media, autoclave 700 mL of ultrapure water at 120 °C for 20 min. Let it cool down before adding 100 mL of each stock solution (*see* **Note 2**).
4. An orbital shaker.
2.3 NBD-Lipid Uptake	1. A water bath at 30 °C.
2. 12-mL conical bottom and round bottom glass tubes.
3. An ice-water bath.
4. SSA + BSA medium: 0.7 % (w/v) Yeast Nitrogen Base (YNB), 2 % (w/v) sorbitol, 20 mM NaN_3, 3 % (w/v) fatty acid-free bovine serum albumin, supplemented with yeast synthetic dropout medium lacking the desired amino acid(s). SSA media is a modification of SG media in which galactose has been substituted for sorbitol. For 1 L of medium, dissolve 20 g of sorbitol in 800 mL of water and add 100 mL of YNB and yeast drop out media stocks as above. Add 1.3 g of NaN_3 (*see* **Note 3**) and stir until completely dissolved. Finally, add 30 g of BSA and let it stand for at least 12 h at 4 °C until the solution is clear. This medium can be kept up to 3 months at 4 °C.
2.4 Flow Cytometry Analysis	1. A Becton Dickinson FACSCalibur (San Jose, CA) flow cytometer or similar model equipped with a 488 nm argon laser for excitation, appropriate filter sets (FL-1, Band pass 530/30 nm; FL-3, Long pass 670 nm) and Cell Quest software.
2. 5-mL round bottom plastic tubes.
3. PBS: 137 mM NaCl, 10 mM Na_2HPO_4, 2 mM NaH_2PO_4, pH 7. For 1 L, dissolve 8 g of NaCl and 1.42 g of Na_2HPO_4 in 800 mL water. While constantly monitoring the pH, add NaH_2PO_4 from a 500 mM stock solution in water until the mix reaches pH 7. Autoclave for long-term storage at room temperature. Filter through a 0.22-μm pore diameter filter right before use (*see* **Note 4**).
4. Cyflogic (CyFlo Ltd., Finland), FlowJo (FlowJo LLC, USA) or similar software for data analysis.

2.5 NucleoCounter® NC-3000™ Analysis

1. Nucleoview® NC-3000™ software and a NucleoCounter® NC-3000™ (Chemometec A/S, Allerod, Denmark) equipped with the following channels: dark field (cell identification), green (ex: 475 nm, em: 560/35 nm for NBD detection) and orange (ex: 530 nm, em: 675/75 nm for PI detection).
2. A2-slides (Chemometec A/S, Allerod, Denmark).

2.6 Other Solutions

1. PI solution: 1 mg/mL propidium iodide in water. Keep at 4 °C in the dark (*see* **Note 5**).

3 Methods

3.1 Preparation of NBD-Lipids

1. Using a glass Hamilton syringe, deposit the appropriate amount of NBD-lipid from a chloroform stock into a disposable 12-mL glass tube.
2. In a fume hood, dry the chloroform using a gentle stream of nitrogen gas so that a dried lipid film is formed at the bottom of the tube (*see* **Note 6**).
3. Solubilize the NBD-lipids in DMSO to a final concentration of 10 mM (*see* **Note 7**).

3.2 Preparation of Yeast Cells

1. Inoculate 5–10 yeast colonies containing the desired genes into 2 mL of SG medium (*see* **Note 8**) and incubate at 30 °C with shaking (100–110 rpm) for 20–24 h (*see* **Note 9**).
2. Measure OD_{600} for the cultures (*see* **Note 10**) and inoculate an appropriate amount of cells into 50 mL of fresh SG medium in 250 mL flasks, so that the cultures will reach a final OD_{600}/mL of 0.2–0.5 after 16–24 h of growth (*see* **Note 11**). Incubate with shaking (100–110 rpm) at 28 °C for the required time.
3. Harvest the cells by centrifugation ($900 \times g$, 15 min, room temperature) and resuspend them in fresh SG medium at a concentration of 10 OD_{600}/mL.
4. For each NBD-lipid uptake test, set up a conical bottom glass tube (*see* **Note 12**) with 125 µL of the cell suspension and keep the remaining cells on ice (*see* **Note 13**).

3.3 NBD-Lipid Uptake

1. Incubate the cells for 5 min at 30 °C using a water bath (*see* **Note 14**).
2. Start labeling each sample by adding 0.75 µL of DMSO lipid suspension under vortexing (*see* **Note 15**) and immediately return the tube to the water bath.
3. Shake gently every 10 min to prevent the cells from sedimenting.
4. After 30 min (*see* **Note 16**), transfer the samples to an ice-water bath and stop labeling by adding 1 mL of ice-cold SSA + BSA medium.

5. Pellet the cells by centrifugation (750×*g*, 5 min, 4 °C) and discard the supernatant (*see* **Note 17**).
6. Resuspend the cells in 1 mL of ice-cold SSA + BSA medium and transfer the suspension to a round bottom glass tube, in order to eliminate residual DMSO lipid suspension sticking to the tubes used for labeling.
7. Repeat the washing (**steps 5** and **6**) once more.
8. Resuspend the cells in 125 µL of ice-cold SSA + BSA medium and keep the samples on ice until analysis by cytometry (*see* **Note 18**).

3.4 Analysis by Flow Cytometry

1. Start up the flow cytometer and the computer, perform the usual fluidics maintenance and launch CellQuest Pro.
2. Open the appropriate acquisition document from the storage folder or, alternatively, create a new one. This document is suggested to include dot plots of forward scattering (FSC) versus side scattering (SSC) (for overall definition of the cell population), FL-3 versus FSC (rough live/dead discrimination), FL-1 versus FL-3 (NBD lipid uptake in relation to PI staining), and a histogram plot of FL-1.
3. Connect to the cytometer and launch or create appropriate instrument settings. The following initial cytometer settings are used (*see* **Note 19**): FSC (voltage E-1, ampgain 1.00, logarithmic mode), SSC (voltage 320, ampgain 1.00, logarithmic mode), FL-1 (voltage 400, ampgain 1.00, logarithmic mode), FL-3 (voltage 550, ampgain 1.00, logarithmic mode).
4. Transfer 50 µL of the cell suspension to a 5-mL round-bottom plastic tube. Add 1 µL PI solution and 1 mL ice-cold PBS and vortex briefly before proceeding directly to sample acquisition on the flow cytometer. Prepare each sample right before measuring.
5. Analyze at least 20,000 cells per sample at low or medium flow rate during the acquisition (*see* **Note 20**).
6. After completing data acquisition and storage, perform the usual fluidics maintenance and turn off the flow cytometer.
7. Proceed to data analysis with your software of choice (Fig. 2).

3.5 Analysis by NucleoCounter® NC-3000™

Preparation of NBD-lipids and yeast cells including the NBD-lipid uptake assay is done according to the procedures presented above (Subheadings 3.1, 3.2 and 3.3).

1. Dilute the samples to 0.2 OD_{600}/mL with ice-cold SSA + BSA medium in 1.5 mL Eppendorf tubes.
2. Start up the NucleoCounter® NC-3000™ and open the Nucleoview® NC-3000™ software.
3. Open the appropriate acquisition document from the storage folder or, alternatively, create a new one. This document is

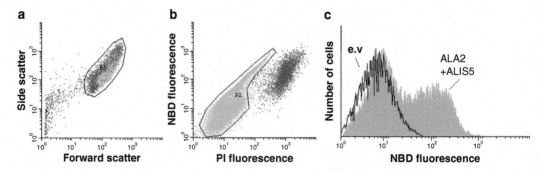

Fig. 2 Monitoring NBD-lipid uptake via flow cytometry. Yeast Δ*drs2*Δ*dnf1*Δ*dnf2* mutant cells expressing the plant P4-ATPase ALA2 and β-subunit ALIS5 were incubated with C6-NBD-PS for 30 min at 30 °C and subsequently analyzed by a FACS cytometer using CellQuest 3.3 software (Becton Dickinson, San Jose, CA). (**a**) Dot plots of side scatter versus forward scatter are used to position gate P1 for selection of the yeast cell population; (**b**) Dot plot of FL-1 (NBD) fluorescence versus FL-3 (PI) fluorescence of the P1 cell population is used to draw gate P2 comprising NBD-labeled living cells and excluding the population of PI-stained dead cells. (**c**) The *green* (NBD) fluorescence of living cells is then plotted on a histogram and the mean fluorescence intensity (geometric mean) of cells showing NBD-lipid uptake is calculated. *Black line*, living cells harboring an empty vector control; *green shadow*, living cells harboring the plant ALA2-ALIS5 complex

suggested to have the following settings: exposure time of 1000 in the green, orange and red channels; exposure time for non-used channels set to 0; number of cells measured set to 10,000.

4. Transfer 50 μL of sample to a new Eppendorf tube, add 1 μL PI solution and vortex briefly. Prepare each sample right before measuring.
5. Load 30 μL of PI-containing sample on an A2-slide (*see* **Note 21**).
6. Name the samples in the Nucleoview® NC-3000™ software window.
7. Load the slide into the NucleoCounter® NC-3000™.
8. Measure the samples.
9. When all data are collected, proceed to analysis using the Nucleoview® NC-3000™ software® (Fig. 3).

4 Notes

1. Any type of high-expression vector could be used with similar results. To optimize the expression in yeast cells, full-length cDNAs must be cloned, eliminating any ATG triplet in the cloning polylinker or in any sequence preceding the first ATG of the open reading frame. These ATGs generate out-of-frame ORFs in the untranslated region (upstream ORFs [uORFs]) of the expressed mRNA, and may substantially reduce the

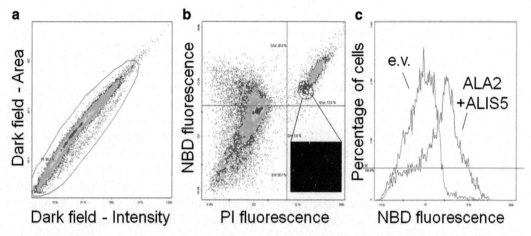

Fig. 3 Monitoring NBD-lipid uptake using NucleoCounter® NC-3000™. Yeast Δdrs2Δdnf1Δdnf2 mutant cells expressing the plant P4-ATPase ALA2 and β-subunit ALIS5 were incubated with C6-NBD-PS and subsequently analyzed using Nucleoview® NC-3000™ software®. (**a**) A bi-exponential dot plot of darkfield area vs. darkfield intensity is used for positioning gate P1 for selection of the cell population; (**b**) Bi-exponential dot plot showing *green* (NBD) against *orange* (PI) fluorescence intensity of the P1 cell population. Quadrants are added to gate for living cells (*lower* and *upper left quadrants*, low PI signal) and exclude dead cells (*lower* and *upper right quadrants*, high PI signal). *Inset*: Optionally, individual cell populations can be presented in an image; (**c**) NBD fluorescence (bi-exponential) of living cells is then plotted on a histogram, and the mean fluorescence intensity (arithmetic mean) of living yeast cells is extracted by setting a marker, M1

expression of the transporter [18]. Moreover, it is also convenient if the sequence around the ATG is as similar as possible to the following sequence: (A/T)A(A/C)A(A/C)A ATG TC(T/C) [19].

2. Synthetic media can be kept at 4 °C for 2–3 months, but we rather prepare it fresh for each experiment.

3. NaN_3 is extremely toxic, especially in the powdered form as it can easily be inhaled. Weigh in a balance placed inside a fume hood or under suction and wear a mask when handling the solid.

4. In case of yeast cells, the PBS can be replaced by freshly tapped ultrapure water.

5. PI is a suspected carcinogen and should be handled with care. The dye must be disposed of safely and in accordance with applicable local regulations.

6. Although trace solvents may remain after this step, this approach has proven to be satisfactory for our purposes. To ensure complete solvent removal, drying under vacuum would be required.

7. DMSO lipid suspensions can be stored at −20 °C and can be used for up to 2 weeks. However, they are prone to precipitation owing their hygroscopic nature. Although we commonly

use 10 mM stocks, a 4 mM stock is enough to provide excess lipid for most transporters.

8. If expression of the target proteins is toxic for the cells or the proteins are unstable upon sustained expression, the first overnight cultures can be started in SD medium. In this case, growth conditions are the same, but incubation times are reduced to 16–18 h. After this time, 4-mL cultures are started at an $OD_{600}/mL = 1$ in SG medium and incubated for 4–5 h under the same temperature and agitation conditions, before re-inoculating into 50 mL fresh SG medium as described in Subheading 3.2, **step 2**. In these 4–5 h, cells will adapt to the galactose-containing medium, but they will not manage to divide, so it is not necessary to measure OD_{600} again.

9. Although agitation might be unimportant in other cases, the ZHY709 strain used in these experiments is very sensitive to excessive shaking. In order to minimize the population of dead cells, it is important to avoid the use of baffled flasks and incubate cultures under soft shaking. Temperature is, however, not that crucial, and it can fluctuate between 28 and 30 °C.

10. We usually prepare a 1/20 dilution of the overnight cultures in water and measure this against a water reference.

11. The starting OD_{600}, as well as the cultivation time, depend on the growth properties of the *S. cerevisiae* mutant strain used. In addition, the expression level of the target protein(s) may also affect growth rates. It is advisable to characterize the growth of the transformed yeast strain prior to these experiments.

12. It is important that labeling with fluorescent lipids is carried out as soon as possible after cells are processed. Therefore, it is advisable to prepare and label all tubes before starting. You will need one conical bottom and one round bottom glass tube for each uptake reaction (i.e., one strain tested against one lipid). We find it especially useful to assign one color to each lipid and one number to each strain. This way, if we are testing uptake of for example NBD-phosphatidylserine in ten different yeast strains carrying different P4-ATPases, all tubes will be labeled with red tape and numbered 1–10. For any other lipid tested for the same strains, tubes will be labeled with a different color and numbered in the same way.

13. To verify lipid flippase expression, an aliquot of the cell suspension may be subjected to western blot analysis.

14. Assays can also be performed on ice or in the presence of 20 µM latrunculin A to prevent lipid uptake by endocytosis.

15. The easiest way to do this is to pipette the lipid directly into the cell suspension, vortex briefly, then wash the pipette tip by pipetting a couple of times up and down into the cell suspension before vortexing briefly again.

16. In our usual set-up we incubate the cells for 30 min, but longer times can be used for transporters that are lowly expressed or not very active. However, lipid flippase activity is dependent on the hydrolysis of cytoplasmic ATP. During prolonged incubation (> 60 min at 30 °C), the medium might get depleted of an energy source (e.g., galactose) required for the cells to maintain a constant ATP level. In this case, the intracellular ATP concentration should be measured at the end of the incubation and the medium adjusted accordingly to ensure that ATP will not decrease with time.

17. At this point, a sample of the medium can be collected to test NBD-lipid metabolism. Lipids can be extracted with a chloroform/methanol (2:1) mixture and separated by thin-layer chromatography using a chloroform/methanol/water mixture (65:25:4, v/v/v).

18. Cells should be analyzed within 1–2 h. A sample of cells (e.g., 50 µL) can be collected, extracted with a chloroform/methanol (2:1) mixture and separated by thin-layer chromatography using a chloroform/methanol/water mixture (65:25:4, v/v/v) to analyze NBD-lipid metabolism within the cells.

19. These settings represent a reference point to roughly define the cell population with respect to size, NBD and PI fluorescence. Further adjustment of detector gains and voltages may be necessary. In addition, individual samples stained with a single fluorophore, i.e., NBD-lipid or PI, can be used to set compensation percentages.

20. During acquisition, it is recommended to gate on cells (R1) in the FSC vs. SSC dot plot and on living cells (R2) in the FL-3 vs. FSC dot plot. Using the Gate List menu option in the Cell Quest Software, a logical gate named G3 (G3 = R1 and R2) can be created to collect at least 20,000 events that follow the requirements for G3.

21. A2 slides allow for loading of two independent samples at once.

References

1. Coleman JA, Quazi F, Molday RS (2013) Mammalian P4-ATPases and ABC transporters and their role in phospholipid transport. Biochim Biophys Acta 1831:555–574

2. López-Marqués RL et al (2014) Structure and mechanism of ATP-dependent phospholipid transporters. Biochim Biophys Acta 1850:461–475

3. Lopez-Marques RL et al (2013) P4-ATPases: lipid flippases in cell membranes. Pflügers Arch 466:1227–1240

4. Menon AK, Iii WE, Hrafnsdóttir S (2000) Specific proteins are required to translocate phosphatidylcholine bidirectionally across the endoplasmic reticulum. Curr Biol 10:241–252

5. Devaux PF, Fellmann P, Hervé P (2002) Investigation on lipid asymmetry using lipid probes: Comparison between spin-labeled lipids and fluorescent lipids. Chem Phys Lipids 116:115–134

6. Gummadi SN, Menon AK (2002) Transbilayer movement of dipalmitoylphosphatidylcholine

in proteoliposomes reconstituted from detergent extracts of endoplasmic reticulum. Kinetics of transbilayer transport mediated by a single flippase and identification of protein fractions enriched in flippase activity. J Biol Chem 277:25337–25343

7. Chang Q, Gummadi SN, Menon AK (2004) Chemical modification identifies two populations of glycerophospholipid flippase in rat liver ER. Biochemistry 44:10710–10718

8. Segawa K et al (2014) Caspase-mediated cleavage of phospholipid flippase for apoptotic phosphatidylserine exposure. Science (New York, N Y) 344:1164–1168

9. Pomorski T et al (1996) Transbilayer movement of fluorescent and spin-labeled phospholipids in the plasma membrane of human fibroblasts: a quantitative approach. J Cell Sci 109(Pt 3):687–698

10. Pomorski T et al (2003) Drs2p-related P-type ATPases Dnf1p and Dnf2p are required for phospholipid translocation across the yeast plasma membrane and serve a role in endocytosis. Mol Biol Cell 14:1240–1254

11. López-Marqués RL et al (2010) Intracellular targeting signals and lipid specificity determinants of the ALA/ALIS P4-ATPase complex reside in the catalytic ALA α-subunit. Mol Biol Cell 21:791–801

12. Poulsen LR, et al. (2008) A Golgi localized Arabidopsis thaliana P4-ATPase involved in root and shoot development requires a β subunit to gain functionality. Plant Cell 20:658–676

13. Fellmann P et al (2000) Transmembrane movement of diether phospholipids in human erythrocytes and human fibroblasts. Biochemistry 39:4994–5003

14. Colleau M et al (1991) Transmembrane diffusion of fluorescent phospholipids in human erythrocytes. Chem Phys Lipids 57:29–37

15. Sleight RG (1994) Fluorescent glycerolipid probes. Synthesis and use for examining intracellular lipid trafficking. Methods Mol Biol (Clifton, N J) 27:143–160

16. Burgers PM (1999) Overexpression of multisubunit replication factors in yeast. Methods (San Diego, Calif) 18:349–355

17. Hua Z, Fatheddin P, Graham TR (2002) An essential subfamily of Drs2p-related P-type ATPases is required for protein trafficking between Golgi complex and endosomal/vacuolar system. Mol Biol Cell 13:3162–3177

18. Bañuelos MA et al (2008) Effects of polylinker uATGs on the function of grass HKT1 transporters expressed in yeast cells. Plant Cell Physiol 49:1128–1132

19. Kozak M (2002) Pushing the limits of the scanning mechanism for initiation of translation. Gene 299:1–34

Chapter 34

Transient Expression of P-type ATPases in Tobacco Epidermal Cells

Lisbeth R. Poulsen, Michael G. Palmgren, and Rosa L. López-Marqués

Abstract

Transient expression in tobacco cells is a convenient method for several purposes such as analysis of protein-protein interactions and the subcellular localization of plant proteins. A suspension of *Agrobacterium tumefaciens* cells carrying the plasmid of interest is injected into the intracellular space between leaf epidermal cells, which results in DNA transfer from the bacteria to the plant and expression of the corresponding proteins. By injecting mixes of *Agrobacterium* strains, this system offers the possibility to co-express a number of target proteins simultaneously, thus allowing for example protein–protein interaction studies. In this chapter, we describe the procedure to transiently express P-type ATPases in tobacco epidermal cells, with focus on subcellular localization of the protein complexes formed by P4-ATPases and their β-subunits.

Key words *Nicotiana benthamiana*, *Agrobacterium tumefaciens*, Epidermal cells, Fluorescent protein, Co-expression, Infiltration

1 Introduction

Transient expression in tobacco epidermal cells has advantageously been applied for the study of membrane proteins. The technique can be used to express fluorescently tagged proteins and thereby determine the subcellular localization of proteins of interest [1, 2]. In combination with the split-GFP system, the method is also commonly used as an in vivo system to detect protein-protein interactions [3].

For introduction of foreign DNA into tobacco, advantage is taken of the natural lifecycle of *Agrobacterium tumefaciens*. *Agrobacterium* is a soil bacterium, which naturally infects plants and causes the tumor-related disease crown gall [4]. For the tumor to develop, a fragment of the bacterial DNA (the so-called transferred DNA; T-DNA) is directly transferred from the bacteria and integrated into the plant genome. The machinery responsible for this process is encoded by a tumor-inducing (Ti) plasmid including the T-DNA and virulence genes (*vir*-genes) important for the transfer of T-DNA. With only relatively minor modifications of this natural plant transformation process, genes of interest can be introduced and expressed in plants [5, 6]. Thus, keeping as little as 25 bp from each side of the T-DNA from the original Ti-plasmid (left and right border (LB, RB), Fig. 1) is enough for integration of the T-DNA into the plant genome, meaning that almost all of the T-DNA region can be exchanged with the gene of interest, in this way removing genes involved in tumor formation. In addition, the Ti-plasmid, which originally consists of approximately 200,000 bp, can be split into two plasmids such that the *vir*-genes and the T-DNA become separated in space while both gene-fragments are

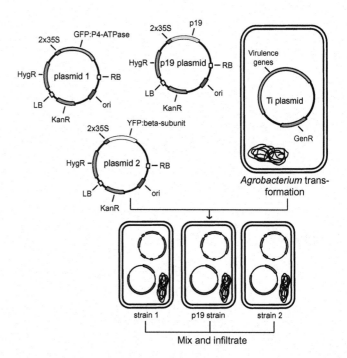

Fig. 1 Using *Agrobacterium tumefaciens* for transient expression. Genes of interest are cloned under the control of the desired promoter into so-called binary plasmids in between DNA sequences denominated left border (LB) and right border (RB) that mark the DNA region, which will be transferred to the plant. Although this is not required for transient expression, most binary plasmids also carry a selection marker for plants between LB and RB. The plasmids also contain origins of replication for *E. coli* and *Agrobacterium* (ori) and a selection marker for bacteria. Laboratory *Agrobacterium* strains carry helper plasmids that contain the virulence genes necessary for transfer of DNA to the plant cell. Plasmids are individually transformed into *Agrobacterium* and strains are mixed prior to infiltration to achieve the desired combination of target proteins. We routinely mix strains carrying the genes of interest with a freshly transformed strain containing the silencing suppressor p19 gene. Other abbreviations: *KanR* kanamycin resistance, *BaR* Basta resistance, *HygR* Hygromycin resistance

fully functional and still work in concert [7, 8]. In such a binary plasmid system employing two vectors the *Agrobacterium* host line is harboring the plasmid containing the *vir*-genes after removal of tumor-inducing genes (Fig. 1), while the small T-DNA fragments (LB and RB) are placed on a smaller plasmid (vector) which can be amplified in both *Escherichia coli* and *Agrobacterium* (Fig. 1) [7, 8]. The gene of interest is then cloned between LB and RB under the control of the desired gene promoter (*see* Subheading 1.3).

1.1 Choice of Agrobacterium Strain and Application System

There are several *Agrobacterium* strains available which can be used for transient expression [7]. They all share the common feature of carrying a modified Ti plasmid (helper plasmid) for transfer of genes of interest into plants. However, each of the strains has different capabilities, which should be taken into account when making a choice. For instance, the widely used *Agrobacterium* strain, GV3101, has been shown to generate higher expression levels compared to another well-known strain named C58C1 (approximately 50 % of GV3101) [9]. In this protocol, strain C5851 will be used, as in our hands this strain provides more consistent results when co-expressing two membrane proteins at the same time.

For gene transfer between *Agrobacterium* and plants, the bacteria need to be infiltrated into the plant tissue. This process can be executed in several ways. The two least invasive methods known to date are vacuum infiltration and direct injection. To vacuum-infiltrate plant tissue, the desired plant area is submerged in an *Agrobacterium* solution, after which the whole plant or leaf disk is exposed to short vacuum treatment. During this treatment the *Agrobacterium* solution is forced into the plant tissue via the stomatal pores. This method can be used for both stable and transient expression [10]. For transient expression in leaves, direct injection can be advantageous. In this case, only the area which has been chosen for expression is being exposed to the *Agrobacterium* solution, which is directly injected into the leaf tissue through stomata without affecting the rest of the plant [11].

1.2 Choice of Tobacco Species

Several tobacco species are commonly used for transient expression, but as for the *Agrobacterium* strains, each species shows specific abilities. As an example, compared to other *Nicotiana* species, *Nicotiana benthamiana* is often used for transient expression and is known to produce a high degree of recombinant protein within a short timeframe [12]. However, due to the small biomass yield from this species, large scale production of recombinant proteins may be facilitated by using species that are more sizeable such as *N. excelsior* [12]. The hybrid species *N. excelsiana* (*N. benthamiana* × *N. excelsior*), which recently has become available, has a higher biomass production than *N. benthamiana* and is easier to infiltrate than *N. excelsior* [9]. The commonly used *N. benthamiana* will be employed as an expression host in this protocol.

1.3 Choice of Promoter

Although in many cases, the native gene promoter can be used for expression in tobacco epidermal cells, it is usual practice to utilize strong constitutive promoters corresponding to other genes of plant (e.g., ubiquitin) [13] or viral origin (e.g., 35S from cauliflower mosaic virus) [14]. This circumvents the limitations derived from low expression or from the presence of tissue-specific or growth condition-specific regulatory sequences in the native promoters. However, high expression derived from constitutive promoters, especially from the viral 35S sequences, can result in posttranslational gene silencing [15]. Co-expression of viral silencing repressors, such as the p19 protein [16], is widely used to overcome this problem. In this protocol, overexpression of membrane proteins will be achieved under the control of a strong double 35S promoter combined with co-expression of a p19 silencing suppressor.

2 Materials

Prepare all solutions using ultrapure water (prepared by purifying deionized water to attain a sensitivity of 18 MΩ cm at 25 °C). Keep all *Agrobacterium* growth media at 4 °C for long-term storage.

2.1 Tobacco Plants

1. *N. benthamiana* seeds.
2. Small round plastic pots 5 cm in diameter.
3. Standard soil.
4. A green house or growth chamber under the following conditions: 16 h light at around 140 µmol photons m^{-2} s^{-1} light intensity followed by 8 h darkness, humidity around 70 %, temperatures ranging between 24 °C during the day and 17 °C during the night.

2.2 Plasmids

1. cDNA fragments (*see* **Note 1**) corresponding to P4-ATPases fused to the coding sequence for green fluorescent protein (GFP) generated in binary plasmids pMDC43 and pMDC84 [17] containing a double 35S promoter (*see* **Note 2**).
2. C-terminally Yellow fluorescent protein (YFP)-tagged versions of the corresponding P4-ATPase β-subunits generated in binary plasmid pEarleyGate101 [18], bearing a single 35S promoter.
3. A plasmid containing the p19 gene encoding the viral silencing suppressor p19 protein [16], to facilitate high expression of recombinant proteins.

2.3 Agrobacterium Preparation

1. YEP plates [20]: 1 % yeast extract, 1 % peptone, 0.5 % NaCl, 1.5 % agar. Sterilize by autoclaving at 120 °C for 20 min. Let the media cool down to around 60–65 °C and add 50 µg/mL

kanamycin and 25 µg/mL gentamicin (*see* **Note 3**), before pouring into round petri dishes. Antibiotic stocks are 5 mg/mL gentamicin and 50 mg/mL kanamycin in water. Filter-sterilize and keep at −20 °C.

2. *Agrobacterium tumefaciens* strain C58C1 [19] (*see* **Note 4**) transformed with each of the plasmids of interest individually and selected on YEP plates.

3. LB liquid medium [21]: 1 % Tryptone, 0.5 % Yeast Extract, 1 % NaCl. Sterilize by autoclaving at 120 °C for 20 min. Let the medium cool down to around 60–65 °C before adding 50 µg/mL kanamycin and 25 µg/mL gentamicin from stock solutions as above.

4. 14-mL round-bottom culture tubes.

5. An orbital shaker and a static incubator at 28 °C.

6. A table centrifuge suitable for 2 mL Eppendorf tubes.

7. A spectrophotometer for measuring absorbance at 600 nm and standard plastic cuvettes of 1.5 mL volume.

2.4 Other Media

1. Infiltration solution: 10 mM $MgCl_2$ and 100 µM 4′-Hydroxy-3′,5′-dimethoxyacetophenone (acetosyringone) in water. Prepare a stock solution of acetosyringone at 10 mM in 50 % ethanol and store at -20 °C (*see* **Note 5**). Keep a 1 M $MgCl_2$ stock solution in water at room temperature. Right before starting your experiment, take equal volumes of acetosyringone and $MgCl_2$ stocks to a 50-mL plastic tube and make a 1:100 dilution in water (i.e., for 50 mL final volume of infiltration solution, you will need to mix 500 µL of each reagent). Keep at room temperature.

2.5 Visualization

1. Leica SP2 UV MP or SP5 II spectral confocal laser scanning microscope (Leica Microsystems, Heidelberg, Germany) with a 63×/1.2 NA water immersion objective and equipped with laser lines at 488 and 514 nm.

2. Standard microscopy slides and cover glasses.

3 Methods

Due to the variability inherent to working with living organisms, each transient expression experiment should be repeated at least three independent times, with a new batch of tobacco plants and fresh *Agrobacterium* transformations, to ensure reproducibility of the results.

3.1 Preparation of Tobacco Plants

1. Plant *N. benthamiana* seeds in individual small pots containing standard soil and grow them in a greenhouse or growth chamber.

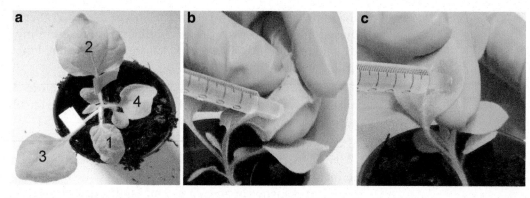

Fig. 2 Infiltration procedure. (**a**) 3-week-old tobacco plant suitable for infiltration. Leaves are numbered starting from the uppermost leave with a reasonable size. Leaves numbered 1, 2, and 3 would be infiltrated in such a plant; (**b**) A syringe without needle is used to force an *Agrobacterium*-containing solution into the intracellular space of tobacco epidermal cells. The lower side of the leaf is used for infiltration. The tip of the syringe is pressed against the leaf and a counter-pressure is applied from the other side by using a finger; (**c**) The *Agrobacterium* solution is released by applying pressure on the syringe piston. A *darker green* area appears where the solution has entered the leaf

2. Remove 3-week old *N. benthamiana* plants at the 4-5 leaf stage from the growth chamber (*see* **Note 6**) and allow them to acclimate to ambient laboratory conditions for at least 1 h before infiltration (*see* **Note 7**).

3. Select and label leaves to be infiltrated (*see* **Note 8**) by hanging a piece of tape from the leaf attachment to the stem (Fig. 2a).

3.2 Preparation of Agrobacterium

1. For each DNA construct freshly transformed in *Agrobacterium* (*see* **Note 9**), transfer one colony to a 14-mL round bottom culture tube containing 2 mL of antibiotic-containing LB medium. For the p19-containing *Agrobacterium*, inoculate 3–4 colonies, as this strain presents a reduced growth due to toxicity of the p19 protein.

2. Incubate cultures at 28 °C for 14–16 h in an orbital shaker with 160–180 rpm agitation.

3. Transfer the cells to 2-mL Eppendorf tubes.

4. Centrifuge at $2500 \times g$ for 4 min at room temperature.

5. Discard supernatant and resuspend in 1 mL infiltration solution by pipetting very gently up and down (*see* **Note 10**).

6. Repeat **steps 4** and **5** once more.

7. Incubate without shaking at 28 °C for 1–2 h (*see* **Note 11**).

8. Measure OD_{600} for each *Agrobacterium* suspension using a 1:10 dilution in water.

9. For each plant to be infiltrated, prepare 1–2 mL (*see* **Note 12**) of infiltration mix by diluting the desired combination of *Agrobacterium* strains to a final $OD_{600} = 0.3$ each in infiltration solution (*see* **Note 13**).

3.3 Infiltration of Epidermal Cells

1. Transfer the cell suspension to a 1- or 2-mL syringe without needle (*see* **Note 14**).
2. Avoiding the veins, place the syringe tip against the lower side of the leaf and apply a gentle pressure against it from the other side of the leaf (*see* **Note 15**) (Fig. 2c).
3. Push slowly the syringe piston to force the cell suspension into the leaf. As the apoplastic space gets filled, the infiltrated area will turn a darker shade of green (*see* **Note 16**) (Fig. 2d). Infiltrate a leaf area of about 0.5–1 cm around the infiltration point.
4. Water the plants and return them to the growth chamber or the greenhouse. Avoid watering during the last 24–30 h prior to visualization (*see* **Note 17**).

3.4 Visualization

1. Cut a piece of about 3 × 3 mm from the desired infiltrated leaf, starting by the area closest to the wound inflicted by the syringe tip (*see* **Note 18**).
2. Place the leaf fragment onto a microscopic slide with the infiltrated side up, add a drop of water on top and cover with a cover slide (*see* **Note 19**).
3. Mount the sample in the confocal microscope and excite GFP at 476 nm (100 % laser intensity) and record its emission spectra between 492 and 500 nm. Excite YFP at 514 nm (40 % laser intensity) and record emission between 545 and 560 nm. Use sequential scanning between lines to follow both fluorescent proteins at once (*see* **Note 20**). Images are scanned with a line average of 2 (*see* **Note 21**).

For each localization experiment, visualize at least 3 independent samples per plant, each corresponding to a single infiltrated leave (*see* **Note 22**).

4 Notes

1. In some cases, expression levels might improve when using genomic DNA fragments.
2. As the role of the protein termini is still relatively unknown for P4-ATPases, we routinely generate and test constructs tagged at both termini.
3. Gentamicin will select for the helper plasmid, while kanamycin will select for the plasmid bearing the gene of interest.
4. Other *Agrobacterium* strains, such as GV3101, have been used with identical positive results.
5. We have used acetosyringone stock solutions up to 1-year old with positive results.

6. It is very important to use healthy unstressed young plants.
7. This acclimation period facilitates subsequent infiltration.
8. Younger leaves tend to express at higher levels than older leaves, as the cells are actively dividing. However, young leaves are also more difficult to infiltrate and to visualize microscopically due to a more evident vascular tissue. Therefore, we infiltrate three leaves per plant, starting by the uppermost leaf with a reasonable size. We routinely infiltrate several plants with the same DNA construct(s) either in the same or in independent experiments, to account for variability of the biological replicates.
9. Transformed *Agrobacterium* cells can also be kept at −80 °C and directly inoculated from the glycerol stock, although final expression levels in the infiltrated tobacco leaves tend to be lower.
10. Vortexing dramatically diminishes the efficiency of the infiltration procedure.
11. Shaking at this step reduces the efficiency of the infiltration procedure.
12. The amount of infiltration solution required will depend on personal skills and plant conditions.
13. Although we have assayed different final OD_{600} for the different *Agrobacterium* strains, an equal concentration of each provides the best results in our hands. However, this should be assessed in a case-to-case basis. Up to four different strains can be successfully combined.
14. Use a syringe with a narrow tip to minimize the wounded area caused during contact with the leave surface.
15. As the leaf will get wet with *Agrobacterium* suspension on both sides, use always gloves when infiltrating.
16. Although it is possible in some cases to infiltrate the whole leaf using one infiltration point, best results are obtained by infiltrating at several different positions within the same leaf. In other cases, especially for very young leaves, the infiltrated areas will be small due to the difficulty to force the cell suspension into the apoplastic space. At this point, it is possible to mark the extent of the infiltrated area by using a black marker pen, if desired. Watch out for squirting *Agrobacterium* suspension.
17. The time required to properly express the P-type ATPase will depend on the *Agrobacterium* strain, the tobacco species, the promoter used and the protein of interest. Typically, overproduction of the desired protein during the first stages of transient expression saturates the cell machinery, causing retention in the endoplasmic reticulum. Progressively, the first wave of high expression subdues and trafficking of the nascent protein

to its target membrane becomes more effective. Times as short as 48 h can be sufficient for proper localization in some cases. For P4-ATPases and their β subunits, the optimal is around 3–5 days in our hands, although longer times might be required for some specific proteins.

18. Do not prepare all your samples at once. Each sample should be cut from the leaf right before visualization.

19. As the leaves are quite thick, it is sometimes useful to use double-sided tape to adhere the cover glass to the slide. That will generate a separation between the two glass pieces that will better accommodate the leaf fragment.

20. In our hands, co-infiltration of a GFP-tagged protein with another YFP-tagged protein, always results in a population of cells expressing either of the two proteins individually. Therefore, it is always possible to find cells that show GFP signal but no YFP signal and *vice versa*. We use these cells to establish the visualization parameters (laser intensity, pinhole aperture, photomultiplier sensitivity) where we cannot detect bleed-through of the fluorescent signals (Fig. 3). These parameters are then used for all the samples in a batch. If any parameter change is necessary, bleed-through should be reassessed.

21. If the signal is too low, it might be necessary to use a higher line average to reduce to background noise. This, however, increases the working time and the concurrent bleaching of the fluorophores, which occurs each time the sample is excited. Therefore, cells with high expression levels are chosen for

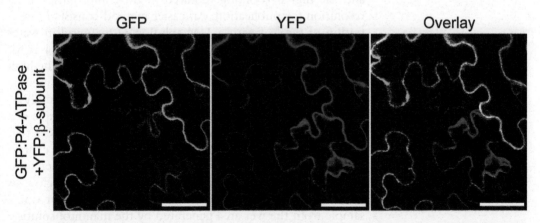

Fig. 3 Bleed-through assessment. Microscopic visualization of tobacco epidermal cells infiltrated with a GFP-tagged P4-ATPase and an YFP-tagged β-subunit. The image shows an example where cells only showing GFP fluorescence (overlay image, *green* signal), only showing YFP fluorescence (overlay image, *magenta* signal) or showing both signals simultaneously (overlay image, *white* signal) are present. Cells only showing the fluorescent signal for one of the proteins are used to optimize acquisition parameters such as photomultiplier sensitivity, pinhole aperture and laser intensity. Once defined, the parameters are kept unchanged throughout visualization of a whole plant set. Scale bar, 50 μm

visualization in applications that require long acquisition times. Bleaching of GFP is faster than YFP. For this reason, it is advisable to use YFP when only one protein of interest is expressed.

22. The expression levels for the desired protein(s) will vary with leaf age, plant health, and even seed batch. Too high expression will result in endoplasmic reticulum retention, while low expression will make the fluorescent signal difficult to distinguish from the background noise. For co-localization experiments, different cells may be expressing each tagged protein at a different level. Therefore, it is important to take a large number of images that represent the biological variability of the sample when analyzing results. This is especially relevant in subcellular localization experiments. In tobacco epidermal leaves, the vacuole occupies most of the cell volume, pressing all other membrane structures against the cell wall. Some organelles, like Golgi or prevacuolar compartments, will appear as easily distinguishable vesicular bodies constantly moving around the cell. The endoplasmic reticulum can be easily spotted as a patchy signal around the nucleus and as membrane threads with a high mobility that cross the cytoplasm (a feature named cytoplasmic streaming). The tonoplast and the plasma membrane are, in contrast, difficult to discern, as both appear as a sharply defined signal along the contour of the cell. One way to distinguish them is to zoom in on the cell nucleus, since the plasma membrane encloses all cellular organelles, while the tonoplast surrounds them. However, this visualization may become an arduous task for an untrained eye and the digital zooming required might compromise image resolution for publication. An easier method consists in infiltration of the leaf with a 1 M solution of mannitol in water 2–5 min before visualization, using the same protocol as for *Agrobacterium* infiltration described in this chapter. The high osmotic pressure of the mannitol solution causes the vacuole to lose water and shrink, in a process termed cell plasmolysis. In our standard procedure, we first take one single sample of a leaf that has been infiltrated at several positions and check that our protein(s) of interest are expressed. If the levels of expression are adequate for visualization, we choose another infiltrated area on the same leaf and inject mannitol as close as possible to the first infiltration wound. Then we cut out a sample from the wet area generated by the mannitol solution entering the apoplast and proceed to visualization. In this case, a drop of mannitol solution is placed between the sample and the cover glass instead of water. After plasmolysis, the plasma membrane remains attached to the cell wall by strips of membrane called Hechtian strands, while the vacuole fragments and appears as large round intracellular membrane structures.

References

1. López-Marqués RL, Poulsen LR, Hanisch S, Meffert K, Buch-Pedersen MJ, Jakobsen MK, Pomorski TG, Palmgren MG (2010) Intracellular targeting signals and lipid specificity determinants of the ALA/ALIS P4-ATPase complex reside in the catalytic ALA alpha-subunit. Mol Biol Cell 21:791–801
2. López-Marqués RL, Poulsen LR, Palmgren MG (2012) A putative plant aminophospholipid flippase, the Arabidopsis P4 ATPase ALA1, localizes to the plasma membrane following association with a β-subunit. PLoS One 7:e33042
3. Jones AM, Xuan Y, Xu M, Wang RS, Ho CH, Lalonde S, You CH, Sardi MI, Parsa SA, Smith-Valle E, Su T, Frazer KA, Pilot G, Pratelli R, Grossmann G, Acharya BR, Hu HC, Engineer C, Villiers F, Ju C, Takeda K, Su Z, Dong Q, Assmann SM, Chen J, Kwak JM, Schroeder JI, Albert R, Rhee SY, Frommer WB (2014) Border control—a membrane-linked interactome of Arabidopsis. Science 344:711–716
4. Smith EF, Townsend CO (1907) A plant-tumor of bacterial origin. Science 25:671–673
5. Herrera-Estrella L, Depicker A, van Montagu M, Schell J (1983) Expression of chimeric genes transferred into plant cells using a Ti-plasmid-derived vector. Nature 303:209–213
6. Murai N, Kemp JD, Sutton DW, Murray MG, Slightom JL, Merlo DJ, Reichert NA, Sengupta-Gopalan C, Stock CA, Barker RF, Hall TC (1983) Phaseolin gene from bean is expressed after transfer to sunflower via tumor-inducing plasmid vectors. Science 222:476–482
7. Hellens R, Mullineaux P, Klee H (2000) A guide to Agrobacterium binary Ti vectors. Trends Plant Sci 5:446–451
8. Zupan J, Muth TR, Draper O, Zambryski P (2000) The transfer of DNA from Agrobacterium tumefaciens into plants: a feast of fundamental insights. Plant J 23:11–28
9. Shamloul M, Trusa J, Mett V, Yusibov V (2014) Optimization and utilization of Agrobacterium-mediated transient protein production in Nicotiana. J Vis Exp 86:e51204
10. Tague BW, Mantis J (2006) In planta Agrobacterium-mediated transformation by vacuum infiltration. Methods Mol Biol 323:215–223
11. Sparkes IA, Runions J, Kearns A, Hawes C (2006) Rapid, transient expression of fluorescent fusion proteins in tobacco plants and generation of stably transformed plants. Nat Protoc 1:2019–2025
12. Sheludko YV, Sindarovska YR, Gerasymenko IM, Bannikova MA, Kuchuk NV (2007) Comparison of several Nicotiana species as hosts for high-scale Agrobacterium-mediated transient expression. Biotechnol Bioeng 96:608–614
13. Garbarino JE, Oosumi T, Belknap WR (1995) Isolation of a polyubiquitin promoter and its expression in transgenic potato plants. Plant Physiol 109:1371–1378
14. Benfey PN, Chua NH (1990) The cauliflower mosaic virus 35S promoter: combinatorial regulation of transcription in plants. Science 250:959–966
15. Vaucheret H, Béclin C, Fagard M (2001) Post-transcriptional gene silencing in plants. J Cell Sci 114:3083–3091
16. Voinnet O, Rivas S, Mestre P, Baulcombe D (2003) An enhanced transient expression system in plants based on suppression of gene silencing by the p19 protein of tomato bushy stunt virus. Plant J 33:949–956
17. Curtis MD, Grossniklaus U (2003) A gateway cloning vector set for high-throughput functional analysis of genes in planta. Plant Physiol 133:462–469
18. Earley KW, Haag JR, Pontes O, Opper K, Juehne T, Song K, Pikaard CS (2006) Gateway-compatible vectors for plant functional genomics and proteomics. Plant J 45:616–629
19. Koncz C, Schell J (1986) The promoter of TL-DNA gene 5 controls the tissue-specific expression of chimeric genes carried by a novel type of Agrobacterium binary vector. Mol Gen Genet 204:383–396
20. An G, Ebert PR, Mitra A, Ha SB (1988) Binary vectors. In: Gelvin SB, Schilperoort RA (eds) Plant molecular biology manual A3. Kluwer Academic Publishers, Dordrecht, pp 1–19
21. Bertani G (1952) Studies on lysogenesis. I The mode of phage liberation by lysogenic Escherichia coli. J Bacteriol 62:293–300

Part VII

Lipid Techniques

Chapter 35

Lipid Exchange by Ultracentrifugation

Nikolaj Düring Drachmann and Claus Olesen

Abstract

Lipids play an important role in maintaining P-type ATPase structure and function, and often they are crucial for ATPase activity. When the P-type ATPases are in the membrane, they are surrounded by a mix of different lipid species with varying aliphatic chain lengths and saturation, and the complex interplay between the lipids and the P-type ATPases are still not well understood. We here describe a robust method to exchange the majority of the lipids surrounding the ATPase after solubilisation and/or purification with a target lipid of interest. The method is based on an ultracentrifugation step, where the protein sample is spun through a dense buffer containing large excess of the target lipid, which results in an approximately 80–85 % lipid exchange. The method is a very gently technique that maintains protein folding during the process, hence allowing further characterization of the protein in the presence of a target lipid of interest.

Key words Native lipids, Synthetic lipids, Lipid exchange, Ultracentrifugation, Density gradient

1 Introduction

P-type ATPases are integral membrane proteins, and are very dependent on the membrane environment both structurally and functionally [1–4]. These studies show that for example the activity of the Ca^{2+}-ATPase varies greatly with the aliphatic chain length of the surrounding lipids, with an optimal activity obtained in bilayers containing lipids with chain lengths of C18-C22. Similar studies on the Na,K-ATPase show comparable results, with optimal activity obtained in bilayers consisting of lipids of C16-C20 [5]. The number of lipid molecules in the annulus of the Ca^{2+}-ATPase is about 32 [6, 7]. It is important to emphasize that the annulus is not a static structure because there is a constant exchange with lipids outside the annulus perimeter with exchange rate in the order of $1-2 \times 10^7$ s^{-1} at 30 °C [7].

In order to test the behavior, structural implications and functionality of different lipid species on P-type ATPases, a lipid exchange procedure has been developed. This technique enables the exchange of the majority of the native lipids present in the mixed protein/lipid/detergent micelle after solubilisation or

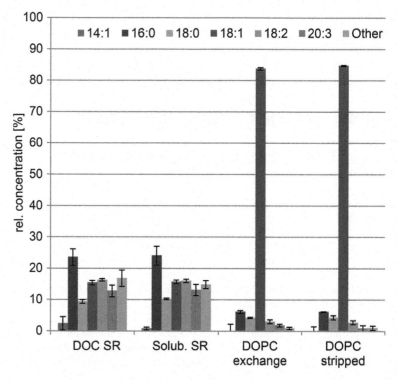

Fig. 1 Relative concentration of fatty acid species before and after LEC of a SERCA preparation. The relative concentrations of fatty acid species determined by gas chromatography of methylated fatty acids from extracted samples show a strong enrichment to ~85 % target lipid after LEC. The following samples were analyzed: DOC SR, deoxycholate-extracted sarcoplasmic reticulum (SR) microsomes before solubilization; Solub. SR, $C_{12}E_8$ solubilized SR microsomes enriched in SERCA; DOPC exchange, solubilized SERCA after LEC with DOPC as target lipid; DOPC strip, DOPC exchanged SERCA after sucrose step-gradient spin to strip the sample from bulk lipids. Error bars correspond to the standard deviation of results from three independent experiments. Adapted figure from [8]

purification, with a target lipid of interest (Fig. 1). We have used this method for investigating the effect of the membrane thickness in relation to structural changes and adaptation using protein X-ray crystallography [8].

The method is developed based on pioneering work by Warren et al. [9, 10]. They showed how the excess addition of a new target lipid to purified SR microsomes containing Ca^{2+}-ATPase would lead to an exchange of the endogenous SR lipids. The major factor of the degree of lipid substitution depends on the ratio of added target lipid and endogenous SR lipids. This property, together with the fact that no preferential substitution of lipids with different head group, chain length or chain saturation is observed [11], has allowed us to develop this technique further.

Upon solubilisation of the SR membranes, lipids pertaining to the annulus of the protein will be part of the content of the solubilized Ca^{2+}-ATPase micelles, while bulk lipids (i.e., lipids not interacting with Ca^{2+}-ATPase), will make up 'empty' micelles containing only lipid and detergent. When spinning this mixture through a lipid buffer, the endogenous lipids will not migrate through the lipid buffer because of the lower density of the lipid/detergent micelles. On the other hand the heavier protein complexes will be pelleted through the buffer and an exchange with lipids in the lipid buffer will take place.

Here we describe an exchange procedure on SERCA1a purified from rabbit fast twitch hind leg muscle with 18:1 DOPC (1,2-dioleoyl-*sn*-glycero-3-phosphocholine) as the target lipid. Appropriate modifications of this procedure should make it applicable to different P-type ATPases and other transmembrane proteins and can be applied for the exchange of a wide range of target lipids after solubilisation.

2 Materials

Prepare all solutions using double-distilled (18.2 MΩ cm) water.

2.1 Solubilized Protein

1. 650 μL of purified SERCA1a isolated from rabbit hind leg muscle (*see* Chap. 3) prepared in the E1 state in 100 mM MOPS-KOH pH 6.8, 80 mM KCl, 10 mM $CaCl_2$, 20 % glycerol, 1 mM AMPPCP, 1.5 mM $MgCl_2$, and 35 mM $C_{12}E_8$ (1:1.5 protein:detergent ratio) (*see* **Note 1**).

2.2 Lipid Buffer

1. $C_{12}E_8$ stock: Prepare 200 mg/mL $C_{12}E_8$ stock by dissolving 20 mg $C_{12}E_8$ (product) in 80.6 μL water (*see* **Note 2**). Dissolve by mixing at room temperature. Store at 4 °C, the solution is stable for a few days.

2. Solubilized lipids: Dissolve 5.6 mg of DOPC lipid in 84 μL of the $C_{12}E_8$ stock solution (resulting in a lipid:$C_{12}E_8$ ratio of 1:3) (*see* **Note 3**). Dissolve by mixing at room temperature. Store at 4 °C, the solution is stable for a few days.

3. 500 μL Lipid buffer: 100 mM MOPS-KOH pH 6.8, 80 mM KCl, 10 mM $CaCl_2$, 25 % glycerol, 1 mM AMPPCP, 1.5 mM $MgCl_2$ and 2.5 % PEG6000. Add 84 μL of the detergent solubilized DOPC lipid from **item 2** above (*see* **Note 4**). Mix well and cool on ice.

2.3 Ultracentrifugation and Sample Collection

1. Ultracentrifuge tubes (Beckman, 11 × 34 mm tubes) (*see* **Note 5**).
2. Swing out rotor (Beckman TLS 55).
3. Ultracentrifuge (Beckmann Optima Max-E).
4. Thin needle (0.2–0.4 mm) and syringe.
5. NanoDrop Spectrophotometer (Thermo Scientific).

3 Methods

Carry out all procedures on ice, unless otherwise specified.

3.1 Sample Preparation and Ultracentrifugation

1. Carefully add 500 μL of ice cold Lipid buffer directly to the bottom of an ultracentrifuge tube. Avoid adding it to the side of the tube.
2. Add 650 μL solubilized protein carefully on top of the dense Lipid buffer in the ultracentrifuge tube. Add the protein by pipetting slowly to the side of the tube holding it in an approximately 40 ° angle position and letting the protein solution slowly build up on top of the Lipid buffer. Make sure to avoid disturbing the interface between the solutions.
3. Carefully place the tubes in the ultracentrifuge rotor. Remember to include a counterweight to the opposite position in the rotor.
4. Start the ultracentrifuge and let it spin overnight at $135,000 \times g$ at 4 °C (approximately 22 h) (*see* **Note 6**).

3.2 Sample Collection and Concentration Measurements

1. After completion of the centrifugation, carefully put the sample centrifugation tube on ice.
2. Use a needle and syringe to carefully take out 200 μL sample from the bottom of the tube. Make sure not to disturb the solution when putting the needle through the liquid.
3. Measure the concentration of the final sample with a Nanodrop photometer at Abs280 using the Lipid buffer as a reference (typically concentrations of the peak fractions are around 12 mg/mL) (*see* **Note 7** and Fig. 2).

4 Notes

1. If another conformational state is desired for the exchange, the components of the solubilisation buffer and the Lipid buffer should be changed in order to change the state of the protein (e.g., add EGTA instead of $CaCl_2$ to induce E2 conformation over E1).
2. In order to account for the volume of the added $C_{12}E_8$ in the stock, the volume of the added water is reduced by 0.97 μL per mg, corresponding to the partial specific volume of $C_{12}E_8$ in the stock. For example, for a 200 mg/mL stock solution, using 20 mg of $C_{12}E_8$, then use: 20 mg/200 mg/mL – 0.97 μL/mg × 20 mg = 80.6 μL H_2O for dissolving.
3. In order to increase the amount of exchanged lipids, add higher amounts of lipid to the Lipid buffer fraction. This will increase the possibility for the lipids to exchange with a target lipid rather than a native lipid during the exchange procedure.

4. To ensure that the lipid phase has a higher density than the solubilized protein, additional 5 % of glycerol (final concentration 25 %) and 2.5 % of PEG6000 are added to the Lipid Buffer. The addition of PEG6000 is included because it is used in the subsequent crystallization and because we would not like to have a glycerol concentration higher than 25 %.

5. In order to optimize the amount of exchanged lipids, the size of the tube can be varied. By using a longer tube with thinner diameter, the protein will stay longer in the Lipid buffer phase, leading to higher degree of exchange. A longer spin is needed in order to pellet the protein through the longer lipid buffer phase. Again, remember to recalculate the centrifugation time, as the sedimentation coefficient will be different if the buffer is changed. We have used the used the following formula for determining the centrifugation time and speed: sedimentation velocity $= \omega^2 rs$ (where, ω is the angular velocity, r is radial distance and s is the sedimentation coefficient).

6. If other buffer composition, protein or lipids are used, it is important to recalculate the sedimentation speed with the new parameters to ensure that the solubilised protein has sedimented to the bottom part of the target lipid phase.

7. A more stringent way of doing this is to take out fractions all the way down through the tube (e.g., 100 μL fractions), starting with the top layer and measure the concentration of all the fractions. By this you would get a total profile of the migration of your protein (Fig. 2). The protein with the highest degree of lipid exchange is found in the bottom of the tube (where the migration through the lipid buffer has been the longest).

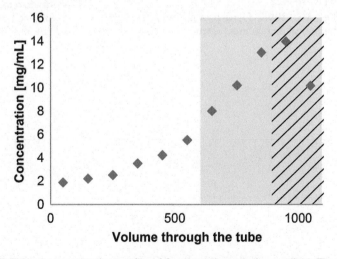

Fig. 2 Typical concentration-profile of fractions through the gradient. The *gray* area corresponds to the part of the sample where there is excess target lipid present, while the *hatched* area is the fractions with the highest degree of lipid exchange

We prefer to do it as described in **steps 2** and **3** of Subheading 3.2, since the risk of disturbing the layers in the tube is low.

Acknowledgement

We would like to thank Jesper Vuust Møller for comments and discussion in the preparation of this book chapter.

References

1. Starling AP, East JM, Lee AG (1993) Effects of phosphatidylcholine fatty acyl chain length on calcium binding and other functions of the (Ca(2+)-Mg2+)-ATPase. Biochemistry 32:1593–1600
2. Cornea RL, Thomas DD (1994) Effects of membrane thickness on the molecular dynamics and enzymatic activity of reconstituted Ca-ATPase. Biochemistry 33:2912–2920
3. Gustavsson M, Traaseth NJ, Karim CB et al (2011) Lipid-mediated folding/unfolding of phospholamban as a regulatory mechanism for the sarcoplasmic reticulum Ca2+-ATPase. J Mol Biol 408:755–765
4. Sonntag Y, Musgaard M, Olesen C et al (2011) Mutual adaptation of a membrane protein and its lipid bilayer during conformational changes. Nat Commun 2:304
5. Johannsson A, Smith GA, Metcalfe JC (1981) The effect of bilayer thikness on the activity of (Na+ + K+)-ATPase. Biochim Biophys Acta 641:416–421
6. Lee AG (2003) Lipid–protein interactions in biological membranes: a structural perspective. Biochim Biophys Acta—Biomembr 1612:1–40
7. Marsh D, Horváth LI (1998) Structure, dynamics and composition of the lipid-protein interface. Perspectives from spin-labelling. Biochim Biophys Acta 1376:267–296
8. Drachmann ND, Olesen C, Møller JV et al (2014) Comparing crystal structures of Ca(2+)-ATPase in the presence of different lipids. FEBS J. doi:10.1111/febs.12957
9. Warren GB, Toon P, Birdsall NJ et al (1974) Reconstitution of a calcium pump using defined membrane components. Proc Natl Acad Sci U S A 71:622–626
10. Warren GB, Toon PA, Birdsall NJM et al (1974) Complete control of the lipid envirenment of membrane-bound proteins: application to a calcium transport system. FEBS Lett 4:160–162
11. Le Maire M, Lind KE, Jørgensen KE et al (1978) Enzymatically Active Ca2+ ATPase from Sarcoplasmic Reticulum Membranes, Solubilized by Nonionic Detergents. J Biol Chem 253:7051–7060

Chapter 36

Reconstitution of Na⁺,K⁺-ATPase in Nanodiscs

Jonas Lindholt Gregersen, Natalya U. Fedosova, Poul Nissen, and Thomas Boesen

Abstract

Nanodiscs are disc-shaped self-assembled lipid bilayers encircled by membrane scaffolding proteins derived from Apolipoprotein A-1 (apo A-1). They constitute a versatile tool for studying membrane proteins since reconstitution into nanodiscs allows studies of the membrane proteins in detergent-free aqueous solutions in a lipid bilayer. Here, we apply the technique to the Na⁺,K⁺-ATPase (NKA) from pig kidney using Membrane Scaffolding Protein 1 D1 (MSP1D1). Contrary to other reports, the nanodiscs obtained by our protocol are built up of the native lipids originally present in the detergent solubilized sample together with the NKA.

Key words Na⁺,K⁺-ATPase, Nanodisc, Self-assembly, Scaffolding protein, Self-assembled lipid bilayer, Native lipids

1 Introduction

Functional investigations of membrane proteins require a suitable model system. Model systems such as membrane proteins solubilized by detergent micelles, bicelles or reconstituted in liposomes all offer great opportunities [1]. However, these models also have disadvantages as for example detergents may induce inactivation or unfolding of the protein, and liposomes provide no immediate access to both sides of the membrane.

Reconstitution of the membrane protein into the nanodiscs enables solubilization in the absence of detergent. Nanodiscs are nanometer-sized disc-shaped lipid-bilayers encircled by membrane scaffolding proteins (MSPs) [2]. The MSPs are derived from apo A-1 by removing its N-terminal globular domain. MSPs exist in different lengths and this allows for preparation of nanodiscs of varying sizes. Furthermore, the nanodisc diameter can determine the oligomeric state of the reconstituted enzyme. Thus, the protein in a nanodisc preparation is provided with a native-like phospholipid bilayer environment, is soluble at a single membrane protein molecule level and homogeneous in terms of oligomeric state [2]. In addition to the advantages listed above, the methodological arsenal applicable for structural and functional studies of nanodisc protein preparations may include all methods developed for soluble proteins.

Here, we present the reconstitution of the NKA from pig kidney outer medulla in nanodiscs, using exclusively lipids from the native cytoplasmic membrane. This yields an active NKA, that is soluble on a single membrane protein molecule level, in a lipid environment similar (or identical) to that in the kidney cell.

2 Materials

All reagents are of analytic grade and solutions are prepared using ultrapure water, i.e., deionized water with at least 18 MΩ resistance at 20 °C. Unless otherwise stated, all reagents and solutions are stored at 4 °C. Waste disposal regulations should be followed for waste materials.

2.1 Na$^+$,K$^+$-ATPase Solubilization

1. Native kidney membrane suspension, e.g., prepared as described in Chap. 2.
2. Buffer A: 20 mM histidine, 250 mM sucrose, 0.9 mM EDTA, pH adjusted to 7.0 using HCl. Weigh 77.5 mg histidine, 2.14 g sucrose, 65.8 mg Na$_2$EDTA·2H$_2$O. Add 20 mL of water and adjust pH to 7.0 using 1 N HCl. Finally, adjust the total volume to 25 mL using water.
3. 150 mM histidine, pH 7.0. Weigh 11.64 g histidine. Add 300 mL of water and adjust pH to 7.0 using KOH. Finally, adjust the total volume to 500 mL using water.
4. 3 M KCl. Weigh 44.8 g KCl and adjust the volume to 200 mL using water.
5. Buffer B: 20 mM histidine, pH 7.0, 1.5 M KCl. Mix 133.2 μL of 150 mM histidine pH 7.0, 500 μL of 3 M KCl and 366,8 μL of water to final volume of 1 mL (*see* **Note 1**).
6. 64 mg/mL octaethylene glycol monododecyl ether (C$_{12}$E$_8$). Weigh 32 mg C$_{12}$E$_8$ and adjust the volume to 500 μL using water.

2.2 Expression and Purification of MSP1D1

1. Purified MSP1D1 at 16 mg/mL, e.g., obtained following Chaps. 3.2 and 3.3 of Ritchie et al. [4].
2. Dialysis buffer: 20 mM Histidine, pH 7.0, 150 mM KCl. Mix 100 mL of 3 M KCl, add 266.7 mL of 150 mM Histidine, pH 7.0 (from Subheading 2.1) and adjust the total volume to 2 L using water (*see* **Note 2**).

2.3 Biobead Preparation

1. Dialysis buffer prepared in Subheading 2.2.
2. Bio-Beads SM-2 (Bio-Rad).

2.4 Nanodisc Purification

1. Dialysis buffer prepared in Subheading 2.2.
2. Superose™ 6 10/300 GL column suitable for HPLC system.
3. HPLC system, e.g., ÄKTA purifier.

3 Methods

Unless stated otherwise, procedures are performed on ice.

3.1 Na⁺,K⁺-ATPase Solubilization

The following protocol, based on Hansen et al. [3], is applicable for obtaining the solubilised preparation of NKA from pig kidney outer medulla (specific Na,K-ATPase activity 30 µmol/mg min).

1. Dilute 6.4 mg native pig kidney membrane suspension to a final concentration of 4 mg/mL in Buffer A.
2. Add 200 µL of Buffer B (*see* **Note 1**).
3. While stirring, add 200 µL 64 mg/mL $C_{12}E_8$ slowly to a final detergent/protein ratio of 2 (w/w) (*see* **Note 3**).
4. Centrifuge the solubilized sample at $130,000 \times g$ for 60 min at 4 °C (55,000 rpm using a TLA-100 rotor in a Beckmann MAX-E ultracentrifuge).
5. Slowly collect the clear supernatant. The expected protein concentration is approx. 2.8 mg/mL estimated by using a NanoDrop 1000 Spectrophotometer assuming that 1 Abs = 1 mg/mL.

3.2 MSP1D1 Preparation

1. Dialyse the purified MSP1D1 against 1 L dialysis buffer overnight at 4 °C.
2. Collect the dialysed MSP1D1. Protein concentration is approx. 13.5 mg/mL estimated by using a NanoDrop 1000 Spectrophotometer assuming an extinction coefficient of $\varepsilon = 21,000$ M^{-1} cm^{-1} and a molecular weight of 24,700 Da for MSP1D1.

3.3 Biobead Preparation

1. Suspend BioBeads SM2 in water at a ratio 1/10 (w/v) and remove the floating beads.
2. Centrifuge briefly at $500 \times g$ or wait for gravitational sedimentation.
3. Remove as much liquid above the beads as possible.
4. Resuspend the beads in 10 mL Dialysis buffer.
5. Repeat **step 2** through **4** three times.

3.4 Nanodisc Reconstitution

1. Mix 1.75 mL 2.8 mg/mL solubilized NKA with 242 µL 13.5 mg/mL MSP1D1 for a molar ratio of 1:4 (*see* **Note 4**) and incubate for 1 h 30 min.

2. Remove the detergent from the samples by BioBeads SM-2 (*see* **Note 5**). For example weigh off 150 mg beads. Do this by weighing 2 × 75 mg in two separate 1.5 mL tubes (*see* **Note 6**). Divide the mix of NKA and MSP1D1 into the two tubes (*see* **Note 7**).

3. Incubate the sample with beads for 20 h under rotation at 4 °C (*see* **Note 8**).

4. Collect the sample for purification (*see* **Note 9**).

3.5 Nanodisc Purification

1. Equilibrate a Superose™ 6 10/300 GL column at 4 °C with filtered (0.2 μm filter pores) and degassed Dialysis buffer.

2. Load 500 μL of the reconstituted nanodiscs on the column using a 1 mL loop (*see* **Note 10**) and collect fractions of 500 μL.

A representative chromatogram is depicted in Fig. 1a. A closer look at the chromatogram (*see* Fig. 1b) reveals three peaks corresponding to either empty (no NKA) discs (~16.5 mL), nanodiscs with one NKA (~14.5 mL) or with two NKA molecules (~13.5 mL) per disc. SDS-PAGE analysis of the fractions is shown in Fig. 2.

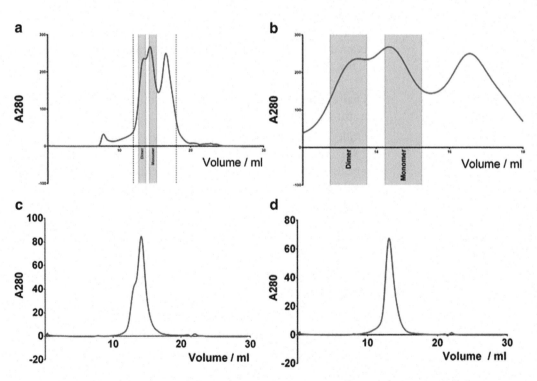

Fig. 1 Chromatograms from size exclusion chromatography of the nanodisc preparation. (**a**) Chromatograms of first run. (**b**) First run, zoom. (**c**) Second run of the fraction corresponding to one NKA molecule per nanodisc. (**d**) Second run of the fraction corresponding to two NKA molecules per nanodisc

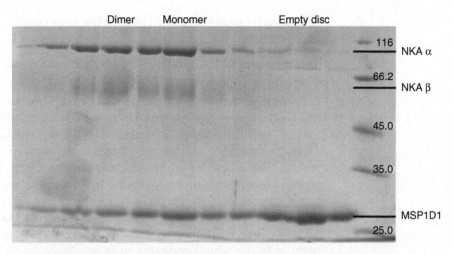

Fig. 2 SDS-PAGE analysis of the collected fractions from the first size exclusion run. The molecular weights of the protein marker on the *right* are given in kDa

The number of NKA/disc in each peak is calculated from the ratio of α-subunit (~100 kDa) to MSP1D1 (~25 kDa).

3. Load the fractions corresponding to one NKA per nanodisc or two NKAs per nanodisc on the column for a second round of size exclusion chromatography (*see* **Note 11**) (*see* Fig. 1c, d).

4 Notes

1. This solution is used to achieve the final buffer composition determining the preferred enzyme conformation. This protocol aims at nanodiscs with NKA stabilized in E2(K2) conformation, applicable for all types of biochemical characterization.

2. The composition of the buffer depends on the targeted final conformational state of the NKA (*see* **Note 1**). The buffer is needed for many subsequent steps, so although 1 L is needed for dialysis we recommend making 2 L.

3. The optimal volume for the added detergent is 1/10 of the final volume of the solubilized enzyme.

4. The molar ratio between MSP1D1 and NKA is optimal at 4:1. The volume of MSP1D1 is calculated following this formula:

$$V_{MSP1D1} = \frac{4 \times V_{NKA} \times c_{NKA} \times M_{MSP1D1}}{M_{NKA} \times c_{MSP1D1}}$$

The volumes are V_{MSP1D1} and V_{NKA} for MSP1D1 and NKA, respectively. The molar masses M_{MSP1D1} and M_{NKA} are estimated to 25,000 Da and 150,000 Da, respectively, and the concentrations c_{MSP1D1} and c_{NKA} are given in mg/mL.

5. The BioBeads to $C_{12}E_8$ ratio is approximately 12:1 (w/w).

6. To pipette the washed beads weigh two 1.5 mL tubes and cut off approx. 0.5 cm of a P1000 tip using a pair of scissors. Use this to transfer an appropriate amount of beads to a 1.5 mL tube. Centrifuge using a regular benchtop centrifuge at $14,000 \times g$ for 15 s to collect the beads at the bottom. Gently remove the excess liquid above the beads by using a P200 pipette tip and subsequently pipette from the bottom to dry out the beads. Weigh the tubes again to get the mass of the beads. This procedure might require several trials. Keep the beads in buffer as much as possible, i.e., only let the beads without buffer when measuring their weight.

7. In the case described here, there should be added 996 μL to each tube. In general, weighing of the beads can be difficult. If the beads are weighed and the total weight is close to the total weight wanted, split the protein mix correspondingly, e.g., if 70 mg and 80 mg are weighed off instead of 2×75 mg, simply add 930 μL to the 70 mg beads and 1062 μL to the 80 mg beads.

8. We use an Intelli-Mixer™ RM-2 from ELMI set to program F1 (simple rotation) at a frequency of 40 rpm placed in a cold room (4 °C).

9. In order to collect all the sample, cut a hole into the lids of two 15 mL tubes (using for example a scalpel) so that a 1.5 mL tube will fit firmLy. Make a hole in the bottom and 5–10 holes in the top of the 1.5 mL tubes containing the samples with a needle (0.4 mm diameter). Place the 1.5 mL tubes into the lids of the 15 mL tubes and collect the samples in these by centrifuging $500 \times g$ at 4 °C for 5 min. The small needle hole will allow only the solution to pass through, leaving the beads in the 1.5 mL tube.

10. The reconstituted nanodiscs can be either concentrated to a volume of 500 μL or loaded in several runs with subsequent pooling of the corresponding fractions. Leaving the sample for too long before loading it on the size exclusion column will result in a larger peak at the void volume. Concentration of the reconstituted sample can be done using a Vivaspin 6 concentrator with a molecular weight cut-off of 100 kDa. Always spin the sample for 10 min at 4 °C using a bench top centrifuge at $14,000 \times g$ before loading it on the size exclusion column in order to pellet excess lipids, aggregated sample, etc.

11. It might be necessary to pool fractions and concentrate them prior to the second round of size exclusion chromatography. The sample is viable for this treatment.

Acknowledgements

This protocol has been developed in collaboration with Linda Reinhard, Department of Molecular Biology and Genetics, Aarhus University. The expert technical assistance of Ms. Birthe Bjerring Jensen, Department of Biomedicine, Aarhus University, is gratefully acknowledged. J.L.G. was supported by a PhD stipend from the Centre for Membrane Pumps in Cells and Disease—PUMPKIN, Danish National Research Foundation, T.B. was supported from the ERC advanced grant—BIOMEMOS awarded to P.N.

References

1. Seddon AM, Curnow P, Booth PJ (2004) Membrane proteins, lipids and detergents: not just a soap opera. Biochim Biophys Acta 1666: 105–117
2. Bayburt TH, Grinkova YV, Sligar SG (2002) Self-assembly of discoidal phospholipid bilayer nanoparticles with membrane scaffold proteins. Nano Lett 2:853–856
3. Hansen AS, Kraglund KL, Fedosova NU et al (2011) Bulk properties of the lipid bilayer are not essential for the thermal stability of Na, K-ATPase from shark rectal gland or pig kidney. Biochem Biophys Res Commun 406:580–583
4. Ritchie TK, Grinkova YV, Bayburt TH et al (2009) Chapter 11—reconstitution of membrane proteins in phospholipid bilayer nanodiscs. Methods Enzymol 464:211–231

Part VIII

Crystallization

Chapter 37

Crystallization of P-type ATPases by the High Lipid–Detergent (HiLiDe) Method

Oleg Sitsel, Kaituo Wang, Xiangyu Liu, and Pontus Gourdon

Abstract

Determining structures of membrane proteins remains a significant challenge. A technique utilizing high lipid–detergent concentrations ("HiLiDe") circumvents the major bottlenecks of current membrane protein crystallization methods. During HiLiDe, the protein–lipid–detergent ratio is varied in a controlled way in order to yield initial crystal hits, which may be subsequently optimized by variation of the crystallization conditions and/or utilizing secondary detergents. HiLiDe preserves the advantages of classical lipid-based methods, yet is compatible with both the vapor diffusion and batch crystallization techniques. The method has been applied with particular success to P-type ATPases.

Key words Membrane proteins, P-type ATPase, Structure, X-ray diffraction, Crystal, HiLiDe

1 Introduction

P-type ATPases are one of the best structurally comprehended families of transmembrane proteins. To date, the structures of five individual family members from four subfamilies have been determined, including the P_{IIA}-type calcium pump SERCA1a from rabbit [1], the P_{IIC}-type Na/K pump from pig or shark [2, 3], the P_{III}-type proton pump from *Arabidopsis* [4], and—most recently—two P_{IB}-type pumps from bacteria, one of which transports copper [5] and the other zinc [6]. The latter two were expressed recombinantly in *Escherichia coli* and the proton pump was produced in *Saccharomyces cerevisiae*, whereas the other proteins were purified from natural source.

Yet regardless of recent progress, structural determination of P-type ATPases as well as other membrane proteins is challenging due to the difficulty of growing well-diffracting crystals. Both obtaining and optimizing crystallization hits remain significant hurdles, and new methods are constantly being developed in order to overcome these barriers.

Crystallization in detergent is the classical technique for obtaining membrane protein crystals [7] and has yielded many structures. Typically, the concentration of detergent after protein purification is minimized, and the sample is then used directly for crystallization. This method is compatible with the highly popular vapor diffusion crystallization setup, as well as with the dialysis and batch setup modes. The downside of using detergent-based methods is that they employ an unnatural detergent environment, which often negatively affects the stability of the protein [8].

The lipid-based methods include lipidic cubic-, sponge-, and bicelle-phase crystallization techniques [9] which all ensure that the protein is kept in a more near-native environment. Lipid-based methods have been successfully applied to a number of challenging membrane proteins [10] and are compatible with specialized robotics. The main overall disadvantage of these methods is their incompatibility with crystallization techniques such as vapor diffusion and the limited selection of compatible lipids. Additionally, the lipidic cubic phase is highly viscous, which makes handling difficult.

HiLiDe is an attempt to combine the simplicity and vapor-diffusion compatibility of detergent-based methods with the power of crystallization in near-native environments of lipid-based methods [11]. HiLiDe is based on using high concentrations of lipid and detergent to relipidate the protein prior to crystallization (similarly to bicelle crystallization), and can be used in a simple and systematic fashion to obtain initial crystal hits and subsequently optimize them. As with the other lipid-based methods, crystals obtained using HiLiDe are of type I, which resemble stacked 2D crystals (or membranes).

HiLiDe-produced crystals are not more technically challenging to harvest than those obtained using detergent-based methods. Furthermore, additional cryoprotection of the crystals prior to flash-freezing is often not necessary if for example using the King Screen (developed at Aarhus University, *see* **Note 4** for details), which has been exploited with great success in P-type ATPase crystallization. A key point is that virtually any lipid/detergent combination deemed necessary by the experimenter may be tested — a luxury that is unavailable in other lipid-based methods. The downsides of HiLiDe are that proper grid screening of lipid-to-detergent ratios is protein-demanding and is better suited to relatively well-expressing targets and that although the relipidated samples are suitable for dispensing using standard robotics, the relipidation procedure itself needs to be done manually, limiting throughput.

2 Materials

All solutions are prepared with ultrapure water.

1. *Purified P-type ATPase solution*: 10 mg/mL protein, with aggregated protein material removed (*see* **Note 1**).
2. *Lipid stock solution*: DOPC in chloroform at 40 mg/mL (*see* **Note 2**).
3. *Detergent stock solution*: 100 mg/mL $C_{12}E_8$ (*see* **Note 3**).
4. *Stock solutions for crystallization screens*: PEG2000MME, PEG6000, KCl, NaCl, sodium acetate, sodium malonate, $MgCl_2$, $(NH_4)_2SO_4$, *tert*-butanol, 2-methyl-2,4-pentanediol, glycerol, and beta-mercaptoethanol (*see* **Note 4**).
5. *Crystal harvesting equipment*: a tabletop dewar with liquid nitrogen, a prechilled storage/transport dewar, and a set for crystal harvesting and cryomanipulation (mounted LithoLoops, CryoCaps, magnetic CryoVials and a CryoWand from Molecular Dimensions would be an example).

3 Methods

3.1 Setting Up the HiLiDe Incubation

1. Transfer the necessary amount of chloroform-dissolved lipid (*see* Fig. 1) to a round-bottom glass tube on ice (*see* **Note 5**).
2. Evaporate the chloroform using a constant stream of nitrogen gas. The stream should not be strong enough to splash the chloroform onto the walls of the tube.
3. Pipette the concentrated protein onto the layer of dried lipid.
4. Pipette the necessary amount of (extra) detergent if any (*see* Fig. 1) onto the layer of protein (*see* **Note 6**).
5. Gently add a small magnetic stirrer bar, (e.g., 5×2 mm^2) so as to avoid splashing of the mixture onto the walls of the tube (*see* **Note 7**).
6. Fill the tube with nitrogen gas and seal with Parafilm.
7. Mix the components of the tube on a magnetic stirrer using 50 rpm for 16–18 h at 4 °C.
8. Remove insoluble material by ultracentrifugation at $100,000 \times g$ for 20 min at 4 °C. Transfer the supernatant into an Eppendorf tube and keep hereafter on ice (*see* **Note 8**).

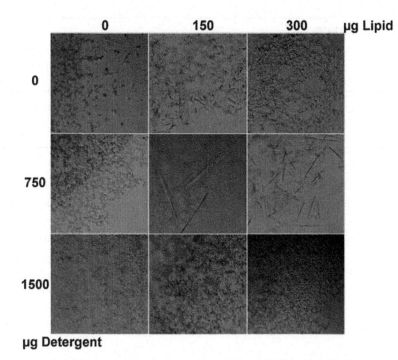

Fig. 1 Effects of using different lipid–detergent ratios during the HiLiDe incubation on crystallization of the Cu+-ATPase *lpg1024* from *Legionella pneumophila*. 100 μL of 10 mg/mL protein were used for this setup. The lipid and detergent used were DOPC and $C_{12}E_8$, respectively. Adapted with permission from 'HiLiDe—Systematic Approach to Membrane Protein Crystallization in Lipid and Detergent' by Pontus Gourdon, Jacob Lauwring Andersen, Kim Langmach Hein, Maike Bublitz, Bjørn Panyella Pedersen, Xiang-Yu Liu, Laure Yatime, Maria Nyblom, Thorbjørn Terndrup Nielsen, Claus Olesen, Jesper Vuust Møller, Poul Nissen, and Jens Preben Morth in Crystal Growth & Design 2011:11(6), 2098–2106. ©2014 American Chemical Society

3.2 Setting Up Crystallization Trials

1. Mix conditions for crystallization setup (*see* **Note 4**). For vapor diffusion setups, reservoir solutions (placed in the well) and precipitant solutions (mixed with the HiLiDe reaction) may differ in composition, however this is generally used in optimization of initial crystal hits.

2. Add substrate and/or inhibitors to the lipidated protein to lock it in a specific conformational state. An extensive list of successful combinations used for P-type ATPase crystallization can be found in [12] (*see* **Note 9**).

3. Set up vapor diffusion crystallization by adding reservoir solution into the crystallization tray well. Pipette and mix 0.5–2 μL of the HiLiDe reaction and precipitant solution on the pedestal and seal with tape (sitting drop) or on a siliconized glass cover slide and seal with immersion oil or vacuum grease (hanging drop) (*see* **Note 10**).

4. Assess presence or absence of crystals by using a light microscope. Examples of various phenomena that can be observed in drops set up using the HiLiDe method are shown in Fig. 1.

3.3 Cryoprotection and Harvesting of Crystals

1. If crystal hits are present, they must be harvested and assessed for X-ray diffraction (*see* **Note 11**).
2. Prepare equipment for harvesting.
3. Uncover the drop containing the crystals.
4. Optional: mix in a cryoprotectant such as glycerol to the drop to be harvested. Detailed protocols for the use of cryoprotectants in macromolecular X-ray crystallography are available elsewhere.
5. Optional: mix in a phasing agent such as tantalum bromide to the drop to be harvested. Detailed protocols for the use of phasing agents in macromolecular X-ray crystallography are available elsewhere.
6. Harvest the crystals.
7. Flash-freeze in liquid nitrogen and store in a storage or transport dewar.
8. Assess diffraction quality using a suitable X-ray source, typically a synchrotron.
9. If diffraction quality is insufficient, then optimization of crystallization conditions is required (*see* **Notes 12–14**).

4 Notes

1. The starting point is purified P-type ATPase after a final gel filtration step for removal of aggregates. Concentrate the protein to a value of 5 mg/mL or higher using a Vivaspin concentrator with a molecular weight cutoff appropriate to the size of the protein. Aliquot, then flash-freeze the concentrated protein in liquid nitrogen and store at –80 °C until further use.
2. Use DOPC as a starting point (polar *E. coli* lipids, POPC or other lipids may be considered as alternatives). Make 40 mg/mL stock solutions of the desired lipids in glass tubes using chloroform as a solvent (caution — volatile and toxic; do procedures under a fume hood while keeping the glass tubes on ice). After preparation of stock solutions, seal glass tubes and store at –20 °C until further use.
3. Use the detergent that the protein was purified in as a starting point. For P-type ATPases this is most often $C_{12}E_8$. Prepare a 10 % stock solution in water and store at –20 °C until further use.

    DDM is one of the most successful detergents for membrane protein purification and crystallization in general.

However for P-type ATPases $C_{12}E_8$ is the most preferred detergent by far, even though $C_{12}E_8$ and DDM have quite similar properties, and DDM can even give more monodisperse and narrow non-void gel filtration peaks comparing to $C_{12}E_8$. An example is the *Arabidopsis* proton pump: although it was purified in DDM, the sample used for crystallization was dialyzed overnight against $C_{12}E_8$ containing buffers.

4. Store reagents for the screen at 4 °C, but adjust to ambient temperature of crystallization setup before starting. The so-called "King Screen" is a particularly successful homemade sparse matrix screen for P-type ATPase crystallization, highly recommended for use as a starting point [13].
   It is made by combining different amounts of:

   (a) PEG2000MME or PEG6000
       with

   (b) KCl, NaCl, sodium acetate, sodium malonate, $MgCl_2$ or $(NH_4)_2SO_4$
       with

   (c) Glycerol
       optionally with

   (d) *Tert*-butanol or 2-methyl-2,4-pentanediol.
       Also consider adding 5 mM beta-mercaptoethanol or another reducing agent.

5. Take care to transfer the total amount of lipid to the bottom of the tube, without spilling it on the walls of the tube. Chloroform dissolves plastic, so transfer quickly if using plastic pipette tips. Use Fig. 1 for inspiration concerning the amount of lipid to use.

6. The protein–lipid–detergent ratio is a crucial factor of successful P-type ATPase crystallization. Protein concentration matters: the higher the protein concentration is, the higher the lipid and detergent concentrations should be during initial screening. Experiments with HiLiDe-based crystallization on the Cu^+-ATPase *lpg1024* from *Legionella pneumophila* indicate that variation in detergent and lipid levels is tolerated over a broad range of concentrations [11]. Experimental determination of detergent/lipid concentration in the sample is also an advantage, as it allows for better control of the relipidation procedure. For these purposes, an approach similar to that used in a study on the crystals of the P-type ATPase SERCA1a [13] could be utilized. Finally, it is very important to keep in mind that batch variation exists. For every protein batch purified, make an initial test using a 3×3 matrix of different lipid/detergent concentrations before optimizing reservoir conditions

during crystal optimization and reproduction. For initial screening, the steps could be ± 0.5–1 mg/mL of lipid/detergent and for established conditions the steps could be reduced to ± 0.1 ~0.2 mg/mL of lipid/detergent. Figure 1 provides an example of how using different ratios of detergent and lipid during the HiLiDe incubation affects the final results of crystallization.

7. In order to avoid relatively large loss of mixture due to evaporation, set up reactions of at least 100 μL for this type of HiLiDe.

8. Measure the concentration of protein in the supernatant after ultracentrifugation on a Nanodrop spectrophotometer. If certain ratios of protein–lipid–detergent cause a strong decrease in protein concentration after HiLiDe, then avoid them in future trials. A high concentration of lipids will raise the baseline in the protein concentration measurement (after relipidation, we have noticed up to a 30 % increase in the baseline level). Make sure to subtract the elevated baseline from the absorbance measurement at 280 nm to obtain more accurate estimations of the protein concentration.

9. An example of an inhibitor/chelator mix used successfully for crystallization of the zinc transporting ATPase SsZntA would be the following stock (to be diluted 24× using purified protein):

   240 mM NaF

   48 mM $AlCl_3$ or $BeSO_4$

   48 mM EGTA

   0.2 mM TPEN

10. HiLiDe is also compatible with batch and dialysis setup modes, although vapor diffusion has been most successful for P-type ATPase crystallization.

11. If too many small crystals are present, performing a twofold or threefold dilution on the relipidated sample before setup is an efficient countermeasure.

12. If no crystal hits are present, three initial options may be explored.

    (a) Modify the crystallization conditions. This is done by expanding the probed crystallization space.

    (b) Optimize the protein–lipid–detergent ratio. Combine with step **a** for greater effect.

    (c) If steps **a** and **b** fail, use a different type of lipid, detergent, or both.

(d) If steps **a–c** fail, redesign construct (if applicable) to remove flexible termini or excessively long loops that may hamper crystallization.

(e) Finally, HiLiDe may be poorly compatible or incompatible with the protein attempted to be crystallized. Try a different method or use a homologous target.

13. If diffraction quality is insufficient, attempt to modify the crystallization conditions by choosing the one providing best-diffracting crystals and using an additive or detergent additive screen (such as those provided by Hampton Research). In case of the detergent additive screen, also try using several different concentrations of detergent additive.

14. Supplying freshly prepared reducing agents to the reservoir just before setup sometimes helps improve crystal quality. Ultracentrifugation of the sample before setup reduces number of crystal seeds and may also help improve crystal quality and size.

Acknowledgements

This work was supported by the Graduate School of Science and Technology at Aarhus University and by grants to P.G. from the Swedish Research Council and the The Lundbeck Foundation.

References

1. Toyoshima C, Nakasako M, Nomura H, Ogawa H (2000) Crystal structure of the calcium pump of sarcoplasmic reticulum at 2.6 A resolution. Nature 405:647
2. Morth JP et al (2007) Crystal structure of the sodium-potassium pump. Nature 450:1043
3. Shinoda T, Ogawa H, Cornelius F, Toyoshima C (2009) Crystal structure of the sodium-potassium pump at 2.4 A resolution. Nature 459:446
4. Pedersen BP, Buch-Pedersen MJ, Morth JP, Palmgren MG, Nissen P (2007) Crystal structure of the plasma membrane proton pump. Nature 450:1111
5. Gourdon P et al (2011) Crystal structure of a copper-transporting PIB-type ATPase. Nature 475:59
6. Wang K et al (2014) Structure and mechanism of Zn-transporting P-type ATPases. Nature 514:518–522
7. Newby ZE et al (2009) A general protocol for the crystallization of membrane proteins for X-ray structural investigation. Nat Protoc 4:619
8. Bowie JU (2001) Stabilizing membrane proteins. Curr Opin Struct Biol 11:397
9. Caffrey M, Cherezov V (2009) Crystallizing membrane proteins using lipidic mesophases. Nat Protoc 4:706
10. Johansson LC, Wohri AB, Katona G, Engstrom S, Neutze R (2009) Membrane protein crystallization from lipidic phases. Curr Opin Struct Biol 19:372
11. Gourdon P et al (2011) HiLiDe-systematic approach to membrane protein crystallization in lipid and detergent. Cryst Growth Des 11:2098
12. Bublitz M, Poulsen H, Morth JP, Nissen P (2010) In and out of the cation pumps: P-Type ATPase structure revisited. Curr Opin Struct Biol 20:431
13. Sorensen TL, Olesen C, Jensen AM, Moller JV, Nissen P (2006) Crystals of sarcoplasmic reticulum Ca(2+)-ATPase. J Biotechnol 124:704

Chapter 38

Two-Dimensional Crystallization of the Ca^{2+}-ATPase for Electron Crystallography

John Paul Glaves, Joseph O. Primeau, and Howard S. Young

Abstract

Electron crystallography of two-dimensional crystalline arrays is a powerful alternative for the structure determination of membrane proteins. The advantages offered by this technique include a native membrane environment and the ability to closely correlate function and dynamics with crystalline preparations and structural data. Herein, we provide a detailed protocol for the reconstitution and two-dimensional crystallization of the sarcoplasmic reticulum calcium pump (also known as Ca^{2+}-ATPase or SERCA) and its regulatory subunits phospholamban and sarcolipin.

Key words Electron crystallography, Two-dimensional crystals, Sarcoplasmic reticulum, Ca^{2+}-ATPase, Phospholamban, Sarcolipin

1 Introduction

Electron crystallography (EC) has long been recognized as a powerful alternative in the structural biology of macromolecular complexes, with particular relevance for membrane protein structure determination. While EC is complementary to other structural biology techniques such as X-ray crystallography and NMR spectroscopy, it offers the unique advantage of a native, membrane-embedded environment. Nonetheless, there are significant technical challenges in obtaining two-dimensional crystals suitable for EC. And, once suitable crystals are obtained, the structural observations may be limited to low or moderate resolution and there can be a long lag time between the production of crystals and the determination of a three-dimensional structure. Despite these limitations, the origins of EC date back to the 1970s and there have since been numerous high-resolution structures (e.g., bacteriorhodopsin, tubulin, acetylcholine receptor [1–3]) and hundreds of reports of moderate-resolution projection or three-dimensional structures (e.g., [4–7]).

While structural biology was once the purview of a select group of practitioners worldwide, the techniques have become more generally accessible to researchers interested in their favorite protein targets. As more researchers turn toward structural biology, it is important to have robust means of structure determination and validation, as well as comprehensive protocols for sample preparation. Herein, we provide an experimental protocol for crystallizing the sarcoplasmic reticulum (SR) calcium pump, complete with pitfalls and moments of frustration and serendipity. The SR calcium pump is also known as Ca^{2+}-ATPase or SERCA, and it plays an essential role in muscle contractility. Two-dimensional crystals of SR membranes have been available since the 1980s [8], and the original insights into SERCA structure were uniquely provided by electron cryo-microscopy of well-ordered helical crystals [9, 10]. At 8 Å resolution, the structural detail provided included the domain architecture of SERCA and the localization of the ten transmembrane helices that coordinate calcium. In 2000, the first high resolution structure of SERCA by X-ray crystallography was published [11], and since then there have been more than 50 structures determined in different states of the enzyme (e.g., [12–19]).

In the modern era of structural biology, SERCA has revealed itself to be one of the most well understood proteins, and uniquely amenable to crystallization and structure determination under numerous conditions and conformational states. Herein we focus on a novel two-dimensional crystal form of SERCA in the presence of its regulatory subunits phospholamban and sarcolipin. The crystals were serendipitously discovered in my laboratory ~8 years ago [20], yet we have encountered significant hurdles in the journey from initial crystals to three-dimensional structure determination. Nonetheless, some interesting general principles have resulted from these studies. For two-dimensional crystallization of SERCA, the factors that control membrane reconstitution and crystallization can be considered separately. The crystallization conditions rely on decavanadate, a large polyanion known to promote linear arrays of SERCA molecules. The lipids used in the reconstitution process are chosen based on maximizing the activity of SERCA, yet the lipid mixture also promotes the formation of well ordered two-dimensional crystals. In this case, the lipid mixture includes phosphatidylcholine, phosphatidylethanolamine and phosphatidic acid and the typical outcome is well ordered helical crystals and small crystalline vesicles. Our serendipitous discovery was that we can switch between different crystal lattices and morphologies (helical versus large two-dimensional crystals) in a process that depends on phosphatidic acid and small changes in magnesium concentration. This lipid-dependent switch in crystal behavior may be applicable to other membrane proteins.

2 Materials

Prepare all solutions using ultrapure water (purified deionized water with a sensitivity of 18 MΩ cm at 25 °C) and analytical grade reagents. Ensure all waste disposal regulations are followed when disposing waste materials. Carefully read and follow Materials Safety Data Sheets for all reagents used.

2.1 Stock Solutions

All stock and buffer solutions should be prepared at room temperature, filtered using a sterile 0.22 μm filter and stored at 4 °C (unless indicated otherwise).

1. 1.5 mL Eppendorf tubes.
2. 15 mL Falcon tubes.
3. Imidazole buffer: 0.5 M imidazole, pH 7.0. Add about 40 mL of water to a glass beaker with a magnetic stir-bar. Weigh 1.7 g imidazole and transfer to the beaker. Mix and adjust pH with HCl (*see* **Note 1**). Make up to 50 mL with water. Filter and store until use.
4. Potassium chloride (KCl): 2 M KCl. Add about 40 mL of water to a glass beaker with a magnetic stir-bar. Weigh 7.46 g KCl and transfer to the beaker. Mix until fully dissolved. Make up to 50 mL with water. Filter and store until use.
5. Magnesium chloride ($MgCl_2$): 1 M $MgCl_2$. Add about 40 mL of water to a glass beaker with a magnetic stir-bar. Weigh 10.16 g $MgCl_2$ ($6 \cdot H_2O$) and transfer to the beaker. Mix until fully dissolved. Make up to 50 mL with water. Filter and store until use.
6. Ethylene glycol tetraacetic acid (EGTA): 50 mM EGTA, pH 8.0. Add about 35 mL of water to a glass beaker with a magnetic stir-bar. Weigh 0.95 g EGTA and transfer to the beaker. Mix and adjust pH with NaOH (*see* **Note 2**). Make up to 50 mL with water. Filter and store until use.
7. Sodium azide: 5 % (w/v) sodium azide. Add about 40 mL of water to a glass beaker with a magnetic stir-bar. Weigh 2.5 g sodium azide and transfer to the beaker. Mix until fully dissolved. Make up to 50 mL with water. Store until use (filtration not required).
8. 3-(*N*-morpholino)propanesulfonic acid (MOPS) buffer: 0.5 M MOPS, pH 7.0. Add about 40 mL of water to a glass beaker with a magnetic stir-bar. Weigh 5.23 g MOPS and transfer to the beaker. Mix and adjust pH with NaOH (*see* **Note 3**). Make up to 50 mL with water. Filter and store until use (*see* **Note 4**).

9. Calcium chloride ($CaCl_2$): 0.5 M $CaCl_2$. Add about 40 mL of water to a glass beaker with a magnetic stir-bar. Weigh 2.77 g $CaCl_2$ (anhydrous) and transfer to the beaker (see **Note 5**). Mix until fully dissolved. Make up to 50 mL with water. Filter and store at room temperature (storage at 4 °C may cause salt precipitation).

10. Dithiothreitol (DTT): 1 M DTT. Add about 8 mL of water to a 15 mL Falcon tube. Weigh 1.54 g DTT and transfer to the tube. Vortex until fully dissolved. Make up to 10 mL with water. The DTT stock solution does not need to be filtered. Transfer 1 mL aliquots into 1.5 mL Eppendorf tubes and store at −20 °C.

11. Glycerol: 80 % (v/v) glycerol. Measure 160 mL of glycerol and transfer it to a glass beaker with a magnetic stir-bar. Make up to 200 mL with water. Transfer to an oversized (greater than 200 mL) screw-top glass bottle and autoclave (see **Note 6**). Store at room temperature.

2.2 Reconstitution

1. 12×75 mm borosilicate glass disposable culture tubes.

2. 8×1 mm (mini) stir-bars.

3. SM2 BioBeads (BioRad): Wash BioBeads in bulk prior to use. Weigh ~5 g BioBeads and transfer to a 50 mL Falcon tube. Add about 30 mL of methanol and gently agitate for 30 min (using a Nutator Mixer, for example). Carefully decant the methanol to waste. Repeat the methanol wash and decant steps. Add about 30 mL of water and gently agitate for 30 min. Carefully decant the water to waste. Repeat the water wash and decant steps twice. Add about 30 mL of 0.05 % sodium azide and store at 4 °C (a 1:100 dilution of the 5 % sodium azide stock will generate the required 0.05 % sodium azide).

4. Detergent—Octaethylene glycol monododecyl ether ($C_{12}E_8$): 10 % (w/v) $C_{12}E_8$. Weigh 0.1 g of solid $C_{12}E_8$ (Barnet Products) and transfer to a 1.5 mL Eppendorf tube. Add 0.92 mL of water. Mix well to dissolve (see **Note 7**) and, if necessary, adjust the final volume to 1 mL. Transfer 100 μL aliquots into 1.5 mL Eppendorf tubes and store at −20 °C.

5. Lipids—Egg yolk phosphatidylcholine (EYPC), egg yolk phosphatidic acid (EYPA), and egg yolk phosphatidylethanolamine (EYPE): Chloroform-solubilized lipids can be obtained at concentrations of 25 mg/mL (Avanti Polar Lipids) and stored at −80 °C.

6. Purified phospholamban (PLN) or sarcolipin (SLN): Purify PLN [21] or SLN [22] as described.

7. Purified Ca^{2+}-ATPase: Purify Ca^{2+}-ATPase as described [23].

8. Reconstitution master-mix: Make from stock solutions on the day of reconstitution and store on-ice. To a 1.5 mL Eppendorf tube, add:

   40 µL of 0.5 M imidazole, pH 7.0.

   25 µL of 2 M KCl.

   10 µL of 1 M $MgCl_2$.

   5 µL of 5 % sodium azide.

   Vortex to mix.

9. Storage buffer: Add about 6.5 mL of water to a glass beaker with a magnetic stir-bar. Add:

   400 µL of 0.5 M MOPS, pH 7.0 stock.

   20 µL of 0.5 M $CaCl_2$ stock.

   10 µL of 1 M $MgCl_2$ stock.

   100 µL of 10 % $C_{12}E_8$ stock.

   Adenosine diphosphate (ADP)—weigh 42.7 mg of solid ADP and transfer to the beaker.

   2.5 µL of 1 M DTT stock.

   2.5 mL of 80 % glycerol stock.

   Mix the components well until the solid ADP has dissolved. Adjust the pH of the solution to pH 7.0 (requires ~100 µL of 1 N NaOH; add dropwise until required pH 7.0 is obtained). Make up to 10 mL with water. Transfer 1 mL aliquots into 1.5 mL Eppendorf tubes and store at −20 °C.

10. 50 % sucrose storage buffer: Add about 40 mL of water to a glass beaker with a magnetic stir-bar. Weigh 25 g of sucrose and transfer to the beaker. Add:

    2.0 mL of stock 0.5 M imidazole, pH 7.0.

    2.5 mL of stock 2 M KCl.

    500 µL of stock 5 % sodium azide.

    Mix while gently heating to dissolve the sucrose (bring the total volume up to 50 mL during this process). Store at 4 °C.

2.3 Crystallization

1. 1.5 mL Screw-cap Eppendorf tubes (*see* **Note 8**).

2. Thapsigargin (TG): 1 mg/mL TG stock solution (Sigma-Aldrich). Add 1 mL of absolute ethanol (> 99.5 % v/v) per mg of TG (obtained as a solid film). Mix to dissolve. Transfer 1 mL aliquots to glass vials. Seal vials tightly with polypropylene-lined caps. Store at −80 °C.

3. Sodium orthovanadate (Na_3VO_4): 50 mM Na_3VO_4 stock solution (*see* **Note 9**). Make fresh on the day of crystallization and store on-ice (*see* **Note 10**). Weigh 46 mg of Na_3VO_4 and transfer to a 15 mL Falcon tube. Add 3.5 mL of water. Mix well to dissolve. Adjust the pH to 2.0 or slightly below (*see*

Note 11). Incubate on ice for 30 min. Adjust the pH to 6.5–7.0 (*see* **Note 12**).

4. Crystallization master-mix: Make from stock solutions on the day of crystallization and store on-ice. To a 15 mL Falcon tube, add:

    4.18 mL of water.

    200 μL of 0.5 M imidazole, pH 7.0.

    250 μL of 2 M KCl. 50 μL of 50 mM EGTA.

    20 μL of 5 % sodium azide.

    Vortex to mix.

5. Crystallization buffer: Make from Crystallization master-mix on the day of crystallization and store on-ice. Add 940 μL of Crystallization master-mix to a 1.5 mL Eppendorf tube. Add the required amount of $MgCl_2$ in a 50 μL volume (*see* **Note 13**). For example, to make 40 mM $MgCl_2$ Crystallization buffer, add 40 μL of 1 M $MgCl_2$ stock solution and 10 μL of ice-cold water. Then add the required amount of Na_3VO_4 in a 10 μL volume (*see* **Note 14**).

3 Methods

3.1 Lipid and Co-reconstituted Peptide Preparation

We describe the lipid and co-reconstituted peptide preparations for four reconstitutions, which represents an appropriate number to become familiar with the experimental setup. Once competent and comfortable with the reconstitution procedure, the researcher may choose the number of reconstitutions per experiment and adjust the components as necessary.

1. Calculate the amount of each lipid required for the reconstitution set to determine the volumes of stock lipids to be added to the lipid pool (*see* **Note 15**). The final lipid ratio will be 8 EYPC:1 EYPA:1 EYPE and lipids will be pooled based on a lipid–protein ratio of 1:1, where the protein is represented solely by the Ca^{2+}-ATPase and not the Ca^{2+}-ATPase and co-reconstituted peptide (*see* **Note 16**). Each reconstitution will require 500 μg Ca^{2+}-ATPase and 500 μg of total lipids. Given a lipid weight ratio of 8 EYPC:1 EYPA:1 EYPE, the calculation of EYPA and EYPE required per reconstitution is straightforward at 50 μg (2 μL of the 25 mg/mL stock lipid solutions). The calculation of the EYPC required per reconstitution is not as straightforward and must take into account the volume containing the 500 μg Ca^{2+}-ATPase to be added will include EYPC at a concentration of 0.25 mg/mL (*see* **Note 17**).

For our calculations here, we will be assuming a purified Ca^{2+}-ATPase concentration of 4 mg/mL and, thus, we will use 125 μL from the purified Ca^{2+}-ATPase aliquot for 500 μg. At 0.25 mg/mL EYPC, this aliquot volume will contribute 31.25 μg EYPC to the reconstitution and, to achieve the 400 μg EYPC required per reconstitution, 368.75 μg EYPC must be added. Thus, 14.75 μL of the 25 mg/mL EYPC stock lipid solution is required per reconstitution.

Finally, the volume of each lipid stock to be added per reconstitution is multiplied by 5 for a set of 4 reconstitutions (*see* **Note 15**). Therefore, the final volume calculations for the lipid pool are 73.75 μL EYPC, 10 μL EYPA, and 10 μL EYPE from their respective 25 mg/mL stock lipid solutions. In this example, the final volume of the lipid pool will be 93.75 μL, representing the required lipids for five reconstitutions, and 18.75 μL of the lipid pool will need to be added to each of the four reconstitution tubes.

2. Add 75 μL of a 2:1 chloroform–trifluoroethanol mixture to 100 μg of dried PLN or dried SLN (*see* **Note 18**). One hundred microgram aliquots of purified PLN or SLN should be dried down, labeled, and stored at −80 °C in a (12×75 mm) glass tube to facilitate this step. Vortex vigorously to dissolve the dried PLN or SLN. Repeat this step for a total of four reconstitution tubes.

3. In a (12×75 mm) glass tube, pool the required lipids for the reconstitution set (based on the calculations in **step 1** above). Measure chloroform-solubilized stock lipids (*see* **Note 19**) using a clean gas-tight Hamilton glass syringe of appropriate size (*see* **Notes 20** and **21**).

4. Transfer the required aliquot of the lipid pool (based on the calculations in **step 1**) into each reconstitution tube. Mix the contents of the reconstitution tubes well by vortexing.

5. Dry the contents of the reconstitution tubes under a gentle stream of nitrogen gas while gently vortexing (*see* **Note 22**). The contents of each reconstitution tube should form a thin film at the bottom of the glass tube (Fig. 1). Desiccate the dried lipid and peptide film overnight in a vacuum desiccator (*see* **Note 23**).

3.2 Ca^{2+}-ATPase Reconstitution

For the following reconstitution step, the reconstitution tubes should be kept at room temperature, whereas stock buffers and other reconstitution components should be kept on ice (unless indicated otherwise). In this section, we describe conditions that have provided the most success for Ca^{2+}-ATPase crystallization. However, it is important to note many reconstitution variables, especially the lipids, can be screened to optimize crystallization (Fig. 1).

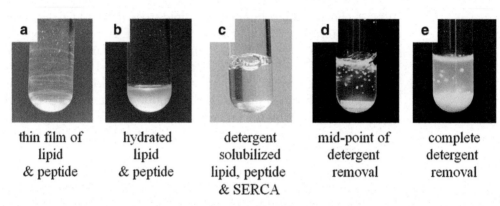

Co-Reconstitution Screen						
	lipids (wt ratio)				peptide	SERCA
	EYPC	EYPE	EYPA	total		
Highest success	8	1	1	500µg	100µg	500µg
Range for screening	6-8	1-2	1-2	300-600µg	25-150µg	300-500µg

Fig. 1 Key steps in the reconstitution of SERCA into proteoliposomes. The table summarizes the conditions that are typically screened to achieve co-reconstitution of SERCA in the presence of phospholamban or sarcolipin. The steps shown are (**a**) thin film of lipid and peptide on the inside wall of a glass test tube; (**b**) hydrated lipids and peptide; (**c**) detergent-solubilized, mixed-micelle solution containing SERCA, lipids and peptide (the magnetic stir bar is visible at the *bottom* of the test tube); (**d**) mid-point of the detergent removal with SM2 Biobeads (the turbidity indicates the beginning of vesicle formation; the *larger white spheres* are Biobeads); (**e**) complete detergent removal with SM2 Biobeads (this solution contains co-reconstituted proteoliposomes of SERCA and phospholamban or sarcolipin)

1. Add 185 µL of water to each reconstitution tube. Without generating bubbles, gently pipette the water to resuspend the lipid and peptide film. The tube may also be gently vortexed, but use care not to generate bubbles. Also, ensure the contents do not exceed the maximum height of the original dried line of the lipid and peptide film.

2. Incubate the reconstitution tubes with resuspended lipid and peptide at 37 °C for 10 min with gentle vortexing every 2 min.

3. Allow the contents of the reconstitution tubes (hydrated lipid and peptide in Fig.1) to cool to room temperature.

4. Add 7 µL of 10 % $C_{12}E_8$ to each reconstitution tube (*see* **Note 24**).

5. Vigorously vortex each reconstitution tube for 3 min wihtout stopping (*see* **Note 25**).

6. Add 8 µL of Reconstitution master mix to each reconstitution tube. Vortex gently to mix.

7. Add the balance volume of storage buffer to each reconstitution tube (see **Note 26**). For this example, we will be adding a balance volume of 175 μL of storage buffer (adding a 125 μL volume of purified Ca^{2+}-ATPase in storage buffer in the next step as we assumed a Ca^{2+}-ATPase concentration of 4 mg/mL). Vortex gently to mix.

8. Add the volume of purified Ca^{2+}-ATPase that contains 500 μg to each reconstitution tube (see **Note 27**).

9. Add 8 × 1 mm (mini) stir-bars to each reconstitution tube.

10. Set up a suitable test-tube rack on a magnetic stir-plate and place the reconstitution tubes in the rack. Stir the reconstitutions gently (see **Note 27**).

11. Add SM2 BioBeads over a 4-h time-course (see **Note 28**).

    First addition: ~1 mg BioBeads (see **Note 29**), incubate for 30 min while stirring gently (see **Note 30**).

    Second addition: ~1 mg BioBeads and 30 min incubation with stirring.

    Third addition: ~1 mg BioBeads and 30 min incubation with stirring.

    Fourth addition: ~2 mg BioBeads and 30 min incubation with stirring.

    Fifth addition: ~5 mg BioBeads and 60 min incubation with stirring.

    Final addition: ~10 mg BioBeads and 60 min incubation with stirring.

3.3 Ca^{2+}-ATPase Crystallization

We describe the Ca^{2+}-ATPase crystallization procedure for eight samples (four reconstitutions screened at two $MgCl_2$ concentrations), which represents an appropriate number to become familiar with the experimental setup. Once competent and comfortable with the crystallization procedure, the researcher may choose the number of crystallizations per experiment and adjust the components as necessary.

During the crystallization steps, the crystallization tubes should be kept on-ice (unless indicated otherwise). All centrifugation steps are carried out at 4 °C. Rotors and centrifuges should be prechilled to 4 °C prior to use. In this section, we describe screening conditions that have provided the most success for Ca^{2+}-ATPase crystallization for both the thin, helical crystals (5 mM $MgCl_2$ Crystal buffer) and the wide, 2D crystals (40 mM $MgCl_2$ Crystal buffer) (Figs. 2 and 3). However, it is important to note that many other crystallization variables can be screened to optimize crystal formation and frequency.

1. Transfer 200 μL of 5 mM $MgCl_2$ Crystallization buffer to a screw-cap Eppendorf tube. Repeat for a total of four crystallization tubes.

2D Crystallization Screen (4 °C)			
Mg^{2+}	Na_3VO_4	Incubation	Crystal type
5 mM	0.5 mM	3-5 days	Helical
30-50 mM	0.25-0.5 mM	1-6 days	Wide 2D
Mn^{2+}			
5 mM	0.5 mM	3-5 days	Wide 2D
20 mM	0.5 mM	3-5 days	Helical

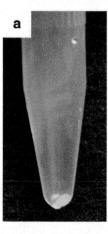

pellet of co-reconstituted
proteoliposomes in
decavanadate
crystallization buffer

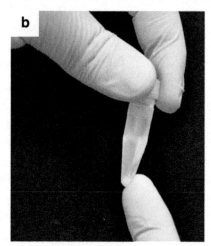

freeze-thaw procedure
used to promote fusion of
proteoliposomes prior to
crystal formation

Fig. 2 Key steps in the two-dimensional crystallization of SERCA starting from reconstituted proteoliposomes. The table summarizes the conditions that are typically screened to achieve crystallization of SERCA in the presence of phospholamban or sarcolipin. (**a**) Pellet of the proteoliposomes collected by centrifugation in the presence of decavanadate crystallization buffer. The *yellow color* is indicative of the presence of decavanadate. (**b**) Demonstration of the freeze–thaw cycle that is used to promote proteoliposome fusion. The small proteoliposomes (~0.2 μm diameter) must fuse to form large two-dimensional crystals (0.5 μm wide, 5–50 μm in length). After rapidly freezing in liquid nitrogen, the pellet is thawed between thumb and forefinger. The solution is gently agitated as it begins to melt, and the process is repeated

2. Transfer 200 μL of 40 mM $MgCl_2$ Crystallization buffer to a screw-cap Eppendorf tube. Repeat for a total of four crystallization tubes.

3. Transfer 50 μL of each reconstitution to a 5 mM $MgCl_2$ crystallization tube and a 40 mM $MgCl_2$ crystallization tube (*see* **Note 31**). Pipette the mixtures gently up and down to mix.

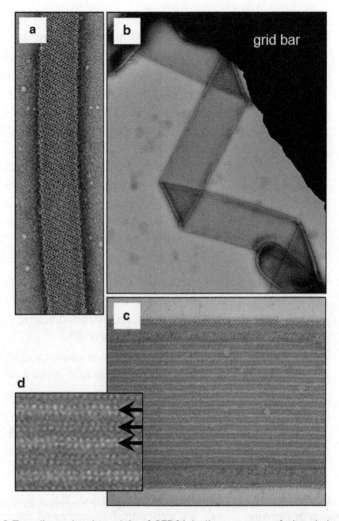

Fig. 3 Two-dimensional crystals of SERCA in the presence of phospholamban (similar crystals are obtained in the presence of sarcolipin). All crystals were imaged by negative-stain electron microscopy. (*A*) A typical helical crystal. The crystal is ~70 nm wide and has p2 symmetry ($a=57$ Å, $b=117$ Å, $\gamma=64$ °). (**B**) A typical two-dimensional crystal. The crystal is ~0.5 μm wide and has $p22_12_1$ symmetry ($a=350$ Å, $b=70$ Å, $\gamma=90$ °). (**C**) Magnified view of a region of a two-dimensional crystal. (**D**) Close-up view of the SERCA dimer ribbons (*arrows*) that make up the lattice. The *b* axis is horizontal and rigidly held together by decanvanadate. The *a* axis is vertical and highly variable, and this variability represents the major hurdle we have encountered in going from initial crystals to a three-dimensional structure

Repeat for each of the 4 reconstitutions. The remaining reconstitution material that will not be used for the crystallization should be frozen and stored as outlined (*see* **Note 32**).

4. Spin the mixture in the crystallization tubes at $40,000 \times g$ for 40 min at 4 °C to pellet the reconstituted proteoliposomes.

5. For each crystallization tube, carefully aspirate the supernatant to waste without disturbing the vesicle pellet.

6. To each crystallization tube, add 20 μL of the appropriate Crystallization buffer on top of the vesicle pellet.

7. To each crystallization tube, add 1 μL of TG to the vesicle pellet (*see* **Note 33**). Tap the screw-cap Eppendorf tube gently to mix its contents without resuspending the vesicle pellet (*see* **Note 34**).

8. Freeze the vesicle pellet and added Crystallization buffer and TG by submerging in liquid nitrogen (*see* **Note 35**).

9. Thaw the frozen vesicle pellet and solution slowly between the thumb and forefinger with gentle mixing just as it thaws (Fig. 2 and *see* **Note 36**).

10. Repeat **steps 8** and **9**.

11. Resuspend the pellet by gently pipetting up and down using a 20 μL pipette tip with the pipette set to 10 μL or less (*see* **Note 37**).

12. Freeze the resuspended vesicle pellet by submerging in liquid nitrogen (*see* **Note 35**).

13. Thaw the frozen resuspension slowly between the thumb and forefinger with gentle mixing just as it thaws (Fig. 2 and *see* **Note 36**).

14. Repeat **steps 12** and **13**.

15. Incubate the crystallization tubes at 4 °C. Also store any remaining master-mix, Crystallization buffers, and Na_3VO_4 solution at 4 °C.

3.4 Crystal Screening by Electron Microscopy

Crystals can be screened after a minimum of 24 h, although crystals are generally best screened after 3–5 days of incubation. To screen the crystals, gently tap the crystallization tubes to resuspend the sedimented material and apply 5 μL of the mixture to a glow-discharged electron microscopy grid for negative-staining. Good crystallization trials identified by negative-stain electron microscopy should have duplicate negative-stain grids prepared, whereas excellent crystallization trials should be prepared for electron cryo-microscopy imaging (*see* **Note 38**). For the successful crystallographer, we have previously provided details of the screening and imaging of crystals by electron microscopy [20, 24, 25], which are beyond the scope of this protocol.

4 Notes

1. Add concentrated HCl (6 N) dropwise to bring the starting pH closer to the required pH, and then add 1 N HCl dropwise to lower the imidazole solution pH to 7.0.

2. EGTA will not go into solution until it nears the required pH. Add concentrated NaOH (10 N) dropwise to bring the starting pH closer to the required pH, and then add 1 N NaOH dropwise to raise the EGTA solution pH to 8.0.

3. Add concentrated NaOH (6 N) dropwise to bring the starting pH closer to the required pH, and then add 1 N NaOH dropwise to raise the MOPS solution pH to 7.0.

4. The 0.5 M MOPS stock solution should be stored in the dark (or wrapped in tinfoil to minimize its exposure to light). The 0.5 M MOPS stock solution should be discarded if it turns yellowish in color and a fresh stock prepared.

5. Dissolving $CaCl_2$ in water is an exothermic reaction and appropriate precautions should be taken. For example, while dissolving the $CaCl_2$ in water, the glass beaker can be placed on ice.

6. Exercise best autoclave practices and take necessary precautions. For example, liquid containers can only be filled to a maximum of one-half to two-thirds capacity, screw-top lids must be loosened to prevent pressure build-up, and there must not be any cracks or vulnerabilities in glass bottles or containers. Autoclave tape, or some other indicator, should be used to ensure that the target temperature has been reached and the autoclave itself should be monitored for operational safety and quality control.

7. It can take several minutes to dissolve the solid $C_{12}E_8$. After sealing the 1.5 mL Eppendorf tube to prevent leakage (use Parafilm M or a screw-cap 1.5 mL Eppendorf), we use a Nutator Mixer to mix the solid detergent and water until dissolution occurs.

8. It is important that the 1.5 mL screw-cap Eppendorf tubes are high quality tubes, such that the tubes will not warp or break during high-speed ($40,000 \times g$) centrifugation.

9. The Na_3VO_4 stock solution is specifically prepared to maximize the decameric species of vanadate (decavanadate), which should be apparent as an intense yellow color when the preparation of the Na_3VO_4 stock solution is completed.

10. If performing the crystallization on the same day as the reconstitution, begin preparing the Na_3VO_4 stock solution during one of the last two 60 min BioBead incubations.

11. Add concentrated HCl (6 N) dropwise to bring the starting pH closer to the required pH, and then add 1 N HCl dropwise to lower the Na_3VO_4 stock solution pH to 2.0 or below (a pH of 1.9 is acceptable).

12. The Na_3VO_4 stock solution should be kept on ice during the pH adjustment. Add ice-cold KOH (1 N) dropwise to bring

the pH between 3.0 and 4.0, and then add ice-cold 0.1 N KOH dropwise to bring the Na_3VO_4 stock solution pH between 6.5 and 7.0. Ensure that the pH does not go above 7.0 during this adjustment as it might cause the Na_3VO_4 to precipitate, at which point the Na_3VO_4 preparation needs to be started again.

13. Determining the $MgCl_2$ concentrations to screen is a key decision in the crystallization experiment. Low $MgCl_2$ concentrations (5–10 mM) will generate thin, helical Ca^{2+}-ATPase crystals, whereas higher $MgCl_2$ concentrations (30–50 mM) will generate wide, 2D Ca^{2+}-ATPase crystals. We have previously provided a detailed discussion surrounding this observed Mg^{2+}-dependent switch in Ca^{2+}-ATPase crystal morphology and the potential underlying mechanisms responsible for the switch [24]. We note here that we have had success screening other divalent cations in place of Mg^{2+} in the Crystallization buffer (J.P. Glaves and H.S. Young, unpublished observations). Interestingly, when screening other divalent cations, the different Ca^{2+}-ATPase crystal morphologies did not necessarily have the same concentration-dependence as that observed for Mg^{2+}. As an example, we observed wide, 2D Ca^{2+}-ATPase crystals when screening low concentrations of $MnCl_2$.

14. Ensure that the Crystallization buffer is ice-cold before adding the required amount of Na_3VO_4. Based on our experience, the amount of Na_3VO_4 should only be varied between 0.25 mM and 0.5 mM (5 or 10 µL, respectively, of the 50 mM Na_3VO_4 stock solution added to 1 mL of Crystallization buffer).

15. The lipid pool should include enough lipids for one extra "reconstitution" for every four reconstitutions (i.e., for four reconstitutions, pool lipids necessary for five reconstitutions; for eight reconstitutions, pools lipids for ten reconstitutions; and so on). Including enough lipids in the lipid pool for these extra "reconstitutions" allows for small amounts of evaporation and minor measurement errors that will occur during the reconstitution setup. The purpose of pooling the required lipids prior to their addition to the individual reconstitution tubes is to minimize the number of very small volume measurements and transfers that will be required during the experiment. It also ensures that the individual reconstitutions draw from a lipid pool with an identical overall composition.

16. The lipid ratio and the lipid-to-protein ratio are among the many reconstitution variables that can be screened to optimize 2D crystallization. We have previously described in detail the effects of the lipid ratio, and the type of lipids used, on the reconstitution, activity, and crystallization of the Ca^{2+}-ATPase [26]. Reference to this work should be made for those interested in the important roles of the lipid composition with

respect to critical steps in the Ca^{2+}-ATPase reconstitution and crystallization processes, including vesicle formation and vesicle fusion, as the in-depth discussion of such details is beyond the scope of this chapter. Regarding the lipid-to-protein ratio, we have experienced the most success working with a lipid-to-protein ratio of 1:1 that either excludes the co-reconstituted peptides in the protein calculation (i.e., 500 μg total lipid to 500 μg Ca^{2+}-ATPase) or includes the co-reconstituted peptides in the protein calculation (i.e., 600 μg total lipid to 500 μg Ca^{2+}-ATPase and 100 μg PLN or SLN). In our experience, varying the lipid-to-protein ratio too far beyond these limits leads to inefficient reconstitution, evidenced by an excess of liposomes at higher lipid-to-protein ratios or poor incorporation of the Ca^{2+}-ATPase and co-reconstituted PLN or SLN at lower lipid-to-protein ratios. Less efficient reconstitution leads to poor subsequent crystallization. As such, although this reconstitution variable can be explored, we recommend working near a 1:1 lipid-to-protein ratio.

17. Detergent-solubilized, affinity-purified aliquots of Ca^{2+}-ATPase are stored at −80 °C. When an aliquot is thawed for the first time, we add EYPC to a final concentration of ~0.25 mg/mL in order to stabilize the enzyme and preserve its activity through subsequent freeze–thaw cycles. As an example, our purified Ca^{2+}-ATPase aliquots are 500 μL in volume and we add 12.5 μL of an aqueous 10 mg/mL EYPC stock to each thawed aliquot. Ca^{2+}-ATPase aliquots that have had EYPC added to them are marked to prevent subsequent over-addition of EYPC. An aqueous 10 mg/mL EYPC stock, for this purpose, is prepared by first drying 10 mg of EYPC (400 μL of the 25 mg/mL chloroform-solubilized EYPC stock) in a (12 × 75 mm) glass tube under a gentle nitrogen gas stream while gently vortexing. Drying the stock EYPC in this manner creates a dried EYPC film on the bottom of the glass tube (Fig. 1). The dried EYPC film is then further dried in a vacuum desiccator overnight. Next, the dried EYPC film is resuspended by adding 1 mL of water to the tube and vortexing. Following resuspension, the aqueous EYPC stock is finally transferred, as 100 μL aliquots, into 1.5 mL Eppendorf tubes and stored at −20 °C.

18. Both the thin, helical Ca^{2+}-ATPase crystals and the wide, 2D Ca^{2+}-ATPase crystals can be grown without PLN or SLN (i.e., Ca^{2+}-ATPase alone). It is critical to note, however, that although the presence of PLN or SLN has a negligible effect on the frequency of thin, helical Ca^{2+}-ATPase crystals, the presence of PLN or SLN dramatically increases the frequency of wide, 2D Ca^{2+}-ATPase crystals. As such, we recommend including PLN or SLN as a variable for wide, 2D Ca^{2+}-ATPase

crystal screens. If PLN or SLN is not to be included in the reconstitution step, the 2:1 chloroform–trifluoroethanol mixture does not need to be added to the glass tube. Alternatively, if PLN or SLN will be included in the reconstitution set, the 2:1 chloroform–trifluoroethanol mixture should be added to an empty glass tube for the Ca^{2+}-ATPase-only control, if such a control is also to be included.

19. Perform all steps including the use of chloroform in a fume hood.

20. For washing the Hamilton syringe between lipid stock measurements (or before measuring aliquots from the lipid pool), rinsing the syringe serially using multiple tubes of chloroform is sufficient to prevent contamination of lipid stocks (or the lipid pool) with the previous lipid stock. Be sure the interior of the syringe is cleared of chloroform from the final rinse step and dry before measuring the next lipid aliquot. The Hamilton syringe should be fully cleaned (*see* **Note 21**) after the lipids have been aliquoted for each experiment and prior to storage of the syringe.

21. Hamilton syringes used for measuring chloroform-solubilized lipids can be cleaned by rinsing with chloroform, followed by multiple rinses with deionized water, and a final rinse with acetone. Be sure to expel all remaining cleaning solvents after each rinse step. The interior of the syringe should be cleared of residual acetone and dry before storing the syringe or using the syringe again. Wipe clean the exterior surfaces of the syringe and the needle with a lint-free tissue. See the Hamilton 'Syringe Care and Use Guide' for more details (www.hamiltoncompany.com).

22. First, determine the vortex speed required during the drying step. Vortex the contents of a reconstitution tube and adjust the vortex speed until the top of the contents mixing line reaches a maximum of approximately 1 cm from the bottom of the tube. Second, determine the nitrogen gas flow-rate required during the drying step. We empirically determine this flow-rate by expelling the gas stream on to the back of the hand (use caution during this step and start with a very low flow-rate that can be adjusted in small increments). If the flow-rate causes an indentation of the skin, it is too high. Next, vortex the contents of the tube under the stream of nitrogen gas. The contents of the tube should not be displaced above the original maximum mixing line approximately 1 cm from the bottom of the tube. If the stream of nitrogen gas is displacing the contents above 1 cm from the bottom of the tube, then lower the flow-rate. Drying of the solubilized lipids and peptide to a thin film using this approach should take approximately 5 min per reconstitution tube.

23. The dried lipid and peptide film can be desiccated for a minimum of 2 h, although we recommend allowing desiccation to proceed overnight.

24. The total amount of $C_{12}E_8$ that will be used in the reconstitution is 1000 µg (representing a detergent–lipid mass ratio of 2:1). The addition of $C_{12}E_8$ in this step adds 700 µg to the reconstitution and the remaining 300 µg will be added with the 300 µL total volume of pure Ca^{2+}-ATPase and additional storage buffer that contains $C_{12}E_8$ at a concentration of 1 mg/mL.

25. After the addition of detergent and vortexing, the contents of the reconstitution tube should be fully solubilized and clear (Fig. 1). If the contents are not clear at this point, the addition of very small volumes of 10 % $C_{12}E_8$ (1 µL per addition) can help with solubility, however, do not exceed an additional 3 µL of 10 % $C_{12}E_8$. The addition of buffer components in the Reconstitution master-mix, particularly $MgCl_2$, in the next step might also improve solubility. Unfortunately, poor solubility or the presence of large aggregates at this stage of the reconstitution could indicate problems with the PLN or SLN peptides and might warrant stopping the experiment altogether (prior to the addition of purified Ca^{2+}-ATPase).

26. The total volume of storage buffer to be added to the reconstitution will be 300 µL. After subtracting the volume of storage buffer containing 500 µg of purified Ca^{2+}-ATPase to be added in the next step from the total storage volume of 300 µL, the remaining volume represents the balance volume of storage buffer added in this step.

27. To ensure enzyme stability and activity after the addition of purified Ca^{2+}-ATPase to the reconstitution, care must be taken to avoid vortexing the mixture and generating air bubbles during the remaining experimental steps.

28. Remove excess storage solution from the stock BioBeads before adding them to the reconstitutions. This prevents excessive dilution of reconstitutions from the BioBead storage solution (0.05 % sodium azide in water). We remove the excess by carefully decanting the storage solution to waste and removing the excess solution from the bulk BioBeads using a vacuum trap. Insert a glass Pasteur pipette into tubing attached to a filtering flask capped with a stopper and inserted vacuum line (use caution to prevent breaking the Pasteur pipette when inserting into the tubing). After turning on the vacuum line, insert the Pasteur pipette into a 200 µL pipette tip (much as you would with an air displacement micropipette) and then insert the capped pipette into the bottom of the BioBead storage tube (i.e., to the bottom of the bulk BioBeads). By using the 200 µL pipette tip, you can quickly replace the pipette tip if it

becomes blocked by the very small BioBeads before the excess storage solution is fully removed. At this point, the tube containing the BioBeads should be capped and kept on ice until the BioBead addition is complete, and storage solution can be added again. It is important that the Biobeads do not become dry at any point before additions, as this will prevent them from adsorbing and removing detergent.

29. Rather than attempting to weigh out very small amounts of BioBeads, we use the general rule-of-thumb that four average-sized BioBeads equals ~1 mg.

30. During BioBead additions and incubation, the reconstitution should be stirred very gently such that the BioBeads are moving around in the tube. However, be careful not to introduce air bubbles or break up or damage the Biobeads (*see* also **Note 27**).

31. The amount of reconstituted material added to the crystallization tube is yet another variable that can be screened to optimize crystallization. Given the reconstitution conditions of 500 μg Ca^{2+}-ATPase in an ~500 μL total reconstitution volume, we have had the most success using 50–75 μg of Ca^{2+}-ATPase (50–75 μL of reconstituted material) following the described crystallization protocol.

32. Following this protocol, there will be ~400 μL of material remaining per reconstitution after this step. The remaining reconstitution can be stored long-term for subsequent Ca^{2+}-ATPase activity measurements or future crystallization trials. First, add 267 μL of the stock 50 % sucrose storage buffer to a screw-cap Eppendorf tube (or an appropriate amount of stock 50 % sucrose storage buffer to achieve a final sucrose concentration of 20 %). The sucrose acts to stabilize the reconstituted vesicles during the freeze–thaw process. Second, add the 400 μL of remaining reconstitution. Third, mix the storage buffer and the reconstitution well by gently pipetting up and down. Once fully mixed, the mixture can be flash-frozen in liquid nitrogen by submerging the sealed screw-cap Eppendorf tube. Store at −80 °C.

The addition of sucrose to the reconstitutions for long-term storage must be addressed when using the vesicles for future crystallization trials, as the sucrose can interfere with the freeze–thaw cycles and vesicle fusion. Also, the dilution of the reconstitution by the addition of the sucrose storage buffer will affect the total volume containing 50–75 μg of Ca^{2+}-ATPase (*see* **Note 31**). To use the frozen reconstituted vesicles in future crystallization trials, the vesicles must first be washed to remove the sucrose. Wash a volume containing 50–75 μg of Ca^{2+}-ATPase by adding it to a screw-cap Eppendorf tube containing 1.2 mL of Crystallization buffer and mixing thoroughly by gently pipetting up and down. Spin the mixture at 40,000 × g for 40 min at 4 °C to pellet the reconstituted vesicles. Carefully aspirate the supernatant to waste without disturbing the vesicle pellet. Add 200 μL of Crystallization buffer to the vesicle pellet and resuspend the pellet by gently pipetting up and down. The crystallization protocol can now proceed as outlined from **step 4** of the Ca^{2+}-ATPase Crystallization section.

33. TG is not required for Ca^{2+}-ATPase crystallization, however, we have had the most success with 2D crystal formation, quality, and frequency using this Ca^{2+}-ATPase inhibitor (Fig. 3). As a crystal-screening variable, TG can be replaced by other Ca^{2+}-ATPase inhibitors, such as cyclopiazonic acid [26], or the crystallization conditions might exclude Ca^{2+}-ATPase inhibitors. Note that this latter crystal screening option of no Ca^{2+}-ATPase inhibitors greatly decreases crystal formation, quality, and frequency.

34. While tapping the screw-cap Eppendorf tube to mix the small volume of Crystallization buffer and added TG, the vesicle pellet may detach from the tube wall. If this occurs, the vesicle pellet will likely remain intact and the proteoliposomes will not disperse at this stage. From experience, we generally consider a vesicle pellet that releases from the tube wall during this step to be a sign of a good reconstitution that will behave and resuspend nicely in the following freeze–thaw steps.

35. When freezing the contents of the crystallization tube, ensure that the liquid volume remains at the bottom of the tube. This can be performed by keeping the crystallization tube upright for the first few seconds of submersion in liquid nitrogen, such that the crystallization volume is at the bottom of the tube (we typically use forceps to submerge and remove the tube, as opposed to simply dropping the tube into liquid nitrogen).

36. To prevent liquid nitrogen burns, use extreme care to ensure that all liquid nitrogen has been gently tapped off of the surface of the crystallization tube before handling. Also, thaw the crystallization tube contents slowly by alternating a few seconds between on-ice and between the thumb and forefinger (to prevent the contents from warming up too much). Gently tap the tube to ensure that the contents are fully thawed (Fig. 2).

37. As with all pipetting in the experiment, do not produce any air bubbles during resuspension. Also, use care to ensure the crystallization tube contents do not warm up too much by alternating between pipetting and placing the tube on-ice. Some pieces of the pellet might not resuspend easily. If some pellet pieces remain after ~1 min of pipetting, it is likely best to discontinue trying to resuspend such pieces and continue to the next step of the experiment.

38. When we prepare frozen-hydrated electron microscopy grids, we often have to dilute the crystals using Crystallization buffer to achieve an optimal amount of material for electron cryo-microscopy screening and imaging. This is why we advise storing the extra Crystallization buffers at 4 °C with the crystallization tubes. As a general procedure, we estimate the number of crystals per grid square during the initial screening by counting the number of crystals on a minimum of 20 grid squares containing "material" (crystals, vesicles, or precipitated material) and then calculating the average number of crystals per grid square [25]. To prepare frozen-hydrated crystal grids, we dilute the crystals using this average as a guide to achieve an approximate frequency of 1 crystal per grid square and immediately apply the diluted crystals to the grid and plunge the grid in liquid ethane.

Acknowledgements

J.P.G. was supported by a Canada Graduate Scholarship from the Canadian Institutes of Health Research. J.O.P. is supported by the NSERC-CREATE International Research Training Group in Membrane Biology. This work was supported by an operating grant from the Canadian Institutes of Health Research (to H.S.Y.).

References

1. Henderson R, Baldwin JM, Ceska TA, Zemlin F, Beckmann E, Downing KH (1990) Model for the structure of bacteriorhodopsin based on high-resolution electron cryo-microscopy. J Mol Biol 213:899–929
2. Nogales E, Wolf SG, Downing KH (1998) Structure of the alpha beta tubulin dimer by electron crystallography. Nature 391:199–203
3. Miyazawa A, Fujiyoshi Y, Unwin N (2003) Structure and gating mechanism of the acetylcholine receptor pore. Nature 423:949–955
4. Jeckelmann JM, Harder D, Ucurum Z, Fotiadis D (2014) 2D and 3D crystallization of a bacterial homologue of human vitamin C membrane transport proteins. J Struct Biol 188(1):87–91
5. Fribourg PF, Chami M, Sorzano CO, Gubellini F, Marabini R, Marco S, Jault JM, Levy D (2014) 3D cryo-electron reconstruction of BmrA, a bacterial multidrug ABC transporter in an inward-facing conformation and in a lipidic environment. J Mol Biol 426:2059–2069
6. Paulino C, Kuhlbrandt W (2014) pH- and sodium-induced changes in a sodium/proton antiporter. eLife 3: e01412
7. Kebbel F, Kurz M, Grutter MG, Stahlberg H (2012) Projection structure of the secondary citrate/sodium symporter CitS at 6 A resolution by electron crystallography. J Mol Biol 418:117–126
8. Dux L, Martonosi A (1983) Two-dimensional arrays of proteins in sarcoplasmic reticulum and purified Ca^{2+}-ATPase vesicles treated with vanadate. J Biol Chem 258:2599–2603
9. Toyoshima C, Sasabe H, Stokes DL (1993) Three-dimensional cryo-electron microscopy of the calcium ion pump in the sarcoplasmic reticulum membrane. Nature 362:469–471
10. Zhang P, Toyoshima C, Yonekura K, Green N, Stokes D (1998) Structure of the calcium pump from sarcoplasmic reticulum at 8 Angstroms resolution. Nature 392:835–839
11. Toyoshima C, Nakasako M, Nomura H, Ogawa H (2000) Crystal structure of the calcium pump of sarcoplasmic reticulum at 2.6 A resolution. Nature 405:647–655
12. Toyoshima C, Nomura H (2002) Structural changes in the calcium pump accompanying the dissociation of calcium. Nature 418:605–611

13. Toyoshima C, Nomura H, Tsuda T (2004) Lumenal gating mechanism revealed in calcium pump crystal structures with phosphate analogues. Nature 432:361–368
14. Toyoshima C, Mizutani T (2004) Crystal structure of the calcium pump with a bound ATP analogue. Nature 430:529–535
15. Toyoshima C, Iwasawa S, Ogawa H, Hirata A, Tsueda J, Inesi G (2013) Crystal structures of the calcium pump and sarcolipin in the Mg^{2+}-bound E1 state. Nature 495: 260–264
16. Winther AM, Bublitz M, Karlsen JL, Moller JV, Hansen JB, Nissen P, Buch-Pedersen MJ (2013) The sarcolipin-bound calcium pump stabilizes calcium sites exposed to the cytoplasm. Nature 495:265–269
17. Olesen C, Picard M, Winther AM, Gyrup C, Morth JP, Oxvig C, Moller JV, Nissen P (2007) The structural basis of calcium transport by the calcium pump. Nature 450:1036–1042
18. Sorensen T, Moller J, Nissen P (2004) Phosphoryl transfer and calcium ion occlusion in the calcium pump. Science 304:1672–1675
19. Olesen C, Sorensen T, Nielsen R, Moller J, Nissen P (2004) Dephosphorylation of the calcium pump coupled to counterion occlusion. Science 306:2251–2255
20. Stokes DL, Pomfret AJ, Rice WJ, Glaves JP, Young HS (2006) Interactions between Ca^{2+}-ATPase and the pentameric form of phospholamban in two-dimensional co-crystals. Biophys J 90:4213–4223
21. Douglas JL, Trieber CA, Afara M, Young HS (2005) Rapid, high-yield expression and purification of Ca^{2+}-ATPase regulatory proteins for high-resolution structural studies. Protein Expr Purif 40:118–125
22. Gorski PA, Glaves JP, Vangheluwe P, Young HS (2013) Sarco(endo)plasmic reticulum calcium ATPase (SERCA) inhibition by sarcolipin is encoded in its luminal tail. J Biol Chem 288:8456–8467
23. Stokes DL, Green NM (1990) Three-dimensional crystals of Ca-ATPase from sarcoplasmic reticulum: symmetry and molecular packing. Biophys J 57:1–14
24. Glaves JP, Fisher L, Ward A, Young HS (2010) Helical crystallization of two example membrane proteins MsbA and the $Ca(2+)$-ATPase. Methods Enzymol 483:143–159
25. Glaves JP, Trieber CA, Ceholski DK, Stokes DL, Young HS (2011) Phosphorylation and mutation of phospholamban alter physical interactions with the sarcoplasmic reticulum calcium pump. J Mol Biol 405:707–723
26. Young HS, Rigaud JL, Lacapère JJ, Reddy LG, Stokes DL (1997) How to make tubular crystals by reconstitution of detergent-solubilized $Ca2(+)$-ATPase. Biophys J 72:2545–2558

Chapter 39

Two-Dimensional Crystallization of Gastric H^+,K^+-ATPase for Structural Analysis by Electron Crystallography

Kazuhiro Abe

Abstract

Electron crystallography of two-dimensional (2D) crystals has provided important information on the structural biology of P-type ATPases. Here, I describe the procedure for making 2D crystals of gastric H^+,K^+-ATPase purified from pig stomach. The 2D crystals are produced by dialyzing detergent-solubilized H^+,K^+-ATPase mixed with synthetic phospholipids. Removal of the detergent induces the reconstitution of H^+,K^+-ATPase molecules into the lipid bilayer. In the presence of fluorinated phosphate analogs, or in combination with transporting cations or the specific antagonist SCH28080, H^+,K^+-ATPase forms crystalline 2D arrays. The molecular conformation and morphology of the 2D crystals vary depending on the crystallizing conditions. Using these 2D crystals, three-dimensional structures of H^+,K^+-ATPase can be generated by data correction from ice-embedded 2D crystals using cryo-electron microscopy, followed by processing the recorded images using electron crystallography methods.

Key words Gastric proton pump, H^+,K^+-ATPase, Electron crystallography, Cryo-electron microscopy, Structural analysis, Membrane protein, Two-dimensional crystal

1 Introduction

Since the first three-dimensional (3D) structure of the membrane protein bacteriorhodopsin [1], electron crystallography of two-dimensional (2D) crystals has been an important method, along with X-ray crystallography and nuclear magnetic resonance (NMR) spectroscopy, for studying the 3D structures of membrane proteins [2, 3]. Its reliance on single- or double-layered 2D-ordered arrays and the ability to obtain structural information from small and moderately disordered crystals make electron crystallography particularly useful for studying membrane proteins in a lipid bilayer environment. Especially in the P-type ATPase field, this method contributed to the determination of the first 3D structure of sarco-endoplasmic reticulum Ca^{2+}-ATPase (SERCA) [4] and Na^+,K^+-ATPase [5]. The use of atomic models of SERCA in its different reaction states determined by X-ray crystallography provided insight into its molecular mechanism [6–10]. Even at medium resolution (6.5–8 Å), however, the 3D structures of H^+,K^+-ATPase determined by electron crystallography have revealed, with the help of atomic models of related ATPases [11–13], its unique and conserved molecular mechanisms relevant to its physiologic roles [14–16].

In the early stages of the structural determination by electron crystallography, 2D crystals were usually obtained in the native membrane taken from a natural source (purple membrane for bacteriorhodopsin, sarcoendoplasmic reticulum vesicles for SERCA, or kidney membrane for Na^+,K^+-ATPase). To obtain much larger and well-ordered crystals, however, recently 2D crystallization has been attempted mainly by reconstituting purified detergent-solubilized proteins into a lipid bilayer comprising synthetic phospholipids [17]. Successful crystallization using this reconstitution method has, however, several requirements:

1. The target protein should be sufficiently stable in the detergent micelle during crystallization.

2. Detergentremoval must be controlled. Because the major driving force for the 2D crystallization is detergent removal, rapid removal of detergent such as using Bio-beads can produce poorly ordered crystals or, in the worst case, protein aggregates. Detergent removal by micro-dialysis is thus highly recommended for 2D crystallization, because it is gradual, and the rate of detergent removal is easy to regulate by changing the dialysis device (i.e., dialysis buttons and membranes), the solution volume, and detergent concentration.

3. The lipid-to-protein ratio (LPR) must be carefully determined. The LPR is critical for crystallization because if the amount of phospholipid is insufficient, the molecules in the membrane are too close to form crystals. Conversely, an excess amount of phospholipid may produce vesicles in which membrane proteins are sparsely reconstituted. The kind of phospholipid used for the reconstitution is also important.

4. Homogeneous molecular conformation is important for well-ordered crystals.

5. Crystal growth should be gradual. Gradual growth of 2D crystal is generally favorable for coherent crystal formation. As crystal formation is mainly driven by reconstitution, the rate of detergent removal by dialysis is an important factor for the production of well-ordered crystals. If detergent removal is too fast, or protein concentration is too high, poorly ordered, multinucleated crystals may be produced.

6. The solution properties must be carefully determined. Solution conditions (e.g., pH, salt, additives) are also critical factors if

the crystal contact is expected to be in the hydrophilic part of the protein.

7. Temperature must be carefully controlled. Temperature impacts the fluidity of phospholipid in the reconstituted membrane. In some cases, temperature annealing (e.g., 20 °C → 37 °C → 20 °C in the case of aquaporin-4) [17] significantly improves crystal quality.

Here, I introduce a protocol for the 2D crystallization of gastric H^+,K^+-ATPase. Using detergent-solubilized membrane fractions purified from pig stomach (*see* Chap. 4), H^+,K^+-ATPase molecules formed an ordered 2D array on the reconstituted lipid bilayer, in the presence of phosphate analogs and/or inhibitors that stabilize the molecular conformation [18]. Refinement of various factors can be achieved by evaluating the quality and frequency of negatively stained 2D crystals. After refining the crystallization conditions, structural information can be extracted from frozen-hydrated 2D crystal images taken by cryo-electron microscopy (cryo-EM), and further image processing and crystallographic analysis. In this chapter, a procedure for the 2D crystallization of $C_{12}E_8$-solubilized H^+,K^+-ATPase in the (SCH)$E2 \cdot$BeF state is provided [15], because this 2D crystal is highly reproducible and much more stable than other 2D crystals with different conformations. Procedures for the sample preparations for cryo-EM and electron crystallographic analysis can be found in references [3, 19, 20].

2 Materials

Prepare all solutions using Milli-Q water. Reagents and materials used for the crystallization should be of the highest available grade. Because free K^+ in the solution affects the molecular conformation of H^+,K^+-ATPase, avoid putting the pH probe directly in the solution. The pH of the solution should be adjusted by measuring small aliquots removed from the solution.

2.1 Solubilization of H^+,K^+-ATPase with $C_{12}E_8$

1. Purified H^+,K^+-ATPase membrane fractions.
   Use sodium dodecyl sulfate (SDS)-treated membrane fractions purified from pig stomach (f2 or f3, *see* Chap. 4) as a source of H^+,K^+-ATPase. These fractions contain highly stable membrane-bound H^+,K^+-ATPase with a purity greater than 80 % of total protein. These purified membrane fractions are suspended in 250 mM sucrose, 0.5 mM EGTA-Tris (pH 7.4), and stored at −80 °C until use (*see* **Note 1**).

2. Detergents
   Octaethyleneglycol dodecylether ($C_{12}E_8$) (Nikko chemicals, Japan): Prepare 10 % (w/v) stock solution dissolved in Milli-Q water, and store at −80 °C until use.

3. Phospholipid

    Dioleoly phosphatidylcholine (DOPC): Use chloroform-dissolved DOPC (Avanti Polar Lipids, USA). Approximately 5 mg of chloroform-dissolved DOPC is dried under nitrogen gas atmosphere on the surface of a glass flask wall with rotation, and lyophilized overnight to completely remove the chloroform. After lyophilization, the DOPC is dissolved in 0.5 mL of 2 % (w/v) $C_{12}E_8$ and 2 mM dithiothreitol (DTT), using an ultrasonic bath with ice-cold water until the solution becomes clear. The phospholipid concentration is determined using colorimetry [21] and is used to calculate the LPR in the following crystallization step. The detergent-solubilized phospholipid can usually be stored for 1 month at 4 °C in the dark.

4. Solubilization buffer: 40 mM MES, 20 mM $Mg(CH_3COO)_2$, 5 mM ATP, 10 % glycerol, and 3 mM DTT, at pH 5.5 adjusted by Tris.

2.2 Dialysis

1. Dialysis button (Hampton Research, USA): 10 μL volume (Fig. 1a, *see* **Note 2**).

2. Dialysis membrane: Spectra/Por dialysis membrane #7 molecular weight cut-off 25,000 (Spectrum Laboratories). Dialysis tubing is opened, cut into small pieces (usually 3 cm × 3 cm square), and washed with Milli-Q water before use (Fig. 1, *see* **Note 2**).

3. Dialysis membrane applicator: Use a conventional golf tee with its tip covered by Parafilm or kitchen wrap. Alternatively, a glass applicator (Hampton Research) is also available.

4. Container for dialysis: Here, we use 500 mL plastic storage bottles and 12-well cell culture plates for dialysis. However, size and materials are depending on experimental conditions. Any containers may be useful, such as glass flasks, beakers, or other suitable plasticware.

5. Dialysis buffer A: 10 mM 2-(*N*-morpholino)ethanesulfonic acid (MES), 10 % (v/v) glycerol, 5 mM $MgCl_2$, 1 mM ADP, 3 mM DTT, 1 mM $BeSO_4$, 4 mM NaF, 1 μM SCH28080 (TOCRIS) (a specific antagonist of H^+,K^+-ATPase), pH adjusted to 5.5 with Tris (*see* **Note 3**, Fig. 2). Prepare 100 mL for each dialysis button.

6. Dialysis buffer B: 20 mM propionate, 10 % (v/v) glycerol, 1 mM $MgCl_2$, 1 mM ADP, 3 mM DTT, 1 mM $BeSO_4$, 4 mM NaF, 100 μM SCH28080, pH adjust to 4.8–4.9 with Tris (*see* **Notes 3 and 4**, Fig. 3). Prepare 5 mL for each dialysis button.

2.3 Electron Microscopy

1. Electron microscope: For screening of 2D crystals, a conventional 100 kV electron microscope (e.g., JEM-1010, JEOL) is sufficient. For recording of negatively stained 2D crystals, a

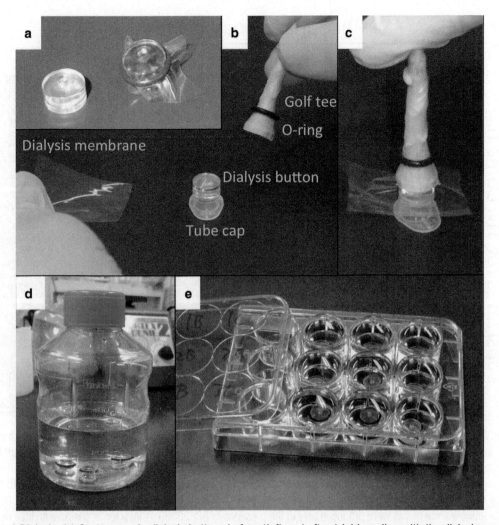

Fig. 1 Dialysis. (**a**) Custom-made dialysis buttons before (*left*) and after (*right*) sealing with the dialysis membrane. (**b**) Sealing process. A golf tee is covered with Parafilm to prevent contamination. The cap of a microcentrifuge tube may be helpful for holding the dialysis button. (**c**) To prevent the inclusion of bubbles, the dialysis membrane should be kept in place immediately after being placed on top of the dialysis button. (**d**) First dialysis against 300 mL dialysis buffer A with three dialysis buttons. (**e**) Second dialysis setup using a 12-well cell culture plate with 5 mL dialysis buffer B in each well

CCD or cMOS detector is preferable because fast Fourier transform (FFT) of the 2D crystal image is an informative and efficient way to evaluate crystal quality (Fig. 4). For data collection from frozen-hydrated specimens, JEM-3000SFF equipped with liquid-helium cooled stage is used [3].

2. Uranium acetate: Prepare 2 % (w/v) uranium acetate solution. This reagent is radioactive and highly toxic, thus diligently follow all waste disposal regulations when disposing the waste materials.

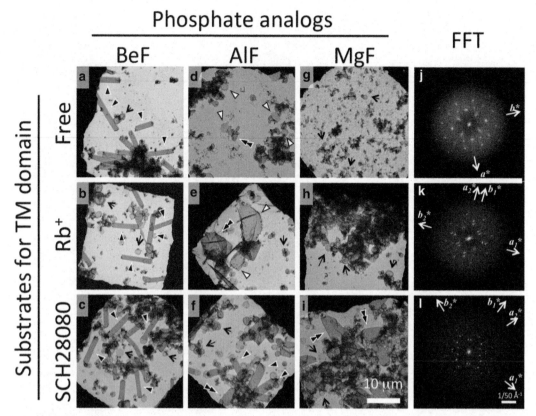

Fig. 2 Gallery of negatively stained 2D crystals with various combinations of phosphate analogs and substrates for the TM domain. Flattened tubular 2D crystals (*black arrowheads*) formed when BeF was used as a phosphate analog regardless of the substrate (**a–c**). On the other hand, flat-sheet 2D crystals (*white arrowheads*) comprising a single-crystalline array were produced in the case of AlF (**d**), and its morphology largely changed when Rb$^+$ (**e**) or SCH28080 (**f**) was added to the dialysis buffer. In the MgF conditions (**g–i**), 2D crystals with vesicular morphologies (*double arrowheads*) were produced only in the presence of SCH28080 (**i**). Vesicles or sheets with little or no H$^+$,K$^+$-ATPase are indicated by arrows. (**j–l**) Fast Fourier transformation (FFT) of sheet (**j**), tubular (**k**), or vesicular crystals (**l**) are shown. Because micrographs of tubular or vesicular crystals always contain two overlapping crystalline lattices, the diffraction spots are overlapped in their respective FFTs (**k, l**). *White arrows* indicate reciprocal lattices in FFTs. *Scale bars* for micrographs and FFTs are shown in panels **i** and **l**, respectively

3. Copper grid with carbon support film [3]: Prepare a continuous carbon film by evaporating carbon rod on the surface of freshly peeled mica under vacuum. Carbon films are peeled off on the water surface, and floating carbon film is scooped up using a 400 mesh copper grid (Nisshin EM, Japan), dry completely, and store under desiccated conditions. Just prior to use, the carbon films are glow-discharged to make the surface hydrophilic.
4. IB-3 ion coater (Eiko, Japan) for glow-discharging.
5. Whatman qualitative filter paper #1 for blotting.
6. Film scanner: SCAI Flat-Bed Scanner (Zeiss, Germany) for digitization of electron micrograph, with 7 μm pixel size.

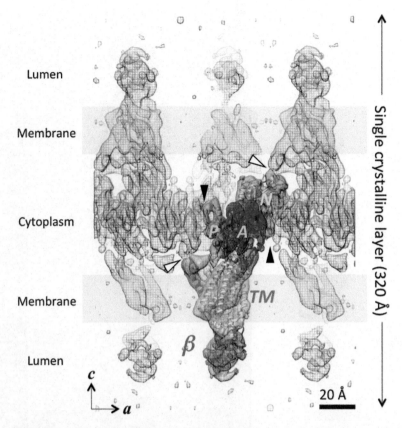

Fig. 3 Molecular packing of H+,K+-ATPase in the 2D crystal. The asymmetric unit corresponds to one H+,K+-ATPase αβ-protomer shown as a colored solid surface, with the superimposed homology model shown as ribbons. Cytoplasmic domains (A, P, N), TM helices, and the β-subunit are indicated in the figure. *Green mesh* represents symmetry-related molecules. A crystal contact occurs between the N domain and the β-subunit N-terminus of the neighboring molecule (*black arrowheads*), and the P domain and A domain of the neighboring molecule (*white arrowheads*). No crystal contact is observed in the membrane or luminal portion of the crystal. A single crystalline layer comprises two membrane layers (indicated as *wheat-colored boxes*), which is responsible for the one-crystalline layer of each crystal in the different morphologies

3 Methods

All procedures are performed on ice or in a cold room unless otherwise stated.

3.1 Solubilization of H+,K+-ATPase with $C_{12}E_8$

1. SDS-treated membrane fraction (8 mg/mL) is solubilized for 10 min on ice with 6.5 mg/mL $C_{12}E_8$ in Solubilization buffer (*see* **Note 5**).

2. Centrifuge at 186,000×*g* for 20 min to remove insoluble materials.

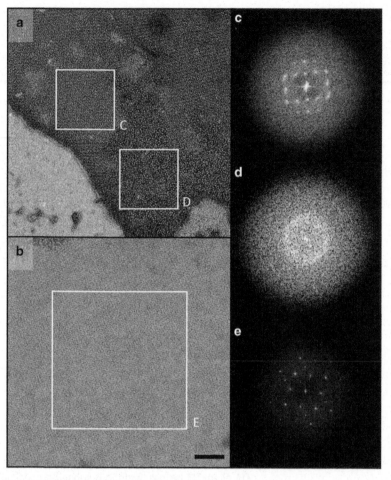

Fig. 4 Evaluation of the quality of negatively stained 2D crystals. Representative poorly ordered (**a**) and well-ordered (**b**) 2D crystals. In contrast to the well-ordered coherent 2D crystal (**b**), the poorly ordered crystal shows the distribution of small lattices (*white box* **c**) and particles of reconstituted H$^+$,K$^+$-ATPase (*white box* **d**). (**c–e**) FFTs of the *square regions* indicated in the figure. *Scale bar* in B = 100 nm

3. Measure protein concentration in the supernatant using the Bradford method [22]. In this step, the protein concentration is expected to be ~6–8 mg/mL.

4. Mix supernatant and DOPC at a LPR (w/w) ranging from 0.95 to 1.15 in C$_{12}$E$_8$-free Solubilization buffer, and adjust the final protein concentration to 0.5–1 mg/mL (*see* **Notes 6** and **7**).

3.2 Dialysis

All steps are performed in a cold room. Wear plastic gloves that have been carefully washed with Milli-Q water to prevent contamination.

1. Place a mixture of H$^+$,K$^+$-ATPase and DOPC (Subheading 3.1, **step 4**) in a 10-μL microdialysis button (Fig. 1a) pre-cooled in the cold room.

2. Seal the dialysis button with dialysis membrane that has equilibrated in water in the cold room before use. Remove the excess water attached to the membrane by wiping out water using a finger with gloves, and then place the membrane over the top of the button (Fig. 1b). Immediately place a golf tee (or glass applicator) on top of the membrane and button (Fig. 1c). Roll the O-ring down the golf tee, and fix the membrane to the button (Fig. 1a, c, see **Note 8**).

3. Place three dialysis buttons, membrane side up, in 300 mL dialysis buffer A cooled on ice, and incubate without stirring for 36 h on ice in the cold room (Fig. 1d).

4. Each single dialysis button is then moved to 5 mL dialysis buffer B in one well of a 12-well cell culture plate, and incubated for 12–18 days at 3 °C (Fig. 1e).

5. The dialysis button is picked up from the dialysis buffer. The dialysis membrane is broken using a 10-µL pipet tip, and the 2D crystal samples are taken out from each of the dialysis buttons. The 2D crystal samples are stable for 1 month as long as they are stored at 3 °C.

3.3 Negative Staining and Observation by Electron Microscope

1. Negative staining is performed in the cold room. A Cu-grid is glow-discharged 20 s with 4 mA current using the ion coater before use, and pre-cooled in the cold room.

2. One microliter of 2D crystal sample solution is placed on the Cu-grid, and mixed well by pipetting.

3. Using a small piece of filter paper, excess buffer is removed by blotting, by putting the filter paper at the side of the grid. Because the 2D crystal of H^+,K^+-ATPase is susceptible to dehydration, 3 µL of 2 % uranium acetate solution is immediately added after blotting.

4. Blot excess uranium acetate, and again add 3 µL uranium acetate for washing.

5. Remove the uranium acetate completely by blotting with filter paper, and leave it on the dry filter paper.

6. After dessicating for a couple of hours at room temperature, the grids can be observed under an electron microscope.

7. Due to the characteristic tubular morphology and large size of more than 5 µm in length, 2D crystals can easily be seen at low magnification (Fig. 2c, 1000× magnification). The number of crystals and their quality are evaluated in this step (Fig. 4, see **Note 9**).

8. If sufficient 2D crystals are obtained, they can be used for the data collection for the structural determination. Briefly, samples

are ice-embedded in a cryo grid using the carbon sandwich method (see **Note 10**, ref. [19]). Electron micrographs are collected from non-tilted and tilted (20°, 45°, and 60°) specimens by liquid helium-cooled cryo-EM operated at 300 keV [3]. The micrographs are digitized using a film scanner and processed with the MRC Image Processing programs [23], which include Fourier filtering to eliminate noise in each micrograph, unbending of the crystalline lattice based on the cross-correlation, and correction of the contrast transfer function, to extract the amplitude and phase information from each micrograph. After collecting and analyzing hundreds of micrographs, the data are merged into a set of lattice lines, and a final 3D density map is calculated [3, 20].

4 Notes

1. In our procedure, poorly ordered 2D crystals are obtained if the G1 fraction without SDS treatment is used. The G2 fraction does not give any 2D crystals. Also, the quality of the 2D crystals varies depending on the lot of the SDS-treated membrane fractions.

2. The diameter and volume of the dialysis button are related to the detergent removal rate, which largely affects the crystal quality. We use a custom-made 10-μL dialysis button with a well size of 2.2 mm diameter and 2.5 mm depth (Fig. 1a). The dialysis rate is also affected by the molecular weight cut-off of the dialysis membrane, dialysis buffer volume, and other factors. Refinement of these factors may improve crystal quality.

3. For 2D crystallization of other conformations, $BeSO_4$ is replaced with other fluorinated phosphate analogs (e.g., 1 mM $AlCl_3$ for the $E2AlF$ state). Interestingly, different phosphate analogs and their combination with Rb^+ or SCH28080 yield 2D crystals with different morphologies, including sheets, tubes, and vesicles, depending on the conformational state of the protein as well as other crystallization conditions (Fig. 2). Only the $E2P$-related conformation, except in the $E2MgF$ state [18], can be crystallized when using $C_{12}E_8$-solubilized H^+,K^+-ATPase. To date, the 2D crystals that are assumed to be in the $E1$ and $E1P$-ADP conformation can only be obtained from octyl glucoside-solubilized H^+,K^+-ATPase, although resolution is limited to 10–14 Å [24].

4. The pH of the solution largely affects the quality of the 2D crystals in this step. This may be related to the unusual crystal packing (Fig. 3), which is essentially the same for all of the H^+,K^+-ATPase 2D crystals we have reported. Single-crystalline

sheets comprise two-membrane layers, and the proteins in the two-membrane layers are related to each other by a two-fold screw axis, resulting in a $p22_12_1$ symmetry. Due to the characteristic crystal packing, all of the intermolecular contacts in the crystals can be found at the cytoplasmic portions of the molecules (Fig. 3), in marked contrast to the relatively hydrophobic membrane proteins such as bacteriorhodopsin [1] or aquaporins [17]. This hydrophilic feature makes 2D crystals of H^+,K^+-ATPase particularly sensitive to solution conditions, including pH, salt concentration, and other additives.

5. Crystal quality relies on the stability of the H^+,K^+-ATPase in the detergent micelle, which can be determined by measuring the residual ATPase activity in the solubilized preparation. We found that H^+,K^+-ATPase activity is preserved in the presence of MgATP at a relatively acidic pH ranging from 5 to 6. The addition of glycerol provides a better preservation effect compared with sucrose or trehalose. Keeping solubilized samples at a low temperature (on ice) is also critical for preserving the activity. The effect of removing ATP from the solubilized sample is negligible in this case, as the ATP concentration is diluted to less than 1 nM in the final sample, based on the volume of the sample and dialysis buffer.

6. Because the protein concentration is related to the rate of reconstitution and the nucleation of the 2D crystals, an excessively high concentration of protein results in the production of incoherent, patchy 2D crystals (Fig. 4a). As a major driving force for 2D crystallization is likely to be detergent removal, the initial concentration of $C_{12}E_8$ also affects the results. In this step, therefore, solubilized H^+,K^+-ATPase mixed with DOPC is diluted approximately 10 times with $C_{12}E_8$-free buffer, resulting in ~0.65 mg/mL $C_{12}E_8$ in the mixture.

7. In our procedure, phospholipid carried from the gastric membrane fraction is also included in the sample. The resulting 2D crystals, however, are indistinguishable when using H^+,K^+-ATPase sample purified by anion-exchange chromatography, in which free phospholipid has been removed. Therefore, we conclude that the reconstituted membrane mainly comprises synthetic DOPC.

8. Because inclusion of bubbles between the membrane and sample prevents reproducible dialysis, care should be taken to prevent bubbles in the button.

9. Usually 2D crystals have characteristic morphologies, and are thus distinguishable from empty vesicles, amorphous membrane, or protein aggregation (Fig. 2). Crystals usually show a higher contrast than empty vesicles or sparsely reconstituted

membranes, because the surface of the 2D crystal is much rougher than that of non-crystalline membranes. The FFT image is helpful for evaluating 2D crystal quality (Fig. 3). If the lattice is distorted, the diffraction spots in the FFT are blurred or smiling (i.e., skewed; Fig.4a, c). Amorphous reconstituted protein produces concentric rings in FFT (Fig. 4d). After refining various factors for crystallization, the diffraction spots become sharper, and are visible as higher resolution shells (Fig. 4e). With coherent 2D crystals, the diffraction spots never change when the region of interest (white squares in Fig. 4) on the micrograph is moved.

10. Because all crystal contacts can be found in the cytoplasmic portion of H^+,K^+-ATPase (Fig. 3), this 2D crystal is susceptible to dehydration and salt concentration changes, which can be preserved by applying a carbon sandwich preparation [19].

Acknowledgements

The author acknowledges Drs. Tomohiro Nishizawa, Kazutoshi Tani, and Yoshinori Fujiyoshi for their contributions to the development of the 2D H^+,K^+-ATPase crystallization procedure. This work was supported by Grants-in-Aid for Young Scientist (A) and Platform for Drug Design, Discovery, and Development from METI, Japan.

References

1. Henderson R, Unwin PN (1975) Three-dimensional model of purple membrane obtained by electron microscopy. Nature 257:28–32
2. Kyogoku Y, Fujiyoshi Y, Shimada I et al (2003) Structural genomics of membrane proteins. Acc Chem Res 36:199–206
3. Fujiyoshi Y (1998) The structural study of membrane proteins by electron crystallography. Adv Biophys 35:25–80
4. Toyoshima C, Sasabe H, Stokes DL (1993) Three-dimensional cryo-electron microscopy of the calcium ion pump in the sarcoplasmic reticulum membrane. Nature 362:467–471
5. Hebert H, Purhonen P, Vorum H et al (2001) Three-dimensional structure of renal Na, K-ATPase from cryo-electron microscopy of two-dimensional crystals. J Mol Biol 314:479–494
6. Toyoshima C, Nakasako M, Nomura H et al (2000) Crystal structure of the calcium pump of sarcoplasmic reticulum at 2.6 Å resolution. Nature 405:647–655
7. Olesen C, Picard M, Winther AM et al (2007) The structural basis of calcium transport by the calcium pump. Nature 450:1036–1042
8. Toyoshima C, Norimatsu Y, Iwasawa S et al (2007) How processing of aspartylphosphate is coupled to luminal gating of the ion pathway in the calcium pump. Proc Natl Acad Sci U S A 104:19831–19836
9. Toyoshima C (2008) Structural aspects of ion pumping by Ca^{2+}-ATPase of sarcoplasmic reticulum. Arch Biochem Biophys 476:3–11
10. Møller JV, Olesen C, Winther A-ML et al (2010) The sarcoplasmic Ca^{2+}-ATPase: design of a perfect chemi-osmotic pump. Q Rev Biophys 43:501–566
11. Morth JP, Pedersen BP, Toustrup-Jensen MS et al (2007) Crystal structure of the sodium-potassium pump. Nature 450:1043–1049
12. Ogawa H, Shinoda T, Cornelius F et al (2009) Crystal structure of the sodium-potassium pump (Na+,K+-ATPase) with bound potassium and ouabain. Proc Natl Acad Sci U S A 106:13742–13747

13. Toyoshima C, Cornelius F (2013) New crystal structures of PII-type ATPases: excitement continues. Curr Opin Struct Biol 23:507–514
14. Abe K, Tani K, Nishizawa T et al (2009) Intersubunit interaction of gastric H+,K+-ATPase prevents reverse reaction of the transport cycle. EMBO J 28:1637–1643
15. Abe K, Tani K, Fujiyoshi Y (2011) Conformational rearrangement of gastric H+,K+-ATPase with an acid suppressant. Nat Commun 2:155
16. Abe K, Tani K, Friedrich T et al (2012) Cryo-EM structure of gastric H+,K+-ATPase with a single occupied cation-binding site. Proc Natl Acad Sci U S A 109:18401–18406
17. Hiroaki Y, Tani K, Kamegawa A et al (2006) Implications of the aquaporin-4 structure on array formation and cell adhesion. J Mol Biol 355:628–639
18. Abe K, Tani K, Fujiyoshi Y (2010) Structural and functional characterization of H+,K+-ATPase with bound fluorinated phosphate analogs. J Struct Biol 170:60–68
19. Fan Y, Abe K, Tani K et al (2013) Carbon sandwich preparation preserves quality of two-dimensional crystals for cryo-electron microscopy. Microscopy 62:597–606
20. Glaeser R, Downing K, Derosier D et al (2007) Electron crystallography of biological macromolecules. Oxford University Press, Oxford
21. Rouser G, Siakotos AN, Fleischer S (1966) Quantitative analysis of phospholipids by thin-layer chromatography and phosphorus analysis of spots. Lipids 1:85–86
22. Bradford MM (1976) A rapid and sensitive method for the quantitation of microgram quantities of protein utilizing the principle of protein-dye binding. Anal Biochem 7:248–254
23. Crowther RA, Henderson R, Smith JM (1996) MRC image processing programs. J Struct Biol 116:9–16
24. Nishizawa T, Abe K, Tani K et al (2008) Structural analysis of 2D crystals of gastric H+,K+-ATPase in different states of the transport cycle. J Struct Biol 162:219–228

Part IX

Computational Approaches in Analyzing P-Type ATPases

Chapter 40

MD Simulations of P-Type ATPases in a Lipid Bilayer System

Henriette Elisabeth Autzen and Maria Musgaard

Abstract

Molecular dynamics (MD) simulation is a computational method which provides insight on protein dynamics with high resolution in both space and time, in contrast to many experimental techniques. MD simulations can be used as a stand-alone method to study P-type ATPases as well as a complementary method aiding experimental studies. In particular, MD simulations have proved valuable in generating and confirming hypotheses relating to the structure and function of P-type ATPases.

In the following, we describe a detailed practical procedure on how to set up and run a MD simulation of a P-type ATPase embedded in a lipid bilayer using software free of use for academics. We emphasize general considerations and problems typically encountered when setting up simulations. While full coverage of all possible procedures is beyond the scope of this chapter, we have chosen to illustrate the MD procedure with the Nanoscale Molecular Dynamics (NAMD) and the Visual Molecular Dynamics (VMD) software suites.

Key words Molecular dynamics simulations, MD simulations, P-type ATPases, Membrane proteins, Protein–lipid interactions, Protein dynamics, Sarco(endo)plasmic reticulum Ca^{2+}-ATPase, SERCA

1 Introduction

1.1 Molecular Dynamics Simulations

In pace with recent technical advances in membrane protein crystallography, the number and quality of deposited crystal structures of P-type ATPases in the Protein Data Bank (PDB) is steadily increasing. Likewise, the number of studies which include MD simulations of P-type ATPases is growing in pace with technical improvements of simulation methods and computational resources as well as with the availability of high-resolution structures.

The amount of information that can be obtained from MD simulations of P-type ATPases is vast and the method can be embraced in a large variety of studies, either stand-alone or as a complimentary method for underscoring experimental observations. Examples of the former include studies of the protonation states of the ion binding sites of the sarco(endo)plasmic reticulum

Ca^{2+}-ATPase (SERCA) [1] and the Na^+,K^+-ATPase [2], as well as studies of the conformational transitions of SERCA [3] just to mention a few, while examples of the latter include studies of the protein–lipid interplay of the detergent solubilized SERCA [4] and the ion pathways of the Zn^{2+}-ATPase [5] and the Na^+,K^+-ATPase [6], respectively. In particular, with the increased use of digital technology, the bridge between experimental and computational scientists within structural biology is growing stronger.

In short, MD simulation is a numerical method which describes the time-dependent dynamics of a system by means of classical mechanics and Newton's laws of motion. In MD simulations, the molecular system is composed of particles (representing atoms or groups of atoms (*see* **Note 1**)) whose potential energy depends only on their position relative to other particles. The particles have different sizes and partial charges, thereby representing different atom types, and are joined by springs of varying lengths and stiffness, representing chemical bonds. Thus, electrons are ignored. As a result of the covalent bonds being represented by springs, bonds cannot be broken or formed during simulations [7, 8]. The potential energy in MD simulations is calculated using force fields (FF). The core of a FF is the potential energy function, which relates the structure of a molecular system to its potential energy. The potential energy function is a sum of energy terms describing bonded and non-bonded interactions within the system. The bonded energy terms describe potential energy associated with covalent bonds (deviation of bond lengths, bond angles, and torsional angles from their equilibrium values). The non-bonded energy terms describe interactions beyond the nearest neighbors, including van der Waals and electrostatic interactions. Electrons are thus implicitly included through both the bonded and the non-bonded terms. In addition to the potential energy function, a FF is also defined by the set of parameters employed in the function (e.g., an equilibrium length for a given covalent bond and a force constant for the same bond which is a measure of how easy the bond extends/contracts). A full review of MD simulation as a method is beyond the scope of this chapter, and the reader is instead referred to books by Andrew Leach and Frank Jensen written on the subject [7, 8].

Different FFs have different functional forms and different sets of parameters. Most importantly, each FF is designed with a certain aim in mind. All-atom FFs optimized for describing organic and biomolecular systems such as protein–ligand complexes include CHARMM [9, 10], GROMOS [11], AMBER [12], and OPLS [13, 14]. The protocol described in this chapter utilizes the CHARMM FF. In the following, important considerations when setting up simulations are presented. Next, a short overview of NAMD and VMD is provided (*see* **Note 2**).

1.2 Important Considerations

Before constructing a system and setting up a simulation of a P-type ATPase, the following questions should be considered:

1. What type of questions do I seek to answer with my simulation?
2. What level of description is required to answer my questions (*see* **Note 1**)?
3. Is the relevant crystal structure available for the P-type ATPase that I wish to study, and is the structure complete or are there any missing residues (*see* **Note 3**)?
4. Are ligands, cofactors, metal ions or other proteins required for stability and/or function of the P-type ATPase and are parameters available in the chosen FF (*see* **Notes 4–6**)?
5. What is the appropriate membrane composition (complex or pure) for the P-type ATPase and the process to be investigated, and are appropriate force field parameters available (*see* **Note 7**)?
6. What is the required time scale for the studied process (*see* **Note 8**)?

1.3 Overview of the NAMD Workflow

In short, most MD simulation procedures involving P-type ATPases can roughly be divided into six steps: (1) Preparation of the P-type ATPase structural model and potential cofactors for simulation, (2) Preparation of an appropriate lipid bilayer and insertion of the ATPase into it, (3) Solvation and ionization, (4) Minimization and equilibration, (5) Production run, and (6) Analysis. A graphical overview of the procedure exemplified in this chapter is illustrated in Fig. 1.

At least five different files are required to set up and run simulations of a molecular system in NAMD (*see* **Note 2**); (1) a coordinate file (PDB), e.g., from the PDB database, (2) a topology file, (3) a Protein Structure File (PSF), (4) a FF parameter file, and (5) a configuration (config) file, *see* Fig. 2.

The coordinate file contains coordinates of the atoms in the system, but does not include any information regarding connectivity; this is provided by the topology file, which defines the atom names, types, masses and partial charges in the chosen FF. The PSF contains all the molecule-specific information needed to apply a particular FF to the molecular system and is generated from the PDB file and the topology file through the psfgen plugin in VMD. The parameter file contains all the numerical constants needed for NAMD to evaluate potential energies of a system using the given FF. Finally, the config file specifies all the settings needed for a simulation. The PDB file can originate from for example X-ray diffraction, NMR studies, homology modeling, or system building in molecular editing programs. The topology file and the parameter file are FF specific and typically available from the developers of the FF. Occasionally, parameters might be available for a

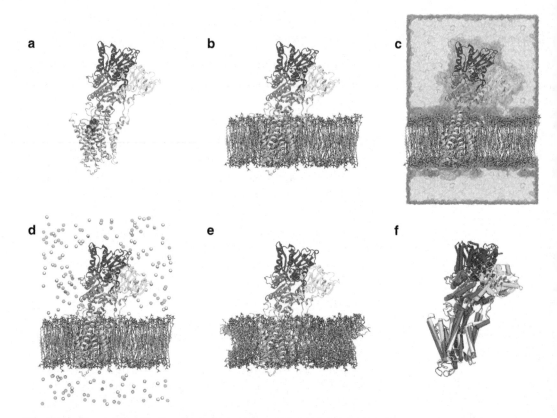

Fig. 1 Procedure covered in the chapter. (**a**) SERCA (PDB ID 2C8K) directly as downloaded from the PDB with bound thapsigargin, ATP, and ions. The protein is shown in *New Cartoon* with the A-domain in *yellow*, N-domain in *red*, P-domain in *blue* and the TM domain in *silver* while ligands and ions are shown as *spheres*. (**b**) SERCA embedded in a POPC bilayer. (**c**) Solvation of the SERCA-bilayer complex with the water shown as surface. (**d**) Ionization with added potassium and chloride ions (water is omitted for clarity). (**e**) System after equilibration of lipid tail groups (water and ions are omitted for clarity). (**f**) SERCA shown in cartoon after 100 ns of MD simulation (transparent) relative to the starting structure (water, ions, and lipids are omitted for clarity).

molecule, while the topology is not. In that case you will need to create the topology file yourself.

2 Materials

The procedure presented in this chapter requires several software components, all of which are freely available for academic purposes. Familiarity with the software is highly recommended in order to perform the entire procedure. The procedure is written to be used in a Linux/Unix environment, although most of the software is also available for Windows. To perform an all-atom MD

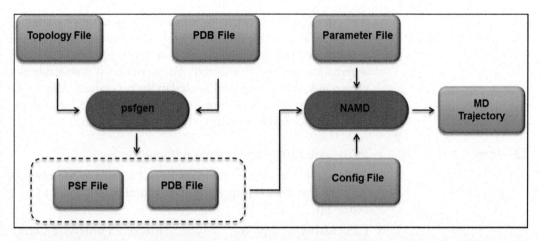

Fig. 2 Flowchart of files used for setting up a MD simulation with NAMD. A coordinate file (PDB file) and a topology file are used to generate a PSF and a new PDB with the psfgen structure tool. The new PDB file contains new segment names and guessed coordinates for hydrogen atoms. These newly generated files are used in the NAMD calculations along with the parameter file with the FF parameters and the config file specifying details about the simulation to be performed.

simulation of a P-type ATPase even at the nanosecond timescale, you will, in addition to the programs below, need access to high-performance computing facilities or a very strong workstation.

2.1 Software

The following programs are used in the chapter:

1. NAMD [15] is a parallel MD code designed for simulations of large biomolecular systems. The NAMD software can be downloaded at http://www.ks.uiuc.edu/Research/namd/ (*see* **Note 2**). Version 2.9 is used in the procedure. Alternatives to NAMD include GROningen MAchine for Chemical Simulations (GROMACS) [16], released under the GNU Lesser General Public License and thus included in most Linux releases. The GROMACS software can be assessed at http://www.gromacs.org.

2. VMD [17] is a molecular visualization program developed for displaying biomolecular systems. The VMD software can be downloaded at http://www.ks.uiuc.edu/Research/vmd/ where a large library of tutorials can also be found (*see* **Note 2**). Version 1.9.1 is used in the procedure. PyMOL (The PyMOL Molecular Graphics System, Schrödinger, LCC) is a popular alternative molecular visualization system, maintained and distributed by Schrödinger, http://www.pymol.org/.

3. PROPKA [18] predicts protonation states of ionizable residues in proteins and protein–ligand complexes based on a 3D struc-

ture (*see* **Note 9**). It can include ligands/ions, which is useful for P-type ATPases. The PROPKA Web Interface and source code can be accessed and downloaded at http://propka.ki.ku.dk/ where a tutorial is also available. Furthermore, it is available as a plugin in VMD. Version 3.1 is used in the procedure. Alternative resources for predicting protonation states include for example H++ [19] accessible at http://biophysics.cs.vt.edu/.

2.2 Files

1. 2C8K.pdb [20] (http://www.pdb.org).
2. 2c8k.pdb file from the Orientations of Proteins in Membranes (OPM) database (http://opm.phar.umich.edu/) [21] (*see* **Note 10**).
3. CHARMM topology file (mackerell.umaryland.edu/charmm_ff.shtml#charmm).
4. CHARMM parameter file (mackerell.umaryland.edu/charmm_ff.shtml#charmm).
5. NAMD config file for minimization (Subheading 3.4).
6. NAMD config files for equilibration (Subheading 3.5).
7. NAMD config file for production run (Subheading 3.6).

3 Methods

We have made the procedure easier to follow step by step by using SERCA in the ATP-bound calcium free E2 state as an example. Commands provided throughout the chapter assume the text-based Tcl/Tk interface in VMD unless otherwise stated and user knowledge of the same is beneficial. Familiarity can be obtained by going through VMD and NAMD tutorials (*see* **Note 2**). It is advised to check for potential errors and warnings reported by the software following every step of the procedure. Furthermore, it is vital to assess the quality of the output from each step to avoid potential waste of time and computational resources.

3.1 Preparation of the Protein Model

In this first step of the procedure, the protein and cofactor models are prepared for simulation by correcting atom and residue names (resname, Tcl terminology) according to those defined in the topology and parameter files. In addition, protonation states of ionizable residues are considered and disulfide bridges are defined. Furthermore, all protons are added in the psfgen step.

1. Downloading the protein structure and viewing it in VMD
Select the specific P-type ATPase and the functional state that you wish to study (E1 or E2) and consider including any cofactors that might be bound to it and could be relevant for the simu-

Table 1
Representations used to inspect the 2C8K PDB in VMD

Selected atoms	Drawing method	Coloring method
protein	NewCartoon	Chain
not protein	VDW	Name

lation (*see* **Notes 4–6**). To follow the example presented here, download 2C8K from the PDB (http://www.pdb.org). Open VMD and load the PDB file you just saved. In the VMD Main window click File→New Molecule and Browse to locate your file. Click Load. To display the contents of the PDB file open the Representations window from the VMD Main window: Graphics→Representations and set up the graphical representations shown in Table 1. In addition to the protein chain, 2C8K contains thapsigargin (residue name TG1), AMPPCP (residue name ACP), and sodium and magnesium ions (*see* Fig. 1a).

2. Predicting pK_a values

The protonation state of ionizable amino acid side chains is determined based on their predicted pK_a value, which ultimately depends on the local environment, i.e., the nearby residues, and the pH value at which the simulation is performed (*see* **Note 9**). As protons cannot be exchanged during simulation (breaking and forming bonds is not possible) we need to ensure that the protons on ionizable side chains are positioned reasonably. In this procedure, pK_a values are estimated using the webserver PROPKA [18].

First, go to the PROPKA website http://propka.ki.ku.dk/. Enter the PDB ID 2C8K in the "a PDB ID" box and press submit. After about 1 min, the output will appear directly in the browser. To familiarize yourself with the output, scroll down until you find GLU 309. The predicted pK_a value for Glu309 is 9.01 (PROPKA v. 3.1) and the side chain should therefore be protonated at physiological pH (7.4). You can click on the residue link to view Glu309 and the surrounding determinants in 3D with Jmol (requires Java installed). Now, you can either go through the rest of the output by viewing it in the browser or analyzing it using the PropKa GUI in VMD (requires that the PropKa plug-in is installed) after saving the PROPKA results file which may be loaded into VMD (see tutorial at http://propka.ki.ku.dk/). Before starting a MD simulation, it is advised to go through all the ionizable side chains for which the predicted pK_a value suggests a nonstandard protonation state and decide how to treat the residues. Nonstandard protonation states of glutamate, aspartate, lysine, and arginine residues will be incorporated into the protein as "patches" in the psfgen step later where all hydrogen atoms are added (*see* Subheading 3.1, **step 4**).

Most histidine residues are predicted to be neutral. However, as neutral histidine residues can exist in two different tautomer forms, you need to inspect each individual histidine residue and decide which tautomer form would be prevalent considering the nearby environment (*see* **Note 11**). You rename the histidine residues in the next step of the procedure when splitting the protein into fragments.

3. Splitting the structure into fragments

    Psfgen requires that non-covalently linked segments are described by separate PDB files. This also includes separate protein segments for P-type ATPases with more than one subunit. We write individual segments of the PDB file by using the Tk Console window (a text console in VMD used to execute Tcl commands), but first we change AMPPCP into ATP by changing the atom type for the C3B carbon atom and the resname (*see* **Note 4**). With the PDB file loaded in VMD, open the Tk Console window from the VMD Main menu under Extensions → Tk console and type the following:

    ```
    set AMPPCP_C3B [atomselect top "resname ACP and name C3B"]
    $AMPPCP_C3B set name O3B
    $AMPPCP_C3B set type O3B
    set AMPPCP [atomselect top "resname ACP"]
    $AMPPCP set resname ATP
    $AMPPCP writepdb ATP.pdb
    ```

    Next, the sodium ion, bound in the potassium ion binding site, is changed into a potassium ion [22] by changing its resname as well as its atom type such as it is described in the topology and parameter files. Finally, write out the coordinates of potassium and magnesium in separate PDB files:

    ```
    set na [atomselect top "name NA"]
    $na set name POT
    $na set resname POT
    $na set type K
    $na writepdb POT.pdb
    set MG [atomselect top "name MG"]
    $MG writepdb MG.pdb
    ```

Before the protein is saved as its own segment, the histidine residues, whose pK_a values you predicted in the previous step, need to have assigned the protonation state that you wish to use in the simulation (*see* **Note 11**). This is done by changing their resnames such that psfgen can recognize where to add hydrogen atoms in the next step (*see* below). For instance, we wish to make sure that His5 is ε-protonated and His32 is δ-protonated:

```
set His5 [atomselect top "protein and resid 5"]
$His5 set resname HSE
set His32 [atomselect top "protein and resid 32"]
$His32 set resname HSD
```

Change the resnames for the remaining 11 histidine residues in the same way. Next, we change the name of the C-terminal oxygen atoms such that psfgen recognizes them as the terminus as described in the topology file, thereby adding the correct charge:

```
set Gly994_O [atomselect top "protein and resid 994
    and name O"]
$Gly994_O set name OT1
$Gly994_O set type OT1
set Gly994_OXT [atomselect top "protein and resid
    994 and name OXT"]
$Gly994_OXT set name OT2
$Gly994_OXT set type OT2
```

Finally, write the protein coordinates in a separate PDB file:

```
set SERCA [atomselect top "protein"]
$SERCA writepdb protein.pdb
```

4. Generating the psf file (psfgen)

    The newly generated PDB files are now combined using psfgen (*see* **Notes 12** and **13** and **Fig. 2**). Considerations regarding protonation states and disulfide bridges are incorporated in this step (*see* **Note 9**). In order to generate the PSF, a topology file describing atom types in the FF is required. Here, the latest version of the CHARMM force field is used [9, 10, 23, 24]. Download the force field package toppar_c36_aug14.tgz (mackerell.umaryland.edu/charmm_ff.shtml#charmm), save and unpack it. All topology and parameter files for the CHARMM36 force field are found in the toppar folder.

    In VMD, the psfgen plugin is called from the Tk Console after deleting all loaded molecules:

```
mol delete all
package require psfgen
```

```
resetpsf
```
Next, the topology files that we wish to use are specified:
```
topology /path/to/topology/toppar/top_all36_prot.rtf
topology /path/to/topology/toppar/top_all36_na.rtf
topology /path/to/topology/toppar/stream/na/toppar_
   all36_na_nad_ppi.str
topology /path/to/topology/toppar/toppar_water_ions.
   str
```
Change the atom name CD1 in isoleucine residues to CD, the name in the topology file:
```
pdbalias atom ILE CD1 CD
```
Read coordinates from the PDB file, matching segment, residue, and atom names for the protein:
```
segment P {pdb protein.pdb}
coordpdb protein.pdb P
```
Apply patches for protonating glutamate residues (GLUP) according to PROPKA predictions and results from the literature [1] and for creating the disulfide bond (DISU) between Cys876 and Cys888 [25]:
```
patch GLUP P:309
patch GLUP P:771
patch GLUP P:908
patch DISU P:876 P:888
```
Again, read coordinates from PDB files, matching segment, residue and atom names for ATP and ions:
```
segment A {pdb ATP.pdb}
coordpdb ATP.pdb A
segment M {pdb MG.pdb}
coordpdb MG.pdb M
segment K {pdb POT.pdb}
coordpdb POT.pdb K
```
Let psfgen regenerate angles and dihedrals and guess coordinates for hydrogen atoms:
```
regenerate angles dihedrals
guesscoord
```
You may see a warning concerning poorly guessed coordinates. This warning usually only concerns hydrogen atoms in which case no further investigations are required as these will be corrected for during the energy minimization (*see* Subheading 3.4). Finally, write the PSF and PDB files of the entire system:

```
writepdb 2c8k_complex.pdb
writepsf 2c8k_complex.psf
```

5. Check the structure in VMD
   Load the PSF- and PDB files in VMD and check that everything looks as expected. First, load the PSF (nothing will appear in the display window) then load the PDB into the PSF (molecule will appear). You will see that the protein now has hydrogen atoms. Pay special attention to the termini, disulfide bridges, residues with nonstandard protonation states, cofactors, ligands, and ions which you just added and inspect each entity one at a time. Now, the protein with cofactor and ions is ready for insertion into the lipid bilayer.

3.2 Preparation of and Insertion into a Lipid Bilayer

In this step of the procedure, the protein is inserted into a lipid bilayer. Before initiating this step, the required membrane composition, e.g., whether it should be complex or homogeneous, as well as the method of protein incorporation should be chosen (*see* **Note 7**). In the procedure described here, a pure 1-palmitoyl-2-oleoylphosphatidylcholine (POPC) bilayer is utilized, as SERCA physiologically is located in a membrane consisting of approximately 65 % of this type of lipid [26]. First, the size of the simulation system is estimated. Next, the membrane is constructed according to its required size using the Membrane Builder plugin in VMD after which SERCA is inserted into it, deleting lipid molecules overlapping with the protein.

1. Orienting the protein and estimating the size of the simulation box
   Before the membrane can be built, the minimum size of the simulation box needs to be determined. The membrane patch should be large enough to accommodate the protein and allow for sufficient space on each side of the protein, as the simulation cell will be treated with periodic boundary conditions (PBC, *see* **Notes 14** and **15**). Consequently, the approximate size of the protein in the x- and y-dimensions will dictate the minimum size of the membrane patch.

   As a starting point, we want to ensure that the transmembrane part of the ATPase is located in the x,y plane, with the long axis of the molecule roughly aligned with the z-axis. Not only is this orientation convenient for embedding the protein into a membrane, it might also facilitate the use of several analysis tools. However, structures obtained from the PDB are often not positioned with the transmembrane part in the x,y plane. For those cases, you can use structure files from the OPM database (*see* **Note 10**). Start by loading the PSF and PDB files generated in the previous step into VMD. Create a

selection with all the components in the structure as well as one which only encompasses the C_α atoms (name CA):

```
mol load psf 2c8k_complex.psf pdb 2c8k_complex.pdb
set mol_SERCA [molinfo top]
set all_SERCA [atomselect top all]
set CA_SERCA [atomselect top "name CA"]
```

Download the OPM coordinates (2c8k.pdb) from the database, save it as 2c8k_opm.pdb, load it into VMD and select the C_α atoms:

```
mol load pdb 2c8k_opm.pdb
set mol_OPM [molinfo top]
set CA_OPM [atomselect $mol_OPM "name CA"]
```

The OPM structure contains dummy atoms representing the position of the inner and outer leaflets (blue and red, respectively) which will appear in the display. Now we want to align our SERCA structure with the OPM structure. This is done by moving the entire SERCA system (with ATP and ions) according to the transformation matrix that gives the best fit between the C_α atoms of the two structures:

```
$all_SERCA move [measure fit $CA_SERCA $CA_OPM]
```

You will see SERCA moves to the position of the OPM structure and now, SERCA has its long axis approximately in the z-direction. The tilt of the protein is a result of the prediction of the TM region by OPM. Now, measure the total size of the protein and write the new coordinates into a PDB file:

```
set M [measure minmax $all_SERCA]
set size [vecsub [lindex $M 1] [lindex $M 0]]
$all_SERCA writepdb 2c8k_complex_move_OPM.pdb
```

Based on the output obtained here for the protein size (approximately 90 Å × 90 Å × 125 Å), an optimal box size would be approximately 125 Å × 125 Å × 155 Å. Thus, the membrane patch should be 125 Å × 125 Å. The z-dimension will be defined when water is added in the solvation step.

2. Build the membrane patch

   The simplest way to obtain a membrane patch is to use the Membrane Builder plugin in VMD which we also do here (*see* **Note 16**). In the previous step, the minimum x- and y-dimensions of the membrane were determined and we now use those values when building the membrane patch.

   Start by selecting the Extensions → Modeling → Membrane Builder menu in the VMD Main window. In the pop-up box, select POPC as the lipid and set both the X and Y length to 125. Set the Output Prefix to */desired/location/*popc, choose c36 topology and click Generate Membrane. The popc.psf and popc.pdb files should be loaded automatically into VMD. The lipid bilayer from the membrane building step looks rather unnatural in its packing, but the lipid tails will be relaxed later in the process. In addition to the POPC bilayer, a layer of solvating water molecules is generated on each side of the membrane.

3. Adjust the protein position

Showing both the membrane patch and the protein in VMD illustrates that the protein is not centrally positioned relative to the membrane in the x- and y-dimensions. For visualization purposes it is nice to have the protein in the middle of the box so we move the protein relative to the membrane by finding the minimum and maximum coordinates for the lipid bilayer and the protein in the x- and y-dimensions and adjust the position of the protein accordingly without changing the z-position. Load 2c8k_complex_move_OPM.pdb and the lipid bilayer into VMD and find minimum and maximum coordinates of the newly built bilayer and of the protein:

```
mol load psf 2c8k_complex.psf pdb 2c8k_complex_move_
   OPM.pdb
set mol_SERCA [molinfo top]
mol load psf popc.psf pdb popc.pdb
set mol_popc [molinfo top]
set membrane [atomselect $mol_popc all]
set membrane_size [measure minmax $membrane]
set SERCA [atomselect $mol_SERCA "protein"]
set protein_size [measure minmax $SERCA]
```

This will output the minimum and maximum x-, y-, and z-coordinates of the bilayer and protein as two lists. For now, the z-direction is ignored but minimum and maximum x- and y-coordinates are extracted in order to calculate how much the protein should be moved in the x- and y-directions:

```
set min_prot [lindex $protein_size 0]
set xmin_prot [lindex $min_prot 0]
set ymin_prot [lindex $min_prot 1]
set max_prot [lindex $protein_size 1]
set xmax_prot [lindex $max_prot 0]
set ymax_prot [lindex $max_prot 1]
set min_mem [lindex $membrane_size 0]
set xmin_mem [lindex $min_mem 0]
set ymin_mem [lindex $min_mem 1]
set max_mem [lindex $membrane_size 1]
set xmax_mem [lindex $max_mem 0]
```

```
set ymax_mem [lindex $max_mem 1]
```

Calculate the adjustment required in the x-direction:

```
set xnegdiff [expr $xmin_prot - $xmin_mem]
set xposdiff [expr $xmax_mem - $xmax_prot]
set xmove [expr (($xnegdiff+$xposdiff)/2) - $xnegdiff]
```

Calculate the adjustment required in the y-direction and adjust the protein position:

```
set ynegdiff [expr $ymin_prot - $ymin_mem]
set yposdiff [expr $ymax_mem - $ymax_prot]
set ymove [expr (($ynegdiff+$yposdiff)/2) - $ynegdiff]
set all_SERCA [atomselect $mol_SERCA all]
$all_SERCA moveby "$xmove $ymove 0"
```

The protein is now located approximately in the middle of the membrane patch and these coordinates are saved (*see* Fig. 1b).

```
$all_SERCA writepdb 2c8k_complex_move2.pdb
```

Finally, check that the z-position of the protein in the membrane looks reasonable. Try for example to display SERCA with VDW representation and color by ResType. The part of the protein embedded in the membrane should be mainly white, corresponding to hydrophobic.

4. Combine membrane and protein structures

   In this step, the two sets of PSF- and PDB files are combined to a single set comprising both SERCA and the membrane.

```
mol delete all
package require psfgen
resetpsf
topology /path/to/topology/toppar/top_all36_prot.rtf
topology /path/to/topology/toppar/top_all36_na.rtf
topology /path/to/topology/toppar/stream/na/toppar_all36_na_nad_ppi.str
topology /path/to/topology/toppar/toppar_water_ions.str
topology /path/to/topology/toppar/top_all36_lipid.rtf
readpsf popc.psf
coordpdb popc.pdb
readpsf 2c8k_complex.psf
coordpdb 2c8k_complex_move2.pdb
writepsf 2c8k_popc_raw.psf
writepdb 2c8k_popc_raw.pdb
```

Load 2c8k_popc_raw.psf and 2c8k_popc_raw.pdb into VMD. Upon inspection of the structure, you will see that several lipids and water molecules overlap with protein atoms and these molecules need to be deleted before we can continue. In order to do so, we identify the lipid and water molecules whose atoms overlap with SERCA (or lie within 0.6 Å of protein atoms).

```
mol load psf 2c8k_popc_raw.psf pdb 2c8k_popc_raw.pdb
set lipid_overlap [atomselect top "resname POPC and
    same residue as within 0.6 of protein"]
set lipid_reslist [$lipid_overlap get resid]
set lipid_seglist [$lipid_overlap get segid]
set water_overlap [atomselect top "water and same
    residue as within 0.6 of protein"]
set water_reslist [$water_overlap get resid]
set water_seglist [$water_overlap get segid]
```

All overlapping molecules are then deleted and new PSF- and PDB files are generated:

```
mol delete all
resetpsf
readpsf 2c8k_popc_raw.psf
coordpdb 2c8k_popc_raw.pdb
foreach segid $lipid_seglist resid $lipid_reslist {
delatom $segid $resid
}
foreach segid $water_seglist resid $water_reslist {
delatom $segid $resid
}
writepsf 2c8k_popc.psf
writepdb 2c8k_popc.pdb
```

Check the generated files by loading them into VMD. You should inspect which lipids and water molecules you deleted (comparing 2c8k_popc_raw.pdb with 2c8k_popc.pdb), and evaluate whether the applied cutoff (0.6 Å) is reasonable or should be different.

3.3 Solvation and Ionization

In this step, the system comprising SERCA and the lipid bilayer is solvated and ionized.

1. Solvate

   The membrane embedded SERCA:ATP structure is now further solvated with water using the Solvate plugin in VMD. The z-dimension of the simulation box should be 155 Å as described in Subheading 3.2. Again, for visualization purposes, it is advised to have approximately the same thickness of water on each side of the protein in the z-direction. Before adding water, we first find the midpoint of the protein along the z-axis:

```
mol delete all
mol load psf 2c8k_popc.psf pdb 2c8k_popc.pdb
set SERCA [atomselect top protein]
set SERCA_minmax [measure minmax $SERCA]
set SERCA_minz [lindex [lindex $SERCA_minmax 0] 2]
set SERCA_maxz [lindex [lindex $SERCA_minmax 1] 2]
set midz [expr ($SERCA_minz+$SERCA_maxz)/2]
```

This gives the midpoint for SERCA along the z axis at −30 Å. The Solvate plugin can only add water as a continuous box surrounding a protein. As water molecules should not be added inside the lipid bilayer, the solvation is performed in two steps; first on one side of the membrane and then on the other side of the membrane. The water molecules solvating the lipid head groups are used to set the limit of where water should be added along z and to define the x-,y-position of the box. Find the minimum and maximum z coordinate for the final box using the protein midpoint:

```
set solv_maxz [expr $midz+(155/2)]
set solv_minz [expr $midz - (155/2)]
```

Find minimum and maximum x- and y-coordinates based on water surrounding the lipid bilayer:

```
set wat [atomselect top "water"]
set water_size [measure minmax $wat]
set solv_minx [lindex [lindex $water_size 0] 0]
set solv_miny [lindex [lindex $water_size 0] 1]
set solv_down_maxz [lindex [lindex $water_size 0] 2]
set solv_maxx [lindex [lindex $water_size 1] 0]
set solv_maxy [lindex [lindex $water_size 1] 1]
set solv_up_minz [lindex [lindex $water_size 1] 2]
```

Add water below the lipid bilayer:

```
set mindown [list $solv_minx $solv_miny $solv_minz]
set maxdown [list $solv_maxx $solv_maxy $solv_down_maxz]
set down [list $mindown $maxdown]
solvate 2c8k_popc.psf 2c8k_popc.pdb -o solv_down_2c8k
    -s WB -minmax $down
```

Add water above the lipid bilayer:

```
set minup [list $solv_minx $solv_miny $solv_up_minz]
set maxup [list $solv_maxx $solv_maxy $solv_maxz]
set up [list $minup $maxup]
```

```
solvate solv_down_2c8k.psf solv_down_2c8k.pdb -o
   solv_2c8k -s WA -minmax $up
```

Water molecules have now been added on both sides of the membrane such that the protein is solvated on all sides (*see* Fig. 1c).

2. Ionize

   Ions are added to mimic the physiological environment and we use KCl as this is more physiologically relevant for SERCA than NaCl. Furthermore, to treat the electrostatic interactions of the system with Particle Mesh Ewald (PME, *see* **Note 17**), the overall charge of the system should be neutral. The Autoionize plugin of VMD is used to add the ions.

   Click on Extensions → Modeling → Add Ions to open the Autoionize plugin interface. In the Autoionize window, add the newly generated solv_2c8k.psf and solv_2c8k.pdb as the input. In Output prefix type ionized_2c8k and choose salt KCl. Under Ion placement mode tick "Neutralise and set KCl concentration…" and set the concentration to 0.20 mol/L. Leave minimum distances at 5 Å and keep the ION segment name. Finally, click Autoionize and ions will be added to the system and the final set of PSF/PDB files generated. Load the files for a final visual check in VMD (*see* Fig. 1d). The system is now ready for simulation.

3.4 Minimization

The first step of the simulation is to energy minimize the system. This step is performed to relax the strain that might be there in the newly built system, moving atoms from unfavorable positions, which could cause the simulation to crash if the actual MD simulation was started. In order to run the minimization, you need a config file which contains all the settings and specifications that NAMD requires for the calculation. Use the example provided below and create a mini.conf file in a text editor or obtain a config file from a NAMD tutorial online. To read more about config files see http://www.ks.uiuc.edu/Training/Tutorials/namd/namd-tutorial-win-html/node27.html or the NAMD documentation on general config file parameters http://www.ks.uiuc.edu/Research/namd/2.9/ug/node9.html.

In the example here, note the lines concerning PBC and wrapping of atoms. These lines are not vital for the minimization but they are included here to avoid potential problems with water molecules moving far away from the rest of the system during the minimization. The dimensions and center of the simulation cell are defined by the cell basis vectors and the cell origin. To get the dimensions of your system, use the Tk Console:

```
mol load psf ionized_2c8k.psf pdb ionized_2c8k.pdb
set all [atomselect top all]
set center [measure center $all]
```

```
set M [measure minmax $all]
set size [vecsub [lindex $M 1] [lindex $M 0]]
```

Use the numbers for the center as the cell origin and use the total size to give the *x*-, *y*- and *z*-components of the three cell basis vectors. Insert them into the file mini.conf. Furthermore, correct the specified file names to match your PSF and PDB.

mini.conf

```
# Input and output
structure ionized_2c8k.psf
coordinates ionized_2c8k.pdb
outputname ionized_2c8k_mini ;# Base name for output
# Specify force field
paraTypeCharmm on
parameters /path/to/parameters/par_all36_prot.prm ;# protein
parameters /path/to/parameters/par_all36_na.prm ;# ATP
parameters /path/to/parameters/toppar_all36_na_nad_ppi.str ;# ATP
parameters /path/to/parameters/par_all36_lipid.prm ;# lipid
parameters /path/to/parameters/toppar_water_ions.str ;# water, ions
exclude scaled1-4
1-4scaling 1.0 ;# Scaling required by CHARMM
Switching on ;# Softens out the potential at the cutoff
cutoff 12. ;# Distance for nonbond cutoff
switchdist 10 ;# Distance for switching function (cutoff - 2)
pairlistdist 14. ;# Atoms move <2A pr cycle (cutoff+2)
# Minimization so temperature is set to 0 K
temperature 0
# System size and origin for PBC - not when using restart files
cellBasisVector1 128 0. 0. ;# max(x)-min(x), 0, 0
cellBasisVector2 0. 128 0. ;# 0, max(y)-min(y), 0
cellBasisVector3 0. 0. 154 ;# 0, 0, max(z)-min(z)
cellOrigin 13.984 14.815 -29.866 ;# Center of box
# Wrap periodic cells
wrapWater on ;# Wrap water to central cell
wrapAll on ;# Wrap other molecules too
# Specify steps pr pairlistcycle
stepspercycle 20 ;# Redo pairlists every 20 timesteps
```

```
# Output
binaryoutput    off    ;# No binary outputs from
    minimization
outputTiming 50 ;# Write timing information every 50
    steps
minimize 5000 ;# 5000 steps of energy minimization
```

For the toppar_water_ions.str file to work with NAMD, you need to comment out a few lines in the file, which would otherwise cause NAMD to report a fatal error. To comment out lines, open the file in a text editor and add an exclamation mark ("!") in front of the lines which should not be read by NAMD. Comment out lines starting with "set", "if", "WRNLEV", "BOMLEV", and "SOD OC2D2", so they now read for example "!set" and "!WRNLEV". As we do not have any atoms in the system with the OC2D2 atom type (check your PSF), we can safely comment out the "SOD OC2D2" line. Run the minimization from a Linux terminal by typing (*see* **Note 18**):

```
namd2 mini.conf > mini.log
```

This might take a couple of hours on a normal workstation. The estimated remaining time is given in the log file. You can check this in a Linux terminal by typing:

grep TIMING mini.log

The minimization produces a new structure file, ionized_2c8k_mini.coor. This file has the PDB format so you can view the energy minimized structure in VMD. Copy the file to a PDB file by changing its extension to .pdb and use it in the next step. Open the original PSF (no new PSF files are generated in simulations) and the output PDB, containing coordinates for the minimized system, in VMD and check the system. As usual, pay special attention to the termini, disulfide bridges, residues with nonstandard protonation states, ligands, and ions. Should the minimization have produced an error (read the log file), much help on the error is searchable online.

3.5 Equilibration

After minimization the system should be equilibrated before running the actual MD simulation, which is also referred to as the production run (*see* Subheading 3.6). Typically, the equilibration consists of several short MD steps in which the restraints of the system are gradually released. With a lipid bilayer system, the first step of the equilibration typically comprises a short simulation where everything in the system but the lipid tails is held fixed in space. You might say that the lipid tails are "melted" to create a more physiological state of the hydrophobic part of the lipid bilayer. Additionally, this step is required to obtain a closer packing of lipid tails around the transmembrane part of the protein than in the initial setup, which is likely to have a large gap of nothing (vacuum) around it directly after the building phase. If this is ignored, water molecules may move into the vacuum around the protein in the MD simulation (water diffuses faster than lipids), creating a nonphysiological state.

Following the lipid tail melting step, the system is further equilibrated while restraining the protein partially. This can be done in one or more steps and allows lipid head groups, water molecules and ions to obtain a closer packing around the protein without the protein changing its conformation. As in the first equilibration step, this removes potential vacuum spots created in the building phase which were not removed in the minimization. In the final equilibration step, every atom is allowed to move freely and the system will obtain the required temperature and pressure before the production run is initiated.

1. Lipid tail melting

    To equilibrate the lipid tails, only lipid tail atoms should be free to move. We specify these atoms through their beta values, creating a file specifying the fixed atoms (beta 1) and free atoms (beta 0). This file is generated from the Tk Console:

    ```
    mol load psf ionized_2c8k.psf pdb ionized_2c8k_mini.pdb
    set all [atomselect top all]
    $all set beta 1
    set tail [atomselect top "resname POPC and not name N C13 C14 C15 C11 C12 P O11 O12 O13 O14 C1 C2 O21 C3 O31"]
    $tail set beta 0
    $all writepdb fix_2c8k_free_lipidtail.pdb
    ```

    When starting the first equilibration step, the dimensions and center of the simulation cell should be specified. These parameters should never be specified when you use restart files. Find the required information in the same way as described in Subheading 3.4 and insert the parameters for cell basis vectors and cell origin at the appropriate lines in the config file (*see* **Note 19**). Change file names in eq_lip.conf as required.

    eq_lip.conf

    ```
    # Input and output
    structure ionized_2c8k.psf
    coordinates ionized_2c8k_mini.pdb
    outputname ionized_2c8k_eq_lip ;# Base name for output

    # Specify Force Field:
    paraTypeCharmm on
    parameters /path/to/parameters/par_all36_prot.prm ;# protein
    parameters /path/to/parameters/par_all36_na.prm ;# ATP
    parameters /path/to/parameters/toppar_all36_na_nad_ppi.str ;# ATP
    ```

```
parameters /path/to/parameters/par_all36_lipid.prm ;#
    lipid
parameters /path/to/parameters/toppar_water_ions.str ;#
    water, ions
exclude scaled1-4
1-4scaling 1 ;# Scaling required by Charmm
switching on ;# Softens out the potential at the
    cutoff
cutoff 12. ;# Distance for nonbond cutoff
switchdist 10. ;# Distance for switching function
    (cutoff - 2)
pairlistdist 14. ;# Atoms move <2A pr cycle (cutoff+2)
# Set initial temperature
temperature 310. ;# Run simulation at 37 C / 310 K
# Use Langevin thermostat:
langevin on ;# langevin dynamics
langevinDamping 0.1 ;# Damping coefficient of 0.1/ps
    for lipid tail equil
langevinTemp 310. ;# Random noise at this level
langevinHydrogen off ;# Do not couple bath to
    hydrogens
# System size and origin for PBC - not if using
    restart files
cellBasisVector1 128 0. 0. ;# max(x)-min(x), 0, 0
cellBasisVector2 0. 128 0. ;# 0, max(y)-min(y), 0
cellBasisVector3 0. 0. 156 ;# 0, 0, max(z)-min(z)
cellOrigin 13.980 14.812 -30.385 ;# Center of box
#Wrap periodic cells
wrapWater on ;# Wrap water to central cell
wrapAll on ;# Wrap other molecules too
# Use Particle Mesh Ewald for electrostatics:
PME yes
PMEGridSpacing 1.0 ;# Allow less than 1 A between
    grid points
# Specify steps pr pairlistcycle
stepspercycle 20 ;# Redo pairlists every 20 steps
# Specify timesteps. Both nonbondedFreq and fEF must
    evenly divide stepspercycle
timestep 1.0 ;# Run simulation with 1 fs timestep
nonbondedFreq 2 ;# Nonbonded forces every 2nd step
fullElectFrequency 4 ;# PME every 4th step
rigidBonds water ;# Always use rigid water molecules
# Output
```

```
outputEnergies 100 ;# Write energies to log file every
    100 step
outputPressure 100 ;# Write pressure to log file every
    100 step
outputTiming 1000 ;# Write timing information every
    1000 step
binaryoutput no
DCDfreq 1000 ;# Write snapshots to dcd file every 1000
    step
DCDUnitCell yes
binaryrestart yes
restartfreq 1000 ;# Write information for restarts
    every 1000 step
XSTfreq 1000 ;# Write cell parameters every 1000 step
# Fix everything but lipid tails using the fix-file
fixedAtoms on
fixedAtomsFile fix_2c8k_free_lipidtail.pdb ;#PDB file
    with fixed atoms in B col
fixedAtomsCol B
fixedAtomsForces on
run 500000 ;# 500 ps of lipid tail equil
```

Run the lipid tail equilibration from a Linux terminal (or using high performance computing facilities) by typing (*see* **Note 18**):

```
namd2 eq_lip.conf > eq_lip.log
```

Once terminated, check the output of the first equilibration step. You can load the trajectory into VMD from the Tk Console. Using 'skip 10' means that only every 10th snapshot is loaded into VMD.

```
mol load psf ionized_2c8k.psf
animate read dcd ionized_2c8k_eq_lip.dcd skip 10
```

Only the lipid tails should have moved and they should now be more disordered, packing more closely around the protein. Now we are ready to equilibrate the environment around the protein (*see* Fig. 1e).

2. Gradual release of the system

    To keep the protein backbone harmonically restrained in the second step of the equilibration, a file specifying the backbone atoms (again through the beta value) and the required force constant (here 7 kcal/mol/Å^2) is needed. It can be generated as follows:

    ```
    mol load psf ionized_2c8k.psf pdb ionized_2c8k_mini.
        pdb
    set all [atomselect top all]
    ```

```
$all set beta 0
set proteinBB [atomselect top "protein and backbone
    and noh"]
$proteinBB set beta 7
$all writepdb fix_2c8k_free_system.pdb
```

The config file below can be used for this second equilibration step. You will see that restart files are specified, ensuring that all positions, velocities and so on are retained from the previous equilibration step. Though coordinates are taken from the restart files, a PDB should always be specified with "coordinates".

eq_lip_wat.conf

```
# Input and output
structure ionized_2c8k.psf
coordinates ionized_2c8k_mini.pdb
outputname ionized_2c8k_eq_lip_wat ;# Base name for
    output
set input ionized_2c8k_eq_lip ;# Base name for input
    restart files
set fstep 0 ;# How long has the simulation run?
#Restart from last run
firsttimestep $fstep
binCoordinates ${input}.restart.coor
binVelocities ${input}.restart.vel
extendedSystem ${input}.restart.xsc
# Specify Force Field:
paraTypeCharmm on
parameters /path-to-parameters/par_all36_prot.prm ;#
    protein
parameters /path-to-parameters/par_all36_na.prm ;#
    ATP
parameters /path-to-parameters/toppar_all36_na_nad_
    ppi.str ;# ATP
parameters /path-to-parameters/par_all36_lipid.prm ;#
    lipid
parameters  /path-to-parameters/toppar_water_ions.
    str ;# water, ions
exclude scaled1-4
1-4scaling 1 ;# Scaling required by Charmm
switching on ;# Softens out the potential at the
    cutoff
cutoff 12. ;# Distance for nonbond cutoff
switchdist 10. ;# Distance for switching function
    (cutoff - 2)
pairlistdist 14. ;# Atoms move <2A pr. cycle (cutoff+2)
```

```
# Do not set temperature when using an input velocity
    restart file
# Use Langevin thermostat:
langevin on ;# langevin dynamics
langevinDamping 0.5 ;# Damping coefficient of 0.5/ps
langevinTemp 310. ;# Random noise at this level
langevinHydrogen off ;# Do not couple bath to
    hydrogens
# Use constant pressure:
useGroupPressure yes ;# needed for rigid bonds
useFlexibleCell yes ;# no for waterbox, yes for
    membrane
useConstantArea no
langevinPiston on
langevinPistonTarget 1.01325 ;# simulate at 1 atm
    pressure (NAMD units: bar)
langevinPistonPeriod 100. ;# Specifies oscillation
    (fs) of P
langevinPistonDecay 50. ;# oscillation decay time
langevinPistonTemp 310 ;# coupled to heat bath
# System size and origin for PBC - do not specify
    when using a restart xsc file
# Wrap periodic cells
wrapWater on ;# Wrap water to central cell
wrapAll on ;# Wrap other molecules too
# Use Particle Mesh Ewald for electrostatics:
PME yes
PMEGridSpacing 1.0 ;# Allow less than 1 A between
    grid points
# Specify steps pr. pairlistcycle
stepspercycle 20 ;# Redo pairlists every 20 steps
# Specify timesteps. Both nonbondedFreq and fEF must
    evenly divide stepspercycle
timestep 1.0 ;# Run simulation with 1 fs timestep
nonbondedFreq 2 ;# Nonbonded forces every 2nd step
fullElectFrequency 4 ;# PME every 4th step
rigidBonds water ;# Always use rigid water
    molecules
# Write output
outputEnergies 100 ;# Write energies to log file every
    100 step
outputPressure 100 ;# Write pressure to log file every
    100 step
outputTiming 1000 ;# Write timing information every
    1000 step
```

```
binaryoutput no
DCDfreq 1000 ;# Write snapshots to dcd file every 1000
    step
DCDUnitCell yes
binaryrestart yes
restartfreq 1000 ;# Write information for restarts
    every 1000 step
XSTfreq 1000 ;# Write cell parameters every 1000 step
# Harmonic constraints
constraints on
consref fix_2c8k_free_system.pdb ;# PDB with refer-
    ence positions
conskfile fix_2c8k_free_system.pdb ;# PBD specifying
    constrained atoms
conskcol B ;# B column gives force constants
run 3000000 ;# Equil water and lipids for 3 ns
```

Run the second equilibration step as you did for the first step and check the output.

As a final step of the equilibration procedure, you can run a 2 ns simulation without any fixed or restrained atoms. The config file will be very similar to the one used for equilibration of lipid molecules and water, but a few things should be changed. Copy the eq_lip_wat.conf file to a new file, eq_all.conf. In this file, change the outputname to ionized_2c8k_eq_all and set input to ionized_2c8k_eq_lip_wat. You can change fstep to 3000000 if you want the time to be continued from the second step. Delete the lines concerning the harmonic constraints (constrainst, consref, conskfile, conskcol) and change run to 2000000. Run this step as before and check the final equilibrated system.

3.6 Production Run

If the equilibrated system looks fine, e.g., no water molecules in the membrane or vacuum artifacts, the production run can be started. The required simulation time is highly dependent on the property you wish to study (*see* **Note 8**). Studying for example protein–ligand interactions may be feasible with a 50 ns simulation or less while a study of protein conformational changes of P-type ATPases will require a much longer simulation. In any case, to prevent wasting computational resources, it is good practice to first run 5–10 ns of simulation and check that everything in the output looks fine before running a longer simulation.

For the production run, you can use a config file very similar to the one used in the final step of the equilibration, but with a few additional variables. Notably, "currenttimestep" is used to get the last time step from the restart files when continuing the simulation. When initiating the production run, "currenttimestep" is set to 0.

"targett" specifies the desired total simulation time. After running the first 10 ns, one might extend to for example 50 ns, requiring "targett" to be adjusted accordingly in addition to changing outputname and input. "steps" specifies how many steps are needed to get to "targett", starting from "currenttimestep". An example of the config file is given below.

prodrun_1.conf

```
# Input and output
structure ionized_2c8k.psf
coordinates ionized_2c8k_mini.pdb
outputname ionized_2c8k_prodrun_1  ;# Base name for output
set input ionized_2c8k_eq_all  ;# Base name for input restart files
# To start from the last time step in the restart files
proc get_first_ts {xscfile} {
    set fd [open $xscfile r]
    gets $fd
    gets $fd
    gets $fd line
    set ts [lindex $line 0]
    close $fd
    return $ts
}
#set currenttimestep [get_first_ts ${input}.restart.xsc]  ;# Delete "#" to use time from restart
set currenttimestep 0  ;# Add "#" if using time from restart
set fstep $currenttimestep  ;# How long has the simulation run?
set targett 10000000  ;# Run to a total time of 10 ns
set steps [expr $targett - $fstep]  ;# How many steps to run now
#Restart from last run
firsttimestep $fstep
binCoordinates ${input}.restart.coor
binVelocities ${input}.restart.vel
extendedSystem ${input}.restart.xsc
# Specify Force Field:
paraTypeCharmm on
parameters /path-to-parameters/par_all36_prot.prm  ;# protein
parameters /path-to-parameters/par_all36_na.prm  ;# ATP
```

```
parameters /path-to-parameters/toppar_all36_na_nad_ppi.
   str ;# ATP
parameters /path-to-parameters/par_all36_lipid.prm ;#
   lipid
parameters /path-to-parameters/toppar_water_ions.str ;#
   water, ions
exclude scaled1-4
1-4scaling 1 ;# Scaling required by Charmm
switching on ;# Softens out the potential at the cutoff
cutoff 12. ;# Distance for nonbond cutoff
switchdist 10. ;# Distance for switching function
   (cutoff - 2)
pairlistdist 14. ;# Atoms move <2A pr. cycle (cutoff+2)
# Do not set temperature when using an input velocity
   restart file
# Use Langevin thermostat:
langevin on ;# langevin dynamics
langevinDamping 0.5 ;# Damping coefficient of 0.5/ps
langevinTemp 310. ;# Random noise at this level
langevinHydrogen off ;# Do not couple bath to hydrogens
# Use constant pressure:
useGroupPressure yes ;# needed for rigid bonds
useFlexibleCell yes ;# no for waterbox, yes for
   membrane
useConstantArea no
langevinPiston on
langevinPistonTarget 1.01325 ;# simulate at 1 atm pres-
   sure (NAMD units: bar)
langevinPistonPeriod 100. ;# Specifies oscillation (fs)
   of P
langevinPistonDecay 50. ;# oscillation decay time
langevinPistonTemp 310 ;# coupled to heat bath
# System size and origin for PBC - do not specify when
   using a restart xsc file
# Wrap periodic cells
wrapWater on ;# Wrap water to central cell
wrapAll on ;# Wrap other molecules too
# Use Particle Mesh Ewald for electrostatics:
PME yes
PMEGridSpacing 1.0 ;# Allow less than 1 A between grid
   points
# Specify steps pr. pairlistcycle
stepspercycle 20 ;# Redo pairlists every 20 steps
```

```
# Specify timesteps. Both nonbondedFreq and fEF must
    evenly divide stepspercycle
timestep 1.0 ;# Run simulation with 1 fs timestep
nonbondedFreq 2 ;# Nonbonded forces every 2nd step
fullElectFrequency 4 ;# PME every 4th step
rigidBonds water ;# Always use rigid water molecules
# Output
outputEnergies 100 ;# Write energies to log file every
    100 step
outputPressure 100 ;# Write pressure to log file every
    100 step
outputTiming 1000 ;# Write timing information every 1000
    step
binaryoutput no
DCDfreq 1000 ;# Write snapshots to dcd file every 1000
    step
DCDUnitCell yes
binaryrestart yes
restartfreq 1000 ;# Write information for restarts every
    1000 step
XSTfreq 1000 ;# Write cell parameters every 1000 step
run $steps ;# Run dynamics for $steps steps
```

Run the production run in the same way as the equilibrations, preferably using high-performance computing facilities.

3.7 Analysis

The analysis depends on the aim of the study and it is beyond the scope of this chapter to describe every possible analysis method. Instead, we give a few general suggestions and refer the reader to useful resources.

Visualize the dynamics in VMD by loading the PSF file and reading the DCD trajectory into this. To concatenate several individual DCD trajectories use the script catdcd (http://www.ks.uiuc.edu/Development/MDTools/catdcd/).

VMD has various analysis tools available (VMD Main window→Extensions→Analysis). For instance, the root mean square deviation (RMSD) can be calculated (VMD Main window→Extensions→Analysis→RMSD Trajectory Tool). When doing so for a P-type ATPase it is advised to calculate the RMSD for the individual domains to check the structural stability. The overall RMSD will often be large due to inter-domain flexibility, in particular between the cytoplasmic domains. For instance, if you align the protein from the final frame of the simulation with the starting structure you can see the proportion of the movements (see Fig. 1f).

A diverse selection of analysis scripts is accessible online in the VMD Script Library (http://www.ks.uiuc.edu/Research/vmd/

script_library/). In particular, bigdcd is recommended for analyzing large trajectory files. The NAMD Plot Plugin (http://www.ks.uiuc.edu/Research/vmd/plugins/namdplot/) is useful for plotting for example pressure, temperature, and other simulation parameters reported by NAMD in the LOG file. Other software packages that might be useful for analyzing the MD simulation include for example GROMACS and MD Analysis [27] (http://www.mdanalysis.org/).

4 Notes

1. When running MD simulations of proteins, the description levels typically considered are all-atom (AA) and coarse-grain (CG). A popular FF utilizing the CG description is MARTINI [28, 29]. On the AA description level, every atom is typically described by an individual particle, while a particle describes a set of atoms (typically 3–4 heavy atoms) on the CG description level. This resolution difference enables CG simulations to assess far longer time scales than typically available with AA simulations, e.g., enabling lipid diffusion and thus self-assembly of a lipid bilayer around a membrane protein without much bias, allowing simulations of complex membranes as well. Although faster than AA simulations, CG simulations lack the atomistic details desired in for example studies of protein–ligand interactions.

2. VMD and NAMD are developed by the Theoretical and Computational Biophysics Group at the University of Illinois at Urbana-Champaign. In-depth tutorials on the general use of VMD, how to create movies and images in VMD as well as setting up MD simulations on membrane proteins using VMD and NAMD is accessible from their homepage http://www.ks.uiuc.edu/Training/Tutorials/. With the tutorials, various examples of config files can furthermore be downloaded. The mailing lists for the two programs are very useful resources for troubleshooting and the archives for these can be accessed at http://www.ks.uiuc.edu/Research/vmd/mailing_list/vmd-l/ and http://www.ks.uiuc.edu/Research/namd/mailing_list/.

3. If there is more than one protein chain in the asymmetric unit (e.g., chain A, B), they might vary and therefore it matters which one you choose for the simulations. Usually, chain A is the most complete chain. Missing residues can be added using MODELLER [30] or another structure modeling software.

4. Cofactors may be important for the simulation. If the functional state studied is an ATP/ADP/P_i-bound state, the relevant cofactor should be included in the cytosolic domains. However, the downloaded PDB file is likely to contain an analog instead of the actual compound, but the analog is helpful

when creating the proper ligand. For instance, the non-hydrolyzable ATP analogs AMPPNP and AMPPCP are easily converted into ATP by changing the N or C atom linking the β- and γ-phosphates to an O atom in the PDB file. There will be small changes in bonding geometries but these are resolved when energy minimizing the full complex before running the MD simulation. Likewise, the phosphate analogs AlF_4^-, BeF_3^-, or MgF_4^{2-} should be converted into either P_i or part of the phosphorylated aspartate residue.

5. The PDB file is likely to contain metal ions and whether or not to include these depends on the functional state studied. The most important metal ions to consider are the magnesium ion in the nucleotide binding site and the ions in the transmembrane (TM) domain. Once you have decided which ions to include, you will need to consider the protonation states of the nearby protein residues (*see* **Note 9**).

6. Ligands which have been co-crystallized with the ATPase are often not relevant to include in the MD simulations and can be skipped in the protein preparation. Such ligands could be phosphatidyl ethanolamine, thapsigargin or ouabain.

7. There are many different strategies for inserting a protein into a lipid bilayer. Here, we simply delete lipid molecules that overlap with the protein and remove any created vacuum gaps with equilibration. Alternatively, you could for instance use self-assembly of a lipid bilayer around the protein at the CG level (*see* **Note 1**), which can be converted into an atomistic representation [31, 32], or grow the protein in the lipid bilayer (g_membed), pushing away lipids as the protein increases in size [33]. Which lipid types to include depends on the aim of the study, e.g., do you merely need a hydrophobic environment to embed the transmembrane part of the protein, are you interested in specific protein–lipid interactions in the physiological state or are you trying to replicate an experimental setup?

8. AA simulations of a P-type ATPase in a lipid bilayer are usually limited to the nanosecond timescale while CG simulations typically allow simulations on the microsecond timescale, both depending on the computational resources available.

9. Protonation states are important to consider and are likely to depend on the functional state of the P-type ATPase. In the vast majority of cases, the acidic residues (aspartic and glutamic acid) are negatively charged, whereas basic residues (lysine and arginine) are modeled with a positive charge. Histidine residues, often with a pK_a of around 6–6.5, are special and often require more attention. Furthermore, there are cases in which basic or acidic residues should be treated in their non-normal state, e.g., for the calcium free states of the Ca^{2+}-ATPase where some of the acidic residues in the calcium

binding site requires protonation to obtain protein stability in a MD simulation [1].

10. The OPM database provides coordinates of the experimental structures of TM proteins for which the position in the membrane has been predicted [21]. In the downloaded structure file the membrane normal coincides with the z-axis, creating a tilt determined by the amino acids predicted to be located within the membrane. Using this information eases the positioning of the protein in the membrane.

11. The CHARMM force field, applied in the setup described here, uses three different residue names for histidine residues depending on the protonation state, namely HSD (δ-protonated neutral form), HSE (ε-protonated neutral form) and HSP (positively charged state).

12. psfgen adds hydrogen atoms according to guessed coordinates as well as specified protonation states of amino acid side chains. It is recommended to set up the psfgen step in a script which can be called by "source /path/to/script/psfgen.tcl". Alternatively, you can use the AutoPSF plugin in VMD which automates the steps described in the generation of PSF and PDB files in Subheading 3.1, **step 4**.

13. In the example described in the protocol, no water molecules are present in the PDB file. If water molecules are present in the PDB file of your choice, it is usually advised to keep them for the simulation. However, these molecules should have their own segment and be included when performing the psfgen step. The oxygen atom in water molecules has the name OH2 and the residue name TIP3 when using the TIP3P water model [34]. You can ensure this when making the PSF by including the following commands:

    ```
    pdbalias atom HOH O OH2
    pdbalias residue HOH TIP3
    ```

14. PBC are used to minimize surface effects at the border of a molecular system. In practice, the simulation system is replicated in all three directions, an indefinite number of times. Should an atom leave the simulation box on one side, it is replaced by an image of the atom entering from the opposite side, ensuring that the total number of atoms in the box remains constant. You can visualize the periodic system using the "Periodic" tab of the VMD Graphical Representations window. If the protein has moved to the side of the box during the simulation, this can be adjusted with the PBCTools plugin (http://www.ks.uiuc.edu/Research/vmd/plugins/pbctools/). It is important that the box is large enough to prevent periodic artifacts, such as the macromolecule interacting with an image of itself or "seeing" the same atom twice.

15. The cutoff for vdW interactions typically used with the CHARMM force field is 12 Å. As a rule of thumb, the minimum distance to the edge of the box from any protein atom should at least be equal to this cutoff value (*see* **Note 14**).

16. The membrane types available using the Membrane builder plug-in of VMD include POPC and 1-palmitoyl-2-oleoylphosphatidylethanolamine (POPE).

17. With PME, an interaction is split into a short and a long-range contribution. For this, a multiple time stepping procedure can be utilized. In this time step procedure, the short-range component is calculated in every time step while the long-range component is evaluated only periodically (*see* examples of conf files in Subheadings 3.5 and 3.6). This procedure is valid, as the long-range forces vary slower compared to the near-range forces [35].

18. Depending on your workstation and on how NAMD was installed, you might be able to run it in parallel (i.e., using several processors). Try to run NAMD with the command "namd2 +p8 mini.conf > mini.log". This should distribute the calculations on eight processors if your NAMD installation and workstation allows it. If in doubt, check the NAMD documentation or ask your local system administrator. For running NAMD using high-performance computing facilities, check the required syntax with the system administrator.

19. When specifying the simulation cell basic vectors at the first equilibration step, the size measured does not correspond to an equilibrated system, as there will be some vacuum between simulation cells at this step (for visualization *see* **Note 14**). Thus, the simulation cell is expected to shrink during the second step of the equilibration (you can check that this happens in VMD and in the xst file produced by NAMD). Furthermore, due to the methodology employed for inserting the protein into the lipid bilayer, the bilayer is expected to shrink in the x- and y-dimensions when a closer packing around the protein is obtained. If the simulation crashes immediately in the first step of the equilibration, it can sometimes be caused by too small cell basic vectors. Thus, you can attempt to increase these by 1–2 Å, however, remember to ensure that the box shrinks and that there is no vacuum between the periodic boxes after equilibration.

References

1. Musgaard M, Thøgersen L, Schiøtt B (2011) Protonation states of important acidic residues in the central Ca(2+) ion binding sites of the Ca(2+)-ATPase: a molecular modeling study. Biochemistry 50:11109–11120

2. Yu H, Ratheal IM, Artigas P, Roux B (2011) Protonation of key acidic residues is critical for the K(+)-selectivity of the Na/K pump. Nat Struct Mol Biol 18:1159–1163

3. Nagarajan A, Andersen JP, Woolf TB (2012) The role of domain: domain interactions versus domain: water interactions in the coarse-grained simulations of the E1P to E2P transitions in Ca-ATPase (SERCA). Proteins 80:1929–1947

4. Sonntag Y et al (2011) Mutual adaptation of a membrane protein and its lipid bilayer during conformational changes. Nat Commun 2:304
5. Wang K et al (2014) Structure and mechanism of Zn-transporting P-type ATPases. Nature 514:518–522
6. Poulsen H et al (2010) Neurological disease mutations compromise a C-terminal ion pathway in the Na(+)/K(+)-ATPase. Nature 467:99–102
7. Leach A (2001) Molecular modelling principles and applications. Prentice Hall, Upper Saddle River, NJ
8. Jensen F (2007) Introduction to computational chemistry. Wiley, New York
9. MacKerell AD Jr, Feig M, Brooks CL 3rd (2004) Improved treatment of the protein backbone in empirical force fields. J Am Chem Soc 126:698–699
10. MacKerell AD Jr et al (1998) All-atom empirical potential for molecular modeling and dynamics studies of proteins. J Phys Chem B 102:3586–3616
11. Oostenbrink C, Villa A, Mark AE, van Gunsteren WF (2004) A biomolecular force field based on the free enthalpy of hydration and solvation: the GROMOS force-field parameter sets 53A5 and 53A6. J Comput Chem 25:1656–1676
12. Cornell WD et al (1996) A second generation force field for the simulation of proteins, nucleic acids, and organic molecules (vol 117, pg 5179, 1995). J Am Chem Soc 118:2309
13. Jorgensen WL, Maxwell DS, TiradoRives J (1996) Development and testing of the OPLS all-atom force field on conformational energetics and properties of organic liquids. J Am Chem Soc 118:11225–11236
14. Kaminski GA, Friesner RA, Tirado-Rives J, Jorgensen WL (2001) Evaluation and reparametrization of the OPLS-AA force field for proteins via comparison with accurate quantum chemical calculations on peptides. J Phys Chem B 105:6474–6487
15. Phillips JC et al (2005) Scalable molecular dynamics with NAMD. J Comput Chem 26:1781–1802
16. Van Der Spoel D et al (2005) GROMACS: fast, flexible, and free. J Comput Chem 26:1701–1718
17. Humphrey W, Dalke A, Schulten K (1996) VMD: visual molecular dynamics. J Mol Graph 14:33–38
18. Li H, Robertson AD, Jensen JH (2005) Very fast empirical prediction and rationalization of protein pKa values. Proteins 61:704–721
19. Anandakrishnan R, Aguilar B, Onufriev AV (2012) H++ 3.0: automating pK prediction and the preparation of biomolecular structures for atomistic molecular modeling and simulations. Nucleic Acids Res 40:W537–W541
20. Jensen AM, Sørensen TL, Olesen C, Møller JV, Nissen P (2006) Modulatory and catalytic modes of ATP binding by the calcium pump. EMBO J 25:2305–2314
21. Lomize MA, Lomize AL, Pogozheva ID, Mosberg HI (2006) OPM: orientations of proteins in membranes database. Bioinformatics 22:623–625
22. Sørensen TL et al (2004) Localization of a K(+)-binding site involved in dephosphorylation of the sarcoplasmic reticulum Ca(2+)-ATPase. J Biol Chem 279:46355–46358
23. Klauda JB et al (2010) Update of the CHARMM all-atom additive force field for lipids: validation on six lipid types. J Phys Chem B 114:7830–7843
24. Best RB et al (2012) Optimization of the additive CHARMM all-atom protein force field targeting improved sampling of the backbone phi, psi and side-chain chi(1) and chi(2) dihedral angles. J Chem Theory Comput 8:3257–3273
25. Daiho T et al (2001) Mutations of either or both Cys876 and Cys888 residues of sarcoplasmic reticulum Ca2+-ATPase result in a complete loss of Ca2+ transport activity without a loss of Ca2+-dependent ATPase activity. Role of the Cys876-Cys888 disulfide bond. J Biol Chem 276:32771–32778
26. Bick RJ, Buja LM, Van Winkle WB, Taffet GE (1998) Membrane asymmetry in isolated canine cardiac sarcoplasmic reticulum: comparison with skeletal muscle sarcoplasmic reticulum. J Membr Biol 164:169–175
27. Michaud-Agrawal N, Denning EJ, Woolf TB, Beckstein O (2011) MDAnalysis: a toolkit for the analysis of molecular dynamics simulations. J Comput Chem 32:2319–2327
28. Marrink SJ, Risselada HJ, Yefimov S, Tieleman DP, de Vries AH (2007) The MARTINI force field: coarse grained model for biomolecular simulations. J Phys Chem B 111:7812–7824
29. Monticelli L et al (2008) The MARTINI coarse-grained force field: extension to proteins. J Chem Theory Comput 4:819–834
30. Sali A, Blundell TL (1993) Comparative protein modelling by satisfaction of spatial restraints. J Mol Biol 234:779–815
31. Scott KA et al (2008) Coarse-grained MD simulations of membrane protein-bilayer self-assembly. Structure 16:621–630
32. Stansfeld PJ, Sansom MS (2011) Molecular simulation approaches to membrane proteins. Structure 19:1562–1572
33. Wolf MG, Hoefling M, Aponte-Santamaria C, Grubmuller H, Groenhof G (2010) g_mem-

bed: efficient insertion of a membrane protein into an equilibrated lipid bilayer with minimal perturbation. J Comput Chem 31: 2169–2174

34. Jorgensen WL, Chandrasekhar J, Madura JD, Impey RW, Klein ML (1983) Comparison of simple potential functions for simulating liquid water. J Chem Phys 79:926–935

35. Darden T, York D, Pedersen L (1993) Particle mesh Ewald: an N·log(N) method for Ewald sums in large systems. J Chem Phys 98:10089–10092

Chapter 41

Computational Classification of P-Type ATPases

Dan Søndergaard, Michael Knudsen,
and Christian Nørgaard Storm Pedersen

Abstract

Analysis of sequence data is inevitable in modern molecular biology, and important information about for example proteins can be inferred efficiently using computational methods. Here, we explain how to use the information in freely available databases together with computational methods for classification and motif detection to assess whether a protein sequence corresponds to a P-type ATPase (and if so, which subtype) or not.

Key words P-type ATPases, Protein sequences, Classification, Motif detection, Databases

1 Introduction

With the increase in new data obtained from genomics projects, the need for automated methods assisting the determination of structural and functional properties of proteins from amino acid sequences alone is becoming apparent.

In this chapter we give an overview of some freely available databases that may be used to assess whether a protein sequence corresponds to a P-type ATPase (and if so, which subtype) or not. We also explain two different methods for classification and discovery of P-type ATPases and present a new web service that makes these methods easily accessible, which we believe will be a useful tool for practitioners in the P-type ATPase field.

2 Databases

In this section we present three popular protein databases: UniProt, PROSITE, and TCDB. Even though neither of them are limited to P-type ATPases in particular (TCDB comes closest by focusing exclusively on transport proteins), they are all of great value for assessing whether a given protein sequence corresponds to a P-type

ATPase or not. We also present PATBox, a new web service designed specifically with the classification and discovery of P-type ATPases in mind.

2.1 The Universal Protein Resource (UniProt)

The UniProt database [1] contains information about proteins obtained from the scientific literature. The information is split into four core databases — UniProtKB/Swiss-Prot, UniProtKB/TrEMBL, UniParc, and UniRef — which we review below.

The UniProtKB/Swiss-Prot database contains manually curated information. The data for a protein may include information about for example function, catalytic activity, pathways involving the protein, and protein-protein interactions. The user may enter a protein sequence on the UniProt web page and BLAST it against the UniProtKB/Swiss-Prot database.

Since manual curation cannot keep up with the rapid arrival of new data, the UniProtKB/TrEMBL database provides an alternative to UniProtKB/Swiss-Prot by also including automatically annotated entries. This gives a much higher coverage but with a higher degree of uncertainty. As of June 11, 2014, the Swiss-Prot (resp. TrEMBL) database contains entries for 545,536 (resp. 69,014,937) protein sequences, and of these approximately 500 (resp. 68,000) correspond to P-type ATPases.

The two UniProtKB databases contain many redundant sequences, which may not only slow down computations but also make interpretation of results challenging. The UniRef database is created with this problem in mind. It contains clusters of sequences from UniProtKB based on sequence similarity and provides three sets — UniRef100, UniRef90, and UniRef50 — where sequences are clustered at the 100, 90, and 50 % identity levels, respectively. Using for example UniRef50 results in a ~70 % reduction in database size compared to UniProtKB. The user may enter a protein sequence on the UniRef web page, which then finds the best matching clusters and provides details about the individual cluster members and links to their entries in UniProtKB.

The last database is UniParc, which gathers protein sequences from various sources, which besides UniProtKB includes the Protein Data Bank (PDB), the Reference Sequence Collection (RefSeq), and many others. The UniParc web page provides a convenient entry point for a search across multiple databases.

The UniProtKB database is accessible from: http://www.uniprot.org/

2.2 PROSITE

The PROSITE database [2] provides two web servers, ScanProsite and ProfileScan, which are capable of identifying protein families and domains. Each method has its own strengths and weaknesses and we discuss both below.

ScanProsite identifies protein families and domains based on short sequence patterns, also known as motifs. Functional or structural similarities between distantly related proteins are not easily picked up by pairwise alignments, but short subsequences (e.g., corresponding to binding sites) may still be very well conserved. Patterns are determined from multiple alignments of structurally and functionally related proteins and domains. Typical patterns are around 10–20 amino acids long. Some positions are fixed, whereas others may vary. For example, a pattern could be A-x-G-[IV]-P which translates to "alanine, any amino acid, glycine, isoleucine or valine, proline."

It is computationally efficient to find all occurrences of a set of patterns in any given amino acid sequence, and the latest version of ScanProsite looks for 1308 patterns. The web server provides links to detailed documentation about all found patterns cross-linked with other databases.

ProfileScan differs from ScanProsite by taking entire sequences into account using position-dependent scoring matrices, also known as profiles, calculated from multiple alignments of functionally and structurally related sequences. When comparing an amino acid sequence to a profile, ProfileScan will always return a score reflecting how well the sequence matches the profile. If sequences are so distantly related that no patterns are conserved (and ScanProsite returns a negative answer), ProfileScan may still be able to detect an overall sequence similarity reflecting a structural, and hence possibly also functional, relationship.

ScanProsite and ProfileScan complement each other. ScanProsite identifies specific regions of sequences and the accompanying documentation provides a valuable starting point for further manual inspection and validation. Even if ScanProsite finds no patterns, distant evolutionary relationships may still be detected by ProfileScan, which may help narrow down the field of possible candidates for further analyses.

The PROSITE database is accessible from: http://prosite.expasy.org/

2.3 The Transport Classification Database (TCDB)

The Transporter Classification Database (TCDB) [3] is a manually curated database containing information about more than 10,000 transport proteins obtained from the scientific literature.

Proteins in TCDB are classified based on both functional and phylogenetic information, and each protein is assigned to a group identified by a name on the form N1.L1.N2.N3.N4, where N1 is the class, L1 the subclass, N2 the family, N3 the subfamily, and N4 the transport system.

As of July 3, 2014, TCDB contains 214 P-type ATPase sequences. P-type ATPases are characterized by having N1. L1.N2 = 3.A.3 and are divided into 32 subfamilies (N3) which are further subdivided based on transport system (N4).

The TCDB web page allows users to enter protein sequences and BLAST them against the database to find the best matching classifications. The results page links to relevant literature and detailed information about the matching groups.

The TCDB database is accessible from: http://www.tcdb.org/

2.4 P-Type ATPase Toolbox

The P-type ATPase Toolbox (PATBox) is a newly developed web service, which uses two computational tools (*see* Subheading 3) for discovery and classification of P-type ATPases. PATBox provides much of the functionality also available in a tool described in [4] and available at http://pumpkin.au.dk. However, it is based upon a more recent dataset and classification methods.

Additionally, the user can browse all sequences from the UniProtKB database containing the D-K-T-G-T-[LIVM]-[TI] motif which is conserved in all P-type ATPases. Each sequence is also annotated with its true subtype, if known, and the subtype predicted by the computational methods.

The web service also provides download access to the manually curated dataset used to build the classifiers and the database. The dataset contains 524 P-type ATPase sequences and their associated subtype collected from [5, 6] and the previously described databases.

The P-type ATPase Toolbox is accessible from: http://services.birc.au.dk/patbox

3 Methods

In this section we provide step-by-step instructions for how to use PROSITE (described in the Subheading 2) to determine whether a protein sequence corresponds to a P-type ATPase or not.

We then present a method for predicting the subtype of a P-type ATPase based on BLAST searches. Finally, we describe a method for finding motifs, which may be of biological importance using a novel pattern discovery and classification technique introduced in [7].

3.1 ScanPROSITE — P-Type ATPase or Not?

The sequence motif D-K-T-G-T-[LIVM]-[TI] occurs in most P-type ATPases but only rarely in other proteins [8, 5], and hence it serves as a good indicator for whether a sequence corresponds to a P-type ATPase or not. Here are step-by-step instructions for how to scan sequences for this motif using ScanPROSITE:

1. Open a browser and navigate to http://prosite.expasy.org/scanprosite/

2. Make sure that Option 3 (*Submit PROTEIN sequences and MOTIFS to scan them against each other*) is checked.

3. In the box labeled "STEP 1", enter the sequences you wish to identify as P-type ATPases or not.

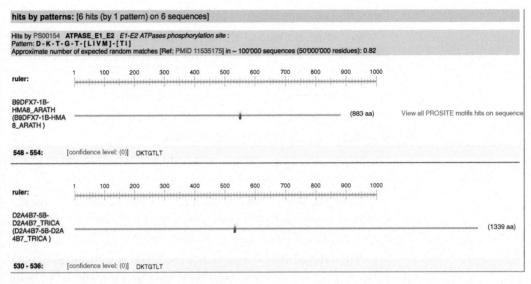

Fig. 1 The *green bar* shows the number of sequences, which contains the motif. Each match is then shown and the position in the sequence at which the motif was located is shown graphically. In this example, six sequences were searched for the D-K-T-G-T-[LIVM]-[TI] motif, and all six of them contained the pattern (referred to as hits)

4. In the box labeled "STEP 2", enter the following PROSITE accession identifier: PS00154.
5. Click the button "START THE SCAN".

When the results are ready, scroll down to see the graphical representation of the results (*see* Fig. 1).

Note that PROSITE reports nine sequences that, despite containing the D-K-T-G-T-[LIVM]-[TI] motif, are not P-type ATPases. Thus, the motif is a good indicator that a sequence is a P-type ATPase, but care should be taken when evaluating the results.

3.2 BLAST+ Method — What Kind of P-Type ATPase?

BLAST+ [9] is an efficient similarity search method for sequence data and a well-known method for searching large databases of sequences such as UniProt. Here we use a local version of BLAST+ with a custom database.

This method is available as a service at The P-type ATPase Toolbox, but for analysis on ones own machine or cluster, follow the instructions below.

This method requires some technical knowledge and experience with use of the command line. When classifying sequences with this approach your query sequence(s) must be saved to a FASTA formatted file.

To use the BLAST+ method you must download and install the BLAST+ software package, which is available for all major platforms from:

ftp://ftp.ncbi.nlm.nih.gov/blast/executables/blast+/LATEST/

You must download the file, which is compatible with your platform. If in doubt whether your platform is 32- or 64-bit, use the 32-bit version.

- Windows (32-bit): ncbi-blast-X.Y.Z+-win32.exe
- Windows (64-bit): ncbi-blast-X.Y.Z+-win64.exe
- Mac OS X: ncbi-blast-X.Y.Z+.dmg
- Linux: Refer to your Linux distribution's package manager or manual to find out which file to download.

When the download is complete, you must follow the installation instructions for your platform in the "Installation" section at:

http://www.ncbi.nlm.nih.gov/books/NBK1762/

When the BLAST+ approach is used, you search against a database of labeled sequences (in this particular case, the label is the subtype of the P-type ATPase). The query sequence is then believed to be of the same class as the labeled sequence with the best score returned by a search against the database.

A database built from our curated data set is available at The P-type ATPase Toolbox and consists of three files which must be placed in your working directory.

To query the database, one must open a command prompt or terminal and navigate to the directory containing the database files. This directory should also contain the FASTA file containing the sequences one wishes to classify. One must then run the command:

```
blastp -db ptype-atpase-blast.db -query input.fa -out
   ptype-atpase-results.txt
```

This creates a file called ptype-atpase-results.txt, which can be opened in a text editor (e.g., Word or Notepad on Windows or TextEdit on Mac). In the output file, locate the line:

```
Sequences producing significant alignments
```

This line signifies the beginning of a listing of all sequences in the database, which were found by BLAST+, sorted by their similarity with the query sequence. Example output is shown below.

```
Score      E
Sequences producing significant alignments: (Bits)  Value

  D2A4B7|5B|D2A4B7_TRICA Putative uncharacterized pro-
tein GLEAN_1...   2693    0.0
  Q16XE5|5B|Q16XE5_AEDAE AAEL008902-PA    [Aedes aegypti]
1239    0.0
  Q7QHN5|5B|Q7QHN5_ANOGA     AGAP011271-PA    (Fragment)
[Anopheles gam...    1216    0.0
```

```
Q7KQN3|5B|Q7KQN3_DROME CG32000, isoform G [Drosophila
melanoga...    1181    0.0
H9KBL9|5B|H9KBL9_APIME Uncharacterized protein    [Apis
mellifera]    1082    0.0
Q9H7F0|5B|AT133_HUMAN Probable cation-transporting ATPase
13A3 ...     901    0.0
H2QP02|5B|H2QP02_PANTR ATPase type 13A3    [Pan troglo-
dytes]    901    0.0
Q5XF89|5B|AT133_MOUSE Probable cation-transporting ATPase
13A3 ...     879    0.0
Q9CTG6|5B|AT132_MOUSE Probable cation-transporting ATPase
13A2 ...     803    0.0
...
```

Each result is presented on a separate line containing the sequence identifier, class, and descriptive name of the sequence. Additionally, a bit-score and an E-value are given. To evaluate the significance of a hit, we will look at the E-value, which represents the number of sequences in the database, which can be expected to be given as a result with the same bit-score by chance. That is, a lower E-value is better since the probability that a similar hit is returned by chance is lower.

For example, consider the first result in the output shown above. An E-value of 0.0 means that, for the given query sequence, we expect to see 0.0 sequences in the database with the bit-score 2693 or higher, meaning that the result is a very significant match. In the output above, the shown results are all very significant and belong to the class 5B. Hence we believe that the query sequence also belongs to the 5B class.

The accuracy of this method depends greatly on the number of results considered. We have experimentally found that considering the first three results provides the highest accuracy. This is a special case ($k=1$) of the k-nearest neighbors method from the field of machine learning, where k is the number of "neighbors" considered.

3.3 Sequence Learner (SeqL) Method — Motif Detection

Sequence Learner (SeqL) [7] is a computational method for classification of sequences. Given two groups of sequences, e.g., sequences that correspond to a particular P-type ATPase subtype and sequences that do not, SeqL computes a set of sequence motifs that separate the two groups. A new sequence of unknown type is then scanned for occurrences of the motifs from each group, and the subtype is predicted to be that of the best matching group.

Note that SeqL only compares two groups. In the case of the 11 P-type ATPase subtypes, we build 11 models (1A versus not 1A, 1B versus not 1B, and so on), and use the model assigning the highest score to the sequence for prediction.

The most substantial benefit of this method is that the model contains the motifs that divide the groups, which may provide insight into the biology of the sequence. We will focus on the motif detection aspect of the SeqL method, but note that it can also be used for classification. For example, PATBox also provides classification of P-type ATPases using SeqL.

SeqL is a command line tool meaning that the user must be comfortable using the command line of the operating system.

3.3.1 Preparing the Input File

Recall that the SeqL method finds motifs, which divide two groups of sequences well. This means that SeqL must be given sequences for each group and for each sequence, one must know which group the sequence belongs to. The input file must be a plain text file where each line contains the class (negative class: -1 or positive class: +1) and the sequence. An example of a SeqL input file is shown below.

```
+1 ACCCGT
+1 CCCGTA
-1 ACACAC
-1 ACATAC
```

Of course, the input file can contain sequences of arbitrary length. Also, an arbitrary number of sequences may be specified for each group.

3.3.2 Installation and Usage

When the input is properly formatted, one must download and install the SeqL classifier. The most recent version of the SeqL source code is available at:

https://github.com/heerme/seql-sequence-learner/archive/master.zip

Since SeqL is distributed as source code, one must compile (translate the source code to machine code) the program before using it. We shall not detail this process as it depends on the operating system and which tools are available.

When SeqL has been compiled and the input file has been prepared and saved as, for example, example.txt, one must first build a preliminary model as follows:

```
./seql_learn -r 0 example.txt example.model
```

This creates a file containing the preliminary model called example.model. To be able to classify sequences using this model, one must first create an optimized model as follows:

```
./seql_mkmodel -i example.model -o example.model.bin -O example.model.predictors
```

This creates two additional files:

- example.model.bin, which can be used for classification of new sequences.

- example.model.predictors, which contains the weighted motifs which divide the two groups and may be interesting in a biological context.

In a motif detection context the predictors file is therefore the most interesting file to look at. The file can be opened in a text editor. A shortened version of the contents of the predictors file for this example is shown below.

```
-0.0372289452055364206151111
0.110800089590139996809093    CG
0.0875246051909023825565725   G
0.0857454245816138133085005   CCG
0.0856232047335227092199617   GT
0.0719889869889090150412159   CGT
0.0719002330870201350476734   CCC
...
-0.0122787123594572476076303  ATAC
-0.0150784793067655267101435  ACAT
-0.0150784793067655267101435  CACAC
-0.0153275287278630562709525  AT
-0.0153275287278630562709525  CAC
-0.0250578076168992865124796  ACACA
...
```

The first line contains the bias of the model and is not relevant in this context. For each remaining line a motif is shown with its associated score. The score signifies how well the motif divides the two groups. In this example, CG is an important motif and indeed it only occurs in the positive class of the example data set. Equivalently, CACAC occurs only in the negative class and therefore has a good negative score.

Acknowledgments

We wish to thank Marco Palos Franco, Bioinformatics Research Centre (BiRC), Aarhus University, for collecting and curating the P-type ATPase dataset.

References

1. The UniProt Consortium (2008) The Universal Protein Resource (UniProt). Nucleic Acids Res 36:D190–D195
2. Sigrist CJA, Cerutti L, Hulo N, Gattiker A, Falquet L, Pagni M, Bairoch A, Bucher P (2002) PROSITE: a documented database using patterns and profiles as motif descriptors. Brief Bioinform 3(3):265–274
3. Saier MH Jr, Tran CV, Barabote RD (2006) TCDB: the transporter classification database for membrane transport protein analyses and information. Nucleic Acids Res 34:D181–D186

4. Pedersen BP, Ifrim G, Liboriussen P, Axelsen KB, Palmgren MG, Nissen P, Wiuf C, Pedersen CNS (2014) Large scale identification and categorization of protein sequences using structured logistic regression. PLoS One 9(1): e85139

5. Axelsen KB, Palmgren MG (1998) Evolution of substrate specificities in the P-type ATPase superfamily. J Mol Evol 46:84–101

6. Møller AB, Asp T, Holm PB, Palmgren MG (2008) Phylogenetic analysis of P5 P-type ATPases, a eukaryotic lineage of secretory pathway pumps. Mol Phylogenet Evol 46(2): 619–634

7. Ifrim G, Wiuf C (2011) Bounded coordinate-descent for biological sequence classification in high dimensional predictor space. Proceedings of the 17th ACM SIGKDD international conference on knowledge discovery and data mining, 708–716

8. Fagan MJ, Saier MH Jr (1994) P-type ATPases of eukaryotes and bacteria: sequence analyses and construction of phylogenetic trees. J Mol Evol 35:57–99

9. Camacho C, Coulouris G, Avagyan V, Ma N, Papadopoulos J, Bealer K, Madden TL (2009) BLAST+: architecture and applications. Bioinformatics 10:421

Chapter 42

Molecular Modeling of Fluorescent SERCA Biosensors

Bengt Svensson, Joseph M. Autry, and David D. Thomas

Abstract

Molecular modeling and simulation are useful tools in structural biology, allowing the formulation of functional hypotheses and interpretation of spectroscopy experiments. Here, we describe a method to construct in silico models of a fluorescent fusion protein construct, where a cyan fluorescent protein (CFP) is linked to the actuator domain of the Sarco/Endoplasmic Reticulum Ca^{2+}-ATPase (SERCA). This CFP-SERCA construct is a biosensor that can report on structural dynamics in the cytosolic headpiece of SERCA. Molecular modeling and FRET experiments allow us to generate new structural and mechanistic models that better describe the conformational landscape and regulation of SERCA. The methods described here can be applied to the creation of models for any fusion protein constructs and also describe the steps needed to simulate FRET results using molecular models.

Key words Molecular modeling, Fluorescent proteins, SERCA, Fluorescence resonance energy transfer, Biosensor

1 Introduction

Molecular modeling and simulations are useful tools for biophysical research. More than just creating pretty pictures for presentations and publications, modeling also assists greatly in the interpretation of experimental results. Fluorescence spectroscopy, including time-resolved anisotropy and FRET (Fluorescence Resonance Energy Transfer), is used in our laboratory to study structural dynamics of the proteins involved in Ca^{2+} transport and its regulation. The aim is to generate new structural and mechanistic models that better describe the conformational landscape and regulation of SERCA.

A biosensor is an analytical biomolecule that detects a molecular event such as ligand binding or protein-protein interaction. The biosensor field is developing rapidly, particularly in fluorescence methods, with new advances in fluorescence detection technologies in microplate instruments and biosensor engineering with genetically encoded probes such as green fluorescent protein

(GFP) and its mutant color derivatives. We have used fluorescent fusion proteins of SERCA for FRET measurements, which were applied to study SERCA ensemble conformation, single-molecule structural dynamics, regulatory subunit interactions, and discovery of enzyme modulators via high-throughput screening [1–6]. Ion-motive ATPases have been tagged with GFP for cellular localization studies, including mammalian, worm, paramecium, plant, and parasite cells [7–12].

Here we describe a procedure to generate models and calculate fluorescence parameters of a fusion protein construct made up by cyan fluorescent protein (**CFP**) linked to the actuator domain of SERCA. X-ray crystal structures suggest that two cytosolic domains of SERCA, the nucleotide-binding and actuator domains, move apart by 3 nm upon Ca^{2+} binding, i.e., undergoing a calcium-induced transition from closed to open conformation of the cytosolic headpiece. We begin with building a fluorescent fusion protein model using the calcium-free (E2·Tg) state of SERCA (PDB ID: **1IWO**) [13]. The procedure to create molecular models for any other SERCA state would be similar; one just needs to select a different SERCA PDB file as starting point. The generated molecular models for the CFP-SERCA fusion proteins were used to predict distances between the fluorophores and fluorophore orientation of CFP-SERCA labeled with fluorescein isothiocyanate (FITC) in the nucleotide-binding domain. These data were used to calculate in silico FRET efficiency between CFP donor on the A domain with FITC acceptor on the N domain, and this predicted FRET efficiency was compared with experimental values measured in ER membranes in physiological solution. Together, experimental and computational results indicate that the time-average conformation of the SERCA headpiece is partly open in both the presence and absence of Ca^{2+}, i.e., not fully open or completely closed as suggested by X-ray crystallography [1]. These results were confirmed by single-molecule fluorescence of a "2-color" SERCA biosensor, indicating that SERCA undergoes a transition from an open, dynamic conformation to a closed, ordered structure and samples several discrete structural sub-states that are Ca^{2+} dependent [3]. The 2-color SERCA biosensor has proved to be a useful tool in identifying SERCA activators and inhibitors in high-throughput drug screens [6].

2 Materials

Molecular modeling, simulation, and molecular graphics software:

2.1 Discovery Studio Visualizer

Accelrys Discovery Studio® is a software suite of life science molecular design solutions for computational chemists and computational biologists. It is a molecular modeling environment for both small molecule and macromolecule applications, targeted mostly towards the needs in drug discovery and pharmaceutical industry. DS Visualizer is a free version of the graphical interface to Discovery Studio, with some features missing compared with the commercial product. For example, most simulation tools have been removed. DS Visualizer is easy to use compared to most molecular graphics software packages and it has useful model building tools for working with small molecules and macromolecules. DS Visualizer is quite well documented and has several useful tutorials, tours, and sample files (available from the program Welcome screen) that will allow the user to get acquainted with the workings of the software quickly. DS Visualizer is available as a free download at http://www.accelrys.com. The current version is 4.1, but the modeling procedure described here works with any version from 3.1 to 4.1. Both Windows and Linux versions of the software are available. Here we describe the use of the Windows version of DS Visualizer, and the Linux version is essentially identical.

2.2 FPMOD

Fusion Protein MODeller (FPMOD), developed in the laboratory of Kevin Truong, University of Toronto, is a modeling software package that is designed to generate an ensemble of models of fusion proteins in order to sample the conformational space [14, 15]. This is done by treating the protein domains as rigid bodies and rotating each of them around the flexible linker connecting them. The FPMOD software can be obtained from the software authors at http://apel.ibbme.utoronto.ca/apel/software. The Windows version includes a graphical user interface (GUI) to the individual programs, which are located in the "scripts" folder. The software can be run from the GUI or the separate programs can be run from the command line. Instructions of how to use the software package can be found at the same website and in a publication by Pham and Truong [16].

2.3 Cygwin

Cygwin is a Unix-like environment and command-line interface for Microsoft Windows. The software package provides a large collection of GNU and Open Source tools which provide functionality similar to a Linux distribution on Windows. The Cygwin package can be downloaded from http://cygwin.com/. These tools are optional but convenient to process the data files generated from the modeling.

3 Methods

The modeling procedure for the CFP-SERCA fusion protein involves the following steps:

1. Customize the DS Visualizer program interface by showing Toolbars.
2. Load the amino acid sequence for the CFP-SERCA construct.

3. Load the protein structures from the PDB database, CFP (**1RM9**) and SERCA (**1IWO**).

4. Align the construct and structure sequences. Identify residues that are missing or different.

5. Build a molecular model of CFP (**1RM9**) by mutating mismatched residues, adding missing residues at N-terminus, and C-terminus.

6. Create a linker at the N-terminus of SERCA (**1IWO**).

7. Join CFP (**1RM9**) to the N-terminus of SERCA (**1IWO**).

8. Generate conformers by rotating backbone torsional angles in the linker region.

9. Save the models and create images.

10. Generate an ensemble of molecular models using FPMOD.

11. Extract fluorophore center and dipole vector coordinate position.

12. Calculate FRET parameters.

3.1 Customize the Program Interface by Showing Toolbars

Start the DS Visualizer program by clicking on the program icon on the computer desktop or on the "Discovery Studio Client" entry in the Start Menu. At the top below the top frame we have the menus as in most Windows programs, below that are the Tools tabs (starting with Macromolecules, Simulation etc.). Below these are Toolbars with commonly used functions. The work area is below these and consists of the Tools area to the left and tabbed display panels to the right (Fig. 1).

The interface is very customizable and can be set up to suit your preferences. For example, the Tools panel can be unpinned so that it is hidden, giving more space to the work area. The window can be undocked from the work area as well. There are help documents

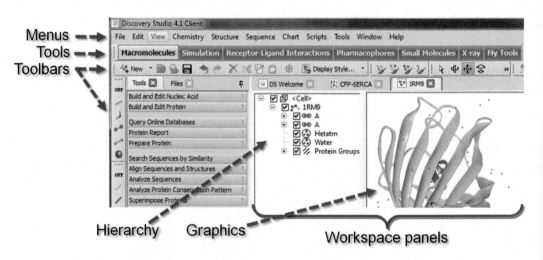

Fig. 1 Overview of the Discovery Studio Visualizer program interface

available on the "DS Welcome" screen or the "Help" menu that describe these features in more detail.

We want to add a few more Toolbars that have functions that we will use often. The easiest way to do that is to go to the "View" menu and go down to the "Toolbars" and when the next level pops up click on "Atom Display" if it does not have a check mark already. The new toolbar will now show above the Tools. The newly added "Toolbars" can be moved to any location. One useful choice is to move them to the left edge of the program window. Drag the toolbar to the left edge and release. Repeat these steps to also show the "ProteinStructure Display", "Sketching", and "Chemistry" toolbars. At this point it is recommended that some time is spent getting used to the program interface. The "Quick Start" tutorials, which can be found on the "DS Welcome" screen, are a good start for anyone unfamiliar with this software.

3.2 Load the Amino Acid Sequence for the CFP-SERCA Construct

To start the modeling procedure, first we want to load the protein sequence for the CFP-SERCA construct. Go to the "File" menu, and to "New" and to "Protein Sequence Window". This opens a new tab in the work area, with a blank sequence section. Copy the sequence data from Table 1 into the sequence section. Save the sequence data by going to the "File" menu and select "Save As…". Save the file as "CFP-SERCA.fa" in an appropriate location on your computer.

Alternatively, the sequence data can be downloaded from our website. Go to the "File" menu, and to "Open URL". In the top text box, type "http://www.msi.umn.edu/~bsven/CFP-SERCA.fa". Click the Open button. A new tab in the work area should appear with the protein sequence.

3.3 Load the Protein Structures from the PDB Database

Now we want to load the CFP and SERCA coordinate files from the Protein Data Bank [17]. We choose the **1RM9** crystal structure [18] as that one has the highest sequence similarity to the CFP protein of our fusion protein construct. Go to the "File" menu and to "Open URL". Type "1RM9", in the ID box and click the Open button. A new tab titled **1RM9** should have opened in the work area, where the CFP protein is shown in ribbon representation. Repeat this for the **1IWO** PDB ID, which is the atomic coordinate file for the structure of SERCA in the E2·Tg state, in which the inhibitor thapsigargin (Tg) locks SERCA in a calcium-free state with closed cytosolic headpiece [13]. Switch back to the **1RM9** tab by clicking on it. On the right side a ribbon representation of the proteins should be visible. On the left side of the work area the "Hierarchy" panel, this shows an overview of what is present in the PDB file (Fig. 1). If the "Hierarchy" panel is not shown already, use the "View" menu to toggle it on. The **1RM9** structure has waters in a second "**A**" chain. We want to delete all atoms except the first protein chain. Select the second "**A**" chain in the "Hierarchy" panel by clicking on it. It should be

Table 1
Sequence for the CFP-SERCA fusion protein construct

VSKGEELFTGVVPILVELDGDVNGHKFSVSGEGEGDATYGKLTLKFICTTGKLPVPWPTL
VTTLTWGVQCFSRYPDHMKQHDFFKSAMPEGYVQERTIFFKDDGNYKTRAEVKFEGDTLV
NRIELKGIDFKEDGNILGHKLEYNYISHNVYITADKQKNGIKANFKIRHNIEDGSVQLAD
HYQQNTPIGDGPVLLPDNHYLSTQSKLSKDPNEKRDHMVLLEFVTAAGITLGMDELYK*GE*
*L*MEAAHSKSTEECLAYFGVSETTGLTPDQVKRHLEKYGHNELPAEEGKSLWELVIEQFED
LLVRILLLAACISFVLAWFEEGEETITAFVEPFVILLILIANAIVGVWQERNAENAIEAL
KEYEPEMGKVYRADRKSVQRIKARDIVPGDIVEVAVGDKVPADIRILSIKSTTLRVDQSI
LTGESVSVIKHTEPVPDPRAVNQDKKNMLFSGTNIAAGKALGIVATTGVSTEIGKIRDQM
AATEQDKTPLQQKLDEFGEQLSKVISLICVAVWLINIGHFNDPVHGGSWIRGAIYYFKIA
VALAVAAIPEGLPAVITTCLALGTRRMAKKNAIVRSLPSVETLGCTSVICSDKTGTLTTN
QMSVCKMFIIDKVDGDFCSLNEFSITGSTYAPEGEVLKNDKPIRSGQFDGLVELATICAL
CNDSSLDFNETKGVYEKVGEATETALTTLVEKMNVFNTEVRNLSKVERANACNSVIRQLM
KKEFTLEFSRDRKSMSVYCSPAKSSRAAVGNKMFVKGAPEGVIDRCNYVRVGTTRVPMTG
PVKEKILSVIKEWGTGRDTLRCLALATRDTPPKREEMVLDDSSRFMEYETDLTFVGVVGM
LDPPRKEVMGSIQLCRDAGIRVIMITGDNKGTAIAICRRIGIFGENEEVADRAYTGREFD
DLPLAEQREACRRACCFARVEPSHKSKIVEYLQSYDEITAMTGDGVNDAPALKKAEIGIA
MGSGTAVAKTASEMVLADDNFSTIVAAVEEGRAIYNNMKQFIRYLISSNVGEVVCIFLTA
ALGLPEALIPVQLLWVNLVTDGLPATALGFNPPDLDIMDRPPRSPKEPLISGWLFFRYMA
IGGYVGAATVGAAAWWFMYAEDGPGVTYHQLTHFMQCTEDHPHFEGLDCEIFEAPEPMTM
ALSVLVTIEMCNALNSLSENQSLMRMPPWVNIWLLGSICLSMSLHFLILYVDPLPMIFKL
KALDLTQWLMVLKISLPVIGLDEILKFIARNYLEG

The CFP part in the sequence, residues 1 through 238, is underlined. The three residues, 239–241 added by sub-cloning, are shown in italics. The SERCA sequence is residues 242–1335

highlighted yellow when selected. Press the Delete key to delete the waters.

Click on the **1IWO** tab to bring that panel to the front. The **1IWO** structure has two SERCA protein molecules in it. We want to delete the second one, i.e., chain "**B**". There is also an inhibitor, thapsigargin, which is bound to the protein in the structure. We want to delete this molecule as well. Select and delete the second "**A**" and "**B**" chains.

3.4 Align the Construct and Structure Sequences

Next we want to align the sequences from the protein structures with the amino acid sequence for the fusion protein construct. The protein structure in PDB needs to be updated for the amino acid sequence of the expressed protein to fill in gaps in the crystal structure of CFP and to match the cDNA variant of CFP.

Click on the **1RM9** tab to bring the panel to the front. Go to the "Sequence" menu and click on "Show Sequence". A new tab with the **1RM9** sequence should be shown. Select the sequence by clicking on the text **1RM9** in the left panel. The sequence should be highlighted black. Copy the sequence by right clicking in the sequence in the right panel and selecting "Copy" from the pop-up menu.

Bring the CFP-SERCA sequence window to front by clicking on that tab. Right click in the right panel and select "Paste" to paste the **1RM9** sequence in with the **1RM9-1IWO** sequence. We can now close the **1RM9** sequence tab.

In the CFP-SERCA tab we will now manually align the sequences by adding two spaces in the beginning of the **1RM9** sequence. Click before the first residue in the sequence to place the cursor at this location. Press the space bar twice to insert spaces. The beginning part of the sequence alignment now shows in a darker color, which means the sequences match.

At position 65 we notice that the sequences are not aligned any more. That is because the CFP fluorophore is made up by residues Thr65, Trp66, and Gly67. The fluorophore is represented in the sequence by one X character. To align the rest of the sequence, click after the X at position 65 to place the cursor there and press the space bar twice to insert two blanks.

Next we want to make sure that when we select a residue in the sequence panel that the residue gets selected in the structure. To do this, go to the "Sequence" menu and select "Link Sequence and Structure". Make sure "**1RM9**" is selected in the top pull down widget and then click Link. Click and drag the sequence tab "CFP-SERCA" to the bottom half of the work area so that we can see the sequence data and the structures at the same time.

3.5 Build a Molecular Model of CFP

We want to change the **1RM9** structure to match the fusion protein construct sequence. Make sure that the "Macromolecule" tools are selected. Pin the tool panel open if needed. Click on "Build and Edit Protein" to show that tool's controls. Set the "Build Action" to "Mutate". Currently only the ribbon representation of the protein is shown. To be able to view and select specific atoms, we want to display all atoms of the protein with the "Line" display style. Click on the "A" chain in the "Hierarchy" panel to select the whole protein then click on the "Line" representation in the "Atom Display" tool panel.

Residue 57 is a fluorinated Trp in the crystal structure and we want to change it to a regular Trp. In the sequence window for the "**1RM9**" sequence, select the X at position 57. The X represents an unknown or non-standard amino acid in the sequence. Click on the "**1RM9**" tab to activate the structure panel again. Alternatively, the residue to be mutated can be selected in the "Hierarchy" panel instead of in the sequence window. We want to display the atoms of this selected residue differently from the rest of the protein. Click on the "Stick" display style in the "Atom Display" toolbar. Click on the "Center Structure" button in the "View" toolbar to display the residue in the center of the panel. Under "Choose Amino Acid" in "Build and Edit Protein" click on "Trp". The fluorinated Trp is now a regular Trp.

Next we will mutate residue 80 from an Arg to a Gln. This mutation is a natural variation that may vary depending on cDNA clone. Repeat the steps as was done for Trp57 to select and display residue 80. Click on "Gln" in the "Build and Edit Protein" tool to mutate the Arg to a Gln.

We can see that the conformation is not optimal because the Gln side chain is too close to other atoms. To adjust its conformation we will use the "Search Side-Chain Rotamers" tool. Make sure residue 80 is selected. In the list of "Macromolecule" tools click on "Search Side-Chain Rotamers" to open the rotamer tools. While residue 80 is selected click on "Set Active Residue" to get the rotamer library for a Gln. Click on rotamer #1 and we should see that the residue side chain changes its position. The steric clashes have been removed, so the side chain structure is now optimized.

The next residue to mutate is Ala206 which should be Lys. This mutation A206K prevents CFP and YFP from self-assembly into dimers and tetramers and is available in most commercially available cDNA clones, thereby preventing aberrant FRET induced by CFP-YFP oligomerization. Repeat the process of residue mutation and side chain rotamer optimization as was done for Arg80. In this case try rotamer #2, which produces a valid conformation with no steric clashes.

We also want to add the two missing residues to the N-terminus. Select the first residue, Lys3, in the "Hierarchy" panel. Click on "Build and Edit Protein". Set the "Build Action" to "Insert". Set "Conformation" to "Extended". Click on "Ser" under "Choose Amino Acid". We have now added a Ser to the N-terminus. Select the Ser2 residue and repeat the process to add Val1. Note that GFP and its mutant color derivatives contain a signal sequence that is cleaved co-translationally, and that the convention for sequence numbering is to start with +1 at remaining N-terminal residue, which is Val1.

We want to optimize the geometry for these two residues by performing an energy minimization. Multiple residue selections can be made by holding down the Ctrl key while clicking a residue. Select both Val1 and Ser2 then click on the "Clean Geometry" button on the "Chemistry" toolbar. We should see the atoms move. Click the "Clean Geometry" button again a few times until there is not much change anymore. We can go back to the sequence window and remove the extra spaces in the beginning of the **1RM9** sequence to make the sequences aligned again.

Finally we want to add 12 residues on the C-terminus of CFP. Nine residues at the C-terminus are not resolved in the crystal structure and an additional three residues that resulted from the subcloning of the two cDNA molecules needs to be added to the CFP structure. Select the last residue in the **1RM9** structure, i.e., Ile229. Center the structure. Click on "Build and Edit Protein". Set the "Build Action" to "Create/Grow Chain". Set "Conformation" to "Extended". Click on "Thr" under "Choose Amino Acid" to add it to the end.

Select the new last residue, i.e., Thr230. Repeat these steps to add the remainder of the sequence "Leu-Gly-Met-Asp-Glu-Leu-Tyr-Lys-Gly-Glu-Leu" to the end of **1RM9**. If you make a mistake and add the wrong residue you can use the Mutate function to correct it.

We have now built the correct CFP and linker regions for the CFP-SERCA construct At this point we should save the modified **1RM9** coordinates. Go to the "File" menu, select "Save As". Set the file type to "Protein Databank Files". Name the file "**1RM9+Linker.pdb**". Select an appropriate location to save and click the Save button.

3.6 Add an N-Terminal Linker to SERCA

Next we need to prepare the **1IWO** structure so that we can attach the **1RM9+Linker** to the N-terminus of SERCA. We could align the sequences as we did for **1RM9** and the fusion protein, but this is not necessary in this case as the sequence does match exactly. We need to add a couple of residues to the N-terminus so that we have an overlapping region between the two proteins. We will add the last three residues, Gly-Glu-Leu, of the linker region to the beginning of the **1IWO** structure.

Display the **1IWO** structure with "Lines" only. Select residue Met1 and center the structure. Click on "Build and Edit Protein". Set the "Build Action" to "Insert". Set "Conformation" to "Extended". Click on "Leu" under "Choose Amino Acid". We have now added a Leu to the N-terminus but we have a problem. The residue extends into the protein. We need make some manual adjustments to turn the N-terminus towards the surface of the protein.

We can adjust the backbone dihedral angles to fix this. Click on the "Torsion" button in the "Sketch" toolbar. With this tool we can adjust the backbone dihedral with the mouse. Click on the bond between the C and CA atoms of residue Glu2. With the left mouse button pressed down slide the mouse left to right to adjust the dihedral. Set this value to about 50°. Next click on the Glu2 CA to N bond and adjust the dihedral to about −160°. Next click on the Met1 C to CA bond and adjust the dihedral to about −145°. The N-terminus should now be facing away from the protein.

Select residues Leu0, Met1, and Glu2. By holding down the Ctrl key while selecting one can add to or remove from a selection.

Click on the "Clean Geometry" button to optimize the geometry for these residues.

We can now continue adding a Glu and Gly to the N-terminus using the "Build and Edit Protein" tools in the same way as was done for Leu0. We can now save the modified 1IWO structure. A suggested filename would be "**1IWO+Linker.pdb**".

3.7 Join CFP to SERCA

Now we will fuse the two structures. Bring the **1RM9+Linker** tab to front and then select the whole chain "**A**". Right click on it and select "Copy". Bring the **1IWO+Linker** tab to front. Right click in molecule panel and select "Paste" to copy the **1RM9+Linker** atoms into the same window.

We now want to superimpose the overlapping regions of the linkers. In the "Hierarchy" panel select the first three residues of **1IWO** and select the last three residues of **1RM9+Linker**. Click on the "Superimpose Protein" in the "Macromolecules" tools to open that tools controls. Select "Tethers" in the "Superimpose by" pull-down list, then click on "Create Protein Tether". If this worked we will see lines from one protein to the other in the molecule panel. Click on "Superimpose", which can be found just below "Create Protein Tether". The overlapping sequences are now superimposed.

Center the view on the three consecutive residues Gly-2, Glu-1, and Leu0 of **1IWO**. Now we want to delete the extra residues. Select residues Gly-2 and Glu-1 of **1IWO** and press the delete key. Select Leu241 of **1RM9** and press the delete key.

To join the two proteins together we want to create a bond between the C atom of residue Glu240 of **1RM9** and the N atom of Leu0 of **1IWO**. Select only these two atoms in the graphics window by first clicking on one atom then holding down the shift key while clicking on the second atom. The atoms can also be selected in the "Hierarchy" panel by clicking one atom then clicking the second while holding down the Ctrl key. Once the two atoms are selected, click on the "Single Bond" button in the "Chemistry" toolbar.

Some tidying up of the chain assignment and residue numbers is recommended. There are now two "**A**" chains as seen in the "Hierarchy" panel. We can fuse these two chains into one by selecting the second "**A**" chain and dragging and dropping it over the first "**A**" chain in the "Hierarchy" panel.

The sequence numbering now goes from 1 to 240 for the **CFP+Linker** part and the last residue of the linker is Leu0 after which SERCA starts at Met1. To be able to do any further simulations on this model we need to correct the residue numbering of the CFP-SERCA construct so that the residue numbers are sequential. Select the residue Leu0 to Gly994 in the "Hierarchy" panel, and then open the "Prepare Protein" tool under the "Macromolecules" tools. Click on "Renumber Sequences…" and select "Renumber residue range from Leu0 to Gly994". Make sure that the "Starting from residue ID" is set to 241 and click OK. The model for the CFP-SERCA fusion protein construct is now correctly built. Save it as "**1RM9+1IWO.pdb**".

3.8 Generate Linker Conformations

This model for the fusion protein with the linker region completely straight does not look very realistic (Fig. 2). In reality the linker region would likely have a much more random conformation. We can apply a few more modeling procedures in DS Visualizer that will generate more realistic conformations of the linker region. The "Build and Edit Protein" has a function to change the backbone conformation.

Open up that tool again and set "Conformation" to "Right-hand Alpha Helix". For example, select residues Thr230 through Leu241. Then click on "Apply Conformation". We can randomize it a little bit more by selecting one or two residues in the region between Thr230 and Leu241 and changing the backbone conformation to any other secondary structure setting.

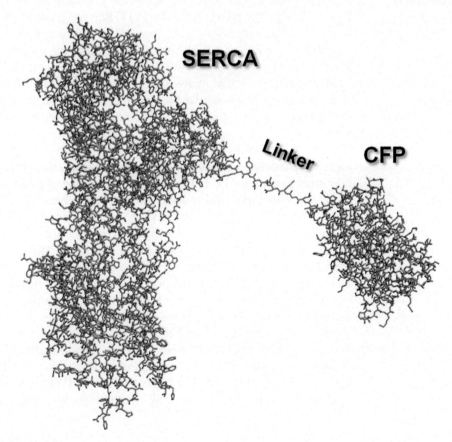

Fig. 2 Initial complete model for the CFP-SERCA fusion protein construct. Note the straight extended conformation of the linker region

Another way of generating conformations of the linker region is by using the "Torsion" tool in the "Sketching" toolbar to rotate dihedrals in the linker. Be careful to only rotate around N-CA or CA-C bonds.

If the CFP clashes with SERCA, i.e., has atom overlaps, one needs to try different dihedral values in the linker. When a conformation has been found that does not have any clashes, then we can select the whole linker region Thr230 trough Leu241 and use the "Clean Geometry" tool to optimize the structure of the selected region. This minimizes the energy by allowing atoms in both backbone and side chains to move. The linker region should now have a reasonable structure.

The model is now finished and we should save it. A suggested filename would be "**1RM9+1IWO_conf1.pdb**". Generate a few more random conformations for the linker region and save these as well. The model is good enough to use as starting point for simulations or to show in a figure for a presentation.

3.9 Making Publication-Quality Figures

DS Visualizer has the ability to save the molecular graphics view as an image and also the ability to export the graphics view as a POV-Ray scene for ray-tracing with the POV-Ray ray-tracing software, http://www.povray.org/. The molecular graphics view in DS Visualizer is quite good, although other software packages like VMD [19], http://www.ks.uiuc.edu/Research/vmd/, or PyMOL, https://www.pymol.org/, can produce somewhat nicer looking images. PDB format files created by DS Visualizer can be opened and visualized in either of these other molecular graphics software.

Choose a useful view and display it with an appropriate graphics representation, such as ribbons to better show the global protein structure (Fig. 3). Different sections of the protein can be displayed with different colors as shown here where the SERCA domains are shown in red, green, blue, and grey, and the CFP part of the construct is shown in cyan. When you are satisfied with how the display looks go to the "File" menu and select "Save As". Set the "Files of type" to "Image Files", and save the file as "CFP-SERCA.png". A requester will pop up asking what resolution the image should be saved at. Keep the current resolution or select a higher resolution for publication quality images.

Previously, we have saved the created models as PDB format files, but if we want to save both the view setting and the loaded molecule data we should save the data using the "Discovery Studio Files (*.dsv)" format. Save the file as "**1RM9+1IWO.dsv**". When loading the dsv file at a later time all view and display setting should be preserved.

3.10 Generate an Ensemble of Molecular Models

It is assumed that CFP can adopt many possible conformations around SERCA. To simulate the multitude of possible conformations that CFP can adopt with respect to SERCA in solution, we

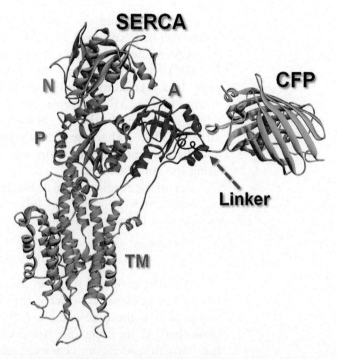

Fig. 3 A ribbon representation of the CFP-SERCA fusion protein construct is shown as an example of a figure suitable for structural analysis. SERCA is shown with the actuator domains in *red*, the nucleotide domain in *green*, the phosphorylation domain in *blue*, and the transmembrane domain in *grey*. The CFP part of the fusion protein is shown in *cyan*

need to create an ensemble of structural models. One way to do that is using the software Fusion Protein MODeller (FPMOD) [14, 15]. In short, SERCA and CFP structures are treated as rigid bodies joined by a flexible linker region that is allowed to rotate freely. The nine C-terminal residues of CFP, which are unresolved and assumed to be flexible in the crystal structure, and the residues of the Gly-Glu-Leu sequence at the N-terminus of SERCA are assigned to be part of this flexible linker. An ensemble of more than 1000 models needs to be generated to sample the conformational space of CFP around SERCA.

The easiest way to generate an ensemble of models using FPMOD is to run the software from the command line. Download the FPMOD software package from http://apel.ibbme.utoronto.ca/apel/software/FPModGUI_v2.rar. Unpack the archive to a location of your choice. Copy the file "conformation_sample_finals.exe" from the "scripts" folder to the location where you saved the PDB format coordinate file for the CFP-SERCA fusion construct. Open a "Command Prompt" from the Windows Start Menu. Change folder to the location where the coordinate files were saved by using the "cd" command and the folder path to where you have saved the PDB files.

Next we need to make a change to the PDB file that will allow FPMOD to read all atoms. Open the file "RM9+1IWO_conf1.pdb" in a text editor, for example Notepad. Search for the text string "HETATM" and replace that with "ATOM" at all placed where it is found in the file. Make sure that there are two spaces after "ATOM", otherwise the columns in the file would get shifted and the file cannot be read correctly. Save the file. It can be saved using the same filename.

To run the conformational sampling and generate an ensemble of models type or copy the following command line into the "Command Prompt" window:

```
"conformation_sample_finals.exe 1RM9+1IWO_conf1.pdb 1RM9+
   1IWO_conf1_ 1000 1 A N AGITLGMDELYKGELMEA TLGMDELYKGEL"
```

After the program name in the command line above comes first the input filename, which is the file we created in the previous steps. Next is the base name for the output filename, which we chose here as the input file name without the ".pdb" extension but with an underscore character added to the end. The output file will be saved with a model number and the extension ".pdb" at the end. After that we have the number of models to generate, in this case it is "1000". To get adequate sampling of the fusion protein conformation at least 1000 models needs to be generated. The next value on the command line is the number of linkers in our construct; in this case that is "1". Next is the chain ID, which is "A" in our construct. After that the command line lists either "N" or "C" depending on whether the fluorescent protein is attached to the N-terminal or C-terminal end of our protein. For our construct CFP is attached at the N-terminus, therefore an "N" is put in the command line. At the end of the command line there are two stretches of the sequence around the flexible linker in our construct. The first sequence has a few extra residues at each end compared to the second sequence. The second sequence comprises the residues that we select to be part of a flexible linker. The flexible region is assumed to be starting after the end of the last β-strand of CFP and end at the residue before the start of the SERCA sequence. This is the same sequence stretch of 12 residues that we modeled in at the C-terminus of CFP. The generation of thousands of models could take up to a couple of hours to complete.

A suggestion to reduce the time in half for generating 1000 models is to start multiple instances of the "conformation_sample_finals.exe" program in separate "Command Prompt" windows. For example start a first run of "conformation_sample_finals.exe" generating 500 models and then start another instance of the program also generating 500 models. A different starting conforma-

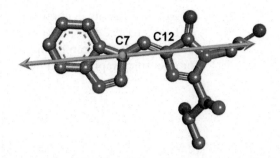

Fig. 4 The CFP fluorophore generated from the sequence Thr65-Trp66-Gly67. The vector centered on the C7 atom and going through the C12 atom represents the excitation and emission transition dipole (*green arrow*)

tion is needed for every run; otherwise the same conformations will be generated. The starting conformation can either be another manually generated conformation or it could also be any of the models created by the first run of the program. Remember to give each output file a different base name so that you will not overwrite the previous data.

3.11 Extracting Fluorophore Center and Dipole Vector Coordinate Positions

For each structure in the ensemble, we can calculate both the distance between the centers of the two fluorophores (R) and the Förster distance (R_0), which is dependent on the relative orientation of the fluorophores. One property that needs to be known is the transition dipole moment for the fluorophore. Transition dipole moments for fluorescent proteins have been determined experimentally [20] and from excited state quantum chemistry calculations [21]. The transition dipole can be approximated quite accurately in the CFP fluorophore by setting the fluorophore center at the C7 atom and the dipole in the direction of the C12 atom (Fig. 4).

We want to extract the X, Y, and Z coordinate values for these atoms from all 1000 pdb files generated in the previous step for further calculations of FRET parameters. One could open every file in a text editor and copy and paste the data, but that would be very tedious work. A better option is to write a program or script to do that and to also calculate the FRET parameters described below. Here we show a simple way to extract the data from the pdb files using Unix command line tools. To get access to these tools one would need a computer running Linux or one could install the Cygwin Unix-like environment for Windows.

The commands below will extract the X, Y, and Z coordinates from all files in the folder that match the "1RM9+1IWO_conf1_*.pdb" pattern, which should be only the models generated by

FPMOD. The final command merges all these temporary files, "x1.txt" to "z2.txt", into a new file "xyz12.txt", which is a Tab delimited file that can be opened in Excel or any other spreadsheet or scientific graphing and data analysis software.

```
grep -h 'C7   4F3  A  66' 1RM9+1IWO_conf1_*.pdb | cut -c
   31-38 > x1.txt
grep -h 'C7   4F3  A  66' 1RM9+1IWO_conf1_*.pdb | cut -c
   39-46 > y1.txt
grep -h 'C7   4F3  A  66' 1RM9+1IWO_conf1_*.pdb | cut -c
   47-54 > z1.txt
grep -h 'C12  4F3  A  66' 1RM9+1IWO_conf1_*.pdb | cut -c
   31-38 > x 2.txt
grep -h 'C12  4F3  A  66' 1RM9+1IWO_conf1_*.pdb | cut -c
   39-46 > y2.txt
grep -h 'C12  4F3  A  66' 1RM9+1IWO_conf1_*.pdb | cut -c
   47-54 > z2.txt
paste x1.txt y1.txt z1.txt x 2.txt y2.txt z2.txt >
   xyz12.txt
```

The first three columns are the X, Y, and Z coordinates for the FRET donor fluorophore center, and the next three columns are X, Y, and Z coordinates representing the FRET donor transition dipole direction.

3.12 Calculating FRET Parameters

Molecular models of fusion protein constructs including both fluorescence donor and acceptor probes can be used to calculate expected FRET efficiency values. Here we have focused on modeling CFP as the donor fluorophore attached to the actuator domain at the N-terminus of SERCA. Similar modeling strategies can be used to add acceptor labeling for intramolecular or intermolecular FRET.

For example, acceptor labeling strategies for CFP-SERCA include intramolecular FRET studies using FITC, a small fluorescent probe that binds in the nucleotide binding pocket of SERCA [1], and for intermolecular FRET using YFP-tagged regulatory subunits phospholamban and sarcolipin [4, 5, 22, 23]. Another example is "2-color-SERCA", that was developed in the laboratories of Seth Robia, Loyola University Chicago, and David Thomas, University of Minnesota, in which GFP is fused to an interior loop of the nucleotide-binding domain and red fluorescent protein (RFP) is fused to the N-terminus in the actuator domain. This GFP-RFP-SERCA construct has been utilized for time-resolved detection of GFP-RFP intramolecular FRET in single-molecule microscopy and high-throughput drug discovery [2, 3, 6].

For the rest of this protocol, we will use a docked conformation of FITC by Winters et al. [1] as the FRET acceptor. The X, Y, and Z coordinates for the fluorophore center position and the

Table 2
Coordinates for FITC docked in the nucleotide pocket of the calcium-free (E2·Tg) structure of SERCA (PDB ID: 1IWO) [13]

	X	Y	Z
FITC center coordinates	−12.692	−18.062	58.889
FITC dipole vector	−13.679	−18.695	58.734

transition dipole vector are shown in Table 2. Copy these values to the spreadsheet. In the spreadsheet we can now calculate the FRET donor to acceptor distance, R, using

$$R = \sqrt{[(x_A - x_D)^2 + (y_A - y_D)^2 + (z_A - z_D)^2]} \tag{1}$$

where, x_D, y_D, and z_D are the donor center fluorophore coordinates and x_A, y_A, and z_A are the acceptor center fluorophore coordinates.

The fluorophore coordinate data can be used to calculate the Förster distance R_0. R_0 is dependent on the orientation factor κ^2, which can be calculated explicitly from the observed probe orientations using

$$\kappa^2 = (\sin\theta_D \sin\theta_A \cos\phi + 2\cos\theta_D \cos\theta_A)^2 \tag{2}$$

θ_D and θ_A are the angles between the inter-probe vector and the donor and acceptor transition moments, respectively, and ϕ is the dihedral angle between the two planes defined by the donor transition moments and the line connecting the centers of the donor and acceptor fluorophores (DR) and the acceptor transition moments and the line connecting the centers of the donor and acceptor fluorophores (AR) [24]. The θ_D and θ_A angles can be calculated from the coordinates using

$$\theta = \text{acos}\,[(A \bullet B)/|A||B|] \tag{3}$$

where A represents the fluorophore transition dipole vector and B represents the fluorophore–fluorophore vector. ϕ can be calculated using

$$\phi = \text{acos}\,([(A_2 - A_1) \times (A_3 - A_1)] \bullet [(B_2 - B_1) \times (B_3 - B_1)] / \\ |(A_2 - A_1) \times (A_3 - A_1)||(B_2 - B_1) \times (B_3 - B_1)|) \tag{4}$$

where A_1, A_2, and A_3 are three coordinate positions that describe the DR plane and B_1, B_2, and B_3 are three coordinate positions that describe the AR plane. Functions to calculate the angles and dihedrals can be found on the web for most spreadsheet and data analysis software.

A κ^2 dependent R_0, $R_0(\kappa)$, can thus be calculated for each model from the simulations using

$$R_0(\kappa) = R_0(2/3) \times (3\kappa/2)^{1/3} \tag{5}$$

where $R_0(2/3)$ is the published R_0 value for the fluorescent probe pairs assuming random orientation of the probes. This value is 55 Å for in case of the CFP-FITC FRET pair [1]. The averaging regime should also be considered for FRET analysis [25, 26]. In the case of fluorescent fusion proteins the static averaging regime is appropriate as the rotational motion is slow for fluorescent proteins in relation to their fluorescence lifetime, which eliminates the need to average κ^2 over time before calculating the FRET efficiency. We can now calculate not only the predicted distribution of distances between the fluorophores (R), but also the predicted values of measured energy transfer efficiency (E) from an ensemble of conformation of fluorophores, using

$$E = 1 / [1 + (R / R_0(\kappa))^6] \tag{6}$$

Next we want to calculate the probability distribution for the distances between the fluorophores (R) and energy transfer efficiency (E) and make plots using graphing software of your choice. Origin, http://originlab.com/, was used to calculate the probability distribution of the data and to make the plots in Fig. 5.

The distance distribution, based on modeling and simulations, is shown to be quite broad. The full-width half-maximum (FWHM) of the distribution is approximately 30 Å (Fig. 5a). The orientation factor κ^2 is 0.6, which is slightly lower than what is predicted from a completely freely moving and isotropic case where κ^2 is $2/3$ (Fig. 5b). The orientation corrected $R_0(\kappa)$ also shows a broad distribution with an average of 51 Å, which is lower that the experimentally determined R_0 of 55 Å for the isotropic case (Fig. 5c). The simulated FRET efficiency distribution show a large population that show low FRET, less than 0.1. However, there are a significant number of conformations that show quite high FRET. The average is 0.290, which is close to the experimentally determined value of 0.336 for CFP-FITC-SERCA [1]. Winters et al. describe more details of this type of FRET analysis as used in the study of the SERCA cytosolic headpiece structural dynamics [1].

In conclusion, we have presented a method for modeling, conformational sampling simulations, and FRET parameter analysis of a fluorescent fusion protein construct of CFP and SERCA. The methods described would be generally applicable for the modeling and simulation of any other fusion protein construct independent of the proteins involved. We hope that this modeling method will be useful for the research community.

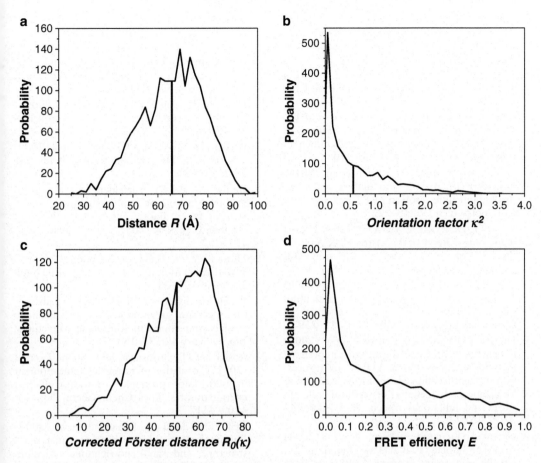

Fig. 5 Simulated FRET parameters based on molecular modeling of CFP-SERCA. Distributions were calculated for (**a**) distance, (**b**) κ^2, (**c**) κ^2-corrected R_0, and (**d**) FRET efficiency. A *vertical line* indicates the mean value for each calculated parameter. The mean value of E is comparable to the experimentally measured steady state FRET

Acknowledgement

This work was supported by NIH grants GM27906 and AR007612. Computational resources were provided by Minnesota Supercomputing Institute. This method chapter was previously presented as a hands-on tutorial at a regional Biophysical Society Networking Symposium at St. Olaf College, Northfield, MN.

References

1. Winters DL, Autry JM, Svensson B et al (2008) Interdomain fluorescence resonance energy transfer in SERCA probed by cyan-fluorescent protein fused to the actuator domain. Biochemistry 47:4246–4256

2. Hou Z, Hu Z, Blackwell DJ et al (2012) 2-Color calcium pump reveals closure of the cytoplasmic headpiece with calcium binding. PLoS One 7:e40369

3. Pallikkuth S, Blackwell DJ, Hu Z et al (2013) Phosphorylated phospholamban stabilizes a compact conformation of the cardiac calcium-ATPase. Biophys J 105:1812–1821

4. Autry JM, Rubin JE, Pietrini SD et al (2011) Oligomeric interactions of sarcolipin and the Ca-ATPase. J Biol Chem 286:31697–31706

5. Gruber SJ, Haydon S, Thomas DD (2012) Phospholamban mutants compete with wild type for SERCA binding in living cells. Biochem Biophys Res Commun 420:236–240

6. Gruber SJ, Cornea RL, Li J et al (2014) Discovery of enzyme modulators via high-throughput time-resolved FRET in living cells. J Biomol Screen 19:215–222

7. Vagin O, Denevich S, Sachs G (2003) Plasma membrane delivery of the gastric H, K-ATPase: the role of beta-subunit glycosylation. Am J Physiol Cell Physiol 285:C968–C976

8. Cho JH, Bandyopadhyay J, Lee J et al (2000) Two isoforms of sarco/endoplasmic reticulum calcium ATPase (SERCA) are essential in Caenorhabditis elegans. Gene 261:211–219

9. Hauser K, Pavlovic N, Klauke N et al (2000) Green fluorescent protein-tagged sarco(endo) plasmic reticulum Ca2+-ATPase overexpression in Paramecium cells: isoforms, subcellular localization, biogenesis of cortical calcium stores and functional aspects. Mol Microbiol 37:773–787

10. Gravot A, Lieutaud A, Verret F et al (2004) AtHMA3, a plant P1B-ATPase, functions as a Cd/Pb transporter in yeast. FEBS Lett 561:22–28

11. Lefebvre B, Batoko H, Duby G et al (2004) Targeting of a Nicotiana plumbaginifolia H+-ATPase to the plasma membrane is not by default and requires cytosolic structural determinants. Plant Cell 16:1772–1789

12. Furuya T, Okura M, Ruiz FA et al (2001) TcSCA complements yeast mutants defective in Ca2+ pumps and encodes a Ca2+-ATPase that localizes to the endoplasmic reticulum of Trypanosoma cruzi. J Biol Chem 276:32437–32445

13. Toyoshima C, Nomura H (2002) Structural changes in the calcium pump accompanying the dissociation of calcium. Nature 418:605–611

14. Pham E, Chiang J, Li I et al (2007) A computational tool for designing FRET protein biosensors by rigid-body sampling of their conformational space. Structure 15:515–523

15. Chiang J, Li I, Pham E et al (2006) FPMOD: a modeling tool for sampling the conformational space of fusion proteins. Conf Proc IEEE Eng Med Biol Soc 1:4111–4114

16. Pham E, Truong K (2010) Design of fluorescent fusion protein probes. Methods Mol Biol 591:69–91

17. Berman HM, Westbrook J, Feng Z et al (2000) The Protein Data Bank. Nucleic Acids Res 28:235–242

18. Budisa N, Pal PP, Alefelder S et al (2004) Probing the role of tryptophans in Aequorea victoria green fluorescent proteins with an expanded genetic code. Biol Chem 385:191–202

19. Humphrey W, Dalke A, Schulten K (1996) VMD: visual molecular dynamics. J Mol Graph 14(33–38):27–38

20. Rosell FI, Boxer SG (2003) Polarized absorption spectra of green fluorescent protein single crystals: transition dipole moment directions. Biochemistry 42:177–183

21. Ansbacher T, Srivastava HK, Stein T et al (2012) Calculation of transition dipole moment in fluorescent proteins – towards efficient energy transfer. Phys Chem Chem Phys 14:4109–4117

22. Kelly EM, Hou Z, Bossuyt J et al (2008) Phospholamban oligomerization, quaternary structure, and sarco(endo)plasmic reticulum calcium ATPase binding measured by fluorescence resonance energy transfer in living cells. J Biol Chem 283:12202–12211

23. Hou Z, Robia SL (2010) Relative affinity of calcium pump isoforms for phospholamban quantified by fluorescence resonance energy transfer. J Mol Biol 402:210–216

24. Dale RE, Eisinger J, Blumberg WE (1979) The orientational freedom of molecular probes. The orientation factor in intramolecular energy transfer. Biophys J 26:161–193

25. Van Der Meer BW, Coker Iii G, Chen S-YS (1994) Resonance energy transfer, theory and data. VCH, New York, NY

26. Vanbeek DB, Zwier MC, Shorb JM et al (2007) Fretting about FRET: correlation between kappa and R. Biophys J 92:4168–4178

Chapter 43

How to Compare, Analyze, and Morph Between Crystal Structures of Different Conformations: The P-Type ATPase Example

Jesper L. Karlsen and Maike Bublitz

Abstract

In the past 15 years, a large body of structural information on P-type ATPases has accumulated in the Protein Data Bank. The available crystal structures cover different enzymes in a variety of conformational states that are associated with the enzymatic activity of ATP-dependent ion translocation across membranes. This chapter provides an overview about the available structural information, along with some practical instructions on how to make meaningful comparisons of structures in different conformations, and how to generate morphs between series of structures, in order to analyze domain movements and structural flexibility.

Key words P-Type ATPase structure, Morphing, Conformational states, Metal fluorides

1 Introduction

The first P-type ATPase crystal structure was determined by Chikashi Toyoshima's laboratory in the year 2000 [1]; it was a structure of the sarco(endo)plasmic reticulum calcium ATPase SERCA1a from rabbit, in a Ca^{2+}-bound state. As of today, a total of 72 P-type ATPase crystal structures have been deposited in the Protein Data Bank. The most represented ATPase is SERCA with 54 structures, followed by the Na^+,K^+-ATPase with 11 structures, the bacterial Cu^+ and Zn^{2+}-ATPases with four and two structures, respectively, and a single structure is available for the plant H^+-ATPase AHA2. These molecular models offer a powerful resource of structural information for the scientific community, but it can be overwhelming and confusing to decide, which structure to look at, in order to address a particular question.

This chapter aims at providing an overview over the currently available structural information on P-type ATPases, and giving some instructions, how to retrieve relevant information from crystallographic data. Furthermore, we describe a novel script called Morphinator, which can generate morphs between structural models, allowing for a visualization of transitional movements between single conformational snapshots.

1.1 Conformational States Around the P-Type Catalytic Cycle and How They Are Stabilized in Crystal Structures

P-Type ATPases obtained their name from the fact that they undergo a transient phosphorylation upon hydrolysis of ATP. Specifically, the γ-phosphate of ATP is transferred to a conserved catalytic aspartate residue, yielding a covalently phosphorylated enzyme (EP). During the cycling between autophosphorylation and dephosphorylation, the ATPase binds and transports specific ion(s) (and in some cases counter-ions) across the membrane. Ion-bound and ion-free (or counter-ion-bound) states are typically called E1 and E2 states, respectively, and in combination with the phosphorylated and unphosphorylated substates, the full catalytic cycle of a P-type ATPase can be described by a scheme like the one shown for SERCA in Fig. 1 (for detailed reviews on P-type ATPase structure and function, see refs. [2–4].

Depending on the question one wants to answer by inspecting one or several crystal structures, it is mandatory to choose the best possible structural model(s) for the given purpose. The plethora of available crystal structures — particularly in the case of SERCA — can easily cause confusion and lead to a suboptimal choice of model. Firstly, it has to be decided, which catalytic state, i.e., which conformation should be inspected. As depicted in Fig. 1 for SERCA, each transition state, MgE1, E1, E1P, E2P, E2.Pi, and E2, represents a distinct structural conformation of the molecule. For example, if one wants to analyze the binding geometry of Ca^{2+} ions in SERCA, a Ca^{2+}-bound E1 or E1>E1P transition state

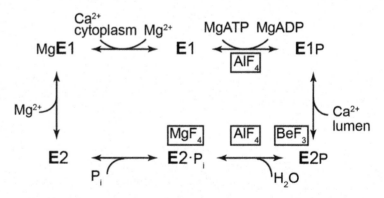

Fig. 1 E1–E2 reaction scheme for P-type ATPases, exemplified for the Ca^{2+}-ATPase SERCA. The Mg^{2+} stabilized E1 state (MgE1) has only recently been identified [2, 3], and it is not yet known, whether this state exists in other P-type ATPases. *Boxes* show the reaction intermediates that can be stabilized by metal fluoride compounds. Combinations with nucleotide analogs, Ca^{2+} or EGTA (a Ca^{2+} chelator) can be used to arrest specific states along the reaction cycle

should be chosen, preferably at the highest available resolution. In contrast, if the mechanism of luminal gate opening and closing is to be analyzed, a comparison of three states would be informative: (1) the calcium-occluded E1>E1P or E1P state, (2) the outward-open E2P state, and (3) the proton-occluded E2P>E2 state.

In Table 1, all currently available crystal structures of P-type ATPases are listed, sorted by the following hierarchy:

1. P-type ATPase.
2. Source organism.
3. Catalytic state represented by the structure.
4. Resolution.

Once the decision for a particular state of interest has been taken, it is constructive to use a representative structure that has been determined at the highest resolution available. However, it is important to always be aware of the fact that, in order to obtain crystal structures of the different conformations, often combinations of ligands and inhibitors have been used to stabilize the given state. In most cases, the inhibited protein has been demonstrated to actually represent a specific arrested catalytic conformation, but still it is worthwhile to keep in mind, that one is looking at an artificially arrested structure in those cases. For this reason, it can be a good idea to compare an inhibitor-bound high-resolution structure with a non-inhibited lower-resolution structure, if available, in order to make sure that the structural details of interest have not been caused by inhibitor binding (for example, thapsigargin (TG)-free E2 structures reveal that there are small side chain movements of residues surrounding the TG binding site in SERCA to accommodate TG binding).

Different metal fluorides have been instrumental in capturing transition states of phosphorylation and dephosphorylation reactions of the ATPase. Aluminum fluoride (AlFx, where x is 3 or 4), together with ADP (and Ca^{2+}) mimics the transition state of phosphorylation, whereas AlFx alone (without Ca^{2+}, i.e., in the presence of EGTA) mimics the transition during dephosphorylation. Beryllium fluoride (BeF_3) mimics a covalently attached phosphate at the catalytic aspartate, and magnesium fluoride (MgFx) mimics a cleaved but still bound inorganic phosphate moiety in a state immediately after the completion of dephosphorylation (*see* also Chap. 19 for detailed descriptions of the biochemical behavior of fluoride analog-bound SERCA).

1.2 What's in the Structural Data?

A structural model deposited as a .pdb file in the Protein Data Bank contains the three-dimensional coordinates (x, y, z, relative to an origin) for every atom of the determined molecular structure, along with its occupancy (a value between 0 and 1) and its B-factor as an indicator for the thermal motion of the atom. A

Table 1
P-Type ATPase crystal structures determined by X-ray diffraction in the Protein Data Bank (as of March 2015)

ATPase	State	Res.	Ions	Inh./Lig	Nuc.	PDB ID	Ref.
rSERCA	MgE1	3.01	Mg^{2+}, Na^+	SLN, PE	TNPAMP	3W5A	[5]
rSERCA	MgE1	3.1	Mg^{2+}, K^+	SLN	(AMP)PCP	4H1W	[6]
rSERCA	MgE1	3.2	Mg^{2+}, Na^+	PE	TNPAMP	3W5B	[5]
rSERCA	E1	2.4	Ca^{2+}, Na^+			1SU4[a]	[1]
rSERCA	E1	3.0	Ca^{2+}, K^+, Cl^-			2C9M	[7]
rSERCA (E309Q)	E1	3.5	Ca^{2+}, K^+	PC		4NAB	[8]
rSERCA	E1	2.5	Ca^{2+}, Na^+	PC	AMPPCP[b]	3AR2	[9]
rSERCA	E1	2.59	Ca^{2+}, K^+		AMPPCP[b]	3N8G	[10]
rSERCA	E1	2.6	Mg^{2+}, Ca^{2+}, K^+		AMPPCP[b]	1T5S	[11]
rSERCA	E1	2.9	Mg^{2+}, Ca^{2+}		AMPPCP[b]	2VFP	[12]
rSERCA	E1P	2.8	Ca^{2+}, K^+		AMPPN[b]	3BA6	[13]
rSERCA	E1>E1P	2.4	Mg^{2+}, Ca^{2+}	AlF_4, PC	ADP[b]	2ZBD[a]	[12]
rSERCA	E1>E1P	2.9	Mg^{2+}, Ca^{2+}, K^+	AlF_4	ADP[b]	1T5T	[11]
rSERCA	E2P	2.65	Mg^{2+}, Na^+	BeF_3		3B9B	[13]
rSERCA	E2P	3.8	Mg^{2+}	BeF_3		2ZBE[a]	[14]
rSERCA	E2P	2.4	Mg^{2+}	BeF_3, TG		2ZBF[a]	[14]
rSERCA	E2P	2.6	Mg^{2+}, Na^+	BeF_3, TG	TNPAMP	3AR9	[6]
rSERCA	E2~P	3.0	Mg^{2+}, K^+	AlF_4	AMPPCP[c]	3B9R	[13]
rSERCA	E2~P	2.55	Mg^{2+}	AlF_4, TG		2ZBG[a]	[14]
rSERCA	E2~P	2.6	Mg^{2+}, Na^+	AlF_4, TG	TNPAMP[c]	3AR8	[9]
rSERCA	E2~P	3.0	Mg^{2+}, K^+	AlF_4, TG		1XP5	[15]
rSERCA	E2.Pi	2.3	Mg^{2+}, Na^+	MgF_4, TG	ADP[c]	1WPG[a]	[12]
rSERCA	E2.Pi	2.5	Mg^{2+}, K^+, Mn^{2+}	MgF_3, CPA	AMPPCP[c]	4BEW	–
rSERCA	E2.Pi	2.5	Mg^{2+}, K^+, Mn^{2+}	MgF_4, CPA	AMPPCP[c]	3FGO	[16]
rSERCA	E2.Pi	2.55	Mg^{2+}, K^+	MgF_4, CPA	ATP[c]	3FPB	[16]
rSERCA	E2.Pi	2.65	Mg^{2+}, Na^+	MgF_4, CPA		2O9J	[17]
rSERCA	E2.Pi>E2	2.45	Na^+, SO_4^{2-}	PE		3W5D	[5]
rSERCA	E2	2.5	Na^+	PE		3W5C	[5]
rSERCA	E2	2.15	Mg^{2+}, Na^+	TG, PE	ATP[c]	3AR4	[6]
rSERCA	E2	2.15	Na^+	TG, PE	TNPATP[c]	3AR7	[6]

(continued)

Table 1
(continued)

ATPase	State	Res.	Ions	Inh./Lig	Nuc.	PDB ID	Ref.
rSERCA	E2	2.2	Mg^{2+}, Na^+	TG, PE	TNPADP[c]	3AR6	[6]
rSERCA	E2	2.2	Na^+	TG, PE	TNPAMP[c]	3AR5	[6]
rSERCA	E2	2.3	Mg^{2+}, Na^+	TG, PE	ADP[c]	3AR3	[6]
rSERCA	E2	2.4	Na^+	BHQ, TG, PE		2AGV[a]	[18]
rSERCA	E2	2.5	Mg^{2+}, K^+, SO_4^{2-}	TG, GOL, TBU		4UU0	[19]
rSERCA	E2	2.5	Mg^{2+}, Na^+	TG, PE	AMPPCP[c]	2DQS[a]	–
rSERCA	E2	2.65	Mg^{2+}, K^+	DTB		3NAL	[20]
rSERCA	E2	2.8		CPA, PE		2EAU[a]	[14]
rSERCA	E2	2.8	Mg^{2+}, Na^+	TG	AMPPCP[c]	2C8K	[7]
rSERCA	E2	2.8	Mg^{2+}, K^+	TG, GOL, PC	AMPPCP[c]	4UU1	[19]
rSERCA	E2	2.9		CPA, TG		2EAT[a]	[14]
rSERCA	E2	3.1		TG		2EAR[a]	[14]
rSERCA	E2	3.1		TG		1IWO[a]	[21]
rSERCA	E2	3.1	Na^+	TG		2C8L	[7]
rSERCA	E2	3.1	Mg^{2+}, K^+	DTG		2YFY	[22]
rSERCA	E2	3.1	Mg^{2+}, Na^+	OTK, PE		3NAM	[20]
rSERCA	E2	3.1	Mg^{2+}, K^+	HZ1, PE		3NAN	[20]
rSERCA	E2	3.1	Mg^{2+}, Na^+	TG	AMPPCP[c]	2C88	[7]
rSERCA	E2	3.2	K^+	1HT, PE		4J2T	[23]
rSERCA	E2	3.2	Mg^{2+}	CPA	ADP[c]	3FPS	[16]
rSERCA	E2	3.25	Mg^{2+}, K^+	CPA		4YCL	[14]
rSERCA	E2	3.3	Mg^{2+}, Na^+	BOC	AMPPCP[c]	2BY4	[24]
rSERCA	E2	3.4	Mg^{2+}	CPA	ADP[c]	2OA0	[17]
bSERCA	E1	2.95	Mg^{2+}, Ca^{2+}, K^+		AMPPCP[b]	3TLM	[25]
pNKA	E1>E1P	2.8	Mg^{2+}, Na^+	AlF_4, COL, NAG, PC	ADP[b]	3WGU	[26]
pNKA	E1>E1P	2.8	Mg^{2+}, Na^+	AlF_4, COL, NAG, PC, OLI	ADP[b]	3WGV	[26]
pNKA	E1>E1P	4.3	Mg^{2+}, Na^+	AlF_4, COL	ADP[b]	4HQJ	[27]
pNKA	E2P	3.41	Mg^{2+}, K^+	COL, NAG, BUF, SUC, PS		4RES	[28]

(continued)

Table 1
(continued)

ATPase	State	Res.	Ions	Inh./Lig	Nuc.	PDB ID	Ref.
pNKA	E2P	4.0	Mg^{2+}	COL, NAG, DGX, SUC, PS		4RET	[28]
pNKA	E2P	3.4	Mg^{2+}	COL, NAG, OBN, PS, $C_{12}E_8$, 1DS, 1AT		4HYT	[28]
pNKA	E2P	4.6	Mg^{2+}	OBN		3N23	[29]
pNKA	E2.Pi	3.5	Mg^{2+}, Rb^+	MgF_4, PC		3B8E	[30]
pNKA	E2.Pi	3.5	Mg^{2+}, Rb^+	MgF_4, COL		3KDP	[30]
sNKA	E2.Pi	2.4	Mg^{2+}, K^+	MgF_4, COL, NAG, NDG		2ZXE	[31]
sNKA	E2.Pi	2.8	Mg^{2+}, K^+	MgF_4, COL, NAG, OBN		3A3Y	[32]
AHA2	E1	3.6	Mg^{2+}		AMPPCP[b]	3B8C	[33]
LpCopA	E2P	2.75	Mg^{2+}	BeF_3, NAD, PC, $C_{12}E_8$		4BBJ	[34]
LpCopA	E2~P	3.2	Mg^{2+}, K^+	AlF_4		3RFU	[35]
LpCopA	E2~P	2.85	Mg^{2+}, K^+	AlF_4, PEG		4BYG	–
LpCopA	E2.Pi	3.58	Mg^{2+}	MgF_3		4BEV	–
SsZntA	E2P	3.2	Mg^{2+}	BeF_3		4UMV	[36]
SsZntA	E2~P	2.7	Mg^{2+}	AlF_4		4UMW	[36]

rSERCA sarco(endo)plasmic reticulum Ca^{2+} ATPase from rabbit hind leg muscle, *pNKA* Na^+,K^+-ATPase from pig kidney, *sNKA* Na^+,K^+-ATPase from shark rectal gland, *AHA2* plasma membrane H^+-ATPase from *Arabidopsis thaliana*, *LpCopA* copper-transporting ATPase from *Legionella pneumophila*, *SsZntA* Zn^{2+}-transporting ATPase from *Shigella sonnei*, *SLN* sarcolipin, *PE* phosphatidylethanolamine, *PC* phosphatidylcholine, AlF_4 aluminum tetrafluoride, BeF_3 beryllium trifluoride, *TG* thapsigargin, MgF_4 magnesium tetrafluoride, MgF_3 magnesium trifluoride, *CPA* cyclopiazonic acid, *BHQ* 2′,5′-di(*tert*-butyl)-1,4-benzohydroquinone, *GOL* glycerol, *TBU* *tert*-butanol, *DTB* thapsigargin derivative DTB, *DTG* debutanoyl thapsigargin, *OTK* thapsigargin derivative dOTg, *HZ1* thapsigargin derivative Boc-(phi)Tg, *1HT* thapsigargin derivative 6a, *BOC* thapsigargin derivative Boc-12ADT, *COL* cholesterol, *NAG* N-acetyl D-glucopyranose, *OLI* oligomycin A, *BUF* bufalin, *SUC* sucrose, *PS* phosphatidylserine, *DGX* digoxin, $C_{12}E_8$ octaethyleneglycol mono-*n*-dodecyl ether, *1DS* 1-O-decanoyl-beta-D-tagatofuranosyl beta-D-allopyranoside, *1AT* beta-d-fructofuranosyl 6-O-decanoyl-alpha-D-glucopyranoside, *OBN* ouabain, *NDG* 2-(acetylamino)-2-deoxy-α-D-glucopyranose, *NAD* nicotinamide adenine dinucleotide, *PEG* polyethylene glycol
[a]This model has been deposited without structure factor data, therefore no electron density can be generated for this entry
[b]The nucleotide is bound in a catalytic binding mode
[c]The nucleotide is bound in a regulatory binding mode

second file typically deposited along with the .pdb model contains the structure factor data, representing the dataset derived from processing of the X-ray diffraction images of the corresponding protein crystal(s). With both model and structure factors at hand, one can generate and inspect the electron density maps, into which the model has been built by manual or automatic interpretation. It is important to understand that the electron density — and not the

.pdb model — is the primary result of a structure determination, and the .pdb model is already an interpretation of the electron density. For this reason, it is always recommended to inspect the electron density along with the model region of interest, in order to evaluate, if the crystallographic data actually support the structural detail that is being looked at (e.g., the side chain orientation of a particular amino acid or inferred hydrogen bond networks). Here, resolution is another important factor, and it is recommended to take the estimated coordinate error of the model into account, which can typically be found in the header of the .pdb file.

PDB files can be opened with any text editor for inspection of the header and the raw coordinate, occupancy and B-factor information, but they can also be interpreted by molecular graphics programs, which allow for a three-dimensional visualization of the molecular model in different representations.

If two or more structures of the same or very similar proteins are to be compared, they have to be superposed first, preferably using an invariant part of the molecule, a "reference domain". If another part of the molecule adopts different positions (relative to the reference domain) in the two models, a superposition will reveal the change, if the reference domain has been appropriately chosen. However, superpositions of more than two different molecules can be difficult to interpret. Particularly in P-type ATPases, structural changes involving the movement of several domains against each other can be too complex to be grasped easily by looking at two still images. For this reason, creating smooth morphs between structural models has proven very helpful in deciphering the conformational changes between different models.

1.3 Generating Morphs with the Morphinator Program

The *Morphinator* program has been developed as a straightforward and user-friendly tool to generate morphs between .pdb files, which can then be inspected as three-dimensional graphical animations.

Morphinator creates molecular morphs by employing well-known programs in the background. The programs that *Morphinator* wraps are *superpose_pdbs* and *geometry_minimization* from the *PHENIX* software suite and *PyMOL* from Schrödinger, although the internal version of *PyMOL* (phenix.pymol) from *PHENIX* is sufficient for full functionality of *Morphinator*.

Morphinator uses an accelerated bidirectional approach in creating the morph steps, which gives the morphs a smooth appearance.

The bidirectional workflow is accomplished by defining the structure files of the two targets as separate origins. Then the two models are moved towards each other in cycles of simultaneous steps of first a linear morph step followed by an energy minimization run. The two PDBs generated from the minimization are now used to make a new linear morph step, after each cycle the number of steps decreases by two, until the two ends meet in the middle (Fig. 2).

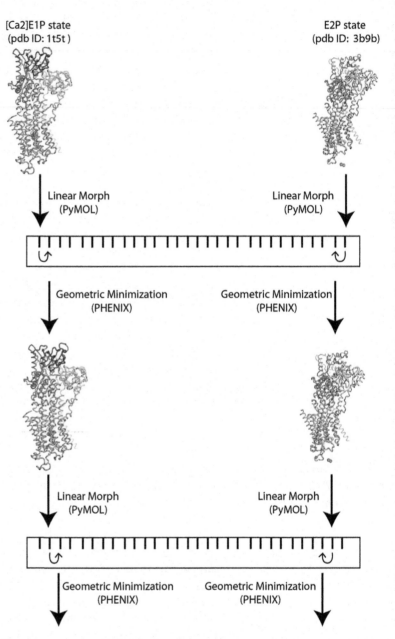

Fig. 2 An example of the bidirectional approach of *Morphinator*. The beginning of a morph between SERCA's [Ca$_2$]E1P state and the E2P state is shown. *Morphinator* will continue to cycle through parallel steps of linear morphing and geometry minimization until the two morphed intermediate structures (represented by *black bars* along a similarity scale in the figure) meet in the middle

By moving in small linear increments away from the two known structures into transitional space, and repeatedly "inflating" the structures by minimization after each linear step, seems to conserve the structural integrity of the molecule during the morph.

The accelerated part of the "accelerated bidirectional approach" is mainly applied to deal with the curved hinge movement of molecular domains. The hinge movement can be looked upon as a partial circular movement, where the hinge is the origin of each involved atom. The linear path between the involved atoms in the two 'endpoint' PDBs can be seen as the baseline.

The projected position of the curved hinge movement to its linear baseline will not be equally distributed, as they would be more dens at the ends. Like cosine and sine describe the projected angular position on the circle's periphery on the x-axis and y-axis. *Morphinator* uses an acceleration factor "e" that is applied at each cycle as an exponent to the remaining number of steps. Thus, at each real step, a virtual amount of steps is created, which the linear morph is split into.

The acceleration factor "e" is a variable that can be set to any value > 0.

A value of "e" > 1 results in smaller linear steps at the end points of the morph than in the middle, much like the behavior of sine and cosine. And the opposite applies for $0 < $ "e" < 1.

The default value of acceleration factor "e" is 1.2, which seems to be a reasonable value for most cases.

The accelerated bidirectional linear incremental steps of *Morphinator* can be described by these two equations that show the increment at each end:

$t =$ total steps

$c =$ current step

$e =$ acceleration factor

$d =$ distance

$l =$ linear morph

$$\{\Delta x_{c+1}, \Delta y_{c+1}, \Delta z_{c+1}\} = \frac{\{x_{(t-c)}, y_{(t-c)}, z_{(t-c)}\} - \{x_c, y_c, z_c\}}{(t-c)^e} \times 2 \quad (1)$$

$$\{\Delta x_{t-c-1}, \Delta y_{t-c-1}, \Delta z_{t-c-1}\} = \frac{\{x_{(t-c)}, y_{(t-c)}, z_{(t-c)}\} - \{x_c, y_c, z_c\}}{(t-c)^e} \quad (2)$$

$\times ((t-c)^e - 1)$

2 Materials

2.1 Software

1. *Coot* [37] is available for download from http://www2.mrc-lmb.cam.ac.uk/personal/pemsley/coot/.
2. *PyMOL* (Schrödinger, LCC) is available for download from https://www.pymol.org/.
3. *Morphinator* can be downloaded from http://morphinator.au.dk.

2.2 Computing Requirements

1. *Coot* runs on Windows, OS X, Linux and UNIX. It requires an OpenGL capable graphics card.
2. *PyMOL* runs on Windows, OS X and Linux. It requires an OpenGL capable graphics card.
3. *Morphinator* runs on OS X and Linux and requires the *PHENIX* software suite installed, and *PyMOL* is recommended.

3 Methods

3.1 Viewing Structures and Electron Densities

Structural data can be inspected with both *Coot* and *PyMOL*. *Coot* is somewhat easier to use with electron densities, whereas *PyMOL* is straightforward for showing several structures in defined representations and for preparing rendered figures. The following points are not meant as step-by-step instructions, neither do they want to replace the plethora of documentation sources available for both programs. They are rather meant as a collection of useful tips and considerations for non-crystallographers in the P-type ATPase field, who want to exploit structural information available from the Protein Data Bank.

3.1.1 Viewing a Molecular Model and Electron Density in Coot

1. Coot is a program for molecular model building and manipulation, rather than representation, but it can be very useful for structural analysis, especially for the display of electron density maps. Extensive documentation is available from https://www2.mrc-lmb.cam.ac.uk/Personal/pemsley/coot/web/docs/coot.html.

2. Start *Coot* and load the structure of interest via *File > Open Coordinates…* or load both the model and the electron density together via *File > Fetch PDB & Map using EDS* (*see* **Note 1**). By default, an all-atom representation of the model is shown, with carbon atoms in yellow, nitrogen in blue, oxygen in red and sulfur in green. Furthermore, two electron density maps open: a 2Fo-Fc map in blue (default contour level 1.0 sigma) and an Fo-Fc map in green (positive) and red (negative) (default contour levels 3.0 and −3.0 sigma, respectively). The Display Manager lists all open objects and allows the user to

control, which objects should be displayed, deleted or active for modification.

3. If we want to inspect for example the calcium-binding site in SERCA, we can click on *Draw > Go To Atom...*, there we choose chain A and residue 800, and click on *Apply*. The view is now centered around the C-alpha atom of the calcium-binding residue Asp800.

4. By pressing and holding the left mouse button while we move the mouse around, we can rotate the view around the centered atom. The right mouse button together with a forward or backward movement zooms in or out, respectively. The center of the view can also be moved, by holding down the Ctrl key while using the left mouse button.

5. The scroll wheel allows us to modify the contour level of the displayed electron density map (use the Display Manager to switch the scrolling activity for different maps). This can be useful for checking if the electron density for an amino acid side chain starts appearing at a lower sigma level, e.g., 0.8, indicating its most likely position. If, on the other hand, a component scatters very strongly, e.g., a phosphate group, it can be helpful to increase the sigma level of the map, in order to localize the point of the strongest scattering, indicating the most likely position of the respective ion.

6. The default clipping (also called 'slab' or the 'depth' of the three-dimensional view) is typically good for viewing coordinate data, but it is often too "thick" for viewing electron density maps. The clipping can be reduced or increased by pressing the 'D' and 'F' keys, respectively.

7. If we now open a second entry, the two structures will likely not overlap, which makes it impossible to compare them directly. *Coot* has two options for superposing two models: (1) *SSM Superpose* using secondary structure information in the specified protein chains, and (2) *LSQ Superpose* applying a least squares fit to a defined residue range. Both approaches ask the user to define, which structure is to be used as reference, and which is to be moved (*see* **Note 2**). In order to look at the complex domain movements in P-type ATPases, it has proven useful to use one consistent sub-domain as a reference, against which all other domain movements are put into relation. In SERCA, the C-terminal region (residues 760–994) is a relatively 'stiff' part of the molecule, which can be used as such a reference. Therefore, the method of choice in Coot is *LSQ Superpose*, since it allows to define the residue range on which to superpose the two structures (*see* **Note 3**).

8. Switching the model representation to C-alphas/Backbone in the Display Manager makes it easier to see conformational differences between the two superposed models.

3.1.2 Viewing a Molecular Model and Electron Density in PyMOL

1. *PyMOL* allows the user to very intuitively load pdb models and represent them as ribbons, cartoons, ball-and-stick, etc. An exhaustive documentation is available at http://www.pymol-wiki.org/index.php/Main_Page.

2. Superpositions of structures on a particular reference section can be done with the *align* command, e.g., for SERCA: *align ref_structure and resi 760-994, moving_structure and resi 760-994*.

3. For visualizing an electron density map in *PyMOL*, a map file in CCP4 format has to be generated first (e.g., with *FFT* or *Coot*) (*see* **Note 4**).

4. After loading the map file, it appears as an object in the list at the top right, but no contour mesh is shown by default. The mesh is added as a separate object by typing a command with the following syntax: *isomesh meshname, mapname, maplevel, selection, carve=desired_radius*. As an example, if the map object is called '2fofc', and we want *PyMOL* to show a mesh (as a new object named *mesh_1*) at a contour level of 1.0 sigma centered around the Ca^{2+} binding residue Asp800 in object 2C9M and with a radius of 5 Å, the command would be: *isomesh mesh_1, 2fofc, 1.0, 2C9M and resi 800, carve=5*.

3.2 Morphing with Morphinator

Morphinator is currently a command-line-only program and can be given the following options:

-1 From PDB. It can be a local file or PDB code, which then will be fetched.

-2 To PDB. It can be a local file or PDB code, which then will be fetched.

-s Steps. The number of intermediates between the two PDBs. Default is 30.

-a Align to PDB. Can be just a domain, where the morph should be anchored. Important to align to the same PDB if a series of morphs is to be combined.

-i Initial minimization on the starting models. This can remove jerks between starting model and first interpolation.

-w Allow files/folders from earlier runs to be overwritten.

-e Exponential number that can make morph look more natural. Values >1.2 slow at ends, <1.2 slow in middle. Reasonable values seem to be between 0.8 and 2.0. Default is 1.2.

-o Output template. Default is "morphinator_PDB-ID1_to_PDB-ID2".

-d Open and display morph in PyMOL after run.

-h Shows this message.

Only the from/to PDB flags (-1 and -2) are obligatory. The other flags are optional.

3.2.1 Creating a Basic Morph with Morphinator

We will take the example from Fig. 2, where a morph between the E1P (1T5T) and E2P (3B9B) states of SERCA is created by *Morphinator* (*see* **Note 5**).

1. One first needs to consider which parts of the molecule are the moving parts and which parts are fixed. In SERCA this is an easy choice, as one would expect the C-terminal transmembrane region to be the fixed part (as explained above), and it will therefore be a good idea to align the morph to this region.

2. Download the 1t5t.pdb file from http://www.rcsb.org/pdb and edit the pdb file manually so it only contains the residues A760–A994 which represent the transmembrane helices. The truncated PDB is called 1t5t_760-994pdb.

3. Now run *Morphinator* by using 1t5t_760-994pdb as alignment target.

morphinator -1 1t5t -2 3b9b –a 1t5t_760-994.pdb

The command above will download the biological assembly of 1T5T and 3B9B, align both pdbs to the transmembrane helices in 1t5t_760-994.pdb and create a morph of 30 steps (default) between them.

4. The output generated by *Morphinator* will be:

   - A multi-state pdb file.
   - A pml (PyMOL script) file.
   - A working folder, all with the root name "morphinator_1t5t_to_3b9b".

5. View the morph in PyMOL by typing in a terminal:

pymol morphinator_1t5t_to_3b9b.pml

or by loading morphinator_1t5t_to_3b9b.pdb directly into PyMOL (*see* **Note 6**).

6. The .pml file will by default make the morph go forth and back, where it stops for five frames at the ends. If this is not desired, the *mset* line in the .pml file can be deleted, or the .pdb file can be loaded directly into *PyMOL*. The morph can be stopped and started with the *mstop* and *mplay* commands in *PyMOL*.

3.2.2 Including Ions and Ligands into the Morph

Morphinator only includes the common residues between the two target pdbs and discards the rest. But it keeps track of the residues that it discards from each pdb file. In our example, the working folder morphinator_1t5t_to_3b9b holds the files 1t5t_residues_removed.txt and 3b9b_residues_removed.txt, which holds this information.

The file 1t5t_residues_removed.txt shows that the Ca^{2+} ions and the ADP ligand have been removed from the morph. Which is not surprising, as the E2P state (3B9B) does not bind either of them. However, we do want to include both the Ca^{2+} ions and ADP in the morph, so they have to be added to the E2P state (*see* **Note 7**).

1. Copy the 3b9b.pdb1 file that we downloaded during the first *Morphinator* run, to a new file called for example 3b9b_mod.pdb.
2. Copy over the Ca^{2+} atoms and the ADP from 1t5t.pdb1 to the new 3b9b_mod.pdb file, using a text editor.
3. When SERCA goes from the [Ca_2]E1P to the E2P state, the Ca^{2+} ions and the ADP dissociate from SERCA. It is therefore necessary to pull the ligands away from SERCA in the E2P model.

   It is recommended to use *Coot* for this rearrangement of the ligands. Open 3b9b_mod.pdb with *Coot* and pull the ADP out of the ATP binding pocket and pull the Ca^{2+} ions out to the luminal side, using the command *Calculate > Model/Fit/Refine... > Rotate/Translate zone*. Save the modified PDB as 3b9b_mod.pdb and exit *Coot*.
4. Now run *Morphinator* again with the modified 3b9b_mod.pdb file:

   morphinator -1 1t5t.pdb1 -2 3b9b_mod.pdb –a 1t5t_760-980.pdb
5. Inspection of the newly created morphinator_1t5t_to_3b9b_mod.pdb in *PyMOL* should now include ADP and Ca^{2+}. To show the Ca^{2+} ions and the ADP as spheres type these commands in *PyMOL*:

   show spheres, resn Ca

   show spheres, resn ADP

3.2.3 Making Circular Morphs

As described above, many of the structures of SERCA's transitional conformations are already solved. This makes SERCA a very good candidate for combining all known stages MgE1 -> E1 -> E1P -> E2P -> E2.Pi into a cyclic morph, showing the pumping of calcium and burning of ATP.

1. Generate a set of morphs between two successive states (*see* **Note 7**), e.g.:

   - MgE1 -> E1
   - E1 -> E1P
   - E1P -> E2P
   - E2P -> E2.Pi
   - E2.Pi -> MgE1
2. Multiple morphs can be combined in *PyMOL* by selecting which states the individual morphs should be loaded at. If one

has two morph PDBs (*morph1.pdb* and *morph2.pdb*) of 30 steps each, they could be loaded into PyMOL like this:

load morph1.pdb, morph1, 1

load morph2.pdb, morph2, 31

The two PDBs would be loaded as objects named morph1 and morph2, starting at state 1 and state 31 respectively.

4 Notes

1. As an example, one might want to look at the effect of nucleotide binding on SERCA in the E1 state. Therefore, one would choose a nucleotide-free (2C9M) and a nucleotide-bound (3N8G) E1 structure. Note that the 1SU4 entry has a higher resolution than 2C9M, but since no structure factor data are available for this entry, we have to use the lower resolution entry instead, if we want to inspect the electron density. Otherwise, the structure with the highest resolution should be used.

2. One a model is moved by a *Superpose* function, it will not match with its original electron density map anymore.

3. Depending on the nature of the structural comparison, it may be necessary to use a different section of the structure as a superposition reference. Not defining any residue range will result in an average best fit over the entire structure, which is not useful in the vast majority of comparisons. The reader is encouraged to try different domains as reference sections, to see the strong (and possibly misleading) influence of the choice of reference on the detected domain movements between two structures.

4. If the map file extension is .ccp4, Pymol can open the map via the *File > Open...* menu. If the extension is different (e.g., .map), the command line has to be used: *load myfile.map, format=ccp4*.

5. If the two target PDBs for morphing are homologs but of two different species, then a sequence substitution is needed for Morphinator to work. We recommend the use of *MODELLER* (https://salilab.org/modeller) for this task.

6. If the -d optional flag is given, *Morphinator* opens the morph in *PyMOL* automatically at the end of the run. The AlF_4 and BeF3-bound phosphate transition analogs in 1T5T and 3B9B, respectively, are ignored, as the complexity of including analog ligands with non-matching atoms counts in the morph is beyond the scope of this tutorial.

7. If the goal is to combine several morphs, then it is important to always use the same alignment file as a superposition reference for all the morphs.

References

1. Toyoshima C, Nakasako M, Nomura H, Ogawa H (2000) Crystal structure of the calcium pump of sarcoplasmic reticulum at 2.6 A resolution. Nature 405:647–655
2. Bublitz M, Poulsen H, Morth JP, Nissen P (2010) In and out of the cation pumps: P-type ATPase structure revisited. Curr Opin Struct Biol 20(431–9)
3. Møller JV, Olesen C, Winther A-ML, Nissen P (2010) The sarcoplasmic Ca^{2+}-ATPase: design of a perfect chemi-osmotic pump. Q Rev Biophys 43(501–66)
4. Kühlbrandt W (2004) Biology, structure and mechanism of P-type ATPases. Nat Rev Mol Cell Biol 5(282–95)
5. Toyoshima C et al (2013) Crystal structures of the calcium pump and sarcolipin in the Mg^{2+}-bound E1 state. Nature 495(260–4)
6. Winther A-ML et al (2013) The sarcolipin-bound calcium pump stabilizes calcium sites exposed to the cytoplasm. Nature 495(265–9)
7. Jensen A-ML, Sørensen TL-M, Olesen C, Møller JV, Nissen P (2006) Modulatory and catalytic modes of ATP binding by the calcium pump. EMBO J 25(2305–2314)
8. Clausen JD et al (2013) SERCA mutant E309Q binds two Ca^{2+} ions but adopts a catalytically incompetent conformation. EMBO J 32(3231–43)
9. Toyoshima C, Yonekura S-I, Tsueda J, Iwasawa S (2011) Trinitrophenyl derivatives bind differently from parent adenine nucleotides to Ca^{2+}-ATPase in the absence of Ca^{2+}. Proc Natl Acad Sci U S A 108(1833–1838)
10. Bublitz M et al (2013) Ion pathways in the sarcoplasmic reticulum Ca^{2+}-ATPase. J Biol Chem 288(10759–65)
11. Sørensen TL-M, Møller JV, Nissen P (2004) Phosphoryl transfer and calcium ion occlusion in the calcium pump. Science 304(1672–5)
12. Toyoshima C, Nomura H, Tsuda T (2004) Lumenal gating mechanism revealed in calcium pump crystal structures with phosphate analogues. Nature 432(361–8)
13. Olesen C et al (2007) The structural basis of calcium transport by the calcium pump. Nature 450(1036–1042)
14. Takahashi M, Kondou Y, Toyoshima C (2007) Interdomain communication in calcium pump as revealed in the crystal structures with transmembrane inhibitors. Proc Natl Acad Sci U S A 104(5800–5805)
15. Olesen C, Sørensen TL-M, Nielsen RC, Møller JV, Nissen P (2004) Dephosphorylation of the calcium pump coupled to counterion occlusion. Science 306(2251–2255)
16. Laursen M et al (2009) Cyclopiazonic acid is complexed to a divalent metal ion when bound to the sarcoplasmic reticulum Ca^{2+}-ATPase. J Biol Chem 284(13513–8)
17. Moncoq K, Trieber CA, Young HS (2007) The molecular basis for cyclopiazonic acid inhibition of the sarcoplasmic reticulum calcium pump. J Biol Chem 282(9748–9757)
18. Obara K et al (2005) Structural role of countertransport revealed in Ca(2+) pump crystal structure in the absence of Ca(2+). Proc Natl Acad Sci U S A 102:14489–14496
19. Drachmann ND et al (2014) Comparing crystal structures of Ca(2+)-ATPase in the presence of different lipids. FEBS J 281(18):4249–4262
20. Winther AML et al (2010) Critical roles of hydrophobicity and orientation of side chains for inactivation of sarcoplasmic reticulum Ca^{2+}-ATPase with thapsigargin and thapsigargin analogs. J Biol Chem 285:28883–28892
21. Toyoshima C, Nomura H (2002) Structural changes in the calcium pump accompanying the dissociation of calcium. Nature 418(605–611)
22. Sonntag Y et al (2011) Mutual adaptation of a membrane protein and its lipid bilayer during conformational changes. Nat Commun 2(304)
23. Paulsen ES et al (2013) Water-mediated interactions influence the binding of thapsigargin to sarco/endoplasmic reticulum calcium adenosinetriphosphatase. J Med Chem 56(3609–19)
24. Søhoel H et al (2006) Natural products as starting materials for development of second-generation SERCA inhibitors targeted towards prostate cancer cells. Bioorg Med Chem 14:2810–2815
25. Sacchetto R et al (2012) Crystal structure of sarcoplasmic reticulum Ca^{2+}-ATPase (SERCA) from bovine muscle. J Struct Biol 178:38–44
26. Kanai R, Ogawa H, Vilsen B, Cornelius F, Toyoshima C (2013) Crystal structure of a Na^+-bound Na^+, K^+-ATPase preceding the E1P state. Nature 502(201–6)
27. Nyblom M et al (2013) Crystal structure of Na^+, $K(+)$-ATPase in the $Na(+)$-bound state. Science 342(123–7)
28. Laursen M, Gregersen JL, Yatime L, Nissen P, Fedosova NU (2015) Structures and characterization of digoxin- and bufalin-bound Na^+, K^+-ATPase compared with the ouabain-bound complex. Proc Natl Acad Sci U S A 112(1755–60)
29. Yatime L et al (2011) Structural insights into the high affinity binding of cardiotonic steroids to the Na^+, K^+-ATPase. J Struct Biol 174:296–306

30. Morth JP et al (2007) Crystal structure of the sodium-potassium pump. Nature 450(1043–1049)
31. Shinoda T, Ogawa H, Cornelius F, Toyoshima C (2009) Crystal structure of the sodium-potassium pump at 2.4 A resolution. Nature 459:446–450
32. Ogawa H, Shinoda T, Cornelius F, Toyoshima C (2009) Crystal structure of the sodium-potassium pump (Na+, K+-ATPase) with bound potassium and ouabain. Proc Natl Acad Sci 106:13742–13747
33. Pedersen BP, Buch-Pedersen MJ, Morth JP, Palmgren MG, Nissen P (2007) Crystal structure of the plasma membrane proton pump. Nature 450(1111–4)
34. Andersson M et al (2014) Copper-transporting P-type ATPases use a unique ion-release pathway. Nat Struct Mol Biol 21(43–8)
35. Gourdon P et al (2011) Crystal structure of a copper-transporting PIB-type ATPase. Nature 475(59–64)
36. Wang K et al (2014) Structure and mechanism of Zn(2+)-transporting P-type ATPases. Nature 514(7523):518–522
37. Emsley P, Lohkamp B, Scott WG, Cowtan K (2010) Features and development of Coot. Acta Crystallogr D Biol Crystallogr 66:486–501

INDEX

A

Absolute quantitation 337
ACMA. *See* 9-amino-6-chloro-2-methoxyacridine (ACMA)
Activity 90, 103, 107–108, 116, 328, 329, 342
Additive-protein interactions
Adenosine diphosphate (ADP) 90, 106, 108, 121, 196, 198, 200–203, 206, 211–214, 218, 219, 225, 243, 244, 306, 425, 446, 452, 487, 525, 535, 536
   insensitive phosphorylated intermediate 196, 211, 213, 218
   sensitive phosphorylated intermediate 211
Adenosine triphosphate (ATP) 11, 14, 16, 19, 21, 24, 38, 42, 44, 51, 52, 57, 60, 64, 68, 69, 89–102, 105, 106, 108, 111, 113, 114, 118, 119, 121–128, 131, 132, 135–140, 143, 144, 146–150, 152, 153, 157, 159, 161, 163, 166, 168, 171, 174, 176–179, 181–183, 185, 187–189, 196, 198, 199, 201, 203–206, 211–214, 217–219, 221, 233–237, 240–256, 263, 265, 267, 294, 299–301, 305, 306, 323, 324, 327–329, 331, 332, 371, 381, 446, 453, 462, 464, 466, 468, 470, 473, 476, 478, 481, 484, 485, 487, 524, 536
   affinity determination 234
   hydrolysis 19, 20, 60, 89, 90, 105, 106, 157, 159, 181, 196, 212, 328
Affinity chromatography 31
Agrobacterium tumefaciens 383, 384, 387
Alexa Fluor 367, 368
Alkanethiol/phospholipid bilayer 59, 60, 294, 295, 301–302, 403
All-atom MD simulations 462–463
Aluminum fluoride (AlFx) 197–199, 206, 207, 525
9-amino-6-chloro-2-methoxyacridine (ACMA) 172–176, 178
Antibodies 76, 275, 323, 327, 355, 365
Aspartyl phosphate 195, 199, 211, 212, 217, 224, 225
Assay 96, 107–108, 123, 124, 174, 187–188, 330, 373–374
   colorimetric 89–103
   complementation 45, 46
   dithionite 183
   drug 91
   enzyme-coupled 64

ATP. *See* Adenosine triphosphate (ATP)
ATPase activity 1, 9, 20, 25–26, 59, 64, 68, 69, 72, 76, 91, 93, 94, 98, 100, 105, 109, 116, 121, 124–126, 133, 135, 136, 143, 203, 328, 330, 331, 354, 438, 453

B

Back-exchange 183
Baculovirus 71–77
Balance beam 342
Barnes maze 342, 344–347, 349
Beryllium fluoride (BeFx) 197–199, 202, 206, 207, 525

C

Ca^{2+} 1, 11, 12, 15, 37, 42, 50, 53, 57–59, 63–65, 67–69, 79, 82, 106–109, 157–159, 161–163, 165–168, 196, 197, 199, 201–207, 212, 213, 218–221, 224, 225, 228–230, 234–237, 239, 240, 248, 250–255, 261–266, 268, 283, 285, 293, 323, 326, 397–399, 422, 424, 426, 427, 429, 434, 435, 437–439, 443, 460, 489, 503, 504, 523–528, 534–536
   affinity 11
   ATPase 234, 235, 237, 239, 248, 250, 253, 254
   H^+ exchange 59–60
   pump 413, 422
   transport 12, 261
   uptake 161–163, 165, 167
$^{45}Ca^{2+}$ 159, 203, 261, 263–266, 268
Calcium 1, 11, 12, 93, 116, 117, 157–159, 161–168, 201, 205, 215, 224, 263–266, 422, 464, 489, 523, 536
Calibration curve 115, 338
Calmodulin 58, 61, 66, 67
Cardiac glycosides 99
Cdc50 subunits 181
Charcoal 122
Charge displacement 299
Classification 495–496
Clone 32
Co-expression 49, 52, 386
   in yeast 37
Coloring solution 112
Competitive inhibition 233–257
Concentration jump method 275, 298, 299

Conformational
  dynamics 281, 282
  states 31, 98, 281, 400, 422, 524
Coot 532–534, 536
Copper 448
Coupling ratio 159
CPM 80–85, 158
Cryo-EM 445, 452
Crystal 198, 415, 416, 429, 444, 453
Crystal structure 198
Crystallization 414, 416, 425–426, 429, 430, 432, 434, 438–440
Cu⁺-ATPase 268, 273–275, 293, 416, 418
Current 149, 152, 299
  measurement 286
  pre-steady-state 307
  pump 89, 147

D

Danio rerio. *See* Zebrafish
Databases 493–496
Decavanadate 422, 430, 433
Density gradient 22, 23, 26, 183
Detergent 58, 93, 131, 135, 415, 416, 424, 435, 444
  removal 444, 452
Dialysis 62, 64–65, 127, 129, 131, 134, 135, 137–141, 153, 404–406, 446, 447, 450–451
DIG-labeled RNA probe 360

E

EGTA. *See* Ethylene glycol tetraacetic acid (EGTA)
Electrochemical detection 334
Electron 421, 446, 452
  crystallography 444
  density 528, 532, 534
  microscopy 2, 20, 432, 440
Electrophysiology 89, 293–302
Embryos 357–360, 367–368
Enzyme-coupled assay 105
Epidermal cells 383–392
Ethylene glycol tetraacetic acid (EGTA) 20, 21, 108, 109, 117, 123, 158, 164, 168, 200, 201, 205, 207, 215, 216, 218, 219, 221, 224, 229, 230, 240, 261–266, 309, 400, 419, 423, 426, 433, 445, 524, 525
Expression 31, 38, 76, 213, 361

F

Flotation 187
Flow cytometry 374, 375, 377, 378
Fluorescence 136, 137, 139, 142, 143, 145, 149, 150, 178, 227, 283, 286, 289, 503
Fluorescence dyes 122
Fluorescent fusion protein 504, 520
Fluorescent protein 369, 386, 389, 503, 504, 516–518, 520
Fluorescence resonance energy transfer (FRET) 289, 503, 504, 506, 510, 517, 518, 520, 521
FXYD 5, 90, 322

G

[γ-³²P]TNP-8N₃-ATP photolabeling 235, 239
[γ-³²P]TNP-8N₃-ATP synthesis 240, 242
Gastric proton pump 445
H⁺,K⁺-ATPase 2, 19–26, 77, 235, 282, 443–454
Grip strength 343, 345

H

High performance liquid chromatography
  (HPLC) 334–336, 338, 339, 405
H⁺,K⁺-ATPase 19, 21, 23–26, 77, 235, 282, 293, 444–446, 448–454
HiLiDe 414, 416–420

I

In situ hybridization 353–362, 369
Infiltration 387, 388
Inorganic phosphate 24, 92, 93, 100, 105, 111, 119, 121, 159, 196, 328, 525
Insect cells 71, 73, 75, 77
Ion pump 1, 2, 127, 128, 134, 137, 145–149, 151, 282, 286, 289, 333
Isoforms 57
Isomeric transition 211, 212, 219

L

Lipid(s)
  annular
    bulk 399
    endogenous 399
    exchange 397
    flippases 181, 182
    native 397, 400
    NBD-labeled 372
    self-assembled bilayer 487, 488
    synthetic 397
  Liposomes 134, 153, 184
Lpg1024 31

M

Magnesium fluoride (MgFx) 197, 199, 206, 207, 525
Mammalian cell lines 296
Mammalian kidney 122
Membrane 53, 75–76, 171, 174, 178, 183, 257, 294, 296, 409, 416, 440, 469, 470, 490
  fragments 133, 151, 296, 298
  isolation 33, 75–76

potential 128, 133, 136, 142, 144, 148, 149, 154, 173, 175, 178, 286, 307
preparation................... 44, 47, 48, 53, 89, 93, 98, 173, 294, 405
protein 1, 19, 29, 30, 37, 42, 50, 59, 67, 79, 80, 93, 102, 181, 182, 205, 228, 267, 281, 282, 322, 327, 383, 385, 403, 414, 417, 459, 487
transport 293, 295
Metal
affinity 276
binding 32, 267, 268, 274
fluorides 197, 201–204, 206, 525
Metallochaperones..267
MgFx. See Magnesium fluoride (MgFx)
Molecular dynamics (MD) simulations...................459–463, 465, 475, 477, 487–489, 503, 504
Molecular modeling 503–521
Molecular dynamics (MD) simulations.....................459–460
Motif detection..499–500
Morphing 530, 534, 537
Mouse model................................... 334, 341–350
Myosin heavy chain..366

N

Na,K-ATPase.......................... 5, 9, 10, 71, 72, 76, 77, 89–91, 93–96, 98, 99, 101, 102, 117, 122, 125, 128, 133, 134, 136–138, 141–143, 145, 147–150, 152, 294, 305–307, 312–317, 321, 322, 327–331, 397, 405
Nanodisc..403–409
Nanoscale molecular dynamics (NAMD) 460–464, 475, 477, 482, 485, 487, 490
Native Source Purification 1
Neurotransmitter 333–339, 354
Nickel affinity chromatography........................33
Nicotiana benthamiana 385
Nucleotide binding..211

O

Oocytes....................................282, 309–310, 312, 317
Open filed
Orientation factor................................ 289, 519, 520
Outer medulla 6, 7, 79, 133, 404, 406
Ovaries 317
Overproduction 29, 31, 32
Oxalate 163
Oxonol VI 128, 130, 133, 136–138, 141, 143–145, 148–150, 173

P

P-type-ATPase dynamics. See P-type ATPase
P_{1B}-ATPases 267, 274
Passive avoidance................................. 342, 344, 346
Phosphate analog............... 195–207, 240, 445, 448, 452, 489
Phospholamban................ 161, 422, 424, 428, 430, 431, 518
Phospholipid probes 371

Phosphoryl transfer 195
Phosphorylated intermediate 38, 51, 196, 197, 211–213
Phosphorylation 52, 211, 218, 219
Photoaffinity labeling
P_i 9, 90, 94, 95, 111, 112, 114, 116–119, 121, 124, 133–136, 198, 199, 203, 205, 212, 214, 216–218, 220, 223, 225, 329, 487, 524, 526, 528, 536
Plasma membrane Ca^{2+} ATPase...57
Plasmid... 308, 310–311, 322
POPC bilayer 462, 471
Potassium antimony (III) oxide tartrate 112
Prediction 468, 470
Protein 2, 5, 44, 93, 113, 199, 241, 269, 369, 399, 405, 416, 418, 459, 461, 464–469, 494, 505, 507, 509–513, 515, 523, 525–528
Dynamics
lipid dynamics
lipid interactions...488
stability 79, 489
Proteoliposomes 176, 177, 187
Proton pumping 174, 178
P-type ATPase 1, 2, 5, 11, 37, 38, 42, 44, 51, 72, 73, 76, 79, 122, 125, 127, 128, 132, 143, 171, 179, 181, 189, 195, 201, 203, 207, 211, 212, 217, 221, 228, 233, 235, 237, 261, 267, 281, 282, 293, 371, 390, 397, 399, 413–419, 443, 459, 461, 463, 464, 466, 483, 486, 488, 493–501, 523–525, 529, 532, 533
Purification......................................31–32, 221–223, 269, 283, 284, 308, 310–311
Purification from cells 64
Pymol...537

R

Rabbit......................... 11–17, 37, 42, 107, 133, 199, 213, 229, 235, 263, 293, 399, 413, 523, 528
Radiolabeled..42, 121–126, 183, 372
Radioactive labeling..261
Reconstitution 59, 185, 403, 424–425, 428, 437

S

Saccharomyces cerevisiae...30, 38, 413
Sarcoplasmic reticulum Ca^{2+}-ATPase (SERCA)..........................1, 2, 11, 161–163, 165, 167, 196, 199, 201, 207, 228, 237, 239, 262, 265, 398, 422, 428, 430, 431, 443, 444, 460, 462, 464, 467, 469–475, 503–509, 511–516, 518–521, 523–525, 530, 533–537
Sarcolipin....................161, 422, 424, 428, 430, 431, 518, 528
Sarcoplasmic reticulum..199, 213
Scaffolding protein 403
SDS. See Sodium dodecyl sulfate (SDS)
SDS-polyacrylamide gel electrophoresis (SDS-PAGE) 15, 16, 21, 22, 33, 34, 49, 50, 53, 64, 135, 212, 217–220, 225, 271, 272, 275, 406, 407

Sections 62, 63, 201, 203, 213, 214, 216, 218–222, 361, 427, 429, 439, 493, 496, 498, 507, 534, 537
Sedation 309
Self-assembly 487, 488, 510
Sequence analysis
SERCA. *See* Sarcoplasmic reticulum Ca^{2+}-ATPase (SERCA)
Sf9 cells 71, 74–77
Single molecule studies 2
Site-specific fluorophore labeling 286
Size exclusion chromatography 30, 31, 34, 79, 83, 406, 407, 409
Sodium dodecyl sulfate (SDS) 8–10, 15, 16, 19, 21, 22, 24, 25, 33, 34, 41, 46, 48–50, 53, 64, 67, 68, 73, 91–94, 96–98, 100, 102, 135, 152, 212, 216–220, 225, 241, 272, 275, 323, 327, 328, 406, 407, 445, 449, 452
Solid supported membrane 293–302
Solubilised protein 401
Spectrophotometry 106
Specific activity 101, 329
Structure 461, 507, 509, 510
Superposition 529, 534, 537
Subtypes 493, 496, 498, 499

T

TEVC. *See* Two-electrode voltage clamping (TEVC)
Thermostability 79–84
TNP-AMP 204
Transfection 325

Transition state 143, 195, 198, 199, 201–203, 237, 240, 250, 524, 525
Transmembrane peptides 58, 59
Tryptophan fluorescence 204, 227–230
Two-dimensional crystals 422
Two-electrode voltage clamping (TEVC) 307–309, 311

U

Ultracentrifugation 399, 400, 420
UV-VIS spectrophotometry

V

Vector 31, 42, 46, 71–73, 76, 221, 239, 275, 276, 282, 324, 325, 335, 357, 372, 373, 375, 378, 475, 490, 506, 517, 519
Vesicles 448
Visualization 387, 389
Voltage clamp fluorometry 281

W

Whole-mount 368

X

Xenopus laevis 286, 305, 307–309
X-ray diffraction 417, 461, 526–528

Z

Zebrafish 353–355, 357–359, 367

Printed by Printforce, the Netherlands